AF753357

$98.00

2026 NATIONAL BUILDING COST MANUAL

Edited by Ben Moselle

50th Edition

Craftsman Book Company
6058 Corte del Cedro, Carlsbad, CA 92011

Photos: iStock by Getty Images™
Illustrations by Laura Knight, Devona Quindoy

ISBN 978-1-57218-407-7
Published August 2025 for the year 2026

Contents of This Manual

Explanation of the Cost Tables

This manual shows construction or replacement costs for a wide variety of residential, commercial, industrial, public, agricultural and military buildings. For your convenience and to minimize the chance of an error, all the cost and reference information you need for each building type is brought together on two or three pages. After reading pages 4 to 6, you should be able to turn directly to any building type and create an error-free estimate or appraisal of the construction or replacement cost.

The costs are per square foot of floor area for the basic building and additional costs for optional or extra components that differ from building to building. Building shape, floor area, design elements, materials used, and overall quality influence the basic structure cost. These and other cost variables are isolated for the building types. Components included in the basic square foot cost are listed with each building type. Instructions for using the basic building costs are included above the cost tables. These instructions include a list of components that may have to be added to the basic cost to find the total cost for your structure.

The figures in this manual are intended to reflect the amount that would be paid by the first user of a building completed in mid-2026.

Costs in the tables include all construction costs: labor, material, equipment, plans, building permit, supervision, overhead and profit. Cost tables do not include land value, site development costs, government mandated fees (other than the building permit) or the cost of modifying unusual soil conditions or grades. Construction expense may represent as much as 60% or as little as 40% of the cost to the first building owner. Site preparation, utility lines, government fees and mandates, finance cost and marketing are not part of the construction cost and may be as much as 20% of the cost to the first building owner.

Building Quality

Structures vary widely in quality and the quality of construction is the most significant variable in the finished cost. For estimating purposes the structure should be placed in one or more quality classes. These classes are numbered from 1 which is the highest quality generally encountered. Each section of this manual has a page describing typical specifications which define the quality class.

Each number class has been assigned a word description (such as best, good, average or low) for convenience and to help avoid possible errors.

The quality specifications do not reflect some design features and construction details that can make a building both more desirable and more costly. When substantially more than basic design elements are present, and when these elements add significantly to the cost, it is appropriate to classify the quality of the building as higher than would be warranted by the materials used in construction.

Many structures do not fall into a single class and have features of two quality classes. The tables have "half classes" which apply to structures which have some features of one class and some features of a higher or lower class. Classify a building into a "half class" when the quality elements are fairly evenly divided between two classes. Generally, quality elements do not vary widely in a single building. For example, it would be unusual to find a top quality single family residence with minimum quality roof cover. The most weight should be given to quality elements that have the greatest cost. For example, the type of wall and roof framing or the quality of interior finish are more significant than the roof cover or bathroom wall finish. Careful evaluation may determine that certain structures fall into two distinct classes. In this case, the cost of each part of the building should be evaluated separately.

Building Shapes

Shape classification considers any cost differences that arise from variations in building outline. Shape classification considerations vary somewhat with different building types. Where the building shape often varies widely between buildings and shape has a significant effect on the building cost, basic building costs are given for several shapes. Use the table that most closely matches the shape of the building you are evaluating. If the shape falls near the division between two basic building cost tables, it is appropriate to average the square foot cost from those two tables.

Explanation of the Cost Tables

Area of Buildings

The basic building cost tables reflect the fact that larger buildings generally cost less per square foot than smaller buildings. The cost tables are based on square foot areas which include the following:

1. All floor area within and including the exterior walls of the main building.
2. Inset areas such as vestibules, entrances or porches outside of the exterior wall but under the main roof.
3. Any enclosed additions, annexes or lean-tos with a square foot cost greater than three-fourths of the square foot cost of the main building.

Select the basic building cost listed below the area which falls closest to the actual area of your building. If the area of your building falls nearly mid-way between two listed building areas, it is appropriate to average the square foot costs for the listed areas.

Wall Heights

Building costs are based on the wall heights given in the instructions for each building cost table. Wall height for the various floors of a building are computed as follows: The basement is measured from the bottom of floor slab to the bottom of the first floor slab or joist. The main or first floor extends from the bottom of the first floor slab or joist to the top of the roof slab or ceiling joist. Upper floors are measured from the top of the floor slab or floor joist to the top of the roof slab or ceiling joist. These measurements may be illustrated as follows:

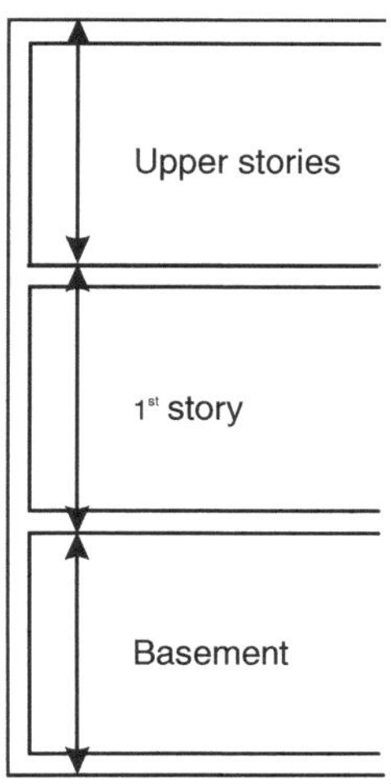

Square foot costs of most building design types must be adjusted if the actual wall height differs from the listed wall height. Wall height adjustment tables are included for buildings requiring this adjustment. Wall height adjustment tables list square foot costs for a foot of difference in perimeter wall height of buildings of various areas. The amount applicable to the actual building area is added or deducted for each foot of difference from the basic wall height.

Buildings such as residences, medical-dental buildings, funeral homes and convalescent hospitals usually have a standard 8-foot ceiling height except in chapels or day room areas. If a significant cost difference exists due to a wall height variation, this factor should be considered in establishing the quality class.

Other Adjustments

A common wall exists when two buildings share one wall. Common wall adjustments are made by deducting the in-place cost of the exterior wall finish plus one-half of the in-place cost of the structural portion of the common wall area.

If an owner has no ownership in a wall, the in-place cost of the exterior wall finish plus the in-place cost of the structural portion of the wall should be deducted from the total building costs. Suggested common wall and no wall ownership costs are included for many of the building types.

Some square foot costs include the cost of expensive veneer finishes on the entire perimeter wall. When these buildings butt against other buildings, adjustments should be made for the lack of this finish. Where applicable, linear foot cost deductions are provided.

The square foot costs in this manual are based on composite costs of total buildings including usual work room or storage areas. They are intended to be applied on a 100% basis to the total building area even though certain areas may or may not have interior finish. Only in rare instances will it be necessary to modify the square foot cost of a portion of a building.

Multiple story buildings usually share a common roof structure and cover, a common foundation and common floor or ceiling structures. The costs of these components are included in the various floor levels as follows:

Explanation of the Cost Tables

The first or main floor includes the cost of a floor structure built at ground level, foundation costs for a one-story building, a complete ceiling and roof structure, and a roof cover. The basement includes the basement floor structure and the difference between the cost of the first floor structure built at ground level and its cost built over a basement. The second floor includes the difference between the cost of a foundation for a one-story building and the cost of a foundation for a two-story building and the cost of the second story floor structure.

Location Adjustments

The figures in this manual are intended as national averages for metropolitan areas of the United States. Use the information on page 7 to adapt the basic building costs to any area listed. Frequently building costs outside metropolitan areas are 2% to 6% lower if skilled, productive, lower cost labor is available in the area. The factors on page 7 can be applied to nearly all the square foot costs and some of the "additional" costs in this book.

Temporary working conditions in any community can affect construction and replacement costs. Construction which must be done under deadline pressure or in adverse weather conditions or after a major fire, flood, or hurricane or in a thin labor market can temporarily inflate costs 25% to 50%. Conditions such as these are usually temporary and affect only a limited area. But the higher costs are real and must be considered, no matter how limited the area and how transient the condition.

Depreciation

Depreciation is the loss in value of a structure from all causes and is caused primarily by three forms of obsolescence: (1) physical (2) functional, and (3) economic.

Physical obsolescence is the deterioration of building components such as paint, carpets or roofing. Much of this deterioration is totally curable. The physical life tables on pages 43, 235 and 269 assume normal physical obsolescence. Good judgment is required to evaluate how deferred maintenance or rehabilitation will reduce or extend the anticipated physical life of a building.

Functional obsolescence is due to some deficiency or flaw in the building. For example, too few bathrooms for the number of bedrooms or an exceptionally high ceiling can reduce the life expectancy of a residence. Some functional obsolescence can be cured. The physical life tables do not consider functional obsolescence.

Economic obsolescence is caused by conditions that occur off site and are beyond control of the owner. Examples of economic obsolescence include a store in an area of declining economic activity or obsolescence caused by governmental regulation (such as a change in zoning). Because this kind of obsolescence is particularly difficult to measure, it is not considered in the physical life tables.

"Effective age" considers all forms of depreciation. It may be less than chronological age, if recently remodeled or improved, or more than the actual age, if deterioration is particularly bad. Though effective age is not considered in the physical life tables, it may yield a better picture of a structure's life than the actual physical age. Once the effective age is determined, considering physical, functional and economic deterioration, use the percent good tables on pages 43, 235 or 269 to determine the present value of a depreciated building. Present value is the result of multiplying the replacement cost (found by using the cost tables) by the appropriate percent good.

Limitations

This manual will be a useful reference for anyone who has to develop budget estimates or replacement costs for buildings. Anyone familiar with construction estimating understands that even very competent estimators with complete working drawings, full specifications and precise labor and material costs can disagree on the cost of a building. Frequently exhaustive estimates for even relatively simple structures can vary 10% or more. The range of competitive bids on some building projects is as much as 20%. Estimating costs is not an exact science and there's room for legitimate disagreement on what the "right" cost is. This manual can not help you do in a few minutes what skilled estimators may not be able to do in many hours. This manual will help you determine a reasonable replacement or construction cost for most buildings. It is not intended as a substitute for judgment or as a replacement for sound professional practice, but should prove a valuable aid to developing an informed opinion of value.

Area Modification Factors

Construction costs are higher in some cities than in other cities. Add or deduct the percentage shown on this page or page 8 to adapt the costs in this book to your job site. Adjust your estimated total project cost by the percentage shown for the appropriate city in this table to find your total estimated cost. Where 0% is shown it means no modification is required. Factors for Canada adjust to Canadian dollars.

These percentages were compiled by comparing the construction cost of buildings in nearly 600 communities throughout North America. Because these percentages are based on completed projects, they consider all construction cost variables, including labor, equipment and material cost, labor productivity, climate, job conditions and markup.

Modification factors are listed alphabetically by state and city, followed by the first three digits of the postal zip code.

These percentages are composites of many costs and will not necessarily be accurate when estimating the cost of any particular part of a building. But when used to modify costs for an entire structure, they should improve the accuracy of your estimates.

Alabama Average		**-3%**
Anniston	362	-3%
Auburn	368	-3%
Bellamy	369	-6%
Birmingham	350-352	6%
Dothan	363	-5%
Evergreen	364	-10%
Gadsden	359	-5%
Huntsville	358	3%
Jasper	355	-9%
Mobile	365-366	-1%
Montgomery	360-361	-1%
Scottsboro	357	0%
Selma	367	-2%
Sheffield	356	-1%
Tuscaloosa	354	-4%

Alaska Average		**16%**
Anchorage	995	21%
Fairbanks	997	23%
Juneau	998	12%
Ketchikan	999	8%
King Salmon	996	18%

Arizona Average		**-5%**
Chambers	865	-14%
Douglas	855	-10%
Flagstaff	860	-8%
Kingman	864	-9%
Mesa	852	5%
Phoenix	850	5%
Prescott	863	-4%
Show Low	859	-10%
Tucson	856-857	-5%
Yuma	853	4%

Arkansas Average		**-7%**
Batesville	725	-10%
Camden	717	-5%
Fayetteville	727	-4%
Fort Smith	729	-7%
Harrison	726	-15%
Hope	718	-11%
Hot Springs	719	-13%
Jonesboro	724	-3%
Little Rock	720-722	-2%
Pine Bluff	716	-12%
Russellville	728	-7%
West Memphis	723	3%

California Average		**8%**
Alhambra	917-918	8%
Bakersfield	932-933	-2%
El Centro	922	-1%
Eureka	955	1%
Fresno	936-938	0%
Herlong	961	2%
Inglewood	902-905	8%
Irvine	926-927	13%
Lompoc	934	3%
Long Beach	907-908	9%
Los Angeles	900-901	8%
Marysville	959	1%
Modesto	953	0%
Mojave	935	5%
Novato	949	12%
Oakland	945-947	17%
Orange	928	12%
Oxnard	930	2%
Pasadena	910-912	9%
Rancho Cordova	956-957	7%
Redding	960	0%
Richmond	948	17%
Riverside	925	3%
Sacramento	958	6%
Salinas	939	2%
San Bernardino	923-924	2%
San Diego	919-921	6%
San Francisco	941	26%
San Jose	950-951	29%
San Mateo	943-944	19%
Santa Barbara	931	3%
Santa Rosa	954	6%
Stockton	952	3%
Sunnyvale	940	26%
Van Nuys	913-916	8%
Whittier	906	8%

Colorado Average		**0%**
Aurora	800-801	8%
Boulder	803-804	5%
Colorado Springs	808-809	0%
Denver	802	9%
Durango	813	-7%
Fort Morgan	807	-3%
Glenwood Springs	816	3%
Grand Junction	814-815	-5%
Greeley	806	5%
Longmont	805	2%
Pagosa Springs	811	-8%
Pueblo	810	-5%
Salida	812	-4%

Connecticut Average		**5%**
Bridgeport	66	6%
Bristol	60	7%
Fairfield	64	7%
Hartford	61	5%
New Haven	65	4%
Norwich	63	0%
Stamford	068-069	11%
Waterbury	67	4%
West Hartford	62	0%

Delaware Average		**1%**
Dover	199	-3%
Newark	197	4%
Wilmington	198	3%

District of Columbia Average		
Washington	200-205	12%

Florida Average		**-5%**
Altamonte Springs	327	-2%
Bradenton	342	-5%
Brooksville	346	-7%
Daytona Beach	321	-10%
Fort Lauderdale	333	2%
Fort Myers	339	-6%
Fort Pierce	349	-10%
Gainesville	326	-7%
Jacksonville	322	0%
Lakeland	338	-7%
Melbourne	329	-6%
Miami	330-332	-2%
Naples	341	-3%
Ocala	344	-12%
Orlando	328	0%
Panama City	324	-9%
Pensacola	325	-7%
Saint Augustine	320	-2%
Saint Cloud	347	-1%
St Petersburg	337	-5%
Tallahassee	323	-7%
Tampa	335-336	0%
West Palm Beach	334	1%

Georgia Average		**0%**
Albany	317	-8%
Athens	306	0%
Atlanta	303	17%
Augusta	308-309	-6%
Buford	305	0%
Calhoun	307	-1%
Columbus	318-319	-4%
Dublin/Fort Valley	310	-7%
Hinesville	313	4%
Kings Bay	315	-9%
Macon	312	4%
Marietta	300-302	7%
Savannah	314	-1%
Statesboro	304	4%
Valdosta	316	-5%

Hawaii Average		**18%**
Aliamanu	968	20%
Ewa	967	17%
Halawa Heights	967	17%
Hilo	967	17%
Honolulu	968	20%
Kailua	968	20%
Lualualei	967	17%
Mililani Town	967	17%
Pearl City	967	17%
Wahiawa	967	17%
Waianae	967	17%
Wailuku (Maui)	967	17%

Idaho Average		**-7%**
Boise	837	1%
Coeur d'Alene	838	-7%
Idaho Falls	834	-8%
Lewiston	835	-11%
Meridian	836	-6%
Pocatello	832	-11%
Sun Valley	833	-8%

Illinois Average		**6%**
Arlington Heights	600	17%
Aurora	605	17%
Belleville	622	2%
Bloomington	617	1%
Carbondale	629	-8%
Carol Stream	601	18%
Centralia	628	-8%
Champaign	618	-1%
Chicago	606-608	21%
Decatur	623	-3%
Galesburg	614	-2%
Granite City	620	0%
Green River	612	0%
Joliet	604	17%
Kankakee	609	2%
Lawrenceville	624	-5%
Oak Park	603	24%
Peoria	615-616	8%
Peru	613	6%
Quincy	602	22%
Rockford	610-611	4%
Springfield	625-627	-2%
Urbana	619	-2%

Indiana Average		**-2%**
Aurora	470	1%
Bloomington	474	-3%
Columbus	472	-4%
Elkhart	465	-3%
Evansville	476-477	-2%
Fort Wayne	467-468	1%
Gary	463-464	4%
Indianapolis	460-462	5%
Jasper	475	-4%
Jeffersonville	471	-6%
Kokomo	469	-4%
Lafayette	479	-3%
Muncie	473	-9%
South Bend	466	-2%
Terre Haute	478	-2%

Iowa Average		**-3%**
Burlington	526	-5%
Carroll	514	-4%
Cedar Falls	506	-3%
Cedar Rapids	522-524	0%
Cherokee	510	1%
Council Bluffs	515	-1%
Creston	508	-8%
Davenport	527-528	5%
Decorah	521	-5%
Des Moines	500-503	2%
Dubuque	520	-1%
Fort Dodge	505	-2%
Mason City	504	-1%
Ottumwa	525	-8%
Sheldon	512	5%
Shenandoah	516	-16%
Sioux City	511	-1%
Spencer	513	-6%
Waterloo	507	-4%

Kansas Average		**-7%**
Colby	677	-4%
Concordia	669	-16%
Dodge City	678	-8%
Emporia	668	-9%
Fort Scott	667	-4%
Hays	676	-12%
Hutchinson	675	-9%
Independence	673	-5%
Kansas City	660-662	6%
Liberal	679	-13%
Salina	674	-5%
Topeka	664-666	-4%
Wichita	670-672	-5%

Kentucky Average		**-6%**
Ashland	411-412	-8%
Bowling Green	421	-3%
Campton	413-414	-12%
Covington	410	2%
Elizabethtown	427	-6%
Frankfort	406	-7%
Hazard	417-418	-15%
Hopkinsville	422	-4%
Lexington	403-405	3%
London	407-409	-6%
Louisville	400-402	1%
Owensboro	423	-4%
Paducah	420	-6%
Pikeville	415-416	-14%
Somerset	425-426	-12%
White Plains	424	-4%

Louisiana Average		**1%**
Alexandria	713-714	-7%
Baton Rouge	707-708	15%
Houma	703	6%
Lafayette	705	4%
Lake Charles	706	-4%
Mandeville	704	5%
Minden	710	-8%
Monroe	712	-7%
New Orleans	700-701	9%
Shreveport	711	-6%

Maine Average		**-3%**
Auburn	42	-3%
Augusta	43	-1%
Bangor	44	-2%
Bath	45	-5%
Brunswick	039-040	1%
Camden	48	-7%
Cutler	46	-9%
Dexter	49	-3%
Northern Area	47	-9%
Portland	41	4%

Maryland Average		**1%**
Annapolis	214	7%
Baltimore	210-212	4%
Bethesda	208-209	10%
Church Hill	216	-4%
Cumberland	215	-9%
Elkton	219	-1%
Frederick	217	3%
Laurel	206-207	7%
Salisbury	218	-7%

Massachusetts Average		**12%**
Ayer	015-016	10%
Bedford	17	19%
Boston	021-022	29%
Brockton	023-024	20%
Cape Cod	26	6%
Chicopee	10	7%
Dedham	19	15%
Fitchburg	14	15%
Hingham	20	20%
Lawrence	18	16%
Nantucket	25	12%
New Bedford	27	8%
Northfield	13	-5%
Pittsfield	12	1%
Springfield	11	6%

Michigan Average		**2%**
Battle Creek	490-491	0%
Detroit	481-482	8%
Flint	484-485	0%
Grand Rapids	493-495	5%
Grayling	497	-2%
Jackson	492	-1%
Lansing	488-489	3%
Marquette	498-499	-5%
Pontiac	483	8%
Royal Oak	480	6%
Saginaw	486-487	-1%
Traverse City	496	-3%

Minnesota Average		**1%**
Bemidji	566	-3%
Brainerd	564	-2%
Duluth	556-558	1%
Fergus Falls	565	-4%
Magnolia	561	-7%
Mankato	560	0%
Minneapolis	553-555	12%
Rochester	559	0%
St Cloud	563	4%
St Paul	550-551	13%
Thief River Falls	567	0%
Willmar	562	2%

Mississippi Average		**0%**
Clarksdale	386	0%
Columbus	397	5%
Greenville	387	-5%
Greenwood	389	0%
Gulfport	395	2%
Jackson	390-392	4%
Laurel	394	3%
McComb	396	-2%
Meridian	393	-2%
Tupelo	388	-4%

Area Modification Factors

Missouri Average		**-3%**
Cape Girardeau	637	-5%
Caruthersville	638	-3%
Chillicothe	646	-4%
Columbia	652	-4%
East Lynne	647	-6%
Farmington	636	-9%
Hannibal	634	-7%
Independence	640	6%
Jefferson City	650-651	-3%
Joplin	648	-2%
Kansas City	641	7%
Kirksville	635	-13%
Knob Noster	653	-7%
Lebanon	654-655	-12%
Poplar Bluff	639	-9%
Saint Charles	633	0%
Saint Joseph	644-645	6%
Springfield	656-658	-7%
St Louis	630-631	7%

Montana Average		**0%**
Billings	590-591	1%
Butte	597	3%
Fairview	592	3%
Great Falls	594	4%
Havre	595	-2%
Helena	596	-1%
Kalispell	599	-5%
Miles City	593	-6%
Missoula	598	-2%

Nebraska Average		**-6%**
Alliance	693	-5%
Columbus	686	-5%
Grand Island	688	-8%
Hastings	689	-2%
Lincoln	683-685	-4%
McCook	690	-10%
Norfolk	687	-6%
North Platte	691	-9%
Omaha	680-681	0%
Valentine	692	-14%

Nevada Average		**1%**
Carson City	897	0%
Elko	898	2%
Ely	893	-7%
Fallon	894	4%
Las Vegas	889-891	2%
Reno	895	4%

New Hampshire Average		**4%**
Charlestown	36	-5%
Concord	34	1%
Dover	38	9%
Lebanon	37	-1%
Littleton	35	2%
Manchester	032-033	8%
New Boston	030-031	11%

New Jersey Average		**9%**
Atlantic City	080-084	5%
Brick	87	2%
Dover	78	9%
Edison	088-089	13%
Hackensack	76	9%
Monmouth	77	12%
Newark	071-073	7%
Passaic	70	9%
Paterson	074-075	7%
Princeton	85	9%
Summit	79	14%
Trenton	86	6%

New Mexico Average		**-8%**
Alamogordo	883	-12%
Albuquerque	870-871	-1%
Clovis	881	-15%
Farmington	874	2%
Fort Sumner	882	-1%
Gallup	873	-9%
Holman	877	-8%
Las Cruces	880	-10%
Santa Fe	875	-7%
Socorro	878	-15%
Truth or Consequences	879	-7%
Tucumcari	884	-11%

New York Average		**5%**
Albany	120-123	8%
Amityville	117	8%
Batavia	140	1%
Binghamton	137-139	2%
Bronx	104	10%
Brooklyn	112	2%
Buffalo	142	1%
Elmira	149	-2%
Flushing	113	10%
Garden City	115	10%
Hicksville	118	9%
Ithaca	148	-6%
Jamaica	114	9%
Jamestown	147	-7%
Kingston	124	-1%
Long Island	111	28%
Montauk	119	8%
New York (Manhattan)	100-102	29%
New York City	100-102	29%
Newcomb	128	4%
Niagara Falls	143	-5%
Plattsburgh	129	0%
Poughkeepsie	125-126	2%
Queens	110	12%
Rochester	144-146	1%
Rockaway	116	4%
Rome	133-134	-4%
Staten Island	103	6%
Stewart	127	10%
Syracuse	130-132	3%
Tonawanda	141	-1%
Utica	135	-5%
Watertown	136	-4%
West Point	109	2%
White Plains	105-108	10%

North Carolina Average		**-1%**
Asheville	287-289	-5%
Charlotte	280-282	5%
Durham	277	5%
Elizabeth City	279	-7%
Fayetteville	283	-5%
Goldsboro	275	3%
Greensboro	274	1%
Hickory	286	-3%
Kinston	285	-7%
Raleigh	276	6%
Rocky Mount	278	-3%
Wilmington	284	-3%
Winston-Salem	270-273	0%

North Dakota Average		**2%**
Bismarck	585	1%
Dickinson	586	10%
Fargo	580-581	2%
Grand Forks	582	0%
Jamestown	584	-2%
Minot	587	0%
Nekoma	583	-11%
Williston	588	14%

Ohio Average		**0%**
Akron	442-443	1%
Canton	446-447	0%
Chillicothe	456	-7%
Cincinnati	450-452	5%
Cleveland	440-441	1%
Columbus	432	9%
Dayton	453-455	3%
Lima	458	-3%
Marietta	457	-8%
Marion	433	0%
Newark	430-431	8%
Sandusky	448-449	2%
Steubenville	439	-3%
Toledo	434-436	4%
Warren	444	-3%
Youngstown	445	0%
Zanesville	437-438	0%

Oklahoma Average		**-8%**
Adams	739	-8%
Ardmore	734	-10%
Clinton	736	-8%
Durant	747	-8%
Enid	737	-11%
Lawton	735	-12%
McAlester	745	-7%
Muskogee	744	-6%
Norman	730	-4%
Oklahoma City	731	-2%
Ponca City	746	-7%
Poteau	749	-13%
Pryor	743	-6%
Shawnee	748	-8%
Tulsa	740-741	-4%
Woodward	738	-7%

Oregon Average		**-2%**
Adrian	979	-9%
Bend	977	1%
Eugene	974	-4%
Grants Pass	975	-6%
Klamath Falls	976	-10%
Pendleton	978	-1%
Portland	970-972	12%
Salem	973	-2%

Pennsylvania Average		**-3%**
Allentown	181	0%
Altoona	166	-8%
Beaver Springs	178	-8%
Bethlehem	180	4%
Bradford	167	-9%
Butler	160	-1%
Chambersburg	172	-4%
Clearfield	168	-3%
DuBois	158	-8%
East Stroudsburg	183	-8%
Erie	164-165	-8%
Genesee	169	-8%
Greensburg	156	-3%
Harrisburg	170-171	-1%
Hazleton	182	-5%
Johnstown	159	-10%
Kittanning	162	-6%
Lancaster	175-176	-1%
Meadville	163	-13%
Montrose	188	-5%
New Castle	161	-4%
Philadelphia	190-191	8%
Pittsburgh	152	2%
Pottsville	179	-6%
Punxsutawney	157	-11%
Reading	195-196	-1%
Scranton	184-185	-3%
Somerset	155	-8%
Southeastern	193	7%
Uniontown	154	-9%
Valley Forge	194	13%
Warminster	189	8%
Warrendale	150-151	2%
Washington	153	0%
Wilkes Barre	186-187	-4%
Williamsport	177	-8%
York	173-174	2%

Rhode Island Average		**3%**
Bristol	28	3%
Coventry	28	3%
Cranston	29	3%
Davisville	28	3%
Narragansett	28	3%
Newport	28	3%
Providence	29	3%
Warwick	28	3%

South Carolina Average		**0%**
Aiken	298	2%
Beaufort	299	-1%
Charleston	294	3%
Columbia	290-292	-1%
Greenville	296	4%
Myrtle Beach	295	-5%
Rock Hill	297	-4%
Spartanburg	293	-2%

South Dakota Average		**-6%**
Aberdeen	574	-4%
Mitchell	573	-7%
Mobridge	576	-13%
Pierre	575	-11%
Rapid City	577	-6%
Sioux Falls	570-571	2%
Watertown	572	-6%

Tennessee Average		**2%**
Chattanooga	374	5%
Clarksville	370	7%
Cleveland	373	3%
Columbia	384	-6%
Cookeville	385	-7%
Jackson	383	5%
Kingsport	376	-1%
Knoxville	377-379	3%
McKenzie	382	-8%
Memphis	380-381	9%
Nashville	371-372	9%

Texas Average		**-1%**
Abilene	795-796	-7%
Amarillo	790-791	-5%
Arlington	760	7%
Austin	786-787	10%
Bay City	774	8%
Beaumont	776-777	4%
Brownwood	768	-6%
Bryan	778	-4%
Childress	792	-8%
Corpus Christi	783-784	2%
Dallas	751-753	7%
Del Rio	788	-10%
El Paso	798-799	-12%
Fort Worth	761-762	6%
Galveston	775	6%
Giddings	789	7%
Greenville	754	2%
Houston	770-772	8%
Huntsville	773	5%
Longview	756	-8%
Lubbock	793-794	-4%
Lufkin	759	-9%
McAllen	785	-12%
Midland	797	5%
Palestine	758	-6%
Plano	750	7%
San Angelo	769	-2%
San Antonio	780-782	1%
Texarkana	755	-8%
Tyler	757	-7%
Victoria	779	-4%
Waco	765-767	-3%
Wichita Falls	763	-5%
Woodson	764	-3%

Utah Average		**-4%**
Clearfield	840	0%
Green River	845	-10%
Ogden	843-844	-6%
Provo	846-847	-6%
Salt Lake City	841	2%

Vermont Average		**-3%**
Albany	58	-5%
Battleboro	53	-3%
Beecher Falls	59	-5%
Bennington	52	-8%
Burlington	54	3%
Montpelier	56	-2%
Rutland	57	-7%
Springfield	51	-4%
White River Junction	50	1%

Virginia Average		**-2%**
Abingdon	242	-7%
Alexandria	220-223	11%
Charlottesville	229	-2%
Chesapeake	233	0%
Culpeper	227	-6%
Farmville	239	-8%
Fredericksburg	224-225	-5%
Galax	243	-10%
Harrisonburg	228	-6%
Lynchburg	245	-2%
Norfolk	235-237	4%
Petersburg	238	-2%
Radford	241	-8%
Reston	201	10%
Richmond	232	3%
Roanoke	240	-2%
Staunton	244	-5%
Tazewell	246	-13%
Virginia Beach	234	-1%
Williamsburg	230-231	2%
Winchester	226	-6%

Washington Average		**1%**
Clarkston	994	-2%
Everett	982	4%
Olympia	985	1%
Pasco	993	1%
Seattle	980-981	13%
Spokane	990-992	-3%
Tacoma	983-984	3%
Vancouver	986	1%
Wenatchee	988	-3%
Yakima	989	-7%

West Virginia Average		**-8%**
Beckley	258-259	-10%
Bluefield	247-248	-12%
Charleston	250-253	4%
Clarksburg	263-264	-4%
Fairmont	266	-10%
Huntington	255-257	-2%
Lewisburg	249	-19%
Martinsburg	254	-7%
Morgantown	265	-4%
New Martinsville	262	-12%
Parkersburg	261	-9%
Romney	267	-8%
Sugar Grove	268	-13%
Wheeling	260	-5%

Wisconsin Average		**0%**
Amery	540	-1%
Beloit	535	2%
Clam Lake	545	-7%
Eau Claire	547	-1%
Green Bay	541-543	3%
La Crosse	546	-4%
Ladysmith	548	-5%
Madison	537	6%
Milwaukee	530-534	6%
Oshkosh	549	7%
Portage	539	0%
Prairie du Chien	538	-4%
Wausau	544	-3%

Wyoming Average		**-3%**
Casper	826	0%
Cheyenne/ Laramie	820	-2%
Gillette	827	-1%
Powell	824	-9%
Rawlins	823	-2%
Riverton	825	-8%
Rock Springs	829-831	-1%
Sheridan	828	-5%
Wheatland	822	0%

UNITED STATES TERRITORIES	
Guam	18%
Puerto Rico	-21%

VIRGIN ISLANDS (U.S.)	
St. Croix	2%
St. John	20%
St. Thomas	5%

CANADIAN AREA MODIFIERS

These figures assume an exchange rate of $1.00 Canadian to $.76 U.S.

Alberta Average	**13%**
Calgary	14%
Edmonton	14%
Fort McMurray	12%

British Columbia Average	**7%**
Fraser Valley	6%
Okanagan	6%
Vancouver	9%

Manitoba Average	**0%**
North Manitoba	0%
Selkirk	0%
South Manitoba	0%
Winnipeg	0%

New Brunswick Average	**-13%**
Moncton	-13%
Newfoundland/Labrador	-3%

Nova Scotia Average	**-8%**
Amherst	-8%
Nova Scotia	-7%
Sydney	-8%

Ontario Average	**7%**
London	7%
Thunder Bay	6%
Toronto	7%

Quebec Average	**-1%**
Montreal	-1%
Quebec City	-1%

Saskatchewan Average	**4%**
La Ronge	3%
Prince Albert	2%
Saskatoon	5%

Building Cost Historical Index

Use this table to find the approximate current dollar building cost when the actual cost is known for any year since 1959. Multiply the figure listed below for the building type and year of construction by the known cost. The result is the estimated 2026 construction cost.

Year	Masonry Buildings	Concrete Buildings	Steel Buildings	Wood-Frame Buildings	Agricultural Buildings	Year of Construction
1959	17.65	18.04	15.02	13.18	13.65	1959
1960	17.24	17.70	14.78	12.99	13.38	1960
1961	16.89	17.64	14.53	12.75	13.33	1961
1962	16.51	17.12	14.18	12.60	13.14	1962
1963	16.26	16.67	14.01	12.36	11.92	1963
1964	15.79	16.48	13.82	11.93	12.52	1964
1965	15.29	16.05	13.34	11.68	12.18	1965
1966	14.59	15.59	12.83	11.17	11.84	1966
1967	14.26	14.84	12.00	10.62	11.37	1967
1968	13.67	14.02	11.45	10.05	10.86	1968
1969	12.91	13.40	11.06	9.67	10.25	1969
1970	12.40	12.82	10.51	9.20	9.74	1970
1971	11.62	11.73	9.76	7.92	9.08	1971
1972	10.81	10.86	9.12	7.94	8.44	1972
1973	9.87	10.29	8.10	7.33	7.93	1973
1974	8.78	9.44	7.61	6.85	7.35	1974
1975	7.98	8.34	6.83	6.44	6.56	1975
1976	7.48	7.94	6.48	6.20	6.21	1976
1977	6.96	7.45	6.16	5.76	5.85	1977
1978	6.49	6.96	5.67	5.29	5.29	1978
1979	5.96	6.20	5.08	4.85	5.00	1979
1980	5.40	5.63	4.53	4.35	4.53	1980
1981	5.08	5.31	4.15	4.15	4.23	1981
1982	4.93	5.08	4.03	4.01	4.09	1982
1983	4.70	4.93	3.95	3.83	3.85	1983
1984	4.39	4.62	3.76	3.54	3.74	1984
1985	4.26	4.39	3.66	3.43	3.68	1985
1986	4.15	4.36	3.59	3.38	3.60	1986
1987	4.14	4.26	3.56	3.32	3.57	1987
1988	4.06	4.10	3.49	3.28	3.51	1988
1989	3.96	4.03	3.32	3.22	3.40	1989
1990	3.73	3.87	3.15	2.99	3.24	1990
1991	4.03	3.81	3.00	2.83	3.07	1991
1992	3.61	3.77	2.96	2.82	3.05	1992
1993	3.52	3.73	2.85	2.78	3.00	1993
1994	3.43	3.49	2.75	2.67	2.78	1994
1995	3.26	3.18	2.55	2.51	2.63	1995
1996	3.15	3.12	2.48	2.46	2.58	1996
1997	3.04	3.04	2.38	2.40	2.52	1997
1998	2.89	2.89	2.29	2.30	2.48	1998
1999	2.80	2.80	2.23	2.28	2.45	1999
2000	2.72	2.72	2.14	2.20	2.36	2000
2001	2.63	2.63	2.10	2.11	2.30	2001
2002	2.57	2.57	2.05	2.09	2.25	2002
2003	2.53	2.53	2.00	2.07	2.21	2003
2004	2.42	2.42	1.94	2.02	2.15	2004
2005	2.24	2.24	1.74	1.81	2.10	2005
2006	2.12	2.12	1.60	1.62	1.88	2006
2007	2.05	2.05	1.54	1.50	1.75	2007
2008	1.91	1.91	1.46	1.44	1.65	2008
2009	1.90	1.90	1.40	1.44	1.65	2009
2010	1.86	1.86	1.32	1.43	1.64	2010
2011	1.89	1.89	1.36	1.45	1.69	2011
2012	1.86	1.86	1.22	1.40	1.65	2012
2013	1.78	1.78	1.30	1.33	1.54	2013
2014	1.76	1.76	1.28	1.31	1.53	2014
2015	1.74	1.74	1.27	1.30	1.52	2015
2016	1.72	1.72	1.40	1.31	1.48	2016
2017	1.67	1.67	1.42	1.32	1.48	2017
2018	1.59	1.59	1.23	1.20	1.38	2018
2019	1.49	1.49	1.28	1.15	1.32	2019
2020	1.47	1.47	1.23	1.17	1.31	2020
2021	1.43	1.43	1.30	1.16	1.31	2021
2022	1.36	1.36	1.14	1.07	1.22	2022
2023	1.19	1.19	0.92	0.98	1.09	2023
2024	1.11	1.11	1.03	1.06	1.08	2024
2025	1.06	1.06	1.12	1.01	1.03	2025
2026	1.00	1.00	1.00	1.00	1.00	2026

Residential Structures Section

The figures in this section include all costs associated with normal construction:

Foundations as required for normal soil conditions. Excavation for foundations, piers, and other foundation components given a fairly level construction site. Floor, wall, and roof structures. Interior floor, wall, and ceiling finishes. Exterior wall finish and roof cover. Interior partitions as described in the quality class. Finish carpentry, doors, windows, trim, etc. Electric wiring and fixtures. Rough and finish plumbing as described in applicable building specifications. Built-in appliances as described in applicable building specifications. All labor and materials including supervision. All design and engineering fees, if necessary. Permits and fees. Utility hook-ups. Contractors' contingency, overhead and profit.

The square foot costs do not include heating and cooling equipment or the items listed in the section "Additional Costs for Residential Structures" which appear on pages 27 to 31. The costs of the following should be figured separately and added to the basic structure cost: porches, basements, balconies, exterior stairways, built-in equipment beyond that listed in the quality classifications, garages and carports.

Single Family Residences

Single family residences vary widely in quality and the quality of construction is the most significant factor influencing cost. Residences are listed in six quality classes. Class 1 is the most expensive commonly encountered and Class 6 is the minimum required under most building codes. Nearly all homes built from stock plans or offered to the public by residential tract developers will fall into Class 3, 4, 5, or 6. For convenience, these classes are labeled *Best Standard, Good Standard, Average Standard* or *Minimum Standard.* Class 1 residences are labeled *Luxury.* Class 2 residences are labeled *Semi-Luxury.* Class 1 and 2 residences are designed by professional architects, usually to meet preferences of the first owner.

The shape of the outside perimeter also has a significant influence on cost. The more complex the shape, the more expensive the structure per square foot of floor. The shape classification of multiple story or split-level homes should be based on the outline formed by the outer-most exterior walls, including the garage area, regardless of the story level. Most residences that fall into Classes 3, 4, 5 or 6 have 4, 6, 8 or 10 corners, as illustrated below. Small insets that do not require a change in the roof line can be ignored when evaluating the outside perimeter.

Class 1 and 2 (*Luxury* and *Semi-Luxury*) residences have more than ten corners and are best evaluated by counting the "building masses." A building mass is a group of contiguous rooms on one or more levels with access at varying angles from a common point or hallway. The illustration at the right below represents a residence with two building masses. Most Class 1 and Class 2 residences have from one to four building masses, ignoring any attached garage. For convenience, cost tables for Class 1 and 2 single family residences with one, two, three or four building masses have been appended to cost tables for Class 3, 4, 5 and 6 residences with 4, 6, 8 and 10 building corners.

Residences on larger lots often include a separate housekeeping unit, either remote from the main structure (as illustrated below at the right) or joined to the main structure by a hallway (no common wall). Evaluate any separate housekeeping unit as a separate residence. The quality class of separate housekeeping units will usually be the same as the main residence if designed and built at the same time as the main residence.

Residences which have features of two or more quality classes can be placed between two of the six labeled classes. The tables have five half-classes (1 & 2, 2 & 3, etc.) which can be applied to residences with some characteristics of two or more quality classes. If a portion of a residence differs significantly in quality from other portions, evaluate the square footage of each portion separately.

These figures can be applied to nearly all single-family residences built using conventional methods and readily available materials, including the relatively small number of highly decorative, starkly original or exceptionally well-appointed residences.

4 corners

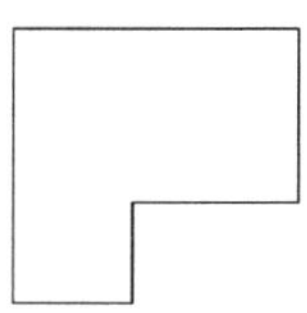

6 corners

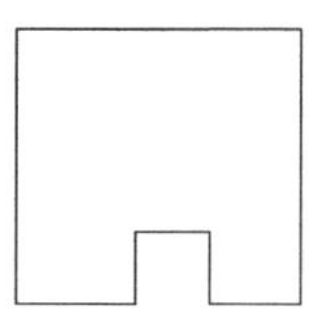

8 corners

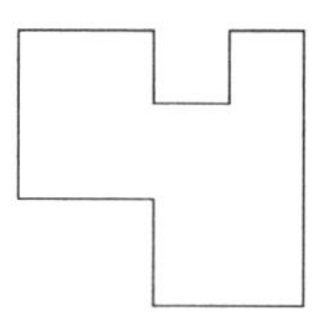

10 corners

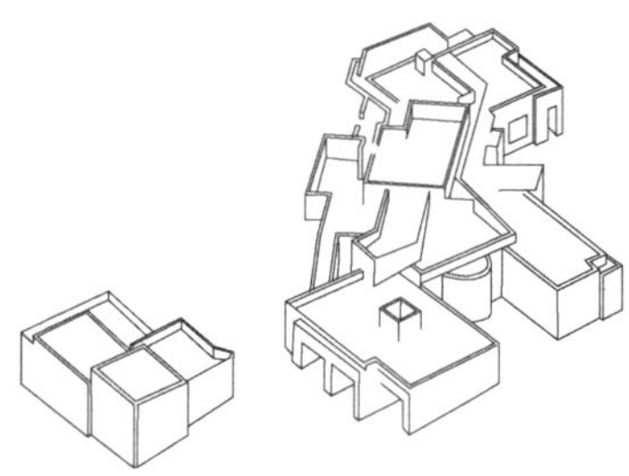

2 building masses and one separate unit

Single Family Residences

Quality Classification

	Class 1 Luxury	Class 2 Semi-Luxury	Class 3 Best Std.	Class 4 Good Std.	Class 5 Average Std.	Class 6 Minimum Std.
Foundation (9% of total cost)	Reinforced concrete.	Reinforced concrete.	Reinforced concrete.	Reinforced concrete or concrete block.	Reinforced concrete or concrete block.	Reinforced concrete.
Floor Structure (12% of total cost)	Engineered wood or steel exceeding code minimums.	Engineered wood or steel or reinforced concrete slab.	Engineered wood or steel or reinforced concrete slab.	Wood frame or slab on grade, changes in shape and elevation.	Standard wood frame or slab on grade with elevation changes.	Slab on grade. No changes in elevation.
Wall Framing and Exterior Finish (14% of total cost)	Wood or steel, very irregular walls, stone veneer, many architectural doors and windows.	Wood or steel, irregular shape, masonry veneer, better grade doors and windows.	Wood or steel, several wall offsets, wood or masonry accents, good grade doors and windows.	Wood or steel, stucco or wood siding, some trim or veneer, average doors and windows.	Wood or steel, stucco or wood siding, few offsets, commodity grade doors and windows.	Wood or steel, stucco or hardboard siding, minimum grade doors and windows.
Roof (10% of total cost)	Complex plan, tile, slate or metal, highly detailed.	Multi-level, slate, tile or flat surface, decorative details.	Multi-pitch, shake, tile or flat surface, large closed soffit.	Wood trusses, tile or good shingles, closed soffit.	Wood frame, shingle or built-up cover, open 24" soffit.	Wood frame, composition shingle cover, open soffit.
Floor Finish (5% of total cost)	Terrazzo, marble, granite, or inlaid hardwood or best carpet throughout.	Marble or granite entry, hardwood, good carpet or sheet vinyl elsewhere.	Simulated marble tile entry, good carpet, hardwood or vinyl elsewhere.	Better sheet vinyl and average carpet, some areas with masonry or tile.	Good sheet vinyl and standard carpet, small area with tile or hardwood.	Composition tile or minimum grade sheet vinyl.
Interior Wall and Ceiling Finish (8% of total cost)	Plaster or gypsum wallboard with artistic finish, many offsets and wall openings, decorative details in nearly all rooms.	Plaster on gypsum or metal lath or 2 layers of 5/8" gypsum wallboard, decorative details, many irregular wall openings.	Gypsum wallboard with putty or texture coat finish, some irregular walls, decorative details in living room, entry and kitchen.	1/2" gypsum wallboard with textured finish, several irregular walls and wall openings, some. decorative details.	1/2" gypsum wallboard with textured finish, most walls are rectangular, doors and windows are the only openings.	1/2" gypsum wallboard, smooth or orange peel finish. Nearly all walls are regular, no decorative details.
Interior Detail (5% of total cost)	Exposed beams or decorative ceiling, 12' to 16' ceiling in great room, many sky widows, built-in shelving and alcoves for art.	Great room has 12' to 16' ceiling, most rooms have windows on two sides, formal dining area, several framed openings.	Cathedral ceiling at entry, one or more floor level changes, several wall openings or pass-throughs, formal dining area.	8' or 9' ceiling throughout, walk-in closet in master bedroom, separate dining area, some decorative wood trim.	8' or 9' ceiling throughout, sliding mirrored closet doors, standard grade molding and trim, breakfast bar or nook.	Drop ceiling in kitchen, other rooms have 7'6" to 8' ceiling, minimum grade molding and trim.
Bath Detail (4% of total cost)	Custom large tile showers, separate elevated spa in master bathroom.	Large tile showers, at least one bathtub, glass block or large window by each bath.	Tile or fiberglass shower, at least one built-in bathtub, window in bathroom.	Good plastic tub and shower in at least one bathroom, one small window in each bath.	Average plastic tub and shower in at least one bathroom.	Minimum plastic tub and shower in one bathroom.
Kitchen Detail (8% of total cost)	Over 30 LF of deluxe wall and base cabinets, stone counter top, island work area, breakfast bar.	Over 25 LF of good custom base and wall cabinets, synthetic stone counter top, desk and breakfast bar.	Over 20 LF of good stock wall and base cabinets, tile or acrylic counter top, desk and breakfast bar or nook.	Over 15 LF of stock standard grade wall and base cabinets, low-cost tile or acrylic counter top, breakfast nook.	Over 10 LF of stock standard grade wall and base cabinets, low-cost acrylic or laminated plastic counter top.	Less than 10 LF of low-cost wall and base cabinets, laminated plastic counter top, space for table.
Plumbing (12% of total cost)	4 deluxe fixtures per bathroom, more bathrooms than bedrooms.	4 good fixtures per bathroom, more bathrooms than bedrooms.	3 good fixtures per bathroom, as many bathrooms as bedrooms.	3 standard fixtures per bathroom, less bathrooms than bedrooms.	3 standard fixtures per bathroom, less bathrooms than bedrooms.	3 minimum fixtures per bathroom, 2 bathrooms.
Special Features (3% of total cost)	10 luxury built-in appliances, wet bar, home theater, pantry, wine cellar.	8 good built-in appliances, wet bar, walk-in pantry, central vacuum.	6 good built-in appliances, walk-in pantry, wet bar, central vacuum.	5 standard built-in appliances, sliding glass or French doors, laundry room.	4 standard grade kitchen appliances.	4 minimum grade kitchen. appliances.
Electrical System (10% of total cost)	Over 100 recessed or track lights, security system, computer network.	80 to 100 recessed lighting fixtures. security system, computer network.	Ample recessed lighting on dimmers, computer network, multiple TV outlets.	Limited recessed lighting on dimmers, multiple TV outlets.	12 lighting fixtures, switch-operated duplex plug outlets in bedrooms.	10 or less lighting fixtures, switch-operated plug outlets in most rooms.
If Exterior Walls are Masonry	Reinforced split face concrete block or brick with face brick veneer.	Reinforced block or brick with masonry veneer or stucco coat.	Textured or coated concrete block or good quality detailed brick.	Colored or coated concrete block or good quality brick.	Colored concrete block or painted common brick.	Painted concrete block or common-brick.

Note: Use the percent of total cost to help identify the correct quality classification.

Single Family Residences

4 Corners (Classes 3, 4, 5 and 6) or One Building Mass (Classes 1 and 2 Only)

Estimating Procedure

1. Establish the structure quality class by applying the information on page 11.
2. Multiply the structure floor area (excluding the garage) by the appropriate square foot cost below.
3. Multiply the total from step 2 by the correct location factor listed on page 7 or 8.
4. Add, when appropriate, the cost of a porch, garage, heating and cooling equipment, basement, fireplace, carport, appliances and plumbing fixtures beyond that listed in the quality classification. See the cost of these items on pages 27 to 31.

Single Family Residence, Class 4

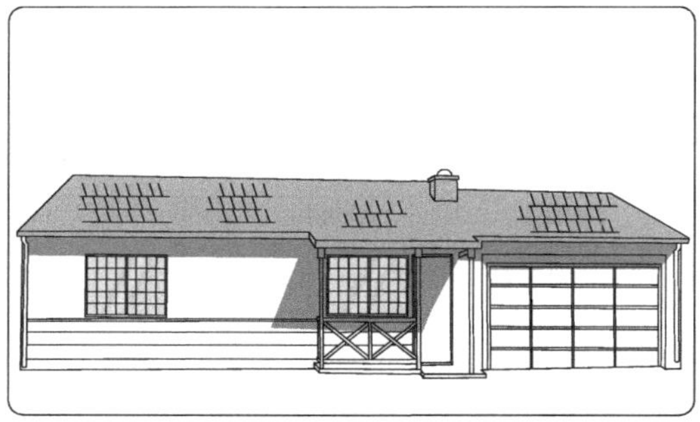

Single Family Residence, Class 6

Square Foot Area

Quality Class	700	800	900	1,000	1,100	1,200	1,300	1,400	1,500	1,600	1,700	1,800	2,000
1, Luxury	657.62	630.12	607.37	587.73	572.42	558.82	546.70	535.81	527.71	519.73	512.43	506.27	494.75
1, & 2	571.87	547.93	528.17	511.10	497.80	485.84	475.40	465.93	458.84	451.97	445.52	440.15	430.19
2, Semi-Luxury	399.67	382.96	369.13	357.17	347.90	339.62	332.29	325.69	320.71	315.73	311.43	307.72	300.58
2 & 3	293.35	281.14	270.95	262.24	255.43	249.27	243.89	239.05	235.38	231.82	228.53	225.90	220.71
3, Best Std.	256.01	245.37	236.45	228.85	222.79	217.53	212.86	208.64	205.41	202.30	199.49	197.06	192.59
3 & 4	218.95	209.65	202.14	195.69	190.47	185.97	181.98	178.30	175.64	172.81	170.58	168.47	164.70
4, Good Std.	188.62	180.60	174.17	168.58	164.19	160.29	156.76	153.62	151.22	148.99	146.92	145.00	141.89
4 & 5	169.92	162.77	156.95	151.85	147.86	144.27	141.12	138.46	136.29	134.19	132.37	130.77	127.69
5 Avg. Std.	152.91	146.60	141.29	136.75	133.26	130.01	127.17	124.56	122.67	120.82	119.13	117.73	115.02
5 & 6	132.78	127.23	122.65	118.71	115.58	112.82	110.34	108.08	106.50	104.84	103.58	102.17	99.86
6, Min. Std.	120.71	115.62	111.47	107.89	105.07	102.53	100.34	98.33	96.83	95.30	94.07	92.84	90.71

Square Foot Area

Quality Class	2,200	2,400	2,600	2,800	3,000	3,200	3,400	3,600	4,000	4,200	4,400	4,600	5,000+
1, Luxury	486.11	477.96	471.40	465.53	461.38	457.52	453.31	450.29	443.93	439.89	436.39	433.36	429.00
1, & 2	422.83	415.64	409.90	404.79	401.17	397.84	394.18	391.52	386.06	382.53	379.47	376.83	373.06
2, Semi-Luxury	295.61	290.48	286.54	282.93	280.36	277.98	275.44	273.62	269.78	267.34	265.18	263.37	260.73
2 & 3	216.90	213.25	210.33	207.72	205.76	203.99	202.26	200.86	198.06	196.28	194.68	193.33	191.40
3, Best Std.	189.28	186.06	183.46	181.26	179.64	178.11	176.46	175.22	172.79	172.81	171.45	170.24	168.55
3 & 4	161.84	159.11	156.94	155.03	153.55	152.20	150.97	149.91	147.81	146.48	145.29	144.28	142.83
4, Good Std.	139.45	137.04	135.23	133.47	132.37	131.16	130.05	129.04	127.31	126.15	125.10	124.24	123.00
4 & 5	125.57	123.54	121.63	120.26	119.12	118.19	117.00	116.31	114.70	113.65	112.81	111.98	110.87
5 Avg. Std.	113.08	111.21	109.68	108.19	107.33	106.39	105.41	104.70	103.26	101.78	101.52	100.84	99.86
5 & 6	98.17	96.53	95.17	93.96	93.19	92.27	91.44	90.78	89.66	88.72	88.16	87.47	86.67
6, Min. Std.	89.12	87.69	86.53	85.50	84.71	83.92	83.18	82.56	81.47	80.65	80.09	79.51	78.76

Note: Tract work and highly repetitive jobs may reduce the cost 8 to 12%. Add 4% to the square foot cost of floors above the second floor level. Work outside metropolitan areas may cost 2 to 6% less. When the exterior walls are masonry, add 9 to 10% for class 2 and 1 structures and 5 to 8% for class 3, 4, 5 and 6 structures. The building area includes all full story (7'6" to 9' high) areas within and including the exterior walls of all floor areas of the building, including small inset areas such as entrances outside the exterior wall but under the main roof. For areas with a ceiling height of less than 80", see the section on half-story areas on page 30.

Single Family Residences

6 Corners (Classes 3, 4, 5, and 6) or Two Building Masses (Classes 1 and 2 Only)

Estimating Procedure

1. Establish the structure quality class by applying the information on page 11.
2. Multiply the structure floor area (excluding the garage) by the appropriate square foot cost below.
3. Multiply the total from step 2 by the correct location factor listed on page 7 or 8.
4. Add, when appropriate, the cost of a porch, garage, heating and cooling equipment, basement, fireplace, carport, appliances and plumbing fixtures beyond that listed in the quality classification. See the cost of these items on pages 27 to 31.

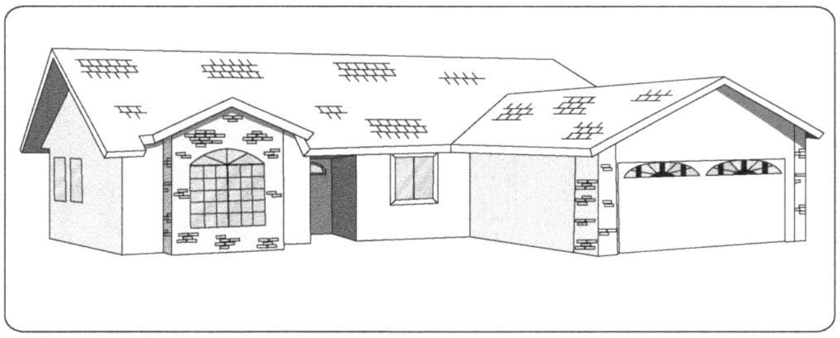

Single Family Residence, Class 5

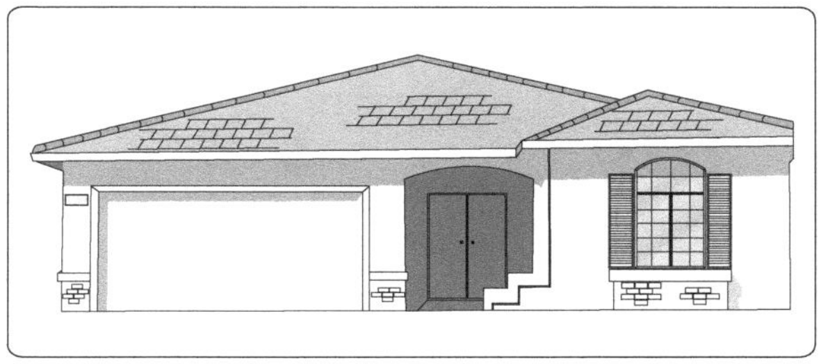

Single Family Residence, Class 5

Square Foot Area

Quality Class	700	800	900	1,000	1,100	1,200	1,300	1,400	1,500	1,600	1,700	1,800	2,000
1, Luxury	670.28	642.19	619.05	599.05	583.48	569.92	558.82	547.74	539.06	531.15	523.89	517.41	506.38
1, & 2	582.86	558.45	538.33	520.88	507.36	495.50	485.84	476.36	468.76	461.86	455.61	449.93	440.31
2, Semi-Luxury	407.44	390.40	375.96	364.62	354.57	346.35	339.62	332.90	327.60	322.74	318.42	314.42	307.79
2 & 3	299.02	286.56	275.99	267.65	260.22	254.17	249.27	244.36	240.44	236.91	233.73	230.74	225.93
3, Best Std.	260.95	250.02	240.86	233.50	227.15	221.84	217.53	213.25	209.90	206.74	203.94	201.45	197.13
3 & 4	223.10	213.85	205.79	199.69	194.27	189.59	186.07	182.29	179.52	176.73	174.41	172.13	168.49
4, Good Std.	192.26	184.24	177.36	172.00	167.34	163.40	160.29	157.09	154.50	152.30	150.35	148.40	145.13
4 & 5	173.25	165.98	159.67	154.93	150.68	147.13	144.27	141.44	139.33	137.15	135.37	133.63	130.83
5 Avg. Std.	155.95	149.49	143.82	139.55	135.75	132.50	130.01	127.57	125.51	123.59	121.90	120.36	117.73
5 & 6	135.36	129.63	124.89	121.14	117.73	114.92	112.82	110.59	108.84	107.21	105.81	104.37	102.24
6, Min. Std.	123.13	117.97	113.55	110.10	107.09	104.59	102.53	100.54	98.87	97.42	96.11	94.90	92.91

Square Foot Area

Quality Class	2,200	2,400	2,600	2,800	3,000	3,200	3,400	3,600	4,000	4,200	4,400	4,600	5,000+
1, Luxury	498.20	490.57	483.73	477.96	473.32	468.74	464.90	461.38	456.55	452.44	448.79	447.40	442.94
1, & 2	433.26	426.66	420.72	415.64	411.63	407.54	404.28	401.17	397.09	393.47	392.13	389.38	385.48
2, Semi-Luxury	302.90	298.06	293.95	290.48	287.59	284.83	282.52	280.36	277.48	274.95	272.76	270.87	268.15
2 & 3	222.26	218.80	215.83	213.25	211.12	209.09	207.39	205.76	203.62	202.10	199.59	198.09	196.69
3, Best Std.	193.92	190.95	188.24	186.06	184.32	182.45	180.95	179.64	177.72	176.14	174.71	173.48	171.76
3 & 4	165.80	163.27	160.93	159.11	157.51	155.95	154.68	153.55	151.85	150.46	149.29	148.26	146.78
4, Good Std.	142.85	140.58	138.63	137.04	135.73	134.34	133.33	132.37	130.92	129.79	128.76	127.81	126.57
4 & 5	128.71	126.62	124.89	123.54	122.16	121.06	120.08	119.12	117.94	116.86	115.93	115.12	113.96
5 Avg. Std.	115.83	114.07	112.57	111.21	110.06	108.99	108.10	107.33	106.19	105.27	104.39	103.67	102.65
5 & 6	100.54	98.87	97.54	96.53	95.50	94.67	93.86	93.19	92.10	91.27	90.54	89.93	88.99
6, Min. Std.	91.41	90.02	88.77	87.69	86.87	86.01	85.30	84.71	83.76	83.01	82.36	81.78	80.97

Note: Tract work and highly repetitive jobs may reduce the cost 8 to 12%. Add 4% to the square foot cost of floors above the second floor level. Work outside metropolitan areas may cost 2 to 6% less. When the exterior walls are masonry, add 9 to 10% for class 2 and 1 structures and 5 to 8% for class 3, 4, 5 and 6 structures. The building area includes all full story (7'6" to 9' high) areas within and including the exterior walls of all floor areas of the building, including small inset areas such as entrances outside the exterior wall but under the main roof. For areas with a ceiling height of less than 80", see the section on half-story areas on page 30.

Single Family Residences

8 Corners (Classes 3, 4, 5, and 6) or Three Building Masses (Classes 1 and 2 only)

Estimating Procedure

1. Establish the structure quality class by applying the information on page 11.
2. Multiply the structure floor area (excluding the garage) by the appropriate square foot cost below.
3. Multiply the total from step 2 by the correct location factor listed on page 7 or 8.
4. Add, when appropriate, the cost of a porch, garage, heating and cooling equipment, basement, fireplace, carport, appliances and plumbing fixtures beyond that listed in the quality classification. See the cost of these items on pages 27 to 31.

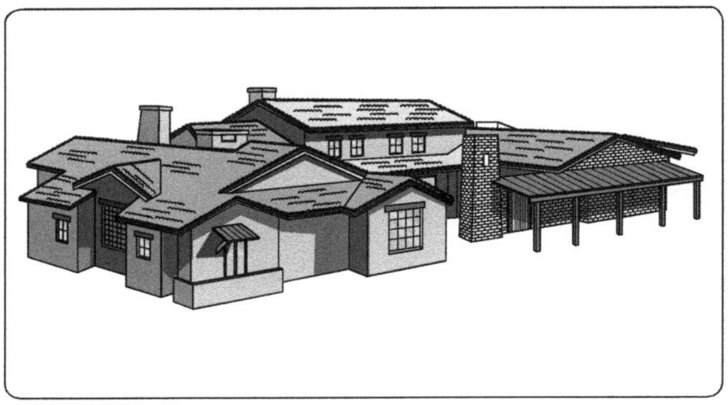

Single Family Residence, Class 1

Single Family Residence, Class 2 & 3

Square Foot Area

Quality Class	700	800	900	1,000	1,100	1,200	1,300	1,400	1,500	1,600	1,700	1,800	2,000
1, Luxury	683.92	655.69	631.42	611.52	595.31	581.90	569.92	559.19	549.59	542.24	535.19	529.03	518.03
1, & 2	594.65	570.29	549.02	531.83	517.62	506.02	495.50	486.30	477.93	471.49	465.38	460.01	450.51
2, Semi-Luxury	414.11	397.22	383.15	371.39	361.77	353.62	346.30	339.82	334.01	329.50	325.29	321.44	314.83
2 & 3	303.98	291.56	281.20	272.58	265.59	259.58	254.17	249.45	245.18	241.91	238.69	236.00	231.12
3, Best Std.	265.27	254.44	245.45	237.92	231.71	226.53	221.84	217.63	213.96	211.11	208.31	206.01	201.68
3 & 4	226.75	217.52	209.73	203.30	198.09	193.71	189.59	186.13	182.86	180.51	178.11	176.11	172.42
4, Good Std.	195.39	187.36	180.74	175.28	170.58	166.88	163.40	160.41	157.55	155.57	153.43	151.69	148.53
4 & 5	176.02	168.86	162.78	157.90	153.62	150.25	147.13	144.54	141.94	140.07	138.22	136.62	133.75
5 Avg. Std.	158.54	152.05	146.62	142.20	138.39	135.37	132.50	130.14	127.71	126.18	124.42	123.15	120.47
5 & 6	137.58	131.96	127.24	123.35	120.12	117.52	114.92	112.93	110.94	109.47	108.03	106.76	104.59
6, Min. Std.	125.04	119.95	115.66	112.14	109.16	106.76	104.59	102.71	100.83	99.51	98.24	94.70	92.97

Square Foot Area

Quality Class	2,200	2,400	2,600	2,800	3,000	3,200	3,400	3,600	4,000	4,200	4,400	4,600	5,000+
1, Luxury	509.30	508.23	494.75	489.62	485.03	480.97	476.16	473.45	467.55	463.32	459.59	456.39	451.83
1, & 2	442.89	433.26	430.19	425.72	421.77	418.27	414.02	411.74	406.66	402.98	399.75	396.94	392.98
2, Semi-Luxury	309.47	302.90	300.58	297.53	294.93	292.31	289.29	287.63	284.18	281.65	279.39	277.48	274.64
2 & 3	227.17	222.26	220.71	218.43	216.36	214.59	212.40	211.22	208.64	206.76	205.10	203.69	201.66
3, Best Std.	198.19	193.92	192.59	190.57	188.80	187.23	185.37	184.33	183.60	181.98	180.58	179.32	177.50
3 & 4	169.31	165.80	164.70	162.90	161.43	160.11	158.54	157.55	155.66	154.25	153.05	151.98	150.42
4, Good Std.	146.00	142.85	141.89	140.42	139.11	138.06	136.62	135.74	134.13	132.87	131.85	130.87	129.60
4 & 5	131.55	128.71	127.69	126.43	125.39	124.27	122.89	122.30	120.82	119.04	118.06	117.23	116.06
5 Avg. Std.	118.45	115.83	115.02	113.92	112.81	111.89	110.80	110.21	108.79	107.84	106.97	106.19	105.15
5 & 6	102.86	100.54	99.86	98.82	97.96	97.11	96.09	95.55	94.46	93.63	92.85	92.20	91.30
6, Min. Std.	91.42	89.50	88.88	88.05	87.24	86.51	85.72	85.15	84.21	83.48	82.82	83.65	81.41

Note: Tract work and highly repetitive jobs may reduce the cost 8 to 12%. Add 4% to the square foot cost of floors above the second floor level. Work outside metropolitan areas may cost 2 to 6% less. When the exterior walls are masonry, add 9 to 10% for class 2 and 1 structures and 5 to 8% for class 3, 4, 5 and 6 structures. The building area includes all full story (7'6" to 9' high) areas within and including the exterior walls of all floor areas of the building, including small inset areas such as entrances outside the exterior wall but under the main roof. For areas with a ceiling height of less than 80", see the section on half-story areas on page 30.

Single Family Residences

10 Corners (Classes 3, 4, 5 and 6) or Four Building Masses (Classes 1 and 2 only)

Estimating Procedure

1. Establish the structure quality class by applying the information on page 11.
2. Multiply the structure floor area (excluding the garage) by the appropriate square foot cost below.
3. Multiply the total from step 2 by the correct location factor listed on page 7 or 8.
4. Add, when appropriate, the cost of a porch, garage, heating and cooling equipment, basement, fireplace, carport, appliances and plumbing fixtures beyond that listed in the quality classification. See the cost of these items on pages 27 to 31.

Single Family Residence, Class 2 & 3

Single Family Residence, Class 1

Square Foot Area

Quality Class	700	800	900	1,000	1,100	1,200	1,300	1,400	1,500	1,600	1,700	1,800	2,000
1, Luxury	698.20	669.01	644.82	625.80	607.78	593.70	582.07	572.23	562.64	554.72	547.74	541.39	530.05
1, & 2	607.16	581.87	560.76	544.20	528.50	516.31	506.12	497.61	489.24	482.96	476.39	470.81	460.94
2, Semi-Luxury	420.66	404.00	390.41	378.46	369.37	360.83	353.76	347.77	341.94	337.12	332.90	328.97	322.09
2 & 3	308.82	296.60	286.58	277.89	271.11	264.82	259.61	255.21	251.00	247.36	244.36	241.46	236.46
3, Best Std.	269.48	258.77	250.07	242.43	236.56	231.18	226.58	222.74	219.07	215.94	213.25	210.77	206.32
3 & 4	230.43	221.10	213.85	207.28	202.30	197.56	193.75	190.41	187.21	184.62	182.29	180.22	176.46
4, Good Std.	198.56	190.63	184.35	178.62	174.18	170.25	166.88	164.15	161.31	159.11	157.09	155.21	152.01
4 & 5	178.85	171.69	165.99	160.91	156.99	153.40	150.35	147.83	145.36	143.36	141.44	139.78	136.96
5 Avg. Std.	161.08	154.51	149.54	144.92	141.38	138.21	135.45	133.12	130.92	128.98	127.57	125.95	123.35
5 & 6	139.71	134.15	129.68	125.76	122.67	119.87	117.58	115.52	113.56	111.95	110.59	109.18	107.09
6, Min. Std.	127.09	121.94	118.00	114.36	111.52	108.96	106.80	105.05	103.20	101.72	100.54	99.31	97.28

Square Foot Area

Quality Class	2,200	2,400	2,600	2,800	3,000	3,200	3,400	3,600	4,000	4,200	4,400	4,600	5,000+
1, Luxury	521.49	519.19	506.99	500.88	496.52	492.43	488.41	485.22	479.00	474.70	470.87	467.58	462.86
1, & 2	453.52	446.64	440.90	435.54	431.89	428.19	424.78	421.91	416.64	411.69	408.45	405.62	401.63
2, Semi-Luxury	316.90	312.16	308.09	304.47	301.79	299.26	296.83	294.97	291.13	288.53	286.16	284.18	281.32
2 & 3	232.62	229.08	226.15	223.45	221.46	219.71	217.85	216.45	213.75	205.35	203.74	202.29	200.29
3, Best Std.	203.01	200.01	197.39	194.95	193.27	191.62	190.13	188.89	186.53	184.85	183.37	182.09	180.26
3 & 4	173.48	170.96	168.74	166.75	165.26	163.82	162.61	161.49	159.35	157.92	156.67	155.55	154.01
4, Good Std.	149.54	147.32	145.53	143.56	142.33	141.15	140.07	139.33	137.37	136.14	134.58	133.21	131.88
4 & 5	134.77	132.65	130.97	129.36	128.24	127.17	126.18	125.42	123.67	122.58	121.59	120.76	119.53
5 Avg. Std.	121.19	119.47	117.97	116.52	115.43	114.59	113.62	112.98	111.37	110.34	109.47	108.70	107.64
5 & 6	105.27	103.69	102.30	101.05	100.13	99.42	98.60	97.97	96.70	95.86	95.04	94.39	93.47
6, Min. Std.	95.73	94.22	93.04	91.90	91.10	90.37	89.66	89.06	87.91	87.12	86.42	85.80	84.99

Note: Tract work and highly repetitive jobs may reduce the cost 8 to 12%. Add 4% to the square foot cost of floors above the second floor level. Work outside metropolitan areas may cost 2 to 6% less. When the exterior walls are masonry, add 9 to 10% for class 2 and 1 structures and 5 to 8% for class 3, 4, 5 and 6 structures. The building area includes all full story (7'6" to 9' high) areas within and including the exterior walls of all floor areas of the building, including small inset areas such as entrances outside the exterior wall but under the main roof. For areas with a ceiling height of less than 80", see the section on half-story areas on page 30.

Manufactured Housing

Quality Classification

	Class 1 Best Quality	Class 2 Good Quality	Class 3 Average Quality	Class 4 Low Quality	Class 5 Lowest Quality
Design	Indistinguishable from site-built construction, good floor plan and sight lines, superior fit and finish	Comparable to site-built construction, good floor plan, shelves and alcoves, good fit and finish	Clearly manufactured housing but with good design and materials, adequate fit and finish	Mobile home design, utilitarian floor plan, commodity-grade materials	Poor design, often sold unfinished, common only in Sun Belt states
Roof (12% of total cost)	Complex roof line, 30-year architectural shingles, roof pitch at least 4" in 12", good overhang on all sides, R-38 insulation	Decorative roof line, gable accents, 25-year shingles, 4" in 12" pitch, 12" overhang on all sides, R-33 insulation	Gable accents, 25-year shingles, 4" in 12" pitch, 8" to 12" overhang front and back, R-21 insulation	Simple roof line, less than 4" in 12" pitch, small overhang front and back, R-19 insulation	Straight roof line, minimum pitch, little or no overhang, minimum roof cover, R-7 insulation
Exterior Walls (18% of total cost)	Good fiber-cement siding, 9' to 10' high, decorative trim, 6" exterior walls, R-19 insulation, 7/16" plywood sheathing	Painted fiber cement siding, 9' high, some trim, 6" exterior walls, R-15 insulation, 7/16" OSB sheathing	Good foam-backed vinyl siding, 8' to 9' high, 4" exterior walls, R-13 insulation, 7/16" OSB sheathing	Vinyl siding, 8' high, 4" exterior walls, R-11 insulation, 3/8" plywood sheathing	Hardboard or economy siding, 7' high, 4" exterior walls, R-7 insulation
Doors and Windows (9% of total cost)	Two 36" wide insulated steel panel exterior doors, solid core wood panel interior doors, good hardware, large insulated low-E vinyl sash windows, recessed entry	Two 36" wide insulated steel exterior doors, hollow core wood interior doors, good hardware, good insulated low-E vinyl sash windows, recessed entry	36" wide steel front door with deadbolt, hollow core wood interior doors, average hardware, insulated vinyl windows, recessed entry	36" wide steel front door, hollow core wood interior doors, economy hardware, smaller dual glazed vinyl windows, 6' sliding bedroom door	34" or 32" wide aluminum exterior doors, hollow core wood interior doors, economy hardware, aluminum windows with storm sash
Interior (5% of total cost)	Hardwood paneling or ½" gypsum board with good workmanship and trim throughout, coffered/ vaulted/beamed ceilings, plank-type acoustical tile, mirrored walls, built-in buffet cabinets, custom drapes, skylights, window sills, good drapes with sheers throughout	Pre-finished hardwood paneling and trim or ½" gypsum board in all rooms, vaulted/beamed, ceiling in main rooms, good floor to ceiling drapes over sheer underlays in living room and dining room, several wall mirrors, some acoustic treatments	Pre-finished and grooved hardwood, plywood paneling or ½" gypsum board, no exposed fasteners, coordinated drapes in all rooms except kitchen and baths, one vaulted ceiling, acoustic tile, pre-finished wood trim	Pre-finished fire rated plywood paneling or 3/8" gypsum board, some exposed fasteners, acoustical tile ceiling, economy drapes in living room, dining room, and bedrooms, vinyl on composition molding.	Stapled 3/8" vinyl-covered wallboard with battens at seams and corners, exposed fasteners or holding strips, unit may have been sold with interior finishing incomplete.
Floors (8% of total cost)	Hardwood or ceramic tile entry, 30-50 oz. carpet, good vinyl in utility and guest bath. Good vinyl or hardwood in kitchen.	26-30 oz. carpet with 1/2" pad in all rooms except guest bath and utility, vinyl in kitchen, utility, and guest bath	22-26 oz. carpet with 1/2" rebond pad in all rooms except baths and kitchen, vinyl in kitchen and baths	16- 22 oz. carpet with 5 lb. pad in living, dining and bedrooms, economy vinyl sheet or tile in other areas	Glued or stapled foam-backed carpet in living room and bedroom, economy vinyl elsewhere
Heating (7% of total cost)	110,000 BTU upflow air-condition-ready forced air furnace with exterior access door, metal ducting to all rooms, fireplace, dual-zone heating	80,000 to 110,000 BTU upflow or downflow air-condition-ready furnace with exterior access door, metal ducting to all rooms, fireplace	80,000 BTU upflow or downflow forced air condition-ready furnace, ducting to all rooms, simulated fireplace	Forced air furnace, fiberglass attic ducting to all rooms, under-door return vents, ready for air conditioning unit.	Forced air furnace, minimum taped fiberglass duct, registers at the room center, return vents under doors
Kitchen (23% of total cost)	18± LF of 25" wide stone or ceramic counter, 4" splash, luxury cabinets, roller drawers, dropped luminous ceiling, island work space, walk-in pantry, name-brand fixtures, cast iron sink, wet bar	16± LF of tile or Corian counter, 4" splash, quality wood cabinets, dropped luminous ceiling, island work space, walk-in pantry, good quality fixtures, stainless or integrated 8" deep sink	14± LF of Corian counter, 2" splash, average quality wood-face cabinets and hardware, built-in range and oven with hood and fan, pantry cabinet, 7" deep stainless or porcelain sink	12± LF laminate counter, smaller commodity-grade cabinets with wood raised panel doors, no lining, built-in range and oven, hood and fan, add for dishwasher if present	10± LF of 24" wide laminate counter, plastic-faced MDF cabinets, stapled and glued, economy range and oven, minimum grade sink and fixtures, add for dishwasher if present
Baths and Plumbing (14% of total cost)	2 to 2¾ baths, 8 fixtures, master bath with two basins, sunken 60" tub, fiberglass shower with glass door, quality medicine cabinets, 6± feet of mirror over 8± feet of cultured marble or ceramic tile lavatory top, decorative faucets, 40-gal. water heater, separate commode closet	2 baths, vent fans, master bath will have two basins, sunken 60" tub and stall shower, quality medicine cabinets and fixtures, cultured marble vanities, good cabinets, 60" one-piece shower in guest bath, 30- to 40-gallon water heater, separate commode closet	2 baths, vent fans, fiberglass shower with glass or plastic door, fiberglass 60" tub, acrylic round toilets, 6 to 8 LF cultured marble vanity in each bath, twin basin master bath with 4± foot mirror, good cabinets, 30- to 40-gallon water heater	1¾ baths, fiberglass shower with plastic door, fiberglass one-piece 54" tub, acrylic round toilets, 4 to 5 linear foot cultured marble vanity with single basin, average quality cabinets and hardware, 30-gallon water heater	1¾ baths, fiberglass 54" one-piece tub and shower with curtain, acrylic round toilets, small 4' plastic marble vanity, minimum quality cabinets and hardware, 20-gallon electric water heater, plastic supply and drain pipe
Bedrooms (4% of total cost)	9 to 14 linear foot floor-to-ceiling sliding mirrored wardrobe doors, or large walk-in closets, phone and cable TV jacks	9 to 14 linear foot floor-to-ceiling mirrored sliding wardrobe doors in master bedroom or walk-in closets, phone and cable TV jacks	10± linear foot wardrobe, floor-to-ceiling mirrored sliding doors in master bedroom, cable TV jacks	8± linear foot wardrobe, pre-finished and grooved plywood doors, mirrored wardrobe door in master bedroom	Five to six linear foot wardrobe, plain plywood sliding doors

Manufactured Housing

A manufactured home is a structure in one or more sections intended to be delivered for erection as a unit on a construction site. No wheels, axles or towbars are included in these costs. Units can be from 8 to 36 feet wide and up to 80 feet long. Manufactured homes assembled from two or three sections are referred to as double wide or triple wide units. The cost FOB the manufacturer is usually about 2/3 of the installed cost. These figures include all costs: typical delivery to the site, setting on piers, finishing ("button up"), connection to utility lines, permits and inspections. Tip-out, expando, or tag-a-long units have one or more telescoping or attached rooms to the side. Include this floor area in your calculations. Do not use area modification factors for manufactured housing.

Estimating Procedure

1. Establish the structure quality class by applying the information on page 16.
2. Multiply the structure floor area (excluding any garage or storage area) by the appropriate square foot cost below.
3. Add, when appropriate, the cost of a permanent foundation, air conditioning, built-ins, porch, skirting, tie-downs, carport, garage or storage building, screen walls and roof snow load rating. See the following page.

Square Foot Area

Quality Class	500	700	900	1100	1300	1500	1700	1900	2100	2300	2500
1, Best	172.32	170.14	168.07	165.87	163.76	161.64	159.55	157.33	155.28	153.16	151.01
1, & 2	162.19	160.07	157.96	155.93	153.70	151.52	149.33	147.34	145.14	143.07	140.89
2, Good	152.03	149.92	147.85	141.62	139.66	137.64	135.44	133.43	131.27	129.28	127.25
2 & 3	142.02	139.80	137.77	129.40	127.34	125.35	123.30	121.33	119.26	117.19	115.24
3, Average	132.39	130.36	128.05	120.29	114.90	112.84	110.97	109.00	107.01	105.07	103.10
3 & 4	123.89	121.73	119.71	112.23	107.01	105.07	103.10	101.09	99.11	97.14	95.13
4, Low Average	115.35	113.32	111.12	104.07	99.11	97.14	95.13	93.14	91.25	89.25	87.28
4 & 5	108.47	106.23	104.21	97.40	92.69	90.73	88.78	86.79	84.86	82.89	80.82
5 Lowest	102.05	99.97	97.84	88.78	86.79	84.86	82.89	80.82	78.84	76.95	74.97

Manufactured Housing, Class 1

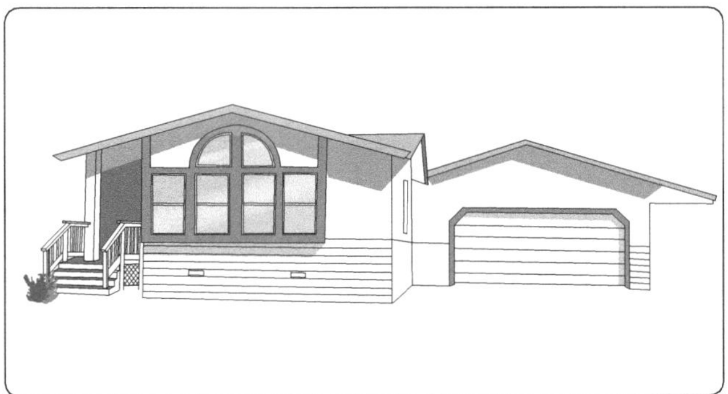

Manufactured Housing, Class 3

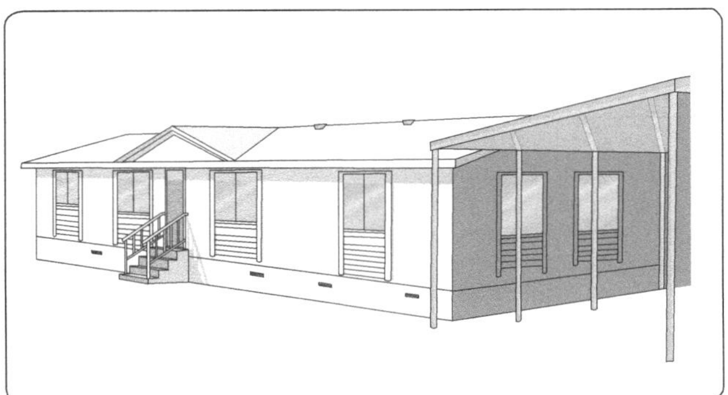

Manufactured Housing, Class 4

Manufactured Housing, Class 5

Manufactured Housing

Additional Costs

Permanent Foundation, in place of setting on piers

Single Story	
Less than 1,000 square feet of floor area	$9,900 to $17,500
Over 1,000 square feet to 1,800 square feet of floor area	$17,500 to $31,800
Over 1,800 square feet to 2,500 square feet of floor area	$31,800 to $52,400

For two-story units, use the footprint of the first floor and select a figure higher in the range of costs. For difficult site conditions, such as a high water table, heavy clay soil, rock, over 3' foundation depth or a sloping site, use a figure in the higher range of costs.

Air Conditioning

Central air for use by existing furnace and ducts	
2 ton, up to 1,100 S.F.	$3,600
2-1/2 to 3 ton, over 1,100 to 1,600 S.F.	$4,130
4 to 5 ton, over 1,600 to 2,500 S.F.	$4,535 to $5,340
Cost per unit	
Thru-wall small unit 1/2 H.P., 6,000 Btu	$1,250
Thru-wall large unit 1 H.P., 12,000 Btu	$1,660
Evaporative cooler, roof mounted	$1,180 to $1,870
Wiring for air conditioning	$227 to $478

Built-Ins

Dishwasher (included in classes 1, 2 & 3)	$970 - $1,290
Garbage disposal (included in all base cost, deduct if missing)	$200 - $1,200
Built-in microwave oven	$540 - $750
Trash compactor	$880 - $1,110
Wet bar (walk-up – if not included in class)	$770 - $930
Wet bar (walk behind – if not included in class)	$2,540 - $2,770
Separate shower in master bath	$880 - $1,110
One-half bath: toilet, sink, and pullman	$1,740 - $1,850
Bathroom sink or laundry sink	$370
Fireplace (permanent – includes flue)	$3,400 - $4,600
Fireplace (free standing – includes flue)	$1,550 - $2,770
Built-in buffet-hutch (included in classes 1 and 2)	$1,170 - $1,475
Whirlpool tub in master bath	$1,420 - $1,740

Porches and Decks (no roofs included)

Wood deck at home floor level with handrail, skirting, steps and outdoor carpet, per square foot of porch or deck	$20.80 to $29.20

Skirting, cost per linear foot of skirt

Lightweight aluminum panels	$7.40
Lap aluminum siding	$13.15
Painted hardboard panels	$17.10
Flagstone-type aluminum panels	$13.20
Concrete composite panels	$22.05 - $27.50
Vinyl panels	$14.70
Brick or stone	$23.10

Storage Buildings, Garages, per S.F. of floor

Aluminum exterior	$21.42
Enameled steel exterior	$16.89
Hardboard panel exterior	$37.54

Figure the garage cost per SF at 2/3 of the home cost per SF.

Tie Downs

Cork screw anchor and straps, per each	$108 - $161

Steps and Rails, per flight to 36" high

Fiberglass steps	$273 - $427
Handrail	$62 - $93

Carport, Porch, or Deck Roof, per S.F. covered

Aluminum supports and roof cover, free standing	$15.35 - $20.60
Aluminum supports and roof cover, attached to house	$9.90 - $14.45
Wood supports and enameled steel cover, free standing	$18.15 - $22.60

Screen Wall Enclosure, per linear foot of 8' wall

Wood frame with screen walls and door	$71.00
Wood or aluminum frame with screen and glass walls, with door	$124.00

Roof Snowload Capability

Cost per square foot of roof	
30 pound design load	$.78 - $1.25
40 pound design load	$1.24 - $2.25
50 pound design load	$2.25 - $2.98
60 pound design load	$2.97 - $3.97
80 pound design load	$3.76 - $5.97
100 pound design load	$4.95 - $6.85
175 pound design load	$6.28 - $7.57

Multi-Family Residences – Apartments

Quality Classification

	Class 1 Best Quality	Class 2 Good Quality	Class 3 High Average Quality	Class 4 Low Average Quality	Class 5 Minimum Quality
Foundation (9% of total cost)	Conventional crawl space built on a sloping site.	Conventional crawl space built on a sloping site.	Conventional crawl space, footing over 40" deep.	Concrete slab or crawl space with 30" footing.	Concrete slab.
Floor Structure (12% of total cost)	Engineered wood, steel or concrete exceeding code requirements, complex plan, changes in elevation.	Engineered wood or steel built to meet code requirements, changes in shape and elevation.	Standard wood frame with irregular shape and changes in elevation.	Standard wood frame or concrete slab, simple floor plan.	Simple slab on grade with no changes in elevation.
Walls and Exterior Finish (12% of total cost)	Complex wood or light steel frame, stone or masonry veneer, 10' average wall height.	Wood or light steel frame, masonry veneer at entrance, good wood or stucco siding.	Wood or light steel frame, decorative trim at entrance, plywood or stucco siding, simple framing plan.	Wood frame, some ornamental details at entrance, plywood or hardboard siding.	Wood frame, little or no ornamentation, inexpensive stucco or hardboard siding.
Roof & Cover (10% of total cost)	Complex roof plan, good insulation, tile or good shake cover.	Good insulation, good shake, tile or 5-ply built-up roof.	4-ply built-up roof, some portions heavy shake or tile.	4-ply built-up roof, some portions shake or composition shingles.	4-ply built-up roof or minimum grade composition single.
Windows and Doors (5% of total cost)	Many large, good quality vinyl or metal windows, architectural grade doors.	Large, good-quality vinyl or metal windows, commercial grade doors.	Good quality vinyl or metal windows, residential grade doors.	Standard residential-grade doors and windows.	Minimum grade doors and windows.
Interior Finish (8% of total cost)	Gypsum board with heavy texture or plaster, some paneled walls, cathedral ceiling at entry, built-in cases, several wall offsets and level changes.	Textured gypsum board, some paneled walls, decorative or stain grade trim at entrance or living room, several irregular walls and wall openings.	Textured 1/2" gypsum board, several irregular walls or wall openings, few ornamental details, standard grade trim and wall molding.	Textured 1/2" gypsum board, some wall-cover or hardboard paneling, most walls are rectangular, standard grade trim and wall molding.	1/2" gypsum board with smooth finish, no ornamental details, doors and windows are the only wall openings.
Floor Finish (5% of total cost)	Masonry or stone tile entry, good hardwood or deluxe carpet in most rooms, good sheet vinyl in other rooms.	Masonry or tile at entry, hardwood or good carpet in most rooms, sheet vinyl in other rooms.	Hardwood or tile at entry, standard carpet in most rooms, sheet vinyl in kitchen and bath.	Average quality carpet or hardwood in most rooms, sheet vinyl or resilient tile in kitchen.	Minimum carpet or resilient tile throughout.
Interior Features (5% of total cost)	Breakfast bar or nook, formal dining room, one walk-in closet, linen closet utility room or pantry.	Formal dining room ample closet space linen closet and utility closet, extra shelving.	Separate dining area, good closet space, linen closet and small utility closet.	Dining area is in the kitchen, small closet in each bedroom, linen closet.	Dining area is part of kitchen, minimum closet space, minimum shelving.
Bath Detail (4% of total cost)	Good tile shower, 8' simulated marble top.	Tile shower, 6' vanity cabinet and top.	Better vanity cabinet and good wall cabinet.	Good vanity cabinet, good medicine cabinet.	Vanity and one small medicine cabinet.
Kitchen (8% of total cost)	16 LF of better hardwood wall and base cabinets, synthetic stone top, 6 very good built-in appliances.	12 LF of good hardwood wall and base cabinets, tile or acrylic top, 5 good built-in appliances.	8 LF of standard hardwood wall and base cabinets, acrylic top, 4 standard grade built-in appliances.	6 LF of low-cost wall and base cabinets, laminate counter top, 4 standard grade appliances.	5 LF of low-cost. wall & base cabinets, laminate counter top, low cost appliances.
Electrical (10% of total cost)	Ample recessed lighting, task lighting in kitchen and bath, security & computer, networks, good chandelier.	Recessed lighting in most rooms, good task lighting in kitchen & bath, security & computer networks.	Recessed lighting in kitchen and living room, switched receptacles in bedrooms, wired for cable TV.	Low-cost recessed lighting in kitchen and living room, switched receptacles in other rooms, cable TV.	Fluorescent ceiling fixture in kitchen, switched receptacles in other rooms.
Plumbing (12% of total cost)	Four excellent fixtures per bathroom, copper supply and drain lines.	Three good fixtures per bathroom, copper supply and drain lines.	Three standard fixtures per bathroom, copper supply and plastic drain lines.	Three low cost fixtures per bathroom, plastic supply and drain lines.	Three minimum-grade fixtures per bathroom, plastic supply & drains.

Plumbing costs assume 1 bathroom per unit. See page 30 for the costs of additional bathrooms.

For Masonry Walls	Good textured block, tile or decorative brick.	Colored or detailed block tile or decorative brick.	Colored concrete block, tile or decorative brick.	Colored concrete block or brick.	Concrete block or common brick.

When masonry walls are used in lieu of wood or light steel frame walls, add 9% to the appropriate S.F. cost.

Note: Use the percent of total cost to help identify the correct quality classification. Exceptional class multi-family residences have architectural details and features uncommon in conventional apartment buildings. Many exceptional class multi-family structures are designed for sale or conversion to condominium ownership.

Multi-Family Residences – Apartments

2 or 3 Units

Estimating Procedure

1. Establish the structure quality class by applying the information on page 19.
2. Multiply the average unit area by the appropriate square foot cost below. The average unit area is found by dividing the building area on all floors by the number of units in the building. The building area should include office and utility rooms, interior hallways and interior stairways.
3. Multiply the total from step 2 by the correct location factor listed on page 7 or 8.
4. Add, when appropriate, the cost of balconies, porches, garages, heating and cooling equipment, basements, fireplaces, carports, appliances and plumbing fixtures beyond that listed in the quality classification. See the cost of these items on pages 27 to 31.
5. Costs assume one bathroom per unit. Add the cost of additional bathrooms from page 30.

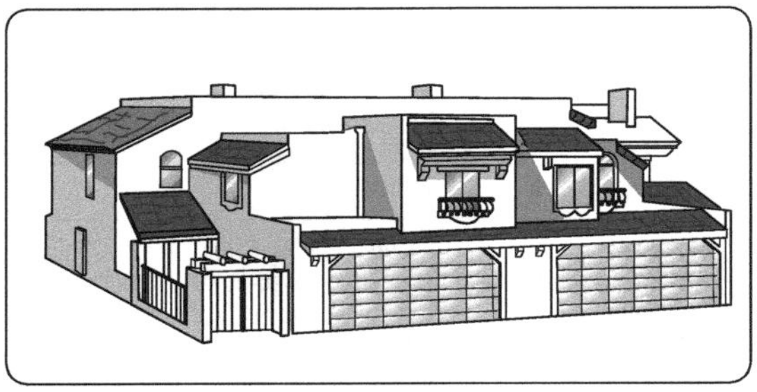

Multi-Family, Class 2

Multi-Family, Class 4

Average Unit Area in Square Feet

Quality Class	400	450	500	550	600	650	700	750	800	900	1,000
Exceptional	289.54	276.84	269.93	263.79	259.10	254.83	251.62	247.95	245.80	241.57	237.62
1, Best	254.30	243.06	237.03	231.66	227.41	223.84	220.99	217.74	215.88	212.03	208.73
1, & 2	223.01	213.19	207.84	203.07	199.51	196.31	193.75	191.08	189.30	185.86	182.96
2, Good	195.12	186.62	181.89	177.82	174.60	171.70	169.60	167.17	165.65	162.68	160.14
2 & 3	178.45	170.58	166.42	162.52	159.65	157.19	155.07	152.97	151.52	148.91	146.48
3, Hi Average	163.31	156.01	152.18	148.85	146.13	143.79	141.81	140.00	138.61	136.11	134.01
3 & 4	150.76	144.11	140.58	137.30	134.85	132.82	131.07	129.18	128.05	125.73	123.75
4, Lo Average	139.30	133.09	129.77	126.79	124.56	122.57	120.90	119.26	118.24	116.14	114.22
4 & 5	128.63	122.89	119.84	117.13	114.95	113.12	111.74	110.13	109.19	107.14	105.41
5 Minimum	118.69	113.56	110.66	108.14	106.29	104.51	103.09	101.83	100.83	98.87	97.41

Average Unit Area in Square Feet

Quality Class	1,100	1,200	1,300	1,400	1,500	1,600	1,700	1,800	1,900	2,000	2,200
Exceptional	234.82	232.29	230.24	228.51	227.04	225.69	224.54	223.52	222.57	221.84	221.15
1, Best	206.10	204.18	202.16	200.73	199.37	198.19	197.20	196.48	195.52	194.82	194.29
1, & 2	180.81	178.95	177.32	175.94	174.97	173.84	172.92	172.24	171.48	171.01	170.43
2, Good	158.15	156.63	155.23	154.05	153.09	152.11	151.39	150.68	150.06	149.49	149.09
2 & 3	144.82	143.14	142.09	140.86	140.01	139.14	138.46	137.97	137.28	136.86	136.40
3, Hi Average	132.43	131.07	129.90	128.83	128.06	127.31	126.62	126.25	125.51	125.15	124.78
3 & 4	122.30	120.92	119.88	118.94	118.29	117.49	117.06	116.39	115.91	115.60	115.21
4, Lo Average	112.93	111.74	110.68	109.86	109.21	108.55	107.95	107.48	107.04	106.70	106.38
4 & 5	104.26	103.20	102.34	101.41	100.87	100.21	99.67	99.34	98.82	98.54	98.24
5 Minimum	96.19	95.30	94.42	93.73	93.08	92.49	92.08	91.61	91.32	90.91	90.71

Note: Work outside metropolitan areas may cost 2 to 6% less. Add 2% to the costs for second floor areas and 4% for third floor areas. Add 9% when the exterior walls are masonry.

Multi-Family Residences – Apartments

4 to 9 Units

Estimating Procedure

1. Establish the structure quality class by applying the information on page 19.
2. Multiply the average unit area by the appropriate square foot cost below. The average unit area is found by dividing the building area on all floors by the number of units in the building. The building area should include office and utility rooms, interior hallways and interior stairways.
3. Multiply the total from step 2 by the correct location factor listed on page 7 or 8.
4. Add, when appropriate, the cost of balconies, porches, garages, heating and cooling equipment, basements, fireplaces, carports, appliances and plumbing fixtures beyond that listed in the quality classification. See the cost of these items on pages 27 to 31.
5. Costs assume one bathroom per unit. Add the cost of additional bathrooms from page 30.

Multi-Family, Class 3 & 4

Multi-Family, Class 3

Average Unit Area in Square Feet

Quality Class	400	450	500	550	600	650	700	750	800	900	1,000
Exceptional	272.57	260.52	253.88	248.53	243.77	239.93	237.09	233.78	231.62	227.16	223.52
1, Best	239.57	228.90	223.04	218.30	214.29	210.81	208.22	205.36	203.50	199.68	196.48
1, & 2	209.96	200.71	195.51	191.48	187.79	184.85	182.65	180.01	178.45	175.11	172.24
2, Good	183.76	175.67	171.19	167.47	164.40	161.78	159.78	157.47	156.01	153.16	150.68
2 & 3	168.13	160.64	156.63	153.22	150.35	147.95	146.21	144.09	142.85	140.04	137.97
3, Hi Average	153.79	147.07	143.19	140.11	137.37	135.32	133.76	131.70	130.71	128.07	126.25
3 & 4	142.09	135.73	132.15	129.35	127.03	125.04	123.41	121.59	120.70	118.38	116.39
4, Lo Average	131.12	125.25	122.24	119.49	117.27	115.37	113.89	112.41	111.44	109.27	107.48
4 & 5	121.06	115.72	112.87	110.33	108.22	106.55	105.33	103.79	102.86	100.91	99.34
5 Minimum	111.80	106.81	104.06	101.91	99.97	98.40	97.28	95.86	94.89	93.13	91.61

Average Unit Area in Square Feet

Quality Class	1,100	1,200	1,300	1,400	1,500	1,600	1,700	1,800	1,900	2,000	2,200
Exceptional	221.31	218.94	216.96	215.21	213.86	212.41	211.36	210.51	209.45	208.74	208.22
1, Best	194.31	192.35	190.56	189.04	187.92	186.63	185.71	184.89	183.95	183.40	182.96
1, & 2	170.45	168.61	167.17	165.77	164.68	163.62	162.86	162.14	161.35	160.87	160.39
2, Good	149.17	147.61	146.29	145.03	144.11	143.22	142.55	141.86	141.22	140.72	140.35
2 & 3	136.48	135.00	133.76	132.62	131.96	131.12	130.39	129.82	129.11	128.68	128.38
3, Hi Average	124.88	123.41	122.34	121.50	120.70	119.84	119.26	118.69	118.13	117.82	117.41
3 & 4	115.31	113.89	112.96	112.03	111.36	110.66	110.13	109.66	109.19	108.64	108.48
4, Lo Average	106.40	105.33	104.32	103.52	102.86	102.19	101.83	101.30	100.83	100.38	100.13
4 & 5	98.26	97.28	96.38	95.52	94.89	94.38	93.85	93.49	92.99	92.69	92.41
5 Minimum	90.74	89.85	88.97	88.19	87.68	87.18	86.72	86.39	85.93	85.51	85.41

Note: Work outside metropolitan areas may cost 2 to 6% less. Add 2% to the costs for second floor areas and 4% for third floor areas. Add 9% when the exterior walls are masonry.

Multi-Family Residences – Apartments

10 or More Units

Estimating Procedure

1. Establish the structure quality class by applying the information on page 19.
2. Multiply the average unit area by the appropriate square foot cost below. The average unit area is found by dividing the building area on all floors by the number of units in the building. The building area should include office and utility rooms, interior hallways and interior stairways.
3. Multiply the total from step 2 by the correct location factor listed on page 7 or 8.
4. Add, when appropriate, the cost of balconies, porches, garages, heating and cooling equipment, basements, fireplaces, carports, appliances and plumbing fixtures beyond that listed in the quality classification. See the cost of these items on pages 27 to 31.
5. Costs assume one bathroom per unit. Add the cost of additional bathrooms from page 30.

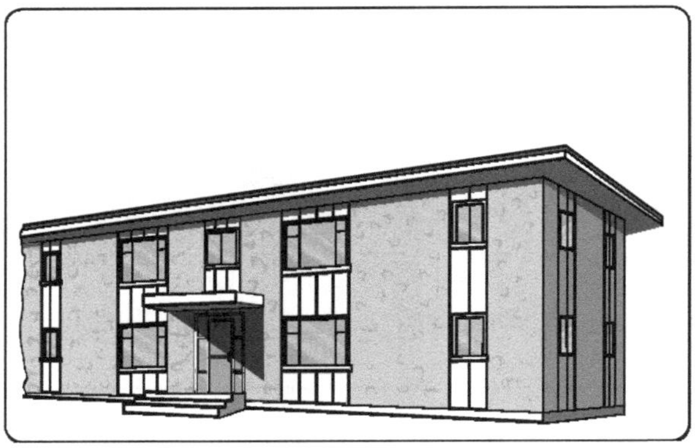

Multi-Family, Class 4

Multi-Family, Class 3 & 4

Average Unit in Square Feet

Quality Class	400	450	500	550	600	650	700	750	800	900	1,000
Exceptional	257.70	245.00	239.93	234.94	230.24	226.64	223.52	220.77	218.52	214.71	211.58
1, Best	226.25	215.25	210.81	206.33	202.16	199.16	196.48	193.96	191.90	188.71	185.83
1, & 2	198.40	188.79	184.85	180.95	177.32	174.69	172.24	169.97	168.41	165.29	163.02
2, Good	173.65	165.16	161.78	158.29	155.23	152.86	150.68	148.86	147.34	144.75	142.57
2 & 3	158.82	151.10	147.95	144.94	142.09	139.85	137.97	136.15	134.84	132.43	130.42
3, Hi Average	145.29	138.13	135.32	132.53	129.90	127.84	126.25	124.56	123.27	121.06	119.33
3 & 4	134.20	127.73	125.04	122.34	119.88	118.01	116.39	114.95	113.80	111.87	110.24
4, Lo Average	123.84	117.89	115.37	112.96	110.68	109.16	107.48	106.29	105.27	103.22	101.85
4 & 5	114.45	108.94	106.55	104.32	102.34	100.69	99.34	97.98	97.05	95.32	93.94
5 Minimum	105.63	100.51	98.40	96.38	94.42	92.98	91.61	90.57	89.57	88.03	86.74

Average Unit in Square Feet

Quality Class	1,100	1,200	1,300	1,400	1,500	1,600	1,700	1,800	1,900	2,000	2,200
Exceptional	208.91	206.84	204.81	203.29	201.67	200.60	199.81	198.79	197.94	197.06	196.65
1, Best	183.50	181.67	180.00	178.45	177.12	176.16	175.53	174.61	173.98	173.10	172.68
1, & 2	160.91	159.28	157.82	156.63	155.34	154.60	153.92	153.17	152.47	151.89	151.35
2, Good	140.85	139.46	138.13	136.96	135.83	135.27	134.60	134.02	133.31	132.82	132.53
2 & 3	128.94	127.60	126.38	125.37	124.20	123.67	123.23	122.57	122.06	121.53	121.20
3, Hi Average	117.89	116.75	115.60	114.64	113.70	113.12	112.72	112.13	111.67	111.18	110.79
3 & 4	108.94	107.74	106.80	105.88	104.93	104.51	103.97	103.61	103.09	102.65	102.43
4, Lo Average	100.51	99.51	98.54	97.74	97.03	96.51	96.02	95.57	95.29	94.77	94.60
4 & 5	92.82	92.00	91.05	90.32	89.55	89.02	88.68	88.34	87.90	87.58	87.28
5 Minimum	85.71	84.81	84.08	83.42	82.67	82.22	81.94	81.52	81.19	80.83	80.66

Note: Work outside metropolitan areas may cost 2 to 6% less. Add 2% to the costs for second floor areas and 4% for third floor areas. Add 9% when the exterior walls are masonry.

Motels

Quality Classification

	Class 1 Best Quality	Class 2 Good Quality	Class 3 Average Quality	Class 4 Low Quality
Foundation (4%)	Concrete slab	Concrete slab	Concrete slab	Concrete slab
Foundation costs will vary greatly with substrate, type, and location.				
Framing* (20% of total Cost)	Wood frame.	Wood frame.	Wood frame.	Wood frame.
Windows (2% of total Cost)	Large, good quality.	Average number and quality.	Average number and quality.	Small, few, low cost.
Roofing (8% of total Cost)	Heavy, shake, tile or slate.	Medium shake or good built-up with large rock, inexpensive tile.	Wood or good composition shingle, light shake, or good built-up with rock.	Inexpensive shingles or built-up with rock.
Overhang (2% of total Cost)	36" open or 24" closed.	30" open or small closed.	16" open.	12" to 16" open.
Exterior Walls (10% of total Cost)	Good wood or stucco, masonry veneer on front.	Good wood siding or stucco with some veneer.	Hardboard, wood shingle, plywood or stucco.	Low cost stucco, hardboard or plywood.
Flooring (5% of total Cost)	Good carpet, good sheet vinyl.	Good carpet, sheet vinyl or inlaid resilient.	Average carpet, average resilient tile in bath.	Minimum tile or low cost carpet.
Interior Finish (23% of total cost including finish carpentry, wiring, lighting, etc.)	Gypsum board with heavy texture or plaster with putty coat. Some good sheet wall cover or paneling.	Gypsum board, taped, textured and painted or plaster. Some wall-paper.	Gypsum board taped and textured or colored interior stucco.	Minimum gyp-sum board.
Baths (15% of total Cost)	Vinyl or foil wall cover, ceramic tile over tub with glass shower door, ample mirrors.	Ceramic tile over tub with glass shower door.	Plastic coated hard-board with low cost glass shower door.	Plastic coated hardboard with one small mirror.
Plumbing** (9% of total Cost)	Copper tube, good quality fixtures.	Galvanized pipe, good fixtures.	Average cost fixtures.	Plastic pipe, low cost fixtures.
Special Features (2% of total Cost)	8' sliding glass door, 8' to 10' tile pullman in bath.	8' sliding glass door, good tile or plastic top pullman in bath.	Small tile or plastic pullman in bath.	None.
***For Masonry Walls**	8" textured face reinforced masonry.	8" colored or detailed reinforced masonry.	8" colored block or common brick, reinforced.	8" painted concrete block.

Note: When masonry walls are used in lieu of wood frame walls add 8% to the appropriate cost

****Add the Following Amounts per Kitchen Unit**

Kitchens	Good sink, 8' to 10' of good cabinets and drainboard - $4,300	Average sink and 6' to 8' average cabinet and drainboard - $3,940	Low cost sink, and 5' of cabinets and drainboard - $2,840	Minimum sink, cabinets and drainboard - $2,410

Add the cost of built-in kitchen fixtures from the table of costs for built-in appliances on page 29.

Note: Use the percent of total cost to help identify the correct quality classification.

Motels

9 Units or Less

Estimating Procedure

1. Establish the structure quality class by applying the information on page 23.
2. Multiply the average unit area by the appropriate cost below. The average unit area is found by dividing the total building area on all floors (including office and manager's area, utility rooms, interior hallways and stairway area) by the number of units in the building.
3. Multiply the total from step 2 by the correct location factor listed on page 7 or 8.
4. Add, when appropriate, the cost of heating and cooling equipment, porches, balconies, exterior stairs, garages, kitchens, built-in kitchen appliances and fireplaces. See pages 23 and 27 to 31.

Motel, Class 3 & 4

Average Unit Area in Square Feet

Quality Class	200	225	250	275	300	330	375	425	500	600	720
1, Best	230.91	222.62	216.15	210.66	206.13	201.66	196.28	191.52	186.20	181.12	176.94
1 & 2	212.12	204.49	198.55	193.54	189.41	185.20	180.19	175.87	171.00	166.43	162.48
2, Good	196.82	189.84	184.24	179.64	175.77	171.96	167.29	163.32	158.72	154.41	150.84
2 & 3	180.86	174.47	169.29	165.04	161.52	157.98	153.66	150.03	145.84	141.94	138.68
3, Average	167.85	161.85	157.15	153.16	149.85	146.56	142.63	139.14	135.32	131.66	128.67
3 & 4	154.06	148.57	144.22	140.56	137.58	134.53	130.86	127.80	124.17	120.87	118.03
4, Low	140.83	135.74	131.74	128.50	125.68	122.99	119.63	116.80	113.49	110.41	107.84

Note: Add 2% for work above the first floor. Work outside metropolitan areas may cost 2 to 6% less. Add 8% when the exterior walls are masonry. Deduct 2% for area built on a concrete slab.

Motels

10 to 24 Units

Estimating Procedure

1. Establish the structure quality class by applying the information on page 23.
2. Multiply the average unit area by the appropriate cost below. The average unit area is found by dividing the total building area on all floors (including office and manager's area, utility rooms, interior hallways and stairway area) by the number of units in the building.
3. Multiply the total from step 2 by the correct location factor listed on page 7 or 8.
4. Add, when appropriate, the cost of heating and cooling equipment, porches, balconies, exterior stairs, garages, kitchens, built-in kitchen appliances and fireplaces. See pages 23 and 27 to 31.

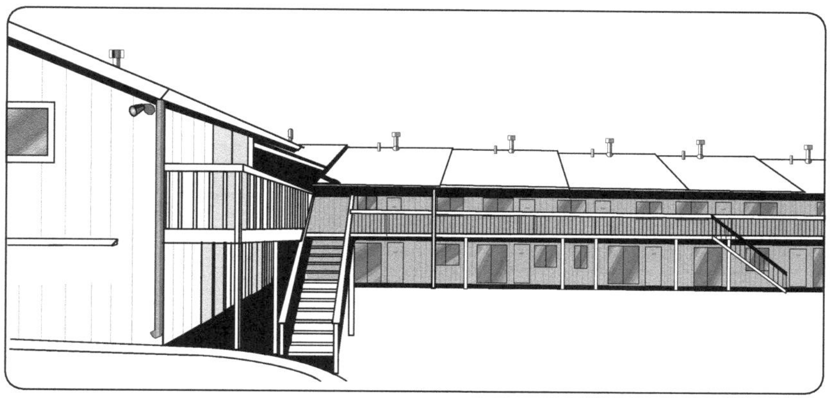

Motel, Class 3

Average Unit Area in Square Feet

Quality Class	200	225	250	275	300	330	375	425	500	600	720
1, Best	223.73	215.79	209.36	204.18	199.82	195.40	190.15	185.61	180.46	175.49	171.48
1 & 2	205.52	198.24	192.33	187.54	183.48	179.49	174.63	170.47	165.73	161.26	157.47
2, Good	190.87	184.12	178.60	174.12	170.43	166.70	162.17	158.25	153.84	149.73	146.29
2 & 3	175.33	169.05	163.95	159.88	156.48	152.93	148.93	145.34	141.27	137.43	134.28
3, Average	162.61	156.83	152.22	148.35	145.16	142.01	138.13	134.84	131.11	127.60	124.58
3 & 4	149.24	143.93	139.71	136.15	133.26	130.33	126.76	123.81	120.27	117.09	114.37
4, Low	136.43	131.52	127.66	124.51	121.80	119.19	115.90	113.15	110.03	107.06	104.57

Note: Add 2% for work above the first floor. Work outside metropolitan areas may cost 2 to 6% less. Add 8% when the exterior walls are masonry. Deduct 2% for area built on a concrete slab.

Motels

Over 24 Units

Estimating Procedure

1. Establish the structure quality class by applying the information on page 23.
2. Multiply the average unit area by the appropriate cost below. The average unit area is found by dividing the total building area on all floors (including office and manager's area, utility rooms, interior hallways and stairway area) by the number of units in the building.
3. Multiply the total from step 2 by the correct location factor listed on page 7 or 8.
4. Add, when appropriate, the cost of heating and cooling equipment, porches, balconies, exterior stairs, garages, kitchens, built-in kitchen appliances and fireplaces. See pages 23 and 27 to 31.

Motel, Class 2 & 3

Average Unit Area in Square Feet

Quality Class	200	225	250	275	300	330	375	425	500	600	720
1, Best	216.35	208.64	202.45	197.42	193.11	188.91	183.82	179.48	174.50	169.75	165.88
1 & 2	198.85	191.70	186.00	181.40	177.46	173.56	168.87	164.88	160.29	155.95	152.37
2, Good	184.57	178.07	172.73	168.42	164.85	161.20	156.93	153.06	148.89	144.82	141.51
2 & 3	169.54	163.44	158.62	154.70	151.29	148.01	144.05	140.50	136.67	133.02	129.98
3, Average	157.24	151.60	147.18	143.48	140.36	137.35	133.64	130.45	126.81	123.36	120.56
3 & 4	144.32	139.12	134.98	131.57	128.81	126.06	122.56	119.64	116.37	113.22	110.60
4, Low	131.95	127.19	123.38	120.29	117.71	115.18	112.05	109.38	106.37	103.56	101.08

Note: Add 2% for work above the first floor. Work outside metropolitan areas may cost 2 to 6% less. Add 8% when the exterior walls are masonry. Deduct 2% for area built on a concrete slab.

Additional Costs for Residential Structures

Covered Porches

Estimate covered porches by applying a fraction of the main building square foot cost.

Porch Description	Suggested Fraction
Ground level floor (usually concrete) without banister, with no ceiling and shed-type roof.	1/4 to 1/3
High (house floor level) floor (concrete or wood) with light banister, no ceiling and shed-type roof.	1/3 to 1/2
Same as above with a finished ceiling and roof like the residence (most typical).	1/2
Same as above but partially enclosed with screen or glass.	1/2 to 2/3
Enclosed lean-to (sleeping porch, etc.) with lighter foundation, wall structure, interior finish or roof than that of house to which it is attached.	1/2 to 3/4
Roofed, enclosed, recessed porch, under the same roof as the main building and with the same type and quality foundation (includes shape costs).	3/4
Roofed, enclosed, recessed porch with the same type roof and foundation as the main building (includes shape costs).	4/4
Good arbor or pergola with floor.	1/4 to 1/3

Uncovered Concrete Decks, cost per square foot, 4" thick concrete

	On Grade	1' High	2' High	3' High	4' High
Less than 100 square feet	$11.71	$16.37	$26.34	$37.01	$53.82
100 to 200 square feet	10.77	14.77	21.35	30.04	40.01
200 to 400 square feet	9.05	11.71	18.32	26.62	34.45
Over 400 square feet	8.78	10.77	16.09	21.37	27.76

Uncovered Wood Decks, cost per square foot, 2" thick deck with typical steps and railing

1' to 4' above ground.	$26.57 to $28.51
Over 4' to 6' above ground	31.24 to 40.32
Over 6' to 9' above ground	32.55 to 42.11
Over 9' to 12' above ground	33.87 to 44.75
Over 12' above ground	35.49 to 46.40

Porch Roofs, cost per square foot based on wood shingle cover

Type	Cost per Square Foot
Unceiled shed roof	$10.20 to $12.10
Ceiled shed roof	17.13 to 19.31
Unceiled gable roof	11.40 to 14.85
Ceiled gable roof	19.30 to 21.50

(See the figures at the right for other roof cover)

Alternate Roof Covers	Cost Difference per S.F.	
Corrugated aluminum	Deduct	$.88 to $1.10
Roll asphalt	Deduct	.87 to .97
Fiberglass shingles	Deduct	1.08 to 1.19
Wood shakes	Add	1.18 to 1.83
Clay or concrete tile	Add	6.83 to 8.36
Slate	Add	7.60 to 10.50

Residential Basements, cost per square foot, including stairs

Size	Unfinished Basements	Finished Basements
Less than 400 square feet	$33 to $54	$50 to $75
400 - 1,000 square feet	25 to 36	41 to 46
Over 1,000 square feet	22 to 26	38 to 44

These basement costs assume normal soil conditions, 7' headroom, no plumbing, partitions or windows. Unfinished basements have reinforced concrete floors and concrete or concrete block walls, a floor drain, stairway with a landing and handrail, open ceilings and one switched light fixture. Finished residential basements have a tile ceiling, resilient flooring, wood panel walls and lighting similar to Class 5 residences. Residential basements are common in climates where footing depths must be 4' or more to prevent frost heaving. These figures assume the residence is in an area where minimum footing depth is 4 feet. Where climate doesn't influence footing depth, unfinished basement costs will be 20% to 50% higher.

Additional Costs for Residential Structures

Balconies, Standard Wood Frame, cost per square foot, including foundations

Supported by 4" x 4" posts, 2" wood floor, open on underside, open 2" x 4" railing.	$24.40 to $26.30
Supported by 4" x 4" posts, 2" wood floor, sealed on underside, solid stucco or wood siding on railing.	29.60 to 31.30
Supported by steel columns, lightweight concrete floor, sealed on underside, solid stucco or open grillwork railing.	43.90 to 48.50

Heating and Cooling Equipment

Prices include wiring and minimum duct work.

Use the higher figures for smaller residences and in more extreme climates where greater heating and cooling density is required. Cost per square foot of heated or cooled area.

Type	Perimeter Outlets	Overhead Outlets
Central Ducted Air Systems, Single Family		
Forced air heating	$5.93 to $6.60	$4.65 to $5.33
Forced air heating and cooling	6.72 to 8.00	6.29 to 6.77
Gravity heat	4.31 to 5.80	—
Central Ducted Air Systems, Multi-Family		
Forced air heating	5.24 to 5.66	4.92 to 5.65
Forced air heating and cooling	7.12 to 7.83	6.23 to 6.70
Motel Units		
Forced air heating	6.04 to 6.51	5.88 to 6.42
Forced air heating and cooling	7.25 to 7.83	7.01 to 7.26
Circulating hot and cold water system	13.86 to 16.78	14.07 to 16.78

Floor and Wall Furnaces, cost each

Single floor unit	$1,140 to $1,320
Dual floor unit	1,980 to 2,160
Single wall unit	765 to 900
Dual wall unit	1,395 to 1,650
Thermostat control, add	126 to 151

Electric Baseboard Units, cost each

500 watts, 3'	$215 to $252
1,000 watts, 4'	331 to 380
1,500 watts, 6'	363 to 410
2,000 watts, 8'	460 to 530
2,500 watts, 10'	540 to 606
3,000 watts, 12'	660 to 720

Outside Stairways, cost per square foot of horizontal step area

Standard wood frame, wood steps with open risers, open on underside, open 2" x 4" railing, unpainted.	$19.19 to $21.12
Standard wood frame, solid wood risers, sealed on underside, solid stucco or wood siding on railing.	23.11 to 27.30
Precast concrete steps with open risers, steel frame, pipe rail with ornamental grillwork.	50.38 to 56.18

Electric vehicle (EV) charging station hookup. 220 volt Level 2 wall mounted NEMA 14-50 receptacle, 60amp breaker and 10' of 6-gauge 3 conductor cable.

32 amp	$270
40 amp	305
50 amp	365
Add for Level 2 charging station, hard wired	870

Ductless mini-split heating and cooling unit. Includes pad-mounted compressor-condenser, 8' of insulated copper refrigerant lines, PVC condensate drain, control wiring, PVC wall chase, clamps, brackets, interior wall-mounted evaporator and wireless control.

9,500 BTU (3/4 ton, 110 volt)	$1,130
18,000 BTU (1-1/2 ton, 230 volt)	1,420
24,000 BTU (2 ton, 230 volt)	1,780
42,000 BTU (3-1/2 ton, 230 volt, 5-zone)	5,730

Window Type or Thru-the-Wall Refrigerated Room Coolers, cost each

1/3 ton	$165 to $205
1/2	590 to 720
3/4	297 to 357
1	363 to 430
1-1/2	515 to 610
2	880 to 1,050

Ton = 12,000 Btu

Electric Wall Heaters, cost each

500 watts	$154 to $186
1,000	156 to 190
2,000	180 to 216
3,000	203 to 244
Add for circulating fan	86 to 126
Add for thermostat	57 to 126

Additional Costs for Residential Structures

Appliances. Add these costs only when the appliance is not included in the quality class. Includes installation.

Built-in single wall oven with broiler	$655	to $790
Built-in double wall oven with microwave	1,185	to 2,240
Drop-in range with single oven, economy	524	to 790
Drop-in range with single oven, excellent	1,309	to 2,460
Range top, four elements		
Residential grade, without grill	589	to 1,110
Residential grade, with grill	974	to 1,650
Commercial grade	4,470	to 7,480
Hot water circulator	726	to 786
Instant hot water dispenser	595	to 840
Range hood and fan	$196	to $464
Franklin or Buck stove		
Steel, cast iron front	1,566	to 2,370
Steel, cast iron front, glass door	2,370	to 3,280
All cast iron, glass panel door	4,070	to 5,870
Under counter 5 CF refrigerator	726	to 1,050
Central vacuum, 3 to 5 outlets	2,235	to 4,470
Dishwasher	357	to 1,320
Garbage disposal	238	to 555
Trash compactor	449	to 753

Fireplaces, cost each, including reinforced foundation, flue, cap, gas line and valve.

	1 Story	2 Story
Freestanding wood burning heat circulating prefab metal fireplace with interior flue, base and cap	$2,245	$2,750
36" wide zero-clearance enclosed metal firebox, brick face, wood mantel	2,600	2,990
48" wide zero-clearance enclosed metal firebox, raised hearth, brick face and mantel	3,590	4,040
Masonry, 5' base, common brick or block on interior face, wood or brick mantle	5,730	6,480
Masonry, 6' base, used brick or natural stone on interior face, raised hearth	11,300	13,500
Masonry, 8' base, used brick or natural stone on interior face, raised hearth	13,600	19,900

Residential Garages and Carports

Attached and detached garages for single family dwellings usually fall in the same quality class as the main structure. Costs are per SF of floor based on wood or light steel construction. Add 8% if exterior walls are masonry. Attached garages assume a common roof and a 20 foot wall in common with the main structure. Multiply the square foot cost below by the correct location factor on page 7 or 8 to find the square foot cost for any garage. Costs include interior finish and one light fixture per 300 SF of floor. Deduct 10% to 18% if interior walls are unfinished. Where dwelling and exterior garage walls are in vertical alignment with second floor walls, the garage cost per SF will be about 2/3 of the main dwelling cost per SF if finished and 1/2 of the main dwelling cost if unfinished. Carports with wood or steel posts, an asphalt floor, and built-up or metal roof will cost $15.80 to $18.30 per SF.

Square Foot Area for Attached Garages for Single Family Dwellings

Quality Class	220	260	280	320	360	400	440	480	540	600	720
1, Luxury	199.09	189.70	185.70	179.88	173.10	168.81	163.75	159.72	155.77	151.88	148.11
1, & 2	172.72	164.72	161.41	156.08	150.74	146.98	142.61	139.07	135.60	132.24	128.95
2, Semi-Luxury	129.82	123.98	121.62	117.66	113.63	110.81	107.51	104.84	102.25	99.69	97.20
2 & 3	105.03	98.38	96.86	95.58	92.29	90.01	87.31	85.18	83.05	80.98	78.98
3, Best Std.	87.30	83.49	81.97	79.36	76.91	74.98	72.76	70.92	69.19	67.45	65.77
3 & 4	73.92	70.93	69.73	67.61	65.15	63.55	61.65	60.11	58.61	57.17	55.75
4, Good Std.	65.46	62.44	61.33	59.68	57.68	56.23	54.56	53.20	51.88	50.59	49.35
4 & 5	61.76	58.44	57.10	55.22	53.26	51.93	50.36	49.12	47.88	46.72	45.55
5 Avg. Std.	57.83	54.29	53.00	50.99	48.81	47.58	46.15	45.02	43.92	42.80	41.76
5 & 6	51.31	48.45	47.37	45.54	43.87	42.75	41.47	40.44	39.46	38.47	37.50
6, Min. Std.	44.96	42.63	41.95	40.57	39.02	38.04	36.93	35.99	35.07	34.24	33.36

Square Foot Area for Detached Garages for Single Family Dwellings

Quality Class	220	260	280	320	360	400	440	480	540	600	720
1, Luxury	226.46	209.30	203.06	192.22	188.41	182.45	174.50	170.18	165.94	161.82	157.81
1, & 2	195.11	180.91	175.13	166.17	163.16	157.97	151.12	147.35	143.69	140.10	136.63
2, Semi-Luxury	145.43	135.22	131.23	124.54	122.39	118.51	113.32	110.53	107.78	105.11	102.47
2 & 3	117.57	109.22	105.89	100.57	98.94	95.77	91.62	89.34	87.10	84.95	82.84
3, Best Std.	97.34	90.39	87.74	83.44	82.09	79.50	75.99	74.13	72.26	70.50	68.74
3 & 4	86.22	80.25	77.85	74.11	73.02	70.64	67.60	65.93	64.26	62.69	61.12
4, Good Std.	76.04	70.81	68.72	65.32	64.35	62.30	59.60	58.09	56.68	55.27	53.89
4 & 5	70.26	65.39	63.49	59.82	59.56	57.68	55.16	53.79	52.46	51.13	49.87
5 Avg. Std.	66.38	60.78	58.61	55.27	54.08	52.35	50.06	48.81	47.61	46.42	45.28
5 & 6	55.86	51.60	49.81	47.25	46.45	44.98	43.01	41.93	40.89	39.89	38.88
6, Min. Std.	48.66	44.95	43.67	41.44	40.87	39.58	37.85	36.93	35.98	35.06	34.24

Additional Costs for Residential Structures

Costs for Multi-Family Residential Bathrooms beyond 1 per unit

	Class 1 Best Quality	Class 2 Good Quality	Class 3 High Average	Class 4 Low Average	Class 5 Minimum Quality
2 or 3 units					
2 fixture bath	$9,846	$7,924	$6,702	$5,584	$4,726
3 fixture bath	14,293	12,185	10,109	8,701	6,966
4 fixture bath	18,187	15,720	14,034	11,498	9,844
4 to 9 units					
2 fixture bath	9,088	7,534	6,297	5,262	4,338
3 fixture bath	12,860	11,170	9,614	7,993	6,431
4 fixture bath	17,796	15,068	12,457	10,380	8,575
10 or more units					
2 fixture bath	8,183	6,966	5,912	4,598	3,793
3 fixture bath	12,602	10,523	8,836	6,963	5,717
4 fixture bath	16,627	14,293	11,431	9,354	7,145

Half Story Areas

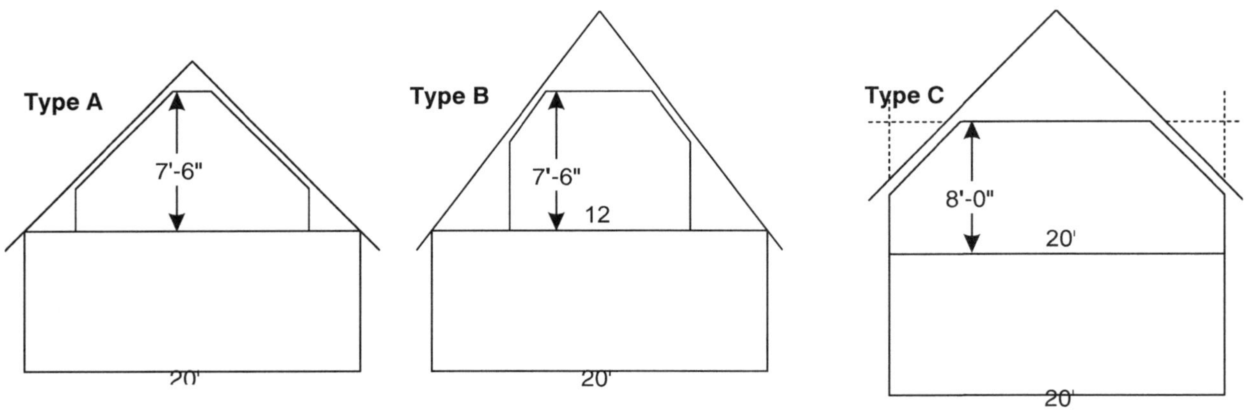

Use a fraction of the basic square foot cost for figuring the reduced headroom floor area. Type "C" includes typical dormers.

Type	Same Finish As Main Area	Lesser Quality Finish
A	1/3	1/4
B	1/2	1/3
C	2/3	1/2

Elevators, per shaft cost for car and machinery

Hydraulic based on two stops

Capacity	100 F.P.M.	200 F.P.M.
2,000 lbs.	$52,300	$86,200
2,500 lbs.	55,600	88,800
3,000 lbs.	58,300	96,600
3,500 lbs.	—	101,600
4,000 lbs.	—	105,600

Add for deluxe car, $10,500. Add for each additional stop over 2: $3,940, baked enamel doors $10,790, stainless steel doors $11,300.

Electric based on six stops

Capacity	200 F.P.M.	250 F.P.M.	300 F.P.M.
2,000 lbs.	$130,900	$138,600	$143,800
2,500 lbs.	138,800	148,300	155,600
3,000 lbs.	148,800	162,600	168,000
3,500 lbs.	162,600	173,300	181,700
4,000 lbs.	172,300	187,200	195,800

Add $9,000 for a deluxe car. Add $10,500 for each additional stop over 6.

Homes Raised on Piles or Columns

	Add per SF of floor
Concrete columns on driven piles	$28.30 plus $1.18 per foot over 5' high
Concrete columns on grade beams	$12.50 plus $0.86 per foot over 5' high
Braced timber piles or poured concrete columns	$4.00 plus $1.16 per foot over 5' high

Multi-Family and Motel Garages Cost Per Square Foot

Garages built at ground level under a multi-family or motel unit. The costs below include the following components:

1. A reinforced concrete floor in all areas.
2. Exterior walls, on one long side and two short sides, made up of a wood frame and good quality stucco, wood siding or masonry veneer.
3. A finished ceiling in all areas.
4. The difference between the cost of a standard wood frame floor structure at second floor level and one at ground level.
5. An inexpensive light fixture for each 600 square feet.

Where no exterior walls enclose the two short sides, use 2/3 of the square foot cost.

Garages built as separate structures for multi-family or motel units. The costs below include the following components:

1. Foundations.
2. A reinforced concrete floor in all areas.
3. Exterior walls on one long side and two short sides, made up of a wood frame and good quality stucco, wood siding or masonry veneer.
4. Steel support columns supporting the roof.
5. A wood frame roof structure with composition tar and gravel, wood shingle or light shake cover. No interior ceiling finish.
6. An inexpensive light fixture for each 600 square feet.

Use the location modifiers on page 7 or 8 to adjust garage costs to any area.

Basement Garages

Costs listed below are per square foot of floor, including the horizontal area of stairs and the approach ramp. These costs assume a single-level garage is built on one level, approximately 5 feet below grade, directly below 2 to 4 story multi-family structure with perimeter walls in vertical alignment. These costs include:

1. Excavation to 5' below ground line.
2. Full wall enclosure.
3. Typical storage facilities.
4. Minimum lighting.
5. Concrete floors.

Use the location modifiers on page 7 or 8 to adjust garage costs to the site.

Ground Level Garages

Area	400	800	1,200	2,000	3,000	5,000	10,000	20,000
Cost	49.64	41.48	39.88	35.11	31.65	31.36	30.52	29.10

Separate Structure Garages

Area	400	800	1,200	2,000	3,000	5,000	10,000	20,000
Cost	56.72	50.66	41.52	44.12	42.28	40.43	38.65	37.83

Basement Garages

Type	5,000	7,500	10,000	15,000	20,000	30,000	40,000	60,000
Reinforced concrete exterior walls and columns. Flat concrete roof slab.	75.60	69.24	66.35	65.33	63.36	62.53	61.67	61.01
Concrete block exterior walls, reinforced concrete columns. Flat concrete roof slab.	75.08	70.48	66.01	64.25	62.74	62.02	61.05	58.79
Concrete block exterior walls, steel posts and beams, light concrete/metal roof fireproofed with spray plaster.	70.55	64.59	61.49	53.21	50.96	56.97	55.23	54.40
Concrete block exterior walls, wood posts and beams, light concrete/metal roof fireproofed with spray plaster.	63.07	60.11	56.00	52.26	50.66	49.89	9.11	48.24
Add for each security gate	4.58	3.35	2.84	2.09	1.76	1.43	1.24	1.09

Cabins and Recreational Dwellings

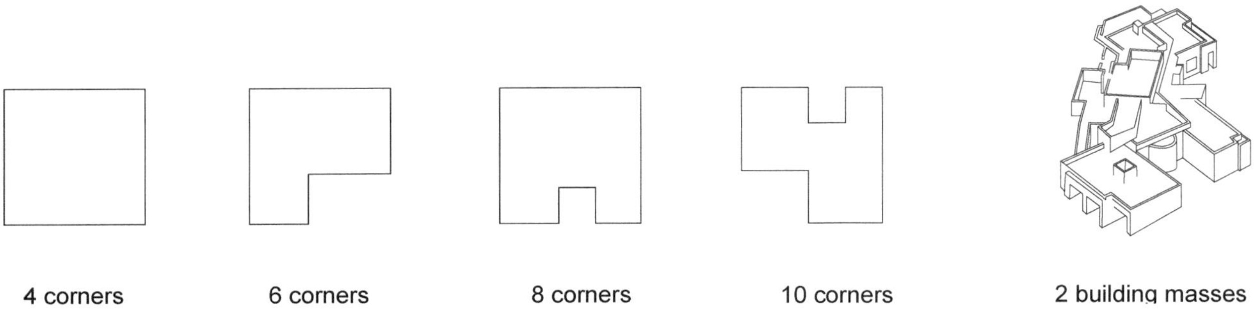

Example of Dwelling Shapes

Cabins and recreational dwellings are designed for single family occupancy, usually on an intermittent basis. These structures are characterized by a more rustic interior and exterior finish and often have construction details which would not meet building requirements in metropolitan areas. Classify these structures into either "conventional type" or "A-frame" construction. Conventional dwellings have an exterior wall which is approximately 8 feet high on all sides. A-frame cabins have a sloping roof which reduces the horizontal area 8 feet above the first floor to between 50% and 75% of the first floor area.

Conventional recreational dwellings vary widely in quality and the quality of construction is the most significant factor influencing cost. Conventional recreational dwellings are listed in six quality classes. Class 1 is the most expensive commonly encountered and Class 6 is the minimum commonly encountered. Nearly all conventional recreational dwellings built from stock plans will fall into Class 3, 4, 5, or 6. For convenience, these classes are labeled *Best Standard, Good Standard, Average Standard* or *Minimum Standard*. Class 1 residences are labeled *Luxury*. Class 2 residences are labeled *Semi-Luxury*. Class 1 and 2 residences are designed by professional architects, usually to meet preferences of the first owner.

The shape of the outside perimeter also has a significant influence on cost: The more complex the shape, the more expensive the structure per square foot of floor. The shape classification of multiple story or split-level conventional recreational dwellings should be based on the outline formed by the outermost exterior walls, including the garage area, regardless of the story level. Most conventional recreational dwellings fall into Classes 3, 4, 5 or 6 and have 4, 6, 8 or 10 corners, as illustrated above. Small insets that do not require a change in the roof line can be ignored when evaluating the outside perimeter.

Class 1 and 2 *(Luxury and Semi-Luxury)* conventional recreational dwellings have more than ten corners and are best evaluated by counting the "building masses." A building mass is a group of contiguous rooms on one or more levels with access at varying angles from a common point or hallway. The illustration at the right above represents a conventional recreational dwelling with two building masses. Most Class 1 and Class 2 conventional recreational dwellings have from one to four building masses, ignoring any attached garage. For convenience, cost tables for Class 1 and 2 conventional recreational dwellings with one, two, three or four building masses have been appended to cost tables for Class 3, 4, 5 and 6 conventional recreational dwellings with 4, 6, 8 and 10 building corners.

Conventional recreational dwellings which have features of two or more quality classes can be placed between two of the six labeled classes. The tables have five half-classes (1 & 2, 2 & 3, etc.) which can be applied to conventional recreational dwellings with some characteristics of two or more quality classes. If a portion of a conventional recreational dwelling differs significantly in quality from other portions, evaluate the square footage of each portion separately.

Cabins and recreational dwellings are often built under difficult working conditions and in remote sites. Individual judgments may be necessary in evaluating the cost impact of the dwelling location. The costs assume construction by skilled professional craftsmen. Where non-professional labor or second quality materials are used, use the next lower quality classification that might otherwise apply. If the structure is assembled from prefabricated components, use costs for the next lower half class.

Conventional Recreational Dwellings

Quality Classification

	Class 1 Luxury	Class 2 Semi-Luxury	Class 3 Best Std.	Class 4 Good Std.	Class 5 Average Std.	Class 6 Minimum Std.
Foundation (8% of total cost)	Reinforced concrete on a sloping site.	Reinforced concrete.	Reinforced concrete.	Reinforced concrete or concrete block.	Reinforced concrete or concrete block.	Wood piers, light concrete or block
Floor Structure (11% of total cost)	Engineered wood or steel, complex plan, elevation changes.	Engineered wood or steel trusses, good floor insulation.	Engineered wood or steel trusses, T&G sub-floor, good floor insulation.	Good wood frame with OSB sub-floor, some floor insulation.	Standard wood frame with OSB sub-floor, some floor insulation.	2" floor joists 16" on center with OSB sub-floor.
Wall Framing and Exterior Finish (14% of total cost)	Wood or steel, irregular walls, wood siding, stone, veneer, top-grade doors and windows.	Wood or steel, irregular walls, wood siding, stone veneer, better doors and windows.	Wood or steel, several wall offsets, plywood or lap siding, good grade doors and windows.	Wood or steel, shingle or plywood siding, some trim or veneer, average doors and windows.	Wood or steel, wood panel siding few or no offsets, commodity grade doors and windows.	Wood or steel, panel hardboard siding, minimum grade doors and windows.
Roof (13% of total cost)	Complex, heavy tile or metal cover, highly detailed.	Multi-pitch, shake, metal or good tile surface.	Dual-pitch, wood single or tile surface, gable over entrances.	Wood trusses, wood or good fiberglass shingle surface.	Simple wood frame, fiberglass shingle surface.	Wood frame, fiberglass shingle or roll roofing cover.
Floor Finish (5% of total cost)	Stone or masonry tile entry, inlaid hardwood or best carpet throughout.	Masonry entry, good hardwood or carpet in most rooms, good sheet vinyl elsewhere.	Hardwood or tile entry, carpet in most rooms sheet vinyl in kitchen and bathrooms.	Good sheet vinyl or average carpet in most areas, some hardwood or tile.	Sheet vinyl or tile on most areas, carpet in living room.	Composition tile or minimum grade sheet vinyl.
Interior Wall and Ceiling Finish (8% of total cost)	Top-grade paneling or wallboard with artistic finish, many offsets and wall openings, decorative details in most rooms.	Good wood paneling or textured wallboard with decorative details in most rooms, many wall openings, several racks and shelves.	Good hardwood veneer paneling or gypsum wallboard, some irregular walls, decorative details in living room, entry and kitchen.	1/2" gypsum wallboard with smooth finish, plywood paneling. at entry and living room, some decorative details.	1/2" gypsum wallboard with smooth finish, most walls are rectangular, doors and windows are the only openings.	Taped 1/2" gypsum wallboard, smooth or orange peel finish. Nearly all walls are regular, few decorative details.
Interior Features (5% of total cost)	Exposed beams or decorative details, 10' to 14' ceiling in great room, many sky widows, built-in shelving.	Great room has exposed beams, most rooms have windows on two sides, several framed openings.	Cathedral ceiling at entry or in master bedroom, floor level changes, several wall openings or pass-throughs.	Cathedral ceiling in master bedroom, sliding glass door, decorative wood molding and trim.	Rustic exposed ceiling beams, sliding closet doors, standard grade wood molding and trim.	Minimum grade molding and trim.
Bath Detail (4% of total cost)	At least 1 large tile shower, good tile counter in master bath.	Tile in 1 bathroom, glass block or good window in each bath, good vanity cabinet.	Tile or fiberglass shower, at least one built-in bathtub, good window in each bath.	Good plastic tub and shower in at least one bathroom, one small window in each bath.	Average plastic tub and shower in at least one bathroom, small vanity cabinet.	Minimum plastic tub and shower in one bathroom, minimum vanity.
Kitchen Detail (8% of total cost)	Over 20 LF of good custom wall & base cabinets, synthetic stone counter top, island work area.	15 to 18 LF of good custom base and wall cabinets, acrylic or tile counter top, desk with book shelf above.	12 to 15 LF of good stock wall and base cabinets, tile or acrylic counter top, desk and shelf or breakfast nook.	10 to 12 LF of stock standard grade wall and base cabinets, low-cost tile or laminated plastic counter top.	8 to 10 LF of stock standard grade wall and base cabinets, laminated plastic or resin coated hardboard top.	Less than 8 LF of low-cost wall and base cabinets, resin-coated hardboard counter top.
Plumbing (11% of total cost)	12 good fixtures, 2 water heaters, laundry room, copper piping.	10 good fixtures large water heater, laundry area, copper piping.	9 average grade fixtures, copper supply and plastic drain piping.	8 standard grade, fixtures, plastic supply and plastic drain lines.	7 low-cost fixtures, plastic supply and plastic drain lines.	6 or less minimum grade fixtures, plastic supply and drain lines.
Special Features (4% of total cost)	10 deluxe built-in appliances, good weather-protection throughout.	7 good built-in appliances, good wall and ceiling insulation.	6 good built-in appliances, good wall and ceiling, insulation.	5 average built-in appliances, adequate wall and ceiling insulation.	4 standard grade kitchen appliances, adequate ceiling insulation.	3 minimum grade built-in kitchen appliances, limited insulation.
Electrical System (9% of total cost)	Ample area and track lighting in most rooms, task light in bathrooms.	Good area and track lighting, simple light fixture in each bathroom.	Good light fixtures in kitchen and baths, limited fixtures in other rooms.	Good light fixture in most rooms, switch-operated outlet in bedrooms.	Simple light fixture in most rooms, switch-operated plugs in bedrooms.	5 or less lighting fixtures, switch-operated plug outlet in most rooms.

Note: Use the percent of total cost to help identify the correct quality classification.

Conventional Recreational Dwellings

4 Corners (Classes 3, 4, 5, and 6) or One Building Mass (Classes 1 and 2 Only)

Estimating Procedure

1. Establish the structure quality class by applying the information on page 33.
2. Multiply the structure floor area by the appropriate cost listed below.
3. Multiply the total from step 2 by the correct location factor listed on page 7 or 8.
4. Add, when appropriate, the cost of a deck or porch, paving, fireplace, garage or carport, heating, extra plumbing fixtures, supporting walls, half story areas, construction on hillside lots, and construction in remote areas.
   See page 42.

Conventional Recreational Dwelling, Class 5

Conventional Recreational Dwelling, Class 3

Square Foot Area

Quality Class	400	500	600	700	800	900	1,000	1,100	1,200	1,300	1,400
1, Luxury	—	—	—	—	515.93	493.53	474.82	459.92	447.16	436.13	426.14
1, & 2	—	—	—	478.06	453.45	433.85	417.34	404.45	392.93	383.56	374.70
2, Semi-Luxury	—	—	448.63	419.60	397.90	380.73	366.26	355.13	344.92	336.62	328.73
2 & 3	—	421.30	389.36	364.11	345.12	330.37	317.66	308.35	299.09	291.92	285.32
3, Best Std.	351.91	315.65	291.76	272.84	258.71	247.51	238.04	230.98	224.22	218.76	213.80
3 & 4	321.62	288.51	266.51	249.24	236.42	226.23	217.72	211.08	205.00	200.00	195.40
4, Good Std.	293.92	263.61	243.71	227.75	216.15	206.73	198.78	192.91	187.24	182.89	178.61
4 & 5	271.24	243.16	224.78	210.27	199.37	190.80	183.40	178.00	172.63	168.61	164.83
5 Avg. Std.	250.13	224.31	207.46	193.88	183.92	175.93	169.21	164.10	159.43	155.63	151.97
5 & 6	230.75	207.03	191.27	178.87	169.61	162.31	156.13	151.35	147.13	143.39	140.21
6, Min. Std.	212.78	190.93	176.55	164.94	156.45	149.83	144.01	139.76	135.53	132.43	129.37

Square Foot Area

Quality Class	1,500	1,600	1,700	1,800	2,000	2,200	2,400	2,600	2,800	3,000	3,200
1, Luxury	419.86	411.35	405.34	399.46	389.09	379.79	373.46	366.46	362.24	356.25	352.82
1, & 2	367.29	361.74	356.27	351.18	342.14	333.67	328.37	322.09	318.57	313.33	310.08
2, Semi-Luxury	323.65	317.59	312.69	308.25	300.39	292.71	288.27	282.74	279.53	275.03	272.15
2 & 3	280.58	275.57	271.31	267.38	260.49	254.00	250.00	245.32	242.54	238.54	236.07
3, Best Std.	210.42	206.40	203.27	200.42	195.29	190.39	187.51	183.71	181.73	178.82	176.81
3 & 4	192.13	188.67	185.81	183.14	178.30	173.94	171.21	168.05	166.08	163.32	161.72
4, Good Std.	175.62	172.48	169.71	167.41	163.03	158.83	156.48	153.59	151.91	149.29	147.74
4 & 5	161.99	159.00	156.80	154.29	150.28	146.60	144.45	141.60	140.13	137.71	—
5 Avg. Std.	149.48	146.78	144.47	142.47	138.65	135.34	133.30	130.66	129.24	—	—
5 & 6	137.92	135.39	133.31	131.50	128.05	124.84	122.92	120.47	—	—	—
6, Min. Std.	127.30	124.89	122.99	121.19	118.05	115.15	113.29	—	—	—	—

Note: Add 4% to the square foot cost for floors above the second floor level.

Conventional Recreational Dwellings

6 Corners (Classes 3, 4, 5, and 6) or Two Building Masses (Classes 1 and 2 Only)

Estimating Procedure

1. Establish the structure quality class by applying the information on page 33.
2. Multiply the structure floor area by the appropriate cost listed below.
3. Multiply the total from step 2 by the correct location factor listed on page 7 or 8.
4. Add, when appropriate, the cost of a deck or porch, paving, fireplace, garage or carport, heating, extra plumbing fixtures, supporting walls, half story areas, construction on hillside lots, and construction in remote areas. See page 42.

Conventional Recreational Dwelling, Class 4 & 5

Conventional Recreational Dwelling, Class 3

Square Foot Area

Quality Class	400	500	600	700	800	900	1,000	1,100	1,200	1,300	1,400
1, Luxury	—	—	—	—	527.28	504.57	485.44	471.40	458.30	447.49	437.76
1, & 2	—	—	—	487.94	463.69	443.71	426.81	414.56	402.81	393.25	384.59
2, Semi-Luxury	—	—	457.25	428.33	406.97	389.47	374.61	363.82	353.52	345.03	337.55
2 & 3	—	429.82	396.82	371.73	353.08	337.89	325.11	315.65	306.64	299.30	292.71
3, Best Std.	358.16	322.18	297.38	278.67	264.56	253.18	243.53	236.47	229.93	224.29	219.46
3 & 4	327.28	294.36	271.87	254.63	241.78	231.52	222.69	216.02	209.98	205.01	200.63
4, Good Std.	299.02	269.07	248.53	232.76	220.96	211.49	203.40	197.34	192.02	187.47	183.33
4 & 5	275.90	248.17	229.09	214.78	203.99	195.10	187.68	182.27	177.10	172.91	169.10
5 Avg. Std.	254.44	228.94	211.34	197.96	188.01	179.95	173.08	168.13	163.32	159.50	155.96
5 & 6	234.76	211.23	194.95	182.69	173.53	165.98	159.70	155.17	150.73	147.15	143.98
6, Min. Std.	216.69	194.80	179.89	168.54	160.11	153.23	147.39	143.07	139.03	135.69	132.75

Square Foot Area

Quality Class	1,500	1,600	1,700	1,800	2,000	2,200	2,400	2,600	2,800	3,000	3,200
1, Luxury	430.44	422.30	416.59	410.65	400.28	390.26	384.81	377.36	373.29	367.52	363.74
1, & 2	378.39	371.38	366.18	360.92	351.69	343.10	338.19	331.88	328.33	323.01	319.49
2, Semi-Luxury	331.99	326.13	321.38	316.86	308.78	301.17	296.71	291.34	288.27	283.59	280.35
2 & 3	287.91	283.14	278.86	274.85	267.76	261.28	257.37	252.73	250.00	245.98	243.06
3, Best Std.	215.87	212.14	209.09	206.09	200.64	195.76	192.86	189.31	187.51	184.35	182.13
3 & 4	197.26	193.77	191.00	188.27	183.40	178.87	176.34	173.10	171.21	168.54	166.48
4, Good Std.	180.27	177.18	174.65	172.12	167.64	163.47	161.11	158.18	156.48	154.10	152.15
4 & 5	166.38	163.32	161.11	158.78	154.68	150.76	148.58	146.00	144.45	142.13	—
5 Avg. Std.	153.38	150.73	148.58	146.53	142.54	139.21	137.12	134.66	133.30	—	—
5 & 6	141.59	139.03	137.12	135.16	131.59	128.32	126.45	124.24	—	—	—
6, Min. Std.	130.60	128.20	126.45	124.68	121.34	118.44	116.72	—	—	—	—

Note: Add 4% to the square foot cost for floors above the second floor level.

Conventional Recreational Dwellings

8 Corners (Classes 3, 4, 5, and 6) or Three Building Masses (Classes 1 and 2 only)

Estimating Procedure

1. Establish the structure quality class by applying the information on page 33.
2. Multiply the structure floor area by the appropriate cost listed below.
3. Multiply the total from step 2 by the correct location factor listed on page 7 or 8.
4. Add, when appropriate, the cost of a deck or porch, paving, fireplace, garage or carport, heating, extra plumbing fixtures, supporting walls, half story areas, construction on hillside lots, and construction in remote areas. See page 42.

Conventional Recreational Dwelling, Class 3

Conventional Recreational Dwelling, Class 1 & 2

Square Foot Area

Quality Class	400	500	600	700	800	900	1,000	1,100	1,200	1,300	1,400
1, Luxury	—	—	—	—	541.17	515.59	496.78	482.70	469.01	458.82	448.20
1, & 2	—	—	—	498.18	475.62	453.45	436.65	424.37	412.31	403.41	394.05
2, Semi-Luxury	—	—	466.65	437.14	417.34	398.02	383.31	372.40	362.06	353.89	345.84
2 & 3	—	437.77	404.82	379.26	362.08	345.35	332.59	322.93	314.16	307.05	300.03
3, Best Std.	363.86	328.00	303.42	284.09	271.33	258.82	249.24	242.00	235.46	230.17	224.83
3 & 4	332.69	299.96	277.25	259.84	248.08	236.51	227.89	221.24	215.13	210.27	205.53
4, Good Std.	303.97	274.07	253.48	237.35	226.70	216.21	208.20	202.14	196.60	192.07	187.83
4 & 5	280.49	252.76	233.82	218.94	209.11	199.43	192.03	186.56	181.52	177.26	173.42
5 Avg. Std.	258.76	233.18	215.65	202.06	192.88	183.94	177.18	172.12	167.41	163.47	159.78
5 & 6	238.65	215.13	198.87	186.33	178.00	169.71	163.37	158.78	154.29	150.77	147.55
6, Min. Std.	220.11	198.46	183.47	171.91	164.10	156.48	150.76	146.53	142.47	139.21	136.14

Square Foot Area

Quality Class	1,500	1,600	1,700	1,800	2,000	2,200	2,400	2,600	2,800	3,000	3,200
1, Luxury	441.44	432.98	426.53	422.26	410.65	401.83	395.01	388.74	384.21	377.87	374.31
1, & 2	387.98	380.73	375.11	371.19	360.98	352.93	347.32	341.70	340.22	332.15	329.14
2, Semi-Luxury	340.61	334.28	329.38	325.64	316.97	309.65	304.84	299.71	296.23	291.54	288.88
2 & 3	295.34	289.91	285.71	282.48	274.95	268.47	264.51	260.05	256.95	252.94	250.65
3, Best Std.	221.34	217.19	214.08	211.70	206.13	201.20	198.24	194.91	192.65	189.64	187.79
3 & 4	202.41	198.57	195.52	193.46	188.54	183.94	181.14	178.20	175.99	173.41	171.72
4, Good Std.	184.94	181.52	178.86	176.81	172.25	168.13	165.54	162.85	160.91	158.21	156.93
4 & 5	170.48	167.41	164.90	163.37	158.82	155.17	152.78	150.19	148.44	146.06	—
5 Avg. Std.	157.30	154.34	152.15	150.46	146.56	143.07	141.03	138.55	136.92	—	—
5 & 6	145.07	142.50	140.52	138.65	135.20	131.90	130.01	127.77	—	—	—
6, Min. Std.	133.82	131.50	129.47	128.05	124.84	121.81	119.94	—	—	—	—

Note: Add 4% to the square foot cost for floors above the second floor level.

Conventional Recreational Dwellings

10 Corners (Classes 3, 4, 5, and 6) or Four Building Masses (Classes 1 and 2 only)

Estimating Procedure

1. Establish the structure quality class by applying the information on page 33.
2. Multiply the structure floor area by the appropriate cost listed below.
3. Multiply the total from step 2 by the correct location factor listed on page 7 or 8.
4. Add, when appropriate, the cost of a deck or porch, paving, fireplace, garage or carport, heating, extra plumbing fixtures, supporting walls, half story areas, construction on hillside lots, and construction in remote areas.
   See page 42.

Conventional Recreational Dwelling, Class 2 & 3

Conventional Recreational Dwelling, Class 1

Square Foot Area

Quality Class	400	500	600	700	800	900	1,000	1,100	1,200	1,300	1,400
1, Luxury	—	—	—	—	550.51	527.28	508.68	494.03	481.25	470.36	459.92
1, & 2	—	—	—	508.39	483.82	463.69	447.11	434.30	422.94	414.16	404.30
2, Semi-Luxury	—	—	475.62	446.23	424.47	406.97	392.40	381.17	371.09	362.59	354.85
2 & 3	—	445.55	412.76	387.19	368.23	353.08	340.28	330.60	321.78	314.44	307.83
3, Best Std.	370.31	333.97	309.28	290.12	276.02	264.56	254.97	247.85	241.18	235.59	230.57
3 & 4	338.43	305.34	282.72	265.18	252.17	241.78	233.18	226.36	220.43	215.59	210.73
4, Good Std.	309.16	278.93	258.34	242.42	230.48	220.96	213.19	207.03	201.51	196.87	192.69
4 & 5	285.33	257.41	238.36	223.56	212.62	203.99	196.48	190.93	185.82	181.65	177.72
5 Avg. Std.	263.33	237.35	219.87	206.21	196.22	188.01	181.26	176.24	171.46	167.47	163.93
5 & 6	242.88	218.94	202.80	190.22	181.00	173.53	167.36	162.44	158.15	154.63	151.18
6, Min. Std.	224.07	202.06	187.14	175.53	166.93	160.11	154.24	150.02	146.00	142.54	139.59

Square Foot Area

Quality Class	1,500	1,600	1,700	1,800	2,000	2,200	2,400	2,600	2,800	3,000	3,200
1, Luxury	453.56	445.24	438.21	432.29	422.61	412.26	406.81	399.25	395.10	386.62	384.50
1, & 2	398.83	391.40	385.27	380.15	371.55	362.28	357.67	351.20	347.53	339.98	338.31
2, Semi-Luxury	349.75	343.54	338.19	333.78	325.97	317.92	313.77	308.29	304.93	298.40	297.11
2 & 3	303.32	297.87	293.35	289.71	282.68	275.85	272.22	267.72	264.66	258.89	257.86
3, Best Std.	227.45	223.35	219.87	217.10	211.83	206.85	203.99	200.51	198.46	193.89	193.11
3 & 4	207.81	204.16	200.80	198.54	193.71	189.09	186.40	183.14	181.26	177.37	176.56
4, Good Std.	189.85	186.56	183.59	181.33	176.95	172.63	170.41	167.44	165.69	161.99	161.41
4 & 5	175.20	172.12	169.39	167.36	163.32	159.29	157.09	154.63	152.82	149.48	—
5 Avg. Std.	161.57	158.78	156.24	154.24	150.69	146.99	144.99	142.50	141.05	—	—
5 & 6	149.09	146.53	144.12	142.46	138.93	135.53	133.81	131.57	—	—	—
6, Min. Std.	137.56	135.16	132.87	131.38	128.12	124.90	123.36	—	—	—	—

Note: Add 4% to the square foot cost for floors above the second floor level.

"A-Frame" Cabins

Quality Classification

	Class 1 Best Quality	Class 2 Good Quality	Class 3 Average Quality	Class 4 Low Quality
Framing (10% of total cost)	Wood frame.	Wood frame.	Wood frame.	Wood frame.
Floor Framing (5% of total cost)	4" x 8" girders 48" o.c. with 2" T&G subfloor, or 2" x 6" to 2" x 8" joists 16" o.c. with 1" subfloor.	4" x 8" girders 48" o.c. with 1-1/4" plywood or 2" T&G subfloor, or 2" x 6" to 2" x 8" joists 16" o.c. with 1" subfloor.	4" x 6" girders 48" o.c. with 1-1/4" plywood or 2" T&G subfloor, or 2" x 6" joists 16" o.c. with 1" subfloor.	4" x 6" girders 48" o.c. with 1-1/4" plywood or 2" T&G subfloor, or 2" x 6" joists 16" o.c. with 1" subfloor.
Roof Framing (8% of total cost)	4" x 8" at 48" o.c. with 2" or 3" T&G sheathing.	4" x 8" at 48" o.c. with 2" or 3" T&G sheathing.	4" x 8" at 48" o.c. with 2" T&G sheathing.	4" x 8" at 48" o.c. with 1-1/4" plywood or 2" T&G sheathing.
Gable End Finish (5% of total cost)	Good plywood, lap board or board and batt.	Average to good plywood, or boards.	Average plywood, board or wood shingle.	Low cost plywood, shingle or composition siding.
Windows (2% of total cost)	Good quality large insulated wood or metal windows.	Average quality insulated wood or metal windows.	Average quality wood or metal windows.	Small glass area of low cost windows.
Roofing (10% of total cost)	Heavy wood shakes.	Medium wood or aluminum shakes.	Wood or composition shingles.	Low cost composition shingles.
Flooring (5% of total cost)	Good carpet or hardwood with sheet vinyl in kitchen and baths.	Average to good quality carpet with good tile or sheet vinyl in kitchen and baths.	Average quality carpet with resilient tile in kitchen and baths.	Composition tile.
Interior Finish (25% of total cost including finish carpentry, wiring, lighting, fireplace, etc.)	Good quality hardwood veneer paneling.	Good textured gypsum wallboard, good plywood or knotty pine paneling.	Textured gypsum wallboard or plywood paneling.	Low cost paneling or wallboard.
Bathrooms (5% of total cost)	Two 3-fixture baths and one 2-fixture bath, good fixtures.	Two 3-fixture baths, good fixtures.	Two 3-fixture baths, average fixtures.	One 3-fixture bath.
Kitchen (5% of total cost)	15' to 18' good quality hardwood veneer base cabinet with matching wall cabinets. 15' to 18' of good quality plastic or ceramic tile drain board.	12' to 16' of hardwood veneer base cabinet with matching wall cabinets. 12' to 16' of plastic or ceramic tile drainboard.	8' to 12' of average quality veneer or painted base cabinets with matching wall cabinets. 8' to 12' of plastic drainboard.	6' to 8' of minimum base cabinets with matching wall cabinets. 6' to 8' of minimum plastic drainboard.
Plumbing (15% of total cost)	Nine good quality fixtures and one larger or two 30 gallon water heaters. Copper supply piping.	Seven good quality fixtures and one water heater.	Seven average quality fixtures and one water heater.	Four low cost fixtures and one water heater. Plastic supply pipe.
Special Features (5% of total cost)	Built-in oven, range, dishwasher, disposer, range hood with good insulation, good lighting fixtures, insulated sliding glass door and ornate entry door.	Built-in range, oven and range hood, some insulation, 8' sliding glass door, average electric fixtures.	Drop-in range and hood, some insulation, low cost electric fixtures.	Minimum electric fixtures.

Note: Use the percent of total cost to help identify the correct quality classification.

"A-Frame" Cabins

4 Corners

Estimating Procedure

1. Establish the structure quality class by applying the information on page 38.
2. Multiply the structure floor area by the appropriate cost listed below.
3. Multiply the total from step 2 by the correct location factor listed on page 7 or 8.
4. Add, when appropriate, the cost of a deck or porch, paving, fireplace, garage or carport, heating, extra plumbing fixtures, supporting walls, half story areas, construction on hillside lots, and construction in remote areas. See page 42.

"A-Frame" Cabin, Class 3 & 4

Square Foot Area

Quality Class	400	500	600	700	800	900	1,000	1,100	1,200	1,300	1,400
1, Best	287.71	259.60	240.07	225.73	214.52	205.65	198.33	192.21	187.03	182.45	178.54
1 & 2	264.36	238.54	220.60	207.37	197.17	188.79	182.24	176.59	171.78	167.68	164.05
2, Good	242.54	218.83	202.51	190.27	180.95	173.37	167.26	162.02	157.71	153.88	150.54
2 & 3	228.99	206.62	191.11	179.65	170.77	163.69	157.82	153.02	148.85	145.25	142.10
3, Average	216.92	195.74	181.04	170.16	161.72	155.03	149.48	144.88	141.03	137.58	134.63
3 & 4	196.87	177.66	164.23	154.44	146.83	140.75	135.75	131.53	127.97	124.85	122.24
4, Low	176.59	159.38	147.43	138.61	131.77	126.26	121.79	118.06	114.76	112.04	109.61

Square Foot Area

Quality Class	1,500	1,600	1,700	1,800	2,000	2,200	2,400	2,600	2,800	3,000	3,200
1, Best	172.20	169.41	166.91	164.59	160.62	157.29	154.51	152.06	149.93	147.99	146.38
1 & 2	158.91	156.35	153.94	151.86	148.24	145.14	142.58	140.32	138.40	136.65	135.13
2, Good	146.62	144.18	142.05	140.08	136.75	133.95	131.54	129.48	127.61	126.04	124.67
2 & 3	139.17	136.88	134.85	133.03	129.79	127.17	124.85	122.90	121.16	119.68	118.30
3, Average	132.20	130.04	128.06	126.41	123.35	120.79	118.59	116.75	115.15	113.68	112.41
3 & 4	121.63	119.66	117.82	116.30	113.50	111.15	109.17	107.43	105.94	104.63	103.41
4, Low	108.83	106.68	105.58	104.08	102.75	100.66	98.83	97.24	95.91	94.70	93.65

"A-Frame" Cabins

6 Corners

Estimating Procedure

1. Establish the structure quality class by applying the information on page 38.
2. Multiply the structure floor area by the appropriate cost listed below.
3. Multiply the total from step 2 by the correct location factor listed on page 7 or 8.
4. Add, when appropriate, the cost of a deck or porch, paving, fireplace, garage or carport, heating, extra plumbing fixtures, supporting walls, half story areas, construction on hillside lots, and construction in remote areas.
   See page 42.

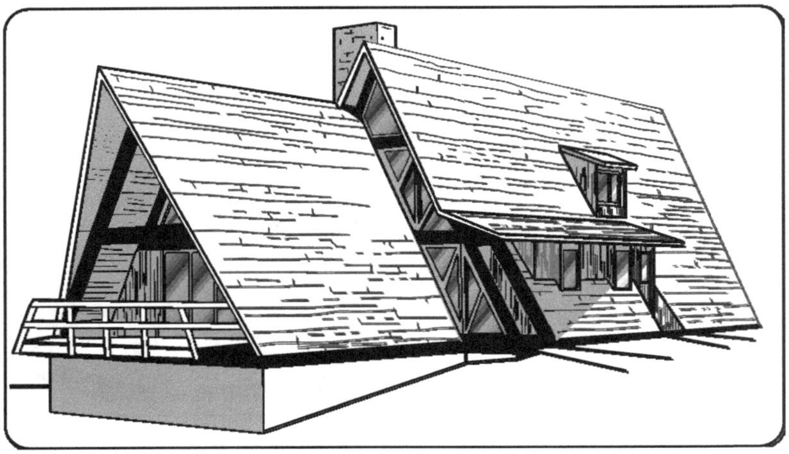

"A-Frame" Cabin, Class 2 & 3

Square Foot Area

Quality Class	400	500	600	700	800	900	1,000	1,100	1,200	1,300	1,400
1, Best	292.34	263.82	244.22	229.71	218.60	209.78	202.51	196.45	191.39	186.87	183.02
1 & 2	268.08	241.99	223.98	210.72	200.55	192.42	185.72	180.18	175.48	171.39	167.68
2, Good	246.27	222.25	205.76	193.53	184.13	176.70	170.63	165.49	161.20	157.47	154.23
2 & 3	232.58	209.83	194.21	182.70	173.94	166.91	161.10	156.25	152.11	148.69	145.60
3, Average	219.24	197.82	183.06	172.25	163.85	157.24	151.82	147.30	143.48	140.07	137.23
3 & 4	200.01	180.52	167.12	157.20	149.59	143.52	138.56	134.50	130.87	127.90	125.25
4, Low	179.03	161.57	149.53	140.75	133.95	128.52	124.03	120.35	117.16	114.43	112.06

Square Foot Area

Quality Class	1,500	1,600	1,700	1,800	2,000	2,200	2,400	2,600	2,800	3,000	3,200
1, Best	177.01	174.09	171.48	169.20	165.18	161.82	158.91	156.45	154.27	152.31	150.67
1 & 2	163.02	160.30	157.98	155.80	152.10	149.02	146.39	144.08	142.06	140.32	138.76
2, Good	150.39	147.91	145.77	143.80	140.38	137.52	135.09	132.98	131.13	129.50	128.04
2 & 3	142.72	140.38	138.29	136.41	133.23	130.49	128.12	126.20	124.47	122.86	121.55
3, Average	135.96	133.77	131.84	130.01	126.92	124.35	122.24	120.25	118.56	117.13	115.77
3 & 4	124.85	122.86	120.99	119.37	116.54	114.18	112.20	110.44	108.86	107.48	106.34
4, Low	111.13	109.46	107.99	105.37	103.25	101.47	99.80	98.49	97.20	96.13	94.68

"A-Frame" Cabins

8 Corners

Estimating Procedure

1. Establish the structure quality class by applying the information on page 38.
2. Multiply the structure floor area by the appropriate cost listed below.
3. Multiply the total from step 2 by the correct location factor listed on page 7 or 8.
4. Add, when appropriate, the cost of a deck or porch, paving, fireplace, garage or carport, heating, extra plumbing fixtures, supporting walls, half story areas, construction on hillside lots, and construction in remote areas. See page 42.

"A-Frame" Cabin, Class 2

Square Foot Area

Quality Class	400	500	600	700	800	900	1,000	1,100	1,200	1,300	1,400
1, Best	297.13	268.69	249.12	234.62	223.48	214.50	207.22	201.04	195.83	191.39	187.37
1 & 2	272.45	246.49	228.50	215.20	204.90	196.70	190.04	184.41	179.58	175.48	171.88
2, Good	249.82	225.93	209.50	197.26	187.90	180.37	174.16	169.01	164.70	160.91	157.51
2 & 3	235.77	213.24	197.70	186.17	177.32	170.18	164.40	159.53	155.42	151.82	148.71
3, Average	223.00	201.73	186.91	176.07	167.73	161.01	155.48	150.98	146.98	143.56	140.71
3 & 4	202.60	183.34	169.88	159.97	152.37	146.26	141.30	137.15	133.57	130.55	127.81
4, Low	181.65	164.25	152.26	143.42	136.62	131.12	126.69	122.90	119.71	116.97	114.59

Square Foot Area

Quality Class	1,500	1,600	1,700	1,800	2,000	2,200	2,400	2,600	2,800	3,000	3,200
1, Best	181.50	178.60	175.97	173.68	169.62	166.23	163.32	160.92	158.73	156.88	155.26
1 & 2	164.11	161.43	159.06	156.94	153.31	150.27	147.66	145.47	143.48	141.84	140.23
2, Good	154.21	151.65	149.44	147.53	144.05	141.18	138.76	136.65	134.80	133.23	131.81
2 & 3	146.00	143.56	141.51	139.65	136.40	133.64	131.44	129.39	127.61	126.18	124.81
3, Average	139.07	136.77	134.77	132.98	129.89	127.35	125.19	123.30	121.63	120.12	118.84
3 & 4	127.57	125.47	123.59	122.00	119.15	116.80	114.81	113.08	111.59	110.25	109.05
4, Low	113.68	112.04	110.56	107.95	105.83	103.98	102.39	101.01	99.80	98.82	97.28

Cabins and Recreational Dwellings

Additional Costs

Half-Story Costs

For conventional recreational dwellings, use the suggested fractions found on page 30 in the section "Additional Costs for Residential Structures." For "A-Frame" cabins, use one of the following costs: A simple platform with low cost floor cover, minimum partitions, and minimum lighting costs $69 to $101 per square foot. Average quality half story area with average quality carpet, average number of partitions finished with gypsum wallboard or plywood veneer and average lighting costs $101 to $112 per square foot. A good quality half story area with good carpet, decorative rustic partitions, ceiling beams and good lighting costs $133 to $155 per square foot.

Decks and Porches, per square foot

2" wood deck with steps and railing (300 S.F. base)	
1' to 4' above ground	$26.85 to $31.51
Over 4' to 6' above ground	31.20 to 40.43
Over 6' to 9' above ground	32.66 to 42.81
Over 9' to 12' above ground	33.85 to 44.82
Over 12' above ground	35.68 to 46.36

Fireplaces, 2-story, including foundation

Metal hood with concrete slab	$3,010 to $3,756
Prefabricated, zero clearance	4,320 to 6,300
Simple concrete block	5,190 to 8,670
Concrete block with stone facing	6,920 to 10,500
Simple natural stone	11,900 to 17,200

Extra Plumbing, cost each

Lavatory	$1,840 to $2,685
Water closet or bidet	2,250 to 2,766
Tub and shower	2,370 to 3,160
Stall shower	1,762 to 2,550
Laundry or utility sink	1,285 to 1,520

Heating, cost each

Wall furnace, 35,000 Btu	$1,438
Wall furnace, 65,000 Btu	1,764
Baseboard hot water, per SF*	5.6
Central heating, perimeter ducts, per S.F.*	7.90

*Cost is per SF of floor area heated.

Garages, Carports and Basements

For garage, carport and basement costs for conventional recreational dwellings, see pages 27 and 29.

Flatwork, per square foot

Asphalt paving	$5.95 to $9.00
4" concrete	6.10 to 9.30
6" concrete	6.45 to 9.50

Reinforced concrete walls, per C.F.

Formed one side only	$25.40 to $29.52
Formed both sides	32.30 to 36.20

Supporting Wall Costs

Cabins and recreational dwellings built on sloping lots cost more than if they are built on level lots. The cost of supporting walls of a building that do not enclose any living area should be estimated by using the figures below. These costs include everything above a normal foundation (12" to 18" above ground) up to the bottom of the next floor structure where square foot costs can be applied. In addition to the cost of supporting walls, add the cost of any extra structural members and the higher cost of building on a slope. A good rule of thumb for this is to add $960 for each foot of vertical distance between the highest and the lowest points of intersection of foundation and ground level.

Wood posts, per foot of height

4" x 4"	$2.59 to $4.20
4" x 6"	4.20 to 7.10
6" x 6"	5.40 to 10.10
8" x 8"	12.10 to 19.90
10" x 10"	22.50 to 32.20
12" x 12"	33.80 to 46.80

Brick, per square foot of wall

8" common brick	$43.50 to $53.10
12" common brick	65.70 to 82.60
8" common brick, 1 side face brick	55.00 to 67.90
12" common brick, 1 side face brick	85.05 to 107.00

Reinforced concrete block, per square foot of wall

8" natural	$12.10 to $14.80
8" colored	16.60 to 19.90
8" detailed blocks, natural	13.70 to 18.10
8" detailed blocks, colored	19.00 to 21.50
8" sandblasted	14.60 to 17.34
8" splitface, natural	12.60 to 14.90
8" splitface, colored	18.50 to 22.30
8" slump block, natural	13.50 to 16.90
8" slump block, colored	18.40 to 21.80
12" natural	23.90 to 26.30

Life in Years and Depreciation for Residences

Quality Class	1	2	3	4	5	6
Single family residences	70	70	70	60	60	55
Manufactured housing	55	50	45	40	30	
Multi-family residences	60	60	55	55	50	
Motels	60	55	55	50		
Conventional recreational dwellings	70	60	60	55	55	50
A-frame cabins	60	55	55	50		

This table shows typical physical lives in years in the absence of unusual physical, functional or economic obsolescence. Raise half classes to the next higher whole class.

To Find the Present Value of an Existing Residence

Present value is the replacement cost less depreciation (inverse of the "% Good" column below). Multiply the appropriate figure in the "% good" column by the current replacement cost developed using this manual to find the present value. For newer residences, the chronological age ("Age" column) is usually the best indicator of percent good. The present value of older residences may be influenced more by physical, functional or economic obsolescence than by age. When physical, functional or economic conditions limit or extend the remaining useful life of a residence, estimate that life in years and use the "Rem. Life" column (rather than the "Age" column) to find the percent good.

	20 Years		25 Years		30 Years		40 Years	
Age	Rem. Life	% Good	Rem. Life	% Good	Rem. Life	% Good	Rem. Life	% Good
0	20	100	25	100	30	100	40	100
1	19	94	24	95	29	96	39	98
2	18	88	23	90	28	93	38	96
3	17	81	22	86	27	89	37	94
4	16	75	21	81	26	86	36	92
5	15	69	20	77	25	82	35	90
6	14	63	19	72	24	79	34	87
7	13	59	18	68	23	75	33	84
8	12	57	17	63	22	71	32	82
9	11	55	16	60	21	67	31	80
10	11	53	16	58	20	64	30	77
11	10	50	15	56	19	60	29	74
12	9	48	14	54	19	59	28	72
13	8	46	13	53	18	57	27	70
14	7	44	12	51	17	56	27	67
15	7	42	11	49	16	54	26	65
16	6	40	11	48	15	53	25	62
17	5	38	10	46	14	52	24	60
18	5	36	9	44	13	50	23	59
19	4	33	8	43	13	49	22	58
20	4	31	7	41	12	47	21	58
21	3	29	7	39	11	46	21	55
22	3	27	6	37	11	44	20	54
23	3	25	6	35	10	43	19	53
24	3	23	5	34	9	42	18	52
25	2	21	5	32	9	40	17	51
26	2	19	4	30	8	39	17	50
27	2	16	4	29	7	37	16	49
28	2	14	4	27	7	36	15	48
29	2	12	3	25	6	34	14	47
30	1	10	3	24	6	33	14	46
31	–	–	3	22	5	31	13	45
32	–	–	3	20	5	30	12	44
33	–	–	2	18	5	29	12	43
34	–	–	2	17	4	27	11	42
35	–	–	2	15	4	26	11	41
36	–	–	2	13	4	24	10	40
38	–	–	1	10	3	21	9	38
40	–	–	–	–	2	19	7	35
42	–	–	–	–	2	16	6	33
46	–	–	–	–	1	10	5	29
50	–	–	–	–	–	–	4	25
55	–	–	–	–	–	–	3	20
60	–	–	–	–	–	–	2	14
64	–	–	–	–	–	–	1	10

	45 Years		50 Years		55 Years		60 Years		70 Years	
Age	Rem. Life	% Good	Rem. Life	% Good	Rem. Life	% Good	Rem. Life	% Good	Rem. Life	% Good
0	45	100	50	100	55	100	60	100	70	100
2	43	97	48	97	53	98	58	98	68	99
4	41	93	46	94	51	96	56	96	66	98
6	39	89	44	91	49	94	54	95	64	98
8	37	85	42	88	47	91	52	92	62	96
10	35	81	39	85	45	88	50	90	60	94
12	33	77	38	82	43	85	48	87	58	92
14	32	73	36	78	41	82	46	85	56	91
16	30	69	35	74	40	79	45	83	54	89
18	28	65	33	70	38	76	43	80	52	87
20	26	60	31	67	36	73	41	77	50	84
22	24	58	29	63	34	70	39	74	48	82
24	23	56	28	60	32	67	37	71	46	80
26	22	54	26	58	31	64	35	68	44	77
28	20	52	24	56	29	61	34	65	42	74
30	18	50	23	54	27	58	32	63	40	73
32	17	48	21	53	26	56	30	61	38	71
34	15	47	20	51	24	55	29	60	36	70
36	14	45	18	49	23	53	27	58	34	68
38	12	43	17	47	21	51	26	56	32	66
40	11	41	16	45	20	50	24	55	30	65
42	10	39	14	44	19	48	23	53	28	63
44	9	37	13	42	17	46	21	51	26	61
46	8	35	12	40	16	45	20	50	25	60
48	7	33	11	38	15	43	19	48	23	58
50	6	31	10	37	14	41	18	46	21	56
52	5	29	9	35	12	40	16	45	19	55
54	5	28	8	33	11	38	15	43	18	53
56	4	26	7	31	10	36	14	41	16	51
58	4	24	6	30	9	35	13	40	15	50
60	3	22	5	28	8	33	12	38	14	47
62	3	20	4	26	7	31	11	36	12	45
64	3	17	4	24	6	30	10	35	11	44
66	2	16	3	22	5	28	9	33	10	42
68	2	14	3	21	5	27	8	32	9	41
70	2	12	3	19	4	25	7	30	9	38
72	1	10	2	17	4	23	6	28	8	36
74	–	–	2	15	4	21	5	26	7	34
76	–	–	2	14	3	19	5	24	7	32
80	–	–	1	10	2	17	4	22	7	28
82	–	–	–	–	2	15	3	18	6	25
84	–	–	–	–	1	13	2	16	5	22
96	–	–	–	–	–	11	1	10	3	14
98	–	–	–	–	–	10	–	–	2	13
100	–	–	–	–	–	–	–	–	1	11

Elementary Schools – Masonry or Concrete

Quality Classification

	Class 1 Best Quality	Class 2 Good Quality	Class 3 Average Quality	Class 4 Low Quality
Foundation (7% of total cost)	Reinforced concrete, depth to 8'	Reinforced concrete, depth 6' to 8'.	Reinforced concrete, depth 4' to 6'.	Reinforced concrete, depth 4' or less.
Floor Structure (4% of total cost)	Concrete on steel beams and deck.	Lightweight concrete on steel beams.	6" thickened edge concrete slab on 6" rock base.	4" reinforced slab on rock base.
Exterior Walls (14% of total cost)	Decorative brick veneer with concrete block backup. Ornate brick or stone trim.	Decorative or colored concrete block. Some brick or stone veneer.	Colored concrete block with some brick or wood trim.	Painted tilt-up concrete wall panels. Few decorative details.
Roof Structure & Cover (29% of total cost)	Glu-lams or steel trusses on steel intermediate columns. Panelized roof system with elastomeric concrete tile or metal roof cover. Engineered for earthquake or high wind zones.	Glu-lams or steel beams on steel intermediate columns. Panelized roof system with 5-ply, built-up or concrete tile roof cover. Good insulation.	Wood or metal trusses on intermediate columns. Panelized roof system with built-up or good composition shingle roofing. Insulated.	Beams or trusses on steel supports. OSB. sheathing. Built-up or low cost membrane roofing. Foil insulation. Not designed for high wind or seismic zones.
Floor Finish (5% of total cost)	Hardwood with good sheet vinyl or carpet in offices. Ceramic tile in restrooms.	Hardwood or sheet vinyl in classrooms. Carpet in offices.	Standard grade sheet vinyl or good composition tile in most areas.	Minimum grade tile.
Windows & Doors (4% of total cost)	Large vinyl-clad or metal insulated low-E windows. Metal doors with glass panels. Institutional grade hardware. LEED certified.	Vinyl-clad insulated windows. Metal exterior doors with glass panels. Solid core interior doors. Commercial grade hardware.	Standard grade insulated windows. Metal exterior doors. Wood interior doors. Standard grade hardware.	Few standard grade metal windows. Low cost metal exterior doors. Low grade hardware.
Interior Wall Finish (13% of total cost)	Gypsum wallboard or lath and plaster finished with good paper or vinyl. Good metal or hardwood veneer wainscot and trim.	Textured gypsum wallboard covered with vinyl or good wallpaper. Good hardwood-veneer wainscot and wood trim in high traffic areas.	Textured and painted interior stucco or gypsum wallboard. Wood trim in high traffic areas.	Painted gypsum wallboard.
Ceiling Finish (2% of total cost)	Suspended good grade acoustical tile with gypsum wallboard backing.	Suspended acoustical tile with concealed grid system.	Suspended acoustical tile with exposed grid system.	Painted gypsum wallboard with acoustic texture.
Specialties (5% of total cost)	Central station alarm. Large chalkboards, cabinets, shelves and cases. Network connection.	Fire alarm and bell system. Network connection. Good chalkboards, map rail & shelving.	Fire alarm, bell and network connections. Some cabinets and shelves.	Minimum alarm system. Few cabinets and shelves.
Plumbing (7% of total cost)	6 good commercial fixtures per classroom. Copper supply & drain pipe. Metal toilet partitions.	5 standard fixtures per classroom. Copper supply & drain pipe. Metal toilet partitions.	4 standard fixtures per classroom. Plastic supply and drain pipe. Composition toilet partitions.	3 minimum fixtures per classroom. Plastic supply and drain pipe. Wood toilet partitions.
Lighting and Power (10% of total cost)	Recessed LED lighting in modular plastic panels. Many task lights or indirect lighting fixtures.	Suspended or panel LED. Some task lighting or indirect light fixtures.	Suspended or panel LED. Some ceiling downlights.	Continuous exposed 2 tube fluorescent strips, 8' O.C.

Note: Use the percent of total cost to help identify the correct quality classification.

Square foot costs include the following components: Foundations as required for normal soil conditions. Floor, wall and roof structures. Interior ceiling, wall and floor finishes (including carpet). Exterior wall finish and roof cover. Interior partitions. Basic lighting and electrical systems. Rough and finish plumbing. Design and engineering fees. Typical permit and hook-up fees. Contractor's mark-up.

Add the cost of: Canopies and canopy lighting. Public address, intercom and security systems. Docks and ramps. Elevators. Draperies. Fire extinguishers and fire sprinklers. Heating and cooling systems. Exterior signs. Walks, paving and curbing. Yard improvements. See the section "Additional Costs for Commercial, Industrial and Public Structures" beginning on page 236.

Elementary Schools – Masonry or Concrete

First Floor

Estimating Procedure

1. Use the tables in this section to estimate the cost of elementary schools (K-6) which are primarily classrooms and lack the extra office space, assembly, library, food service and recreational facilities common in secondary schools.
2. Establish the structure quality class by applying the information on page 44.
3. Calculate the area of the first floor. This should include all area within the building exterior walls and all inset areas outside the main walls but under the main building roof.
4. Find in the table below the square foot cost for the appropriate quality class and the nearest building area.
5. Use figures in the Wall Height Adjustment row to adjust that square foot cost for wall heights more or less than 10 feet.
6. Multiply the adjusted square foot cost by the area of the first floor.
7. Multiply that total by the location factor on page 7 or 8.
8. Add costs from the section Additional Costs for Commercial, Industrial and Public Buildings beginning on page 236.
9. Using figures on the next page, add the cost of any second or higher floors or a basement.

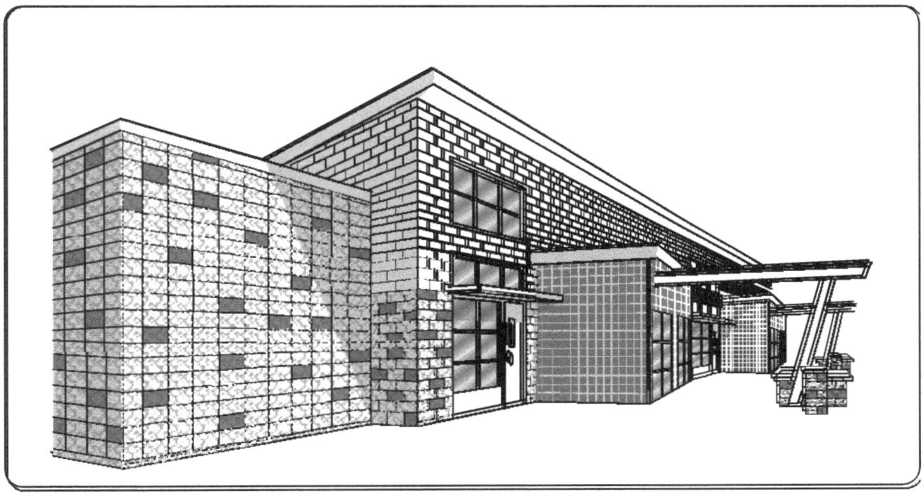

Elementary School, Class 2 & 3

First Floor – Square Foot Area

Quality Class	5,000	6,000	7,500	10,000	12,500	15,000	17,500	20,000	25,000	30,000	40,000
1, Best	560.79	535.97	516.21	500.26	486.65	475.02	465.02	448.43	434.82	423.85	406.50
1 & 2	502.87	480.68	463.04	448.58	436.36	425.91	417.05	401.98	390.03	380.24	364.43
2, Good	452.58	432.71	416.92	403.92	392.83	383.48	375.36	362.01	351.14	342.09	328.26
2 & 3	391.15	373.95	360.09	348.76	339.48	331.38	324.37	312.72	303.44	295.69	283.61
3, Average	363.02	347.04	334.40	324.00	315.04	307.71	301.11	290.30	281.62	274.49	263.27
3 & 4	316.81	302.91	291.78	282.76	275.00	268.60	262.88	253.37	245.70	239.60	229.77
4, Low	270.35	258.35	248.90	241.26	234.68	229.04	224.17	216.17	209.58	204.43	195.93
Wall Height Adjustment*	3.06	2.35	2.12	2.05	1.93	1.88	1.69	1.67	1.63	1.63	1.63

***Wall Height Adjustment:** Add or subtract the amount listed in this row to or from the square foot of floor cost for each foot of wall height more or less than 10 feet.

Elementary Schools – Masonry or Concrete

Upper Floors and Basements

Estimating Procedure

1. Establish the quality class for second and higher floors and the basement. The quality class will usually be the same as the first floor of the building. Square foot costs for unfinished basements will be nearly the same regardless of the structure quality class.
2. Calculate the area of any second or higher floor or a basement.
3. Find in the tables below the square foot cost for the appropriate quality class and the nearest area.
4. Use figures in the Wall Height Adjustment row to adjust the square foot cost for wall heights more or less than 10 feet.
5. Multiply the adjusted square foot cost by the area.
6. Multiply that total by the location factor on page 7 or 8.
7. Add costs from the section Additional Costs for Commercial, Industrial and Public Buildings beginning on page 236.
8. Add totals from this page to the cost for the first floor to find the total building cost.

Second and Higher Floors – Square Foot Area

Quality Class	5,000	6,000	7,500	10,000	12,500	15,000	17,500	20,000	25,000	30,000	40,000
1, Best	515.88	493.03	474.90	460.23	447.72	437.05	427.82	412.56	400.03	389.96	374.00
1 & 2	462.70	442.23	425.99	412.67	401.43	391.84	383.69	369.80	358.81	349.82	335.28
2, Good	416.38	398.08	383.53	371.57	361.40	352.80	345.35	333.04	323.01	314.75	301.98
2 & 3	372.46	356.09	342.90	332.13	323.28	315.54	308.82	297.71	288.92	281.56	270.06
3, Average	334.02	319.29	307.63	298.09	289.78	283.06	277.03	267.08	259.11	252.50	242.16
3 & 4	291.48	278.63	268.49	260.09	252.97	247.08	241.87	233.08	226.03	220.45	211.42
4, Low	248.71	237.67	228.98	221.94	215.90	210.72	206.20	198.81	192.81	188.01	180.27
Wall Height Adjustment*	2.87	2.15	1.97	1.91	1.77	1.69	1.58	1.55	1.53	1.53	1.53

***Wall Height Adjustment:** Add or subtract the amount listed in this row to or from the square foot of floor cost for each foot of wall height more or less than 10 feet.

Finished Basement – Square Foot Area

Quality Class	5,000	6,000	7,500	10,000	12,500	15,000	17,500	20,000	25,000	30,000	40,000
1, Best	315.24	301.29	290.23	281.15	273.56	267.07	261.42	252.10	244.40	238.28	228.51
2, Good	254.42	243.24	234.38	227.02	220.79	215.57	210.99	203.51	197.36	192.31	184.55
3, Average	204.06	195.08	187.93	182.11	177.06	172.97	169.27	163.19	158.29	154.29	147.99
4, Low	151.95	145.23	139.91	135.63	131.96	128.78	126.00	121.53	117.84	114.87	110.13

Unfinished Basements

Area	5,000	6,000	7,500	10,000	12,500	15,000	17,500	20,000	25,000	30,000	40,000
Cost	76.71	73.62	71.26	69.31	67.49	66.10	64.96	62.75	61.11	59.70	57.43

Wall Height Adjustment: Add or subtract the amount listed in this table to or from the square foot of floor cost for each foot of basement wall height more or less than 10 feet.

Area	5,000	6,000	7,500	10,000	12,500	15,000	17,500	20,000	25,000	30,000	40,000
Finished	1.81	1.64	1.62	1.58	1.57	1.53	1.52	1.39	1.37	1.37	1.35
Unfinished	1.64	1.62	1.57	1.53	1.52	1.39	1.39	1.37	1.35	1.34	1.29

Elementary Schools – Wood or Steel Frame

Quality Classification

	Class 1 Best Quality	Class 2 Good Quality	Class 3 Average Quality	Class 4 Low Quality
Foundation (7% of total cost)	Reinforced concrete, depth to 8'	Reinforced concrete, depth 6' to 8'.	Reinforced concrete, depth 4' to 6'.	Reinforced concrete, depth 4' or less.
Floor Structure (4% of total cost)	Concrete on steel beams and deck.	Sheathing on wood or steel floor trusses.	6" thickened edge concrete slab on 6" rock base.	4" reinforced slab on rock base.
Exterior Walls (14% of total cost)	Braced wood or steel studs. Decorative brick or stone veneer with ornate details.	Wood or steel studs. Good wood or composition siding. Some masonry veneer.	Wood studs. Stucco with integral color. Some brick trim.	Wood studs. Painted stucco or inexpensive wood panel siding.
Roof Structure & Cover (29% of total cost)	Glu-lams or steel trusses on steel intermediate columns. Panelized roof system with elastomeric concrete tile or metal roof cover. Engineered for seismic or high wind zones.	Glu-lams or steel beams on steel intermediate columns. Panelized roof system with 5-ply, built-up or concrete tile roof cover. Good insulation.	Glu-lams on steel intermediate columns. Panelized roof system with built-up or good composition shingle roofing. Insulated. Adequate Insulation.	Beams or trusses on steel supports. OSB. sheathing. Built-up or composition shingle roofing. Foil insulation. Not designed for high wind or seismic zones.
Floor Finish (5% of total cost)	Hardwood with good sheet vinyl or carpet in offices. Ceramic tile in restrooms.	Hardwood or sheet vinyl in classrooms. Carpet in offices.	Standard grade sheet vinyl or good composition tile in most areas.	Minimum grade tile.
Windows & Doors (4% of total cost)	Large vinyl-clad or metal insulated low-E windows. Metal doors with glass panels. Institutional grade hardware. LEED certified.	Vinyl-clad insulated windows. Metal exterior doors with glass panels. Solid core interior doors. Commercial grade hardware.	Standard grade insulated windows. Metal exterior doors. Wood interior doors. Standard grade hardware.	Few standard grade metal windows. Low cost metal exterior doors. Low grade hardware.
Interior Wall Finish (13% of total cost)	Gypsum wallboard or lath and plaster finished with good paper or vinyl. Good metal or hardwood veneer wainscot and trim.	Textured gypsum wallboard covered with vinyl or good wallpaper. Good hardwood-veneer wainscot and wood trim in high traffic areas.	Textured and painted interior stucco or gypsum wallboard. Wood trim in high traffic areas.	Painted gypsum wallboard.
Ceiling Finish (2% of total cost)	Suspended good grade acoustical tile with gypsum wallboard backing.	Suspended acoustical tile with concealed grid system.	Suspended acoustical tile with exposed grid system.	Painted gypsum wallboard with acoustic texture.
Specialties (5% of total cost)	Central station alarm. Large chalkboards, cabinets, shelves and cases. Network connection.	Fire alarm and bell system. Network connection. Good chalkboards, map rail & shelving.	Fire alarm, bell and network connections. Some cabinets and shelves.	Minimum alarm system. Few cabinets and shelves.
Plumbing (7% of total cost)	6 good commercial fixtures per classroom. Copper supply & drain pipe. Metal toilet partitions.	5 standard fixtures per classroom. Copper supply & drain pipe. Metal toilet partitions.	4 standard fixtures per classroom. Plastic supply and drain pipe. Composition toilet partitions.	3 minimum fixtures per classroom. Plastic supply and drain pipe. Wood toilet partitions.
Lighting and Power (10% of total cost)	Recessed LED lighting in modular plastic panels. Many task lights or indirect lighting fixtures.	Suspended or panel LED fixtures. Some task lighting or indirect light fixtures.	Continuous 4 tube fluorescent strips with egg crate diffusers, 8' O.C. Some ceiling downlights.	Continuous exposed 2 tube fluorescent strips, 8' O.C.

Note: Use the percent of total cost to help identify the correct quality classification.

Square foot costs include the following components: Foundations as required for normal soil conditions. Floor, wall and roof structures. Interior ceiling, wall and floor finishes (including carpet). Exterior wall finish and roof cover. Interior partitions. Basic lighting and electrical systems. Rough and finish plumbing. Design and engineering fees. Typical permit and hook-up fees. Contractor's mark-up.

Add the cost of: Canopies and canopy lighting. Public address, intercom and security systems. Docks and ramps. Elevators. Draperies. Fire extinguishers and fire sprinklers. Heating and cooling systems. Exterior signs. Walks, paving and curbing. Yard improvements. See the section "Additional Costs for Commercial, Industrial and Public Structures" beginning on page 236.

Elementary Schools – Wood or Steel Frame

First Floor

Estimating Procedure

1. Use the tables in this section to estimate the cost of elementary schools (K-6) which are primarily classrooms and lack the extra office space, assembly, library, food service and recreational facilities common in secondary schools.
2. Establish the structure quality class by applying the information on page 47.
3. Calculate the area of the first floor. This should include all area within the building exterior walls and all inset areas outside the main walls but under the main building roof.
4. Find in the table below the square foot cost for the appropriate quality class and the nearest building area.
5. Use figures in the Wall Height Adjustment row to adjust that square foot cost for wall heights more or less than 10 feet.
6. Multiply the adjusted square foot cost by the area of the first floor.
7. Multiply that total by the location factor on page 7 or 8.
8. Add costs from the section Additional Costs for Commercial, Industrial and Public Buildings beginning on page 236.
9. Using figures on the next page, add the cost of any second or higher floors or a basement.

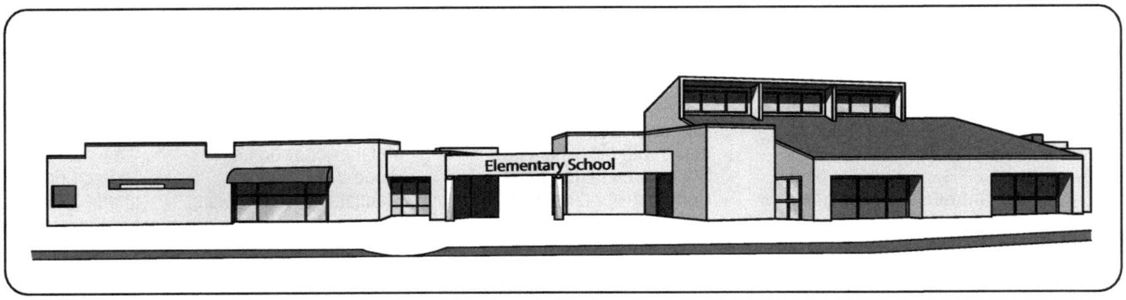

Elementary School, Class 3

Square Foot Area

Quality Class	5,000	6,000	7,500	10,000	12,500	15,000	17,500	20,000	25,000	30,000	40,000
1, Best	352.55	336.99	324.56	314.51	305.97	298.66	292.38	281.94	273.39	266.47	255.57
1 & 2	316.18	302.21	291.13	282.04	274.34	267.79	262.21	252.74	245.21	239.06	229.15
2, Good	284.57	272.04	262.14	253.94	247.00	241.10	236.02	227.60	220.79	215.08	206.38
2 & 3	254.54	243.35	234.32	226.94	220.91	215.64	211.07	203.48	197.46	192.42	184.55
3, Average	228.26	218.17	210.21	203.70	198.06	193.45	189.33	182.49	177.06	172.55	165.49
3 & 4	169.96	162.45	156.49	151.69	147.54	144.02	140.92	135.90	131.78	128.51	123.21
4, Low	199.20	190.45	183.46	177.74	172.87	168.85	165.27	159.30	154.48	150.64	144.45
Wall Height Adjustment*	1.93	1.49	1.34	1.27	1.22	1.15	1.07	1.05	1.03	1.03	1.03

***Wall Height Adjustment:** Add or subtract the amount listed in this table to or from the square foot of floor cost for each foot of wall height more or less than 10 feet.

Upper Floor and Basement

Estimating Procedure

1. Establish the quality class for second floor and the basement. The quality class will usually be the same as the first floor of the building. Square foot costs for unfinished basements will be nearly the same regardless of the structure quality class.
2. Calculate the area of any second floor or basement.
3. Find in the tables below the square foot cost for the appropriate quality class and the nearest area.
4. Use figures in the Wall Height Adjustment row to adjust the square foot cost for wall heights more or less than 10 feet.
5. Multiply the adjusted square foot cost by the area.
6. Multiply that total by the location factor on page 7 or 8.
7. Add costs from the section Additional Costs for Commercial, Industrial and Public Buildings beginning on page 236.
8. Add totals from this page to the cost for the first floor to find the total building cost.

Second Floor – Square Foot Area

Quality Class	5,000	6,000	7,500	10,000	12,500	15,000	17,500	20,000	25,000	30,000	40,000
1, Best	299.13	285.92	275.37	266.86	259.62	253.43	248.08	239.21	231.98	226.13	216.85
1 & 2	268.27	256.42	247.04	239.30	232.79	227.24	222.49	214.42	208.05	202.85	194.46
2, Good	241.44	230.84	222.42	215.46	209.53	204.62	200.24	193.10	187.30	182.48	175.11
2 & 3	215.98	206.49	198.81	192.57	187.42	182.98	179.10	172.64	167.54	163.25	156.59
3, Average	193.67	185.13	178.37	172.82	168.05	164.13	160.63	154.86	150.26	146.41	140.44
3 & 4	169.01	161.58	155.66	150.83	146.71	143.29	140.20	135.14	131.08	127.83	122.57
4, Low	144.22	137.84	132.77	128.69	125.18	122.18	119.59	115.31	111.81	109.02	104.54
Wall height Adjustment*	1.64	1.26	1.13	1.09	1.01	1.00	.92	.90	.90	.90	.90

***Wall Height Adjustment:** Add or subtract the amount listed in this table to or from the square foot of floor cost for each foot of second and higher floor wall height more or less than 10 feet.

Finished Basements – Square Foot Area

Quality Class	5,000	6,000	7,500	10,000	12,500	15,000	17,500	20,000	25,000	30,000	40,000
1, Best	225.22	215.26	207.35	200.90	195.42	190.79	186.80	180.10	174.63	170.20	163.27
2, Good	181.79	173.79	167.44	162.21	157.75	154.03	150.76	145.41	141.02	137.40	131.88
3, Average	145.80	139.40	134.27	130.11	126.52	123.56	120.95	116.60	113.10	110.22	105.74
4, Low	108.59	103.76	99.96	96.89	94.25	92.00	90.04	86.81	84.18	82.07	78.70

Unfinished Basements

Area	5,000	6,000	7,500	10,000	12,500	15,000	17,500	20,000	25,000	30,000	40,000
Cost	54.78	52.64	50.90	49.48	48.23	47.24	46.40	44.86	43.67	42.63	41.03

Wall Height Adjustment: Add or subtract the amount listed in this table to or from the square foot of floor cost for each foot of basement wall height more or less than 10 feet.

Area	5,000	6,000	7,500	10,000	12,500	15,000	17,500	20,000	25,000	30,000	40,000
Finished	1.30	1.15	1.14	1.12	1.11	1.07	1.05	1.01	.99	.99	.96
Unfinished	1.15	1.14	1.11	1.07	1.05	1.01	1.01	.99	.96	.94	.90

Temporary "bungalow" classrooms: One or two-story 24' x 40' prefabricated classrooms, including foundation and utility hookup will cost about $99,000 per classroom for units with standard grade lighting, doors, windows, exterior finish and electric space heating. Prefabricated 24' x 40' classrooms with better grade lighting, doors, windows, exterior finish and packaged A/C will cost about $143,000 each.

Secondary Schools – Masonry or Concrete

Quality Classification

	Class 1 Best Quality	Class 2 Good Quality	Class 3 Average Quality	Class 4 Low Quality
Foundation (7% of total cost)	Reinforced concrete, depth to 8'	Reinforced concrete, depth 6' to 8'.	Reinforced concrete, depth 4' to 6'.	Reinforced concrete, depth 4' or less.
Floor Structure (4% of total cost)	Concrete on steel beams and deck.	Lightweight concrete on steel beams.	6" thickened edge concrete slab on 6" rock base.	4" reinforced slab on rock base.
Exterior Walls (14% of total cost)	Decorative brick veneer with concrete block backup. Ornate brick or stone trim.	Decorative or colored concrete block. Some brick or stone veneer.	Colored concrete block with some brick or wood trim.	Painted tilt-up concrete wall panels. Few decorative details.
Roof Structure & Cover (29% of total cost)	Glu-lams or steel trusses on steel intermediate columns. Panelized roof system with elastomeric, concrete tile or metal roof cover. Engineered for earthquake or high wind zones.	Glu-lams or steel beams on steel intermediate columns. Panelized roof system with 5-ply, built-up or concrete tile roof cover. Good insulation.	Wood or metal trusses on intermediate columns. Panelized roof system with built-up or good composition shingle roofing. Insulated.	Beams or trusses on steel supports. OSB. sheathing. Built-up or low cost membrane roofing. Foil insulation. Not designed for high wind or seismic zones.
Floor Finish (5% of total cost)	Hardwood with good sheet vinyl or carpet in offices. Ceramic tile in restrooms.	Hardwood or sheet vinyl in classrooms. Carpet in offices.	Standard grade sheet vinyl or good composition tile in most areas.	Minimum grade tile.
Windows & Doors (4% of total cost)	Large vinyl-clad or metal insulated low-E windows. Metal doors with glass panels. Institutional grade hardware.	Vinyl-clad insulated windows. Metal exterior doors with glass panels. Solid core interior doors. Commercial grade hardware.	Standard grade insulated windows. Metal exterior doors. Wood interior doors. Standard grade hardware.	Few standard grade metal windows. Low cost metal exterior doors. Low grade hardware.
Interior Wall Finish (13% of total cost)	Gypsum wallboard or lath and plaster finished with good paper or vinyl. Good metal or hardwood veneer wainscot and trim.	Textured gypsum wallboard covered with vinyl or good wallpaper. Good hardwood-veneer wainscot and wood trim in high traffic areas.	Textured and painted interior stucco or gypsum wallboard. Wood trim in high traffic areas.	Painted gypsum wallboard.
Ceiling Finish (2% of total cost)	Suspended good grade acoustical tile with gypsum wallboard backing.	Suspended acoustical tile with concealed grid system.	Suspended acoustical tile with exposed grid system.	Painted gypsum wallboard with acoustic texture.
Specialties (5% of total cost)	Central station alarm. Large chalkboards, cabinets, shelves and cases. Network connection.	Fire alarm and bell system. Network connection. Good chalkboards, map rail & shelving.	Fire alarm, bell and network connections. Some cabinets and shelves.	Minimum alarm system. Few cabinets and shelves.
Plumbing (7% of total cost)	6 good commercial fixtures per classroom. Copper supply & drain pipe. Metal toilet partitions.	5 standard fixtures per classroom. Copper supply & drain pipe. Metal toilet partitions.	4 standard fixtures per classroom. Plastic supply and drain pipe. Composition toilet partitions.	3 minimum fixtures per classroom. Plastic supply and drain pipe. Wood toilet partitions.
Lighting and Power (10% of total cost)	Recessed LED lighting in modular plastic panels. Many task lights or indirect lighting fixtures.	Suspended or panel LED fixtures. Some task lighting or indirect light fixtures.	Continuous 4 tube fluorescent strips with egg crate diffusers, 8' O.C. Some ceiling downlights.	Continuous exposed 2 tube fluorescent strips, 8' O.C.

Note: Use the percent of total cost to help identify the correct quality classification.

Square foot costs include the following components: Foundations as required for normal soil conditions. Floor, wall and roof structures. Interior ceiling, wall and floor finishes (including carpet). Exterior wall finish and roof cover. Interior partitions. Basic lighting and electrical systems. Rough and finish plumbing. Design and engineering fees. Typical permit and hook-up fees. Contractor's mark-up.

Add the cost of: Canopies and canopy lighting. Public address, intercom and security systems. Docks and ramps. Draperies. Fire extinguishers and fire sprinklers. Heating and cooling systems. Exterior signs. Walks, paving and curbing. Elevators. Kitchen equipment. Yard improvements. See the section "Additional Costs for Commercial, Industrial and Public Structures" beginning on page 236.

Secondary Schools – Masonry or Concrete

First Floor

Estimating Procedure

1. Use the tables in this section to estimate the cost of secondary schools and junior colleges which include classrooms, faculty and staff office space, assembly, library-media center, food service and recreational facilities.
2. Establish the structure quality class by applying the information on page 50.
3. Calculate the area of the first floor. This should include all area within the building exterior walls and all inset areas outside the main walls but under the main building roof.
4. Find in the table below the square foot cost for the appropriate quality class and the nearest building area.
5. Use figures in the Wall Height Adjustment row to adjust that square foot cost for wall heights more or less than 12 feet.
6. Multiply the adjusted square foot cost by the area of the first floor.
7. Multiply that total by the location factor on page 7 or 8.
8. Add costs from the section Additional Costs for Commercial, Industrial and Public Buildings beginning on page 236.
9. Using figures on the next page, add the cost of any second or higher floors or a basement.

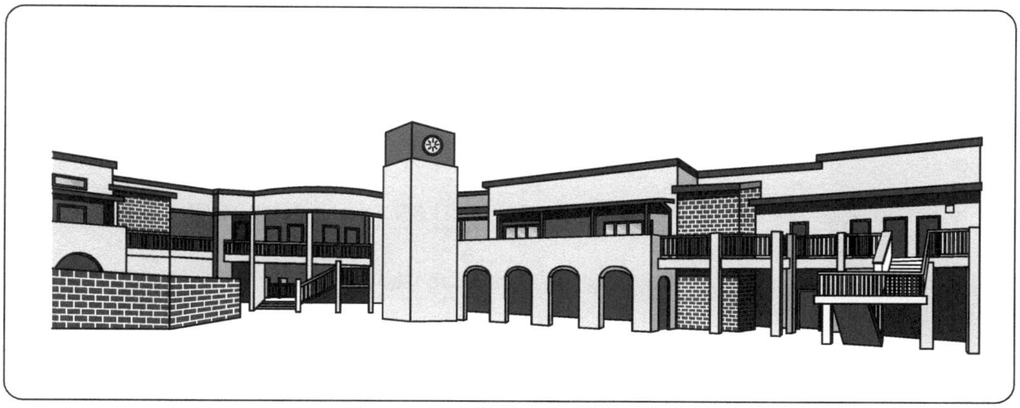

Secondary School, Class 2

First Floor – Square Foot Area

Quality Class	20,000	25,000	30,000	35,000	40,000	45,000	50,000	60,000	70,000	80,000	100,000
1, Best	547.31	523.08	503.83	488.20	474.98	463.62	453.85	437.66	424.42	413.65	396.77
1 & 2	490.81	469.11	451.96	437.84	425.88	415.69	407.05	392.35	380.64	371.12	355.71
2, Good	441.68	422.31	406.91	394.19	383.40	374.31	366.37	353.31	342.68	333.86	320.40
2 & 3	395.15	377.74	363.76	352.33	342.93	334.75	327.65	315.86	306.50	298.70	286.48
3, Average	354.34	338.70	326.33	316.17	307.43	300.29	293.86	283.34	274.86	267.83	256.91
3 & 4	309.20	295.64	284.78	275.97	268.43	262.12	256.56	247.26	239.75	233.83	224.29
4, Low	263.80	252.15	242.91	235.47	229.03	223.54	218.77	210.96	204.57	199.49	191.25
Wall Height Adjustment*	2.62	2.01	1.81	1.74	1.63	1.58	1.48	1.42	1.39	1.39	1.39

***Wall Height Adjustment:** Add or subtract the amount listed in this row to or from the square foot of floor cost for each foot of wall height more or less than 12 feet.

Secondary Schools – Masonry or Concrete

Upper Floors and Basements

Estimating Procedure

1. Establish the quality class for second and higher floors and the basement. The quality class will usually be the same as the first floor of the building. Square foot costs for unfinished basements will be nearly the same regardless of the structure quality class.
2. Calculate the area of any second or higher floor or a basement.
3. Find in the tables below the square foot cost for the appropriate quality class and the nearest area.
4. Use figures in the Wall Height Adjustment row to adjust the square foot cost for wall heights more or less than 10 feet.
5. Multiply the adjusted square foot cost by the area.
6. Multiply that total by the location factor on page 7 or 8.
7. Add costs from the section Additional Costs for Commercial, Industrial and Public Buildings beginning on page 236.
8. Add totals from this page to the cost for the first floor to find the total building cost.

Second and Higher Floors – Square Foot Area

Quality Class	20,000	25,000	30,000	35,000	40,000	45,000	50,000	60,000	70,000	80,000	100,000
1, Best	492.59	470.77	453.44	439.41	427.48	417.29	408.49	393.92	381.96	372.30	357.05
1 & 2	441.72	422.22	406.76	394.00	383.29	374.14	366.35	353.11	342.61	334.02	320.14
2, Good	397.50	380.11	366.18	354.79	345.04	336.83	329.70	317.97	308.42	300.49	288.32
2 & 3	355.62	339.98	327.37	317.08	308.60	301.29	294.89	284.26	275.85	268.82	257.85
3, Average	318.92	304.80	293.67	284.60	276.69	270.26	264.49	254.99	247.37	241.09	231.26
3 & 4	278.32	266.07	256.31	248.37	241.55	235.90	230.91	222.57	215.79	210.49	201.81
4, Low	237.46	226.95	218.64	211.93	206.14	201.21	196.90	189.89	184.12	179.56	172.14
Wall Height Adjustment*	2.73	2.06	1.91	1.77	1.67	1.63	1.53	1.48	1.47	1.47	1.47

***Wall Height Adjustment:** Add or subtract the amount listed in this row to or from the square foot of floor cost for each foot of wall height more or less than 10 feet.

Finished Basement – Square Foot Area

Quality Class	20,000	25,000	30,000	35,000	40,000	45,000	50,000	60,000	70,000	80,000	100,000
1, Best	301.02	287.69	277.08	268.57	261.26	255.01	249.63	240.69	233.43	227.56	218.21
2, Good	242.93	232.27	223.81	216.82	210.89	205.85	201.50	194.33	188.46	183.62	176.21
3, Average	194.90	186.28	179.49	173.90	169.09	165.16	161.64	155.80	151.17	147.36	141.32
4, Low	145.09	138.68	133.60	129.49	125.97	122.95	120.35	116.01	112.51	109.72	105.19

Unfinished Basements

Area	20,000	25,000	30,000	35,000	40,000	45,000	50,000	60,000	70,000	80,000	100,000
Cost	62.87	60.39	58.41	56.81	55.30	54.14	53.24	51.42	50.12	48.94	47.11

Wall Height Adjustment: Add or subtract the amount listed in this table to or from the square foot of floor cost for each foot of basement wall height more or less than 10 feet.

Area	20,000	25,000	30,000	35,000	40,000	45,000	50,000	60,000	70,000	80,000	100,000
Finished	1.55	1.39	1.37	1.35	1.34	1.30	1.29	1.20	1.17	1.17	1.14
Unfinished	1.30	1.26	1.20	1.17	1.14	1.08	1.08	1.06	1.04	1.03	.98

Secondary Schools – Wood or Steel Frame

Quality Classification

	Class 1 Best Quality	Class 2 Good Quality	Class 3 Average Quality	Class 4 Low Quality
Foundation (7% of total cost)	Reinforced concrete, depth to 8'	Reinforced concrete, depth 6' to 8'.	Reinforced concrete, depth 4' to 6'.	Reinforced concrete, depth 4' or less.
Floor Structure (4% of total cost)	Concrete on steel beams and deck.	Sheathing on wood or steel floor trusses.	6" thickened edge concrete slab on 6" rock base.	4" reinforced slab on rock base.
Exterior Walls (14% of total cost)	Braced wood or steel studs. Decorative brick or stone veneer with ornate details.	Wood or steel studs. Good wood or composition siding. Some masonry veneer.	Wood studs. Stucco with integral color. Some brick trim.	Wood studs. Painted stucco or inexpensive wood panel siding.
Roof Structure & Cover (29% of total cost)	Glu-lams or steel trusses on steel intermediate columns. Panelized roof system with elastomeric, concrete tile or metal roof cover. Engineered for seismic or high wind zones.	Glu-lams or steel beams on steel intermediate columns. Panelized roof system with 5-ply, built-up or concrete tile roof cover. Good insulation.	Glu-lams on steel intermediate columns. Panelized roof system with built-up or good composition shingle roofing. Insulated. Adequate Insulation.	Beams or trusses on steel supports. OSB. sheathing. Built-up or composition shingle roofing. Foil insulation. Not designed for high wind or seismic zones.
Floor Finish (5% of total cost)	Hardwood with good sheet vinyl or carpet in offices. Ceramic tile in restrooms.	Hardwood or sheet vinyl in classrooms. Carpet in offices.	Standard grade sheet vinyl or good composition tile in most areas.	Minimum grade tile.
Windows & Doors (4% of total cost)	Large vinyl-clad or metal insulated low-E windows. Metal doors with glass panels. Institutional grade hardware. LEED certified.	Vinyl-clad insulated windows. Metal exterior doors with glass panels. Solid core interior doors. Commercial grade hardware.	Standard grade insulated windows. Metal exterior doors. Wood interior doors. Standard grade hardware.	Few standard grade metal windows. Low cost metal exterior doors. Low grade hardware.
Interior Wall Finish (13% of total cost)	Gypsum wallboard or lath and plaster finished with good paper or vinyl. Good metal or hardwood veneer wainscot and trim.	Textured gypsum wallboard covered with vinyl or good wallpaper. Good hardwood-veneer wainscot and wood trim in high traffic areas.	Textured and painted interior stucco or gypsum wallboard. Wood trim in high traffic areas.	Painted gypsum wallboard.
Ceiling Finish (2% of total cost)	Suspended good grade acoustical tile with gypsum wallboard backing.	Suspended acoustical tile with concealed grid system.	Suspended acoustical tile with exposed grid system.	Painted gypsum wallboard with acoustic texture.
Specialties (5% of total cost)	Central station alarm. Large chalkboards, cabinets, shelves and cases. Network connection.	Fire alarm and bell system. Network connection. Good chalkboards, map rail & shelving.	Fire alarm, bell and network connections. Some cabinets and shelves.	Minimum alarm system. Few cabinets and shelves.
Plumbing (7% of total cost)	6 good commercial fixtures per classroom. Copper supply & drain pipe. Metal toilet partitions.	5 standard fixtures per classroom. Copper supply & drain pipe. Metal toilet partitions.	4 standard fixtures per classroom. Plastic supply and drain pipe. Composition toilet partitions.	3 minimum fixtures per classroom. Plastic supply and drain pipe. Wood toilet partitions.
Lighting and Power (10% of total cost)	Recessed LED lighting in modular plastic panels. Many task lights or indirect lighting fixtures.	Suspended or panel LED fixtures. Some task lighting or indirect light fixtures.	Continuous 4 tube fluorescent strips with egg crate diffusers, 8' O.C. Some ceiling downlights.	Continuous exposed 2 tube fluorescent strips, 8' O.C.

Note: Use the percent of total cost to help identify the correct quality classification.

Square foot costs include the following components: Foundations as required for normal soil conditions. Floor, wall and roof structures. Interior ceiling, wall and floor finishes (including carpet). Exterior wall finish and roof cover. Interior partitions. Basic lighting and electrical systems. Rough and finish plumbing. Design and engineering fees. Typical permit and hook-up fees. Contractor's mark-up.

Add the cost of: Canopies and canopy lighting. Public address, intercom and security systems. Docks and ramps. Elevators. Draperies. Fire extinguishers and fire sprinklers. Heating and cooling systems. Exterior signs. Walks, paving and curbing. Kitchen equipment. Yard improvements. See the section "Additional Costs for Commercial, Industrial and Public Structures" beginning on page 236.

Secondary Schools – Wood or Steel Frame

First Floor

Estimating Procedure

1. Use the tables in this section to estimate the cost of secondary schools and junior colleges which include classrooms, faculty and staff office space, assembly, library-media center, food service and recreational facilities.
2. Establish the structure quality class by applying the information on page 53.
3. Calculate the area of the first floor. This should include all area within the building exterior walls and all inset areas outside the main walls but under the main building roof.
4. Find in the table below the square foot cost for the appropriate quality class and the nearest building area.
5. Use figures in the Wall Height Adjustment row to adjust that square foot cost for wall heights more or less than 12 feet.
6. Multiply the adjusted square foot cost by the area of the first floor.
7. Multiply that total by the location factor on page 7 or 8.
8. Add costs from the section Additional Costs for Commercial, Industrial and Public Buildings beginning on page 236.
9. Using figures on the next page, add the cost of any second or higher floors or a basement.

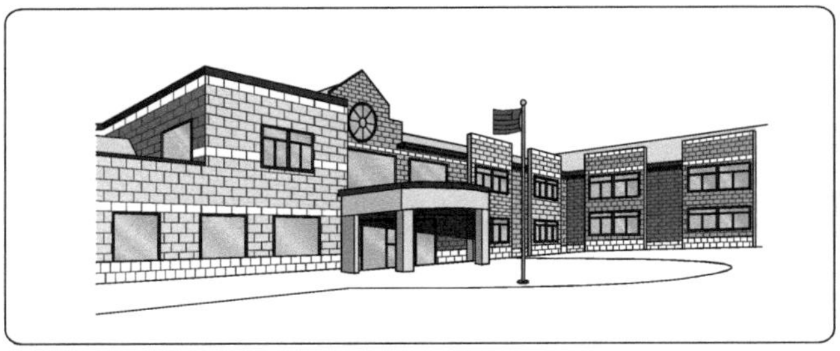

Secondary School, Class 3

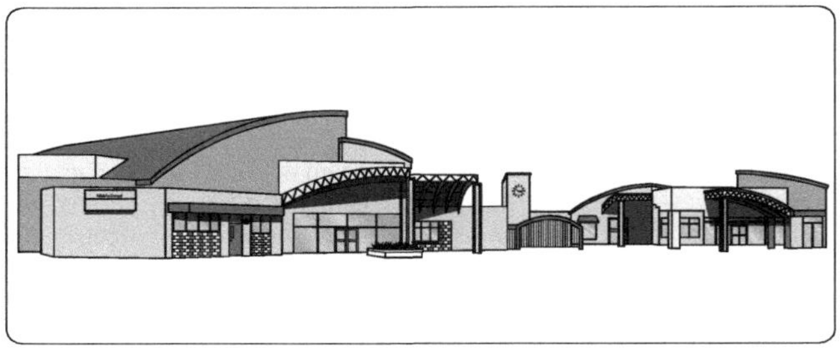

Secondary School, Class 2

Square Foot Area

Quality Class	20,000	25,000	30,000	35,000	40,000	45,000	50,000	60,000	70,000	80,000	100,000
1, Best	344.10	328.84	316.76	306.93	298.63	291.49	285.34	275.15	266.81	260.08	249.44
1 & 2	308.57	294.96	284.14	275.26	267.75	261.34	255.91	246.64	239.30	233.32	223.63
2, Good	277.70	265.51	255.81	247.84	241.07	235.29	230.34	222.14	215.45	209.92	201.42
2 & 3	248.43	237.51	228.67	221.48	215.60	210.45	205.98	198.59	192.67	187.79	180.11
3, Average	222.79	212.95	205.15	198.80	193.30	188.78	184.79	178.11	172.80	168.41	161.56
3 & 4	194.43	185.85	179.05	173.49	168.74	164.79	161.29	155.45	150.76	147.04	141.00
4, Low	165.89	158.53	152.72	148.07	144.01	140.57	137.52	132.65	128.61	125.40	120.25
Wall Height Adjustment*	1.86	1.45	1.32	1.25	1.15	1.13	1.04	1.03	1.01	1.01	1.00

***Wall Height Adjustment:** Add or subtract the amount listed in this table to or from the square foot of floor cost for each foot of wall height more or less than 12 feet.

Secondary Schools – Wood or Steel Frame

Upper Floor and Basement

Estimating Procedure

1. Establish the quality class for second floor and the basement. The quality class will usually be the same as the first floor of the building. Square foot costs for unfinished basements will be nearly the same regardless of the structure quality class.
2. Calculate the area of any second floor or basement.
3. Find in the tables below the square foot cost for the appropriate quality class and the nearest area.
4. Use figures in the Wall Height Adjustment row to adjust the square foot cost for wall heights more or less than 10 feet.
5. Multiply the adjusted square foot cost by the area.
6. Multiply that total by the location factor on page 7 or 8.
7. Add costs from the section Additional Costs for Commercial, Industrial and Public Buildings beginning on page 236.
8. Add totals from this page to the cost for the first floor to find the total building cost.

Second Floor – Square Foot Area

Quality Class	20,000	25,000	30,000	35,000	40,000	45,000	50,000	60,000	70,000	80,000	100,000
1, Best	285.37	272.74	262.71	254.57	247.64	241.74	236.64	228.20	221.28	215.70	206.85
1 & 2	255.91	244.60	235.66	228.28	222.06	216.76	212.22	204.61	198.48	193.51	185.49
2, Good	230.32	220.21	212.16	205.55	199.93	195.16	191.04	184.24	178.68	174.11	167.06
2 & 3	206.01	196.97	189.67	183.70	178.79	174.53	170.87	164.68	159.84	155.73	149.38
3, Average	184.77	176.61	170.13	164.87	160.31	156.58	153.23	147.73	143.34	139.64	133.99
3 & 4	161.23	154.18	148.46	143.88	139.94	136.66	133.78	128.93	125.04	121.97	116.92
4, Low	137.57	131.47	126.65	122.78	119.42	116.57	114.08	109.99	106.67	104.02	99.74
Wall height Adjustment*	1.60	1.17	1.09	1.03	.99	.94	.89	.87	.83	.83	.83

***Wall Height Adjustment:** Add or subtract the amount listed in this table to or from the square foot of floor cost for each foot of second and higher floor wall height more or less than 10 feet.

Finished Basements – Square Foot Area

Quality Class	20,000	25,000	30,000	35,000	40,000	45,000	50,000	60,000	70,000	80,000	100,000
1, Best	215.30	205.75	198.16	192.04	186.84	182.38	178.54	172.15	166.93	162.72	156.07
2, Good	173.72	166.12	160.07	155.04	150.81	147.20	144.09	138.98	134.76	131.33	126.01
3, Average	139.37	133.24	128.36	124.37	120.95	118.09	115.59	111.42	108.11	105.34	101.04
4, Low	103.78	99.16	95.54	92.63	90.09	87.92	86.06	82.97	80.47	78.47	75.23

Unfinished Basements

Area	20,000	25,000	30,000	35,000	40,000	45,000	50,000	60,000	70,000	80,000	100,000
Cost	44.91	43.14	41.74	40.58	39.51	38.71	38.02	36.75	35.79	34.97	33.64

Wall Height Adjustment: Add or subtract the amount listed in this table to or from the square foot of floor cost for each foot of basement wall height more or less than 10 feet.

Area	20,000	25,000	30,000	35,000	40,000	45,000	50,000	60,000	70,000	80,000	100,000
Finished	1.01	.91	.90	.89	.87	.83	.82	.77	.75	.75	.73
Unfinished	.91	.90	.87	.83	.82	.77	.77	.75	.73	.72	.70

Temporary "bungalow" classrooms: One or two-story 24' x 40' prefabricated classrooms, including foundation and utility hookup will cost about $99,000 per classroom for units with standard grade lighting, doors, windows, exterior finish and electric space heating. Prefabricated 24' x 40' classrooms with better grade lighting, doors, windows, exterior finish and packaged A/C will cost about $143,000 each.

Government Offices – Masonry or Concrete

Quality Classification

	Class 1 Best Quality	Class 2 Good Quality	Class 3 Average Quality	Class 4 Low Quality
Foundation (7% of total cost)	Reinforced concrete, depth to 12'	Reinforced concrete, depth 8' to 10'.	Reinforced concrete, depth 6' to 8'.	Reinforced concrete, depth 6' or less.
Floor Structure (5% of total cost)	Concrete on steel beams and deck.	Lightweight concrete on steel beams.	6" thickened edge concrete slab on 6" rock base.	4" reinforced slab on rock base.
Exterior Walls (12% of total cost)	Marble or polished granite, spandrel glass. Metal trim. Decorative atrium entrance.	Glass curtain wall. Textured-block, good brick or stone veneer. Atrium entrance.	Colored concrete block with some brick or wood trim. Decorative entrance.	Tilt-up concrete, brick or concrete block. Few decorative details.
Roof Structure & Cover (16% of total cost)	Glu-lams or steel trusses on steel intermediate columns. Panelized roof system with elastomeric concrete tile or metal roof cover. Engineered for earthquake or high wind zones.	Glu-lams or steel beams on steel intermediate columns. Panelized roof system with 5-ply, built-up or concrete tile roof cover. Good insulation.	Wood or metal trusses on intermediate columns. Panelized roof system with built-up or good composition shingle roofing. Insulated.	Beams or trusses on steel supports. OSB. sheathing. Built-up or low cost membrane roofing. Foil insulation. Not designed for high wind or seismic zones.
Floor Finish (8% of total cost)	Marble or good terrazzo in public rooms. Carpeted offices. Ceramic tile in restrooms.	Terrazzo or marble tile in public rooms. Carpet in offices. Vinyl in restrooms.	Good sheet vinyl in meeting rooms. Carpet in offices. Composition tile in restrooms.	Low cost sheet vinyl in public rooms. Floor Tile elsewhere.
Windows & Doors (6% of total cost)	Low-E glass in metal sash. Metal-frame laminated glass entrance doors. Metal interior doors. Institutional grade hardware. LEED certified.	Colored low-E glass. Decorative metal exterior doors. Solid core 8' high wood interior doors. Commercial grade hardware.	Insulated store front or glazed curtain wall. Metal exterior doors. Wood interior doors. Standard grade hardware.	Few standard grade metal windows. Low cost metal exterior doors. Low grade hardware.
Interior Wall Finish (14% of total cost)	Decorative plaster in public rooms. Plaster or vinyl covered wallboard in private offices. Decorative metal or hardwood veneer wainscot and trim.	Plaster or good paneling in public rooms. Vinyl-covered wallboard in hallways. Textured or vinyl-covered wallboard with good trim elsewhere.	Textured and painted interior stucco or gypsum wallboard. Wood trim in public meeting rooms.	Painted gypsum wallboard.
Ceiling Finish (6% of total cost)	Suspended decorative textured plaster in meeting rooms. Acoustic ceiling tile elsewhere.	Suspended acoustical tile in with concealed grid in meeting rooms. Acoustical tile elsewhere.	Suspended acoustical tile with exposed grid system.	Painted gypsum wallboard with acoustic texture.
Specialties (4% of total cost)	Security, alarm, PA and surveillance systems. Video and network connections.	Alarm and security systems. Network connections. Directory boards and built-in cabinetry.	Fire alarm, bell and network connections. Some cabinets and shelves.	Minimum alarm system. Few cabinets and shelves.
Plumbing (6% of total cost)	Institutional grade fixtures. Copper supply, vent & drain pipe. Metal or synthetic stone toilet partitions.	Good commercial grade fixtures. Copper supply & drain pipe drain pipe. Metal toilet partitions.	Standard grade fixtures. PEX or PVC supply, vent and drain pipe. Composition toilet partitions.	Minimum fixtures. Plastic supply, vent & drain pipe. Plastic-faced toilet partitions.
Lighting and Power (16% of total cost)	Decorative indirect fixtures in public rooms. Many recessed task lights on separate controls. Recessed LED fixtures in offices and hallways.	Indirect lighting fixtures in public rooms. Track lighting or recessed LED fixtures with some task lighting in offices and hallways.	Continuous LED strips with decorative diffusers, 8' O.C. Some recessed ceiling fixtures.	Continuous exposed 2 tube fluorescent strip, fixtures 8' O.C.

Notes: Use the percent of total cost to help identify the correct quality classification. State and federal office buildings are usually Class 1 or Class 2 and rarely Class 3 or Class 4. Municipal office buildings are usually Class 3 or Class 4 and rarely Class 1 or Class 2. The figures in this section apply to government offices designed for public assembly (with committee hearing rooms or council chambers). For government offices serving walk-in clientele (motor vehicle department offices, post offices) use figures from either the section Urban Stores or the section Suburban Stores. For government offices used primarily by administrative staff (police stations, hall of records), use figures from the section General Office Buildings.

Square foot costs include the following components: Foundations as required for normal soil conditions. Floor, wall and roof structures. Interior ceiling, wall and floor finishes (including carpet). Exterior wall finish and roof cover. Interior partitions. Basic lighting and electrical systems. Rough and finish plumbing. Specialties as listed. Design and engineering fees. Typical utility hook-up. Contractor's mark-up.

Add the cost of: Canopies and canopy lighting. Public address, intercom and security systems beyond what appears in the quality classification above. Docks and ramps. Elevators. Draperies. Fire extinguishers and fire sprinklers. Heating and cooling systems. Exterior signs. Walks, paving and curbing. Yard improvements. See the section "Additional Costs for Commercial, Industrial and Public Structures" beginning on page 236.

Government Offices – Masonry or Concrete

First Floor

Estimating Procedure

1. Use the tables in this section to estimate the cost of buildings designed for occupancy by state, federal, or municipal agencies. Many government agencies occupy office space designed for use by commercial tenants rather than built for use by government officials. Use figures from the sections on Urban Stores, Suburban Stores or General Office Buildings when the structure was designed for commercial occupancy.
2. Establish the structure quality class by applying the information on page 56.
3. Calculate the area of the first floor. This should include all area within the building exterior walls and all inset areas outside the main walls but under the main building roof.
4. Find in the table below the square foot cost for the appropriate quality class and the nearest building area.
5. Use figures in the Wall Height Adjustment row to adjust that square foot cost for wall heights more or less than 12 feet.
6. Multiply the adjusted square foot cost by the area of the first floor.
7. Multiply that total by the location factor on page 7 or 8.
8. Add costs from the section Additional Costs for Commercial, Industrial and Public Buildings beginning on page 236.
9. Using figures on the next page, add the cost of any second or higher floors or a basement.

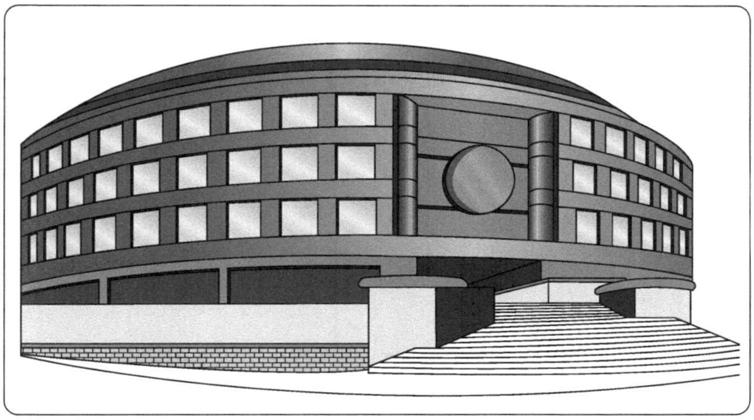

Government Office, Class 1

First Floor – Square Foot Area

Quality Class	5,000	7,500	10,000	12,500	15,000	20,000	25,000	30,000	35,000	40,000	50,000
1, Best	559.16	519.96	496.44	480.43	468.49	451.81	440.32	431.80	425.26	419.94	411.64
1 & 2	518.46	481.94	460.39	445.34	434.24	418.66	408.15	400.30	394.03	389.18	381.79
2, Good	480.01	446.37	426.28	412.50	402.23	388.00	378.01	370.76	365.14	360.61	353.56
2 & 3	447.86	416.38	397.72	384.85	375.11	361.89	352.67	345.74	340.45	336.30	329.70
3, Average	420.96	391.57	373.80	361.64	352.67	340.28	331.28	324.99	320.09	316.07	309.85
3 & 4	381.83	354.90	338.96	328.06	319.91	308.54	300.68	294.79	290.27	286.76	281.15
4, Low	340.96	316.98	302.67	292.95	285.61	275.41	268.32	263.26	259.24	255.95	250.91
Wall Height Adjustment*	6.37	5.06	4.46	4.05	3.76	3.07	2.75	2.45	2.40	2.31	1.94

***Wall Height Adjustment:** Add or subtract the amount listed in this row to or from the square foot of floor cost for each foot of wall height more or less than 10 feet.

Government Offices – Masonry or Concrete

Upper Floors and Basements

Estimating Procedure

1. Establish the quality class for second and higher floors and the basement. The quality class will usually be the same as the first floor of the building. Square foot costs for unfinished basements will be nearly the same regardless of the structure quality class.
2. Calculate the area of any second or higher floor or a basement.
3. Find in the tables below the square foot cost for the appropriate quality class and the nearest area.
4. Use figures in the Wall Height Adjustment row to adjust the square foot cost for wall heights more or less than 10 feet.
5. Multiply the adjusted square foot cost by the area.
6. Multiply that total by the location factor on page 7 or 8.
7. Add costs from the section Additional Costs for Commercial, Industrial and Public Buildings beginning on page 236.
8. Add totals from this page to the cost for the first floor to find the total building cost.

Second and Higher Floors – Square Foot Area

Quality Class	5,000	7,500	10,000	12,500	15,000	20,000	25,000	30,000	35,000	40,000	50,000
1, Best	490.92	456.52	435.86	421.82	411.27	396.71	386.59	379.10	373.32	368.69	361.40
1 & 2	455.13	423.10	404.19	390.96	381.23	367.59	358.30	351.42	345.99	341.67	335.23
2, Good	421.43	391.89	374.24	362.16	353.14	340.66	331.83	325.50	320.55	316.56	310.44
2 & 3	393.21	365.57	349.16	337.90	329.36	317.69	309.65	303.54	298.91	295.24	289.45
3, Average	369.57	343.74	328.14	317.48	309.65	298.75	290.86	285.32	281.02	277.49	272.03
3 & 4	335.25	311.61	297.55	288.01	280.84	270.91	263.94	258.82	254.83	251.73	246.87
4, Low	299.32	278.31	265.71	257.17	250.77	241.81	235.56	231.09	227.61	224.69	220.31
Wall Height Adjustment*	5.59	4.46	3.91	3.60	3.30	2.73	2.39	2.15	2.07	2.02	1.69

***Wall Height Adjustment:** Add or subtract the amount listed in this row to or from the square foot of floor cost for each foot of wall height more or less than 10 feet.

Finished Basement – Square Foot Area

Quality Class	5,000	7,500	10,000	12,500	15,000	20,000	25,000	30,000	35,000	40,000	50,000
1, Best	304.56	283.22	270.38	261.65	255.13	246.05	239.80	235.15	231.58	228.66	224.17
2, Good	261.43	243.11	232.10	224.62	219.05	211.31	205.85	201.95	198.83	196.35	192.60
3, Average	229.27	213.25	203.58	196.95	192.05	185.30	180.45	176.99	174.31	172.15	168.75
4, Low	185.67	172.64	164.80	159.54	155.58	149.96	146.16	143.35	141.17	139.40	136.67

Unfinished Basements

Area	5,000	7,500	10,000	12,500	15,000	20,000	25,000	30,000	35,000	40,000	50,000
Cost	77.42	71.98	68.77	66.53	64.89	62.56	60.97	59.79	58.89	58.16	57.00

Wall Height Adjustment: Add or subtract the amount listed in this table to or from the square foot of floor cost for each foot of basement wall height more or less than 10 feet.

Area	5,000	7,500	10,000	12,500	15,000	20,000	25,000	30,000	35,000	40,000	50,000
Finished	1.65	1.35	1.19	1.08	1.01	.82	.71	.63	.61	.58	.51
Unfinished	1.53	1.20	1.06	.98	.90	.71	.63	.57	.56	.54	.46

Government Offices – Wood Frame

Quality Classification

	Class 1 Best Quality	Class 2 Good Quality	Class 3 Average Quality	Class 4 Low Quality
Foundation (7% of total cost)	Reinforced concrete, depth to 8'	Reinforced concrete, depth 6' to 8'.	Reinforced concrete, depth 4' to 6'.	Reinforced concrete, depth 4' or less.
Floor Structure (5% of total cost)	Concrete on steel beams and deck.	Sheathing on wood or steel floor trusses.	6" thickened edge concrete slab on 6" rock base.	4" reinforced slab on rock base.
Exterior Walls (12% of total cost)	Braced wood or steel studs. Decorative brick or stone veneer with ornate details.	Wood or steel studs. Good wood or composition siding. Some masonry veneer.	Wood studs. Stucco with integral color. Some brick trim.	Wood studs. Painted stucco or inexpensive wood panel siding.
Roof Structure & Cover (16% of total cost)	Glu-lams or steel trusses on steel intermediate columns. Panelized roof system with elastomeric concrete tile or metal roof cover. Engineered for earthquake or high wind zones.	Glu-lams or steel beams on steel intermediate columns. Panelized roof system with 5-ply, built-up or concrete tile roof cover. Good insulation.	Wood or metal trusses on intermediate columns. Panelized roof system with built-up or good composition shingle roofing. Insulated.	Beams or trusses on steel supports. OSB. sheathing. Built-up or low cost membrane roofing. Foil insulation. Not designed for high wind or seismic zones.
Floor Finish (8% of total cost)	Marble or good terrazzo in public rooms. Carpeted offices. Ceramic tile in restrooms.	Terrazzo or marble tile in public rooms. Carpet in offices. Vinyl in restrooms.	Good sheet vinyl in meeting rooms. Carpet in offices. Composition tile in restrooms.	Low cost sheet vinyl in public rooms. Floor Tile elsewhere.
Windows & Doors (6% of total cost)	Low-E glass in metal sash. Metal-frame laminated glass entrance doors. Metal interior doors. Institutional grade hardware. LEED certified.	Colored low-E glass. Decorative metal exterior doors. Solid core 8' high wood interior doors. Commercial grade hardware.	Insulated store front or glazed curtain wall. Metal exterior doors. Wood interior doors. Standard grade hardware.	Few standard grade metal windows. Low cost metal exterior doors. Low grade hardware.
Interior Wall Finish (14% of total cost)	Decorative plaster in public rooms. Plaster or vinyl covered wallboard in private offices. Decorative metal or hardwood veneer wainscot and trim.	Plaster or good paneling in public rooms. Vinyl-covered wallboard in hallways. Textured or vinyl-covered wallboard with good trim elsewhere.	Textured and painted interior stucco or gypsum wallboard. Wood trim in public meeting rooms.	Painted gypsum wallboard.
Ceiling Finish (6% of total cost)	Suspended decorative textured plaster in meeting rooms. Acoustic ceiling tile elsewhere.	Suspended acoustical tile in with concealed grid in meeting rooms. Acoustical tile elsewhere.	Suspended acoustical tile with exposed grid system.	Painted gypsum wallboard with acoustic texture.
Specialties (4% of total cost)	Security, alarm, PA and surveillance systems. Video and network connections.	Alarm and security systems. Network connections. Directory boards and built-in cabinetry.	Fire alarm, bell and network connections. Some cabinets and shelves.	Minimum alarm system. Few cabinets and shelves
Plumbing (6% of total cost)	Institutional grade fixtures. Copper supply, vent & drain pipe. Metal or synthetic stone toilet partitions.	Good commercial grade fixtures. Copper supply & drain pipe drain pipe. Metal toilet partitions.	Standard grade fixtures. PEX or PVC supply, vent and drain pipe. Composition toilet partitions.	Minimum fixtures. Plastic supply, vent & drain pipe. Plastic-faced toilet partitions.
Lighting and Power (16% of total cost)	Decorative indirect fixtures in public rooms. Many recessed task lights on separate controls. Recessed LED fixtures in offices and hallways.	Indirect lighting fixtures in public rooms. Track lighting or recessed LED fixtures with some task lighting in offices and hallways.	Continuous 4 tube LED strips with decorative diffusers, 8' O.C. Some recessed ceiling fixtures.	Continuous exposed 2 tube fluorescent strip, fixtures 8' O.C.

Notes: Use the percent of total cost to help identify the correct quality classification. State and federal office buildings are usually Class 1 or Class 2 and rarely Class 3 or Class 4. Municipal office buildings are usually Class 3 or Class 4 and rarely Class 1 or Class 2. The figures in this section apply to government offices designed for public assembly (with committee hearing rooms or council chambers). For government offices serving walk-in clientele (motor vehicle department offices, post offices) use figures from either the section Urban Stores or the section Suburban Stores. For government offices used primarily by administrative staff (police stations, hall of records), use figures from the section General Office Buildings.

Square foot costs include the following components: Foundations as required for normal soil conditions. Floor, wall and roof structures. Interior ceiling, wall and floor finishes (including carpet). Exterior wall finish and roof cover. Interior partitions. Basic lighting and electrical systems. Rough and finish plumbing. Specialties as listed. Design and engineering fees. Typical utility hook-up. Contractor's mark-up.

Add the cost of: Canopies and canopy lighting. Public address, intercom and security systems beyond what appears in the quality classification above. Docks and ramps. Elevators. Draperies. Fire extinguishers and fire sprinklers. Heating and cooling systems. Exterior signs. Walks, paving and curbing. Yard improvements. See the section "Additional Costs for Commercial, Industrial and Public Structures" beginning on page 236.

Government Offices – Wood Frame

First Floor

Estimating Procedure

1. Use the tables in this section to estimate the cost of buildings designed for occupancy by state, federal, or municipal agencies. Many government agencies occupy office space designed for use by commercial tenants rather than built for use by government officials. Use figures from the sections on Urban Stores, Suburban Stores or General Office Buildings when the structure was designed for commercial occupancy.
2. Establish the structure quality class by applying the information on page 59.
3. Calculate the area of the first floor. This should include all area within the building exterior walls and all inset areas outside the main walls but under the main building roof.
4. Find in the table below the square foot cost for the appropriate quality class and the nearest building area.
5. Use figures in the Wall Height Adjustment row to adjust that square foot cost for wall heights more or less than 10 feet.
6. Multiply the adjusted square foot cost by the area of the first floor.
7. Multiply that total by the location factor on page 7 or 8.
8. Add costs from the section Additional Costs for Commercial, Industrial and Public Buildings beginning on page 236.
9. Using figures on the next page, add the cost of any second or higher floors or a basement.

Government Office, Class 3

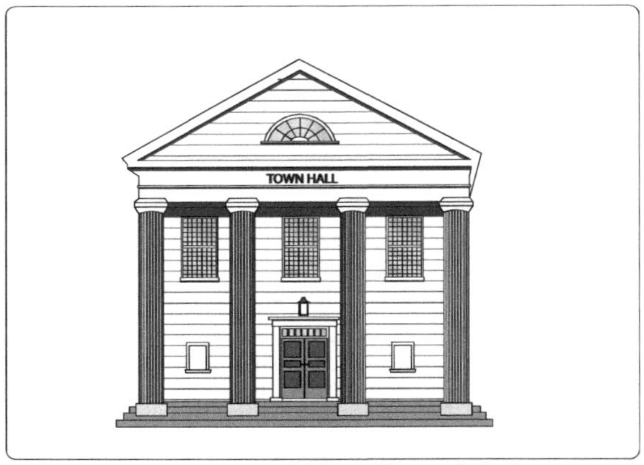

Government Office, Class 3 & 4

Square Foot Area

Quality Class	5,000	7,500	10,000	12,500	15,000	20,000	25,000	30,000	35,000	40,000	50,000
1, Best	351.48	326.86	312.05	301.99	294.51	283.99	276.78	271.44	267.31	263.98	258.76
1 & 2	325.91	302.95	289.37	279.91	272.94	263.21	256.51	251.60	247.70	244.61	240.03
2, Good	301.74	280.59	267.92	259.27	252.82	243.89	237.58	233.06	229.51	226.65	222.26
2 & 3	281.53	261.74	250.01	241.92	235.81	227.49	221.68	217.36	214.01	211.37	207.25
3, Average	264.61	246.13	234.96	227.30	221.68	213.94	208.26	204.29	201.21	198.69	194.75
3 & 4	214.31	199.24	190.26	184.13	179.55	173.13	168.69	165.47	162.95	160.90	157.73
4, Low	240.04	223.07	213.03	206.22	201.08	193.94	188.98	185.29	182.45	180.24	176.77
Wall Height Adjustment*	4.00	3.17	2.79	2.56	2.36	1.91	1.72	1.58	1.50	1.46	1.23

***Wall Height Adjustment:** Add or subtract the amount listed in this table to or from the square foot of floor cost for each foot of wall height more or less than 10 feet.

Government Offices – Wood Frame

Upper Floor and Basement

Estimating Procedure

1. Establish the quality class for second floor and the basement. The quality class will usually be the same as the first floor of the building. Square foot costs for unfinished basements will be nearly the same regardless of the structure quality class.
2. Calculate the area of any second floor or basement.
3. Find in the tables below the square foot cost for the appropriate quality class and the nearest area.
4. Use figures in the Wall Height Adjustment row to adjust the square foot cost for wall heights more or less than 10 feet.
5. Multiply the adjusted square foot cost by the area.
6. Multiply that total by the location factor on page 7 or 8.
7. Add costs from the section Additional Costs for Commercial, Industrial and Public Buildings beginning on page 236.
8. Add totals from this page to the cost for the first floor to find the total building cost.

Second Floor – Square Foot Area

Quality Class	5,000	7,500	10,000	12,500	15,000	20,000	25,000	30,000	35,000	40,000	50,000
1, Best	308.73	287.08	274.09	265.27	258.64	249.48	243.11	238.39	234.80	231.84	227.27
1 & 2	286.23	266.06	254.17	245.86	239.73	231.15	225.32	221.00	217.58	214.87	210.79
2, Good	265.04	246.45	235.33	227.76	222.06	214.22	208.72	204.69	201.61	199.09	195.25
2 & 3	247.27	229.89	219.56	212.49	207.12	199.82	194.73	190.91	187.95	185.71	182.05
3, Average	232.42	216.17	206.37	199.66	194.73	187.88	182.92	179.42	176.71	174.51	171.08
3 & 4	210.83	195.94	187.14	181.13	176.61	170.36	165.99	162.75	160.25	158.31	155.25
4, Low	188.24	175.00	167.08	161.74	157.70	152.04	148.14	145.35	143.13	141.29	138.53
Wall height Adjustment*	3.51	2.79	2.47	2.27	2.08	1.69	1.50	1.36	1.32	1.27	1.07

***Wall Height Adjustment:** Add or subtract the amount listed in this table to or from the square foot of floor cost for each foot of second and higher floor wall height more or less than 10 feet.

Finished Basement – Square Foot Area

Quality Class	5,000	7,500	10,000	12,500	15,000	20,000	25,000	30,000	35,000	40,000	50,000
1, Best	217.58	202.33	193.19	186.93	182.28	175.80	171.35	168.00	165.47	163.40	160.18
2, Good	186.80	173.70	165.86	160.49	156.49	150.95	147.08	144.27	142.07	140.27	137.58
3, Average	163.79	152.33	145.45	140.69	137.23	132.39	128.92	126.45	124.55	123.01	120.58
4, Low	132.68	123.34	117.76	113.97	111.15	107.16	104.41	102.43	100.87	99.59	97.65

Unfinished Basements

Area	5,000	7,500	10,000	12,500	15,000	20,000	25,000	30,000	35,000	40,000	50,000
Cost	55.34	51.44	49.14	47.54	46.37	44.72	43.59	42.70	42.09	41.54	40.71

Wall Height Adjustment: Add or subtract the amount listed in this table to or from the square foot of floor cost for each foot of basement wall height more or less than 10 feet.

Area	5,000	7,500	10,000	12,500	15,000	20,000	25,000	30,000	35,000	40,000	50,000
Finished	1.17	.96	.84	.77	.71	.60	.52	.47	.44	.42	.37
Unfinished	1.07	.87	.75	.70	.64	.52	.47	.41	.40	.38	.33

Public Libraries – Masonry or Concrete

Quality Classification

	Class 1 Best Quality	Class 2 Good Quality	Class 3 Average Quality	Class 4 Low Quality
Foundation (6% of total cost)	Reinforced concrete, depth to 12'.	Reinforced concrete, depth 8' to 10'.	Reinforced concrete, depth 6 to 8'.	Reinforced concrete, depth 6' or less.
Floor Structure (5% of total cost)	Concrete on steel beams and deck.	Lightweight concrete on steel beams.	6" thickened edge concrete slab on 6" rock base.	4" reinforced slab on rock base.
Exterior Walls (15% of total cost)	Marble or polished granite, spandrel glass. Metal trim. Decorative atrium entrance.	Glass curtain wall. Textured-block, good brick or stone veneer. Atrium entrance.	Colored concrete block with some brick or wood trim. Decorative entrance.	Tilt-up concrete, brick or concrete block. Few decorative details.
Roof Structure & Cover (18% of total cost)	Glu-lams or steel trusses on steel intermediate columns. Panelized roof system with elastomeric concrete tile or metal roof cover. Engineered for earthquake or high wind zones.	Glu-lams or steel beams on steel intermediate columns. Panelized roof system with 5-ply, built-up or concrete tile roof cover. Good insulation.	Wood or metal trusses on intermediate columns. Panelized roof system with built-up or good composition shingle roofing. Insulated.	Beams or trusses on steel supports. OSB. sheathing. Built-up or low cost membrane roofing. Foil insulation. Not designed for high wind or seismic zones.
Floor Finish (5% of total cost)	Marble or good terrazzo in public rooms. Carpeted offices. Ceramic tile in restrooms.	Terrazzo or marble tile in public rooms. Carpet in offices. Vinyl in restrooms.	Good sheet vinyl in meeting rooms. Carpet in offices. Composition tile in restrooms.	Low cost sheet vinyl in public rooms. Floor tile elsewhere.
Windows & Doors (6% of total cost)	Low-E glass in metal sash. Metal-frame laminated glass entrance doors. Metal interior doors. Institutional grade hardware. LEED certified.	Colored low-E glass. Decorative metal exterior doors. Solid core 8' high wood interior doors. Commercial grade hardware.	Insulated store front or glazed curtain wall. Metal exterior doors. Wood interior doors. Standard grade hardware.	Few standard grade metal windows. Low cost metal exterior doors. Low grade hardware.
Interior Wall Finish (13% of total cost)	Decorative plaster in public rooms. Plaster- or vinyl-covered wallboard in collection area. Decorative metal or hardwood veneer wainscot and trim.	Plaster or good paneling in public rooms. Vinyl-covered wallboard in hallways. Textured or vinyl-covered wallboard with good trim elsewhere.	Textured and painted interior stucco or gypsum wallboard. Wood trim in public meeting rooms.	Painted gypsum wallboard.
Ceiling Finish (2% of total cost)	Suspended decorative textured plaster in meeting rooms. Acoustic ceiling tile elsewhere.	Suspended acoustical tile in with concealed grid in meeting rooms. Acoustical tile elsewhere.	Suspended acoustical tile with exposed grid system.	Painted gypsum wallboard with acoustic texture.
Specialties (13% of total cost)	Wired or wireless security and surveillance systems, video and network connections, large circulation desk area.	Surveillance and security system. Network connections throughout. Dedicated information and circulation stations.	Data network and AV connections throughout. Some cabinets and shelves.	Minimum alarm system. Few cabinets and shelves.
Plumbing (7% of total cost)	Institutional grade fixtures. Copper supply, vent & drain pipe. Metal or synthetic stone toilet partitions.	Good commercial grade fixtures. Copper supply & drain pipe. Metal toilet partitions.	Standard grade fixtures. PEX or PVC supply, vent and drain pipe. Composition toilet partitions.	Minimum fixtures. Plastic supply, vent & drain pipe. Plastic-faced toilet partitions.
Lighting and Power (10% of total cost)	Decorative indirect fixtures in public rooms. Many recessed task lights on separate controls. Recessed LED fixtures in offices and hallways.	Indirect lighting fixtures in public rooms. Track lighting or recessed LED fixtures with some task lighting in offices and hallways.	Continuous 4 tube fluorescent strips with decorative diffusers, 8' O.C. Some recessed ceiling fixtures.	Continuous exposed 2 tube fluorescent strip, fixtures 8' O.C.

Notes: Use the percent of total cost to help identify the correct quality classification. Includes space for the library collection, seating, public access computer stations, staff work area, meeting rooms, circulation and information desks.

Square foot costs include the following components: Foundations as required for normal soil conditions. Floor, wall and roof structures. Interior ceiling, wall and floor finishes (including carpet). Exterior wall finish and roof cover. Interior partitions. Basic lighting and electrical systems. Rough and finish plumbing. Specialties as listed. Design and engineering fees. Typical utility hook-up. Contractor's mark-up.

Add the cost of: Book and media shelving and storage. Canopies and canopy lighting. Desks, tables and study carrels. Installed AV equipment, computer network and computers. Intercom and security systems beyond what appears in the quality classification above. Docks and ramps. Elevators. Draperies. Fire extinguishers and fire sprinklers. Heating and cooling systems. Exterior signs. Walks, paving and curbing. Yard improvements. See the section "Additional Costs for Commercial, Industrial and Public Structures" beginning on page 236.

Public Libraries – Masonry or Concrete

First Floor

Estimating Procedure

1. Use the tables in this section to estimate the cost of buildings designed for use as a public, academic or school library. Many community libraries occupy space designed for use by commercial tenants rather than built for use as a public library. Use figures from the sections on Urban Stores, Suburban Stores or General Office Buildings when the structure was designed for commercial occupancy.
2. Establish the structure quality class by applying the information on page 62.
3. Calculate the area of the first floor. This should include all area within the building exterior walls and all inset areas outside the main walls but under the main building roof.
4. Find in the table below the square foot cost for the appropriate quality class and the nearest building area.
5. Use figures in the Wall Height Adjustment row to adjust that square foot cost for wall heights more or less than 14 feet.
6. Multiply the adjusted square foot cost by the area of the first floor.
7. Multiply that total by the location factor on page 7.
8. Add costs from the section Additional Costs for Commercial, Industrial and Public Buildings beginning on page 236.
9. Using figures on the next page, add the cost of any second or higher floors or a basement.

Public Library, Class 2

Public Library, Class 1

First Floor – Square Foot Area

Quality Class	4,000	6,000	8,000	10,000	12,000	16,000	20,000	24,000	28,000	32,000	40,000
1, Best	568.27	543.09	523.11	506.87	493.14	481.41	471.21	454.45	440.58	429.44	411.94
1 & 2	509.57	487.08	469.24	454.56	442.18	431.60	422.59	407.31	395.24	385.32	369.31
2, Good	458.60	438.50	422.44	409.31	398.01	388.61	380.37	366.82	355.78	346.62	332.64
2 & 3	410.25	392.18	377.66	365.76	356.02	347.52	340.19	327.94	318.25	310.11	297.41
3, Average	367.88	351.66	338.77	328.30	319.23	311.80	305.10	294.18	285.38	278.12	266.75
3 & 4	321.05	306.93	295.68	286.49	278.63	272.14	266.37	256.69	248.93	242.82	232.83
4, Low	273.97	261.78	252.23	244.48	237.78	232.08	227.13	219.00	212.38	207.10	198.56
Wall Height Adjustment*	3.11	2.39	2.15	2.05	1.95	1.91	1.74	1.67	1.64	1.64	1.64

***Wall Height Adjustment:** Add or subtract the amount listed in this row to or from the square foot of floor cost for each foot of wall height more or less than 14 feet.

Public Libraries – Masonry or Concrete

Upper Floors and Basements

Estimating Procedure

1. Establish the quality class for second and higher floors and the basement. The quality class will usually be the same as the first floor of the building. Square foot costs for unfinished basements will be nearly the same regardless of the structure quality class.
2. Calculate the area of any second or higher floor or a basement.
3. Find in the tables below the square foot cost for the appropriate quality class and the nearest area.
4. Use figures in the Wall Height Adjustment row to adjust the square foot cost for wall heights more or less than 10 feet.
5. Multiply the adjusted square foot cost by the area.
6. Multiply that total by the location factor on page 7.
7. Add costs from the section Additional Costs for Commercial, Industrial and Public Buildings beginning on page 236.
8. Add totals from this page to the cost for the first floor to find the total building cost.

Second and Higher Floors – Square Foot Area

Quality Class	4,000	6,000	8,000	10,000	12,000	16,000	20,000	24,000	28,000	32,000	40,000
1, Best	515.74	492.91	474.77	460.07	447.63	436.90	427.68	412.46	399.95	389.84	373.84
1 & 2	462.49	442.10	425.90	412.59	401.34	391.75	383.59	369.70	358.69	349.76	335.23
2, Good	416.24	397.97	383.43	371.49	361.31	352.72	345.23	332.94	322.94	314.63	301.88
2 & 3	372.36	355.97	342.75	332.02	323.14	315.45	308.75	297.67	288.88	281.49	269.97
3, Average	333.89	319.18	307.48	297.97	289.73	282.98	276.95	267.04	259.05	252.43	242.11
3 & 4	291.39	278.57	268.35	260.02	252.93	247.01	241.75	233.03	225.97	220.36	211.32
4, Low	248.63	237.59	228.92	221.88	215.85	210.64	206.16	198.77	192.77	187.95	180.23
Wall Height Adjustment*	2.83	2.15	1.97	1.88	1.76	1.67	1.57	1.55	1.53	1.53	1.53

***Wall Height Adjustment:** Add or subtract the amount listed in this row to or from the square foot of floor cost for each foot of wall height more or less than 10 feet.

Finished Basement – Square Foot Area

Quality Class	4,000	6,000	8,000	10,000	12,000	16,000	20,000	24,000	28,000	32,000	40,000
1, Best	309.35	295.64	284.77	275.97	268.49	262.02	256.51	247.35	239.88	233.78	224.29
2, Good	249.65	238.71	230.00	222.78	216.71	211.53	207.05	199.69	193.68	188.71	181.06
3, Average	200.24	191.41	184.44	178.75	173.78	169.71	166.09	160.12	155.34	151.40	145.21
4, Low	149.14	142.53	137.31	133.08	129.47	126.36	123.67	119.25	115.61	112.75	108.10

Unfinished Basements

Area	4,000	6,000	8,000	10,000	12,000	16,000	20,000	24,000	28,000	32,000	40,000
Cost	76.35	72.96	70.29	68.09	66.23	64.65	63.30	61.01	59.19	57.69	55.33

Wall Height Adjustment: Add or subtract the amount listed in this table to or from the square foot of floor cost for each foot of basement wall height more or less than 10 feet.

Area	4,000	6,000	8,000	10,000	12,000	16,000	20,000	24,000	28,000	32,000	40,000
Finished	1.63	1.29	1.14	1.08	1.04	1.01	.94	.90	.88	.88	.88
Unfinished	1.48	1.14	1.04	.98	.96	.90	.82	.81	.80	.80	.80

Public Libraries – Wood or Steel Frame

Quality Classification

	Class 1 Best Quality	Class 2 Good Quality	Class 3 Average Quality	Class 4 Low Quality
Foundation (6% of total cost)	Reinforced concrete, depth to 8'.	Reinforced concrete, depth 6' to 8'.	Reinforced concrete, depth 4' to 6'.	Reinforced concrete, depth 4' or less.
Floor Structure (5% of total cost)	Concrete on steel beams and deck.	Sheathing on wood or steel floor trusses.	6" thickened edge concrete slab on 6" rock base.	4" reinforced slab on rock base.
Exterior Walls (15% of total cost)	Braced wood or steel studs. Decorative brick or stone veneer with ornate details.	Wood or steel studs. Good wood or composition siding. Some masonry veneer.	Wood studs. Stucco with integral color. Some brick trim.	Wood studs. Painted stucco or inexpensive wood panel siding.
Roof Structure & Cover (18% of total cost)	Glu-lams or steel trusses on steel intermediate columns. Panelized roof system with elastomeric concrete tile or metal roof cover. Engineered for earthquake or high wind zones.	Glu-lams or steel beams on steel intermediate columns. Panelized roof system with 5-ply, built-up or concrete tile roof cover. Good insulation.	Wood or metal trusses on intermediate columns. Panelized roof system with built-up or good composition shingle roofing. Insulated.	Beams or trusses on steel supports. OSB sheathing. Built-up or low cost membrane roofing. Foil insulation. Not designed for high wind or seismic zones.
Floor Finish (5% of total cost)	Marble or good terrazzo in public rooms. Carpeted offices. Ceramic tile in restrooms.	Terrazzo or marble tile in public rooms. Carpet in offices. Vinyl in restrooms.	Good sheet vinyl in meeting rooms. Carpet in offices. Composition tile in restrooms.	Low cost sheet vinyl in public rooms. Floor Tile elsewhere.
Windows & Doors (6% of total cost)	Low-E glass in metal sash. Metal-frame laminated glass entrance doors. Metal interior doors. Institutional grade hardware. LEED certified.	Colored low-E glass. Decorative metal exterior doors. Solid core 8' high wood interior doors. Commercial grade hardware.	Insulated store front or glazed curtain wall. Metal exterior doors. Wood interior doors. Standard grade hardware.	Few standard grade metal windows. Low cost metal exterior doors. Low grade hardware.
Interior Wall Finish (13% of total cost)	Decorative plaster in public rooms. Plaster- or vinyl-covered wallboard in private offices. Decorative metal or hardwood veneer wainscot and trim.	Plaster or good paneling in public rooms. Vinyl-covered wallboard in hallways. Textured or vinyl-covered wallboard with good trim elsewhere.	Textured and painted interior stucco or gypsum wallboard. Wood trim in public meeting rooms.	Painted gypsum wallboard.
Ceiling Finish (2% of total cost)	Suspended decorative textured plaster in meeting rooms. Acoustic ceiling tile elsewhere.	Suspended acoustical tile in with concealed grid in meeting rooms. Acoustical tile elsewhere.	Suspended acoustical tile with exposed grid system.	Painted gypsum wallboard with acoustic texture.
Specialties (13% of total cost)	Wired or wireless security and surveillance systems, video and network connections, large circulation desk area.	Surveillance and security system. Network connections throughout. Dedicated information and circulation stations.	Data network and AV connections throughout. Some cabinets and shelves.	Minimum alarm system. Few cabinets and shelves.
Plumbing (7% of total cost)	Institutional grade fixtures. Copper supply, vent & drain pipe. Metal or synthetic stone toilet partitions.	Good commercial grade fixtures. Copper supply & drain pipe. Metal toilet partitions.	Standard grade fixtures. PEX or PVC supply, vent and drain pipe. Composition toilet partitions.	Minimum fixtures. Plastic supply, vent & drain pipe. Plastic-faced toilet partitions.
Lighting and Power (10% of total cost)	Decorative indirect fixtures in public rooms. Many recessed task lights on separate controls. Recessed LED fixtures in offices and hallways.	Indirect lighting fixtures in public rooms. Track lighting or recessed LED fixtures with some task lighting in offices and hallways.	Continuous 4 tube LED strips with decorative diffusers, 8' O.C. Some recessed ceiling fixtures.	Continuous exposed 2 tube fluorescent strip, fixtures 8' O.C.

Notes: Use the percent of total cost to help identify the correct quality classification. Includes space for the library collection, seating, public access computer stations, staff work area, meeting rooms, circulation and information desks.

Square foot costs include the following components: Foundations as required for normal soil conditions. Floor, wall and roof structures. Interior ceiling, wall and floor finishes (including carpet). Exterior wall finish and roof cover. Interior partitions. Basic lighting and electrical systems. Rough and finish plumbing. Specialties as listed. Design and engineering fees. Typical utility hook-up. Contractor's mark-up.

Add the cost of: Book and media shelving and storage. Canopies and canopy lighting. Desks, tables and study carrels. Installed AV equipment, computer network and computers. Public address, intercom and security systems beyond what appears in the quality classification above. Docks and ramps. Elevators. Draperies. Fire extinguishers and fire sprinklers. Heating and cooling systems. Exterior signs. Walks, paving and curbing. Yard improvements. See the section "Additional Costs for Commercial, Industrial and Public Structures" beginning on page 236.

Public Libraries – Wood or Steel Frame

First Floor

Estimating Procedure

1. Use the tables in this section to estimate the cost of buildings designed for use as public libraries. Many community libraries occupy space designed for use by commercial tenants rather than built for use as public libraries. Use figures from the sections on Urban Stores, Suburban Stores or General Office Buildings when the structure was designed for commercial occupancy.
2. Establish the structure quality class by applying the information on page 65.
3. Calculate the area of the first floor. This should include all area within the building exterior walls and all inset areas outside the main walls but under the main building roof.
4. Find in the table below the square foot cost for the appropriate quality class and the nearest building area.
5. Use figures in the Wall Height Adjustment row to adjust that square foot cost for wall heights more or less than 14 feet.
6. Multiply the adjusted square foot cost by the area of the first floor.
7. Multiply that total by the location factor on page 7.
8. Add costs from the section Additional Costs for Commercial, Industrial and Public Buildings beginning on page 236.
9. Using figures on the next page, add the cost of any second or higher floors or a basement.

Public Library, Class 4

Public Library, Class 1 & 2

Square Foot Area

Quality Class	4,000	6,000	8,000	10,000	12,000	16,000	20,000	24,000	28,000	32,000	40,000
1, Best	360.69	344.71	332.07	321.78	313.03	305.55	299.12	288.46	279.70	272.62	261.49
1 & 2	323.46	309.18	297.87	288.57	280.69	273.96	268.27	258.58	250.89	244.58	234.44
2, Good	291.12	278.34	268.18	259.83	252.68	246.67	241.45	232.87	225.86	220.04	211.16
2 & 3	260.43	248.97	239.72	232.19	226.02	220.61	215.94	208.18	202.01	196.86	188.79
3, Average	233.53	223.22	215.07	208.40	202.63	197.91	193.70	186.74	181.15	176.54	169.33
3 & 4	203.81	194.84	187.68	181.86	176.90	172.76	169.05	162.95	158.02	154.17	147.80
4, Low	173.89	166.22	160.11	155.21	150.94	147.33	144.20	139.04	134.80	131.47	126.04
Wall Height Adjustment*	1.97	1.53	1.38	1.30	1.24	1.17	1.10	1.07	1.04	1.04	1.04

***Wall Height Adjustment:** Add or subtract the amount listed in this table to or from the square foot of floor cost for each foot of wall height more or less than 14 feet.

Public Libraries – Wood or Steel Frame

Upper Floor and Basement

Estimating Procedure

1. Establish the quality class for the second floor and the basement. The quality class will usually be the same as the first floor of the building. Square foot costs for unfinished basements will be nearly the same regardless of the structure quality class.
2. Calculate the area of any second floor or basement.
3. Find in the tables below the square foot cost for the appropriate quality class and the nearest area.
4. Use figures in the Wall Height Adjustment row to adjust the square foot cost for wall heights more or less than 10 feet.
5. Multiply the adjusted square foot cost by the area.
6. Multiply that total by the location factor on page 7.
7. Add costs from the section Additional Costs for Commercial, Industrial and Public Buildings beginning on page 236.
8. Add totals from this page to the cost for the first floor to find the total building cost.

Second Floor – Square Foot Area

Quality Class	4,000	6,000	8,000	10,000	12,000	16,000	20,000	24,000	28,000	32,000	40,000
1, Best	316.70	302.71	291.55	282.51	274.85	268.29	262.63	253.28	245.57	239.37	229.56
1 & 2	283.99	271.49	261.53	253.37	246.44	240.55	235.54	227.03	220.27	214.73	205.81
2, Good	255.59	244.40	235.46	228.12	221.87	216.59	212.00	204.44	198.30	193.20	185.39
2 & 3	228.64	218.60	210.47	203.87	198.44	193.71	189.59	182.77	177.37	172.83	165.73
3, Average	205.02	195.98	188.81	182.99	177.92	173.75	170.07	163.95	159.06	155.00	148.67
3 & 4	178.93	171.07	164.79	159.68	155.31	151.66	148.44	143.09	138.76	135.35	129.74
4, Low	152.69	145.92	140.57	136.25	132.52	129.37	126.60	122.09	118.37	115.43	110.68
Wall height Adjustment*	1.74	1.33	1.22	1.13	1.09	1.04	.98	.94	.92	.92	.92

***Wall Height Adjustment:** Add or subtract the amount listed in this table to or from the square foot of floor cost for each foot of second and higher floor wall height more or less than 10 feet.

Finished Basements – Square Foot Area

Quality Class	4,000	6,000	8,000	10,000	12,000	16,000	20,000	24,000	28,000	32,000	40,000
1, Best	222.51	212.67	204.86	198.50	193.12	188.53	184.53	177.94	172.55	168.20	161.31
2, Good	179.62	171.73	165.43	160.26	155.86	152.20	148.94	143.65	139.32	135.76	130.26
3, Average	144.04	137.72	132.69	128.57	125.00	122.10	119.48	115.19	111.74	108.93	104.47
4, Low	107.27	102.54	98.77	95.74	93.12	90.89	88.95	85.75	83.18	81.09	77.76

Unfinished Basements

Area	4,000	6,000	8,000	10,000	12,000	16,000	20,000	24,000	28,000	32,000	40,000
Cost	55.64	53.18	51.19	49.65	48.31	47.15	46.14	44.49	43.14	42.07	40.35

Wall Height Adjustment: Add or subtract the amount listed in this table to or from the square foot of floor cost for each foot of basement wall height more or less than 10 feet.

Area	4,000	6,000	8,000	10,000	12,000	16,000	20,000	24,000	28,000	32,000	40,000
Finished	1.17	.90	.82	.77	.73	.71	.66	.65	.65	.65	.65
Unfinished	1.07	.82	.73	.70	.68	.65	.62	.61	.60	.60	.60

Fire Stations – Masonry or Concrete

Quality Classification

	Class 1 Best Quality	Class 2 Good Quality	Class 3 Average Quality	Class 4 Low Quality
Foundation (6% of total cost)	Reinforced concrete, depth to 12'.	Reinforced concrete, depth 8' to 10'.	Reinforced concrete, depth 6 to 8'.	Reinforced concrete, depth 6' or less.
Floor Structure (5% of total cost)	Concrete on steel beams and deck.	Lightweight concrete on steel beams.	6" thickened edge concrete slab on 6" rock base.	4" reinforced slab on rock base.
Exterior Walls (15% of total cost)	Decorative brick with store-front glazing. Metal trim. Decorative public entrance.	Insulated glass with textured-block, good brick or stone veneer. Public entrance.	Colored concrete block with some brick or wood trim. Plain entrance.	Tilt-up concrete, brick or concrete block. Few decorative details.
Roof Structure & Cover (18% of total cost)	Glu-lams or steel trusses on steel intermediate columns. Panelized roof system with elastomeric or metal roof cover. Patio or roof deck. Engineered for earthquake or high wind zones.	Glu-lams or steel beams on steel intermediate columns. Panelized roof system with 5-ply, built-up or elastomeric roof cover. Good insulation.	Wood or metal trusses on intermediate columns. Panelized roof system with built-up or good composition roofing. Adequate insulation.	Beams or trusses on steel supports. OSB sheathing. Built-up or low cost membrane roofing. Foil insulation. Not designed for high wind or seismic zones.
Floor Finish (5% of total cost)	Stone or good terrazzo in public rooms. Carpeted leisure-time rooms. Tile in restrooms.	Terrazzo or hardwood in public rooms. Carpet in offices. Vinyl in restrooms.	Good sheet vinyl in meeting rooms. Carpet in offices. Composition tile in restrooms.	Low cost sheet vinyl in public rooms. Floor tile elsewhere.
Windows & Doors (6% of total cost)	Low-E glass in metal sash. Metal-frame laminated glass entrance doors. Metal interior doors. Institutional grade hardware. LEED certified.	Colored low-E glass. Decorative metal exterior doors. Solid core 8' high wood interior doors. Commercial grade hardware.	Insulated store front or glazed curtain wall. Metal exterior doors. Wood interior doors. Standard grade hardware.	Few standard grade metal windows. Low cost metal exterior doors. Low grade hardware.
Interior Wall Finish (13% of total cost)	Decorative plaster in public rooms. Plaster- or vinyl-covered wallboard in leisure time areas. Decorative metal or hardwood veneer wainscot and trim.	Plaster or good paneling in public rooms. Vinyl-covered wallboard in hallways. Textured or vinyl-covered wallboard with good trim elsewhere.	Textured and painted interior stucco or gypsum wallboard. Wood trim in public meeting rooms.	Painted gypsum wallboard.
Ceiling Finish (2% of total cost)	Suspended decorative textured plaster in administrative rooms. Acoustic ceiling tile elsewhere.	Suspended acoustical tile with concealed grid in meeting rooms. Ceiling tile elsewhere.	Suspended acoustical tile with exposed grid. Wallboard in vehicle bays.	Painted gypsum wallboard with trowel texture.
Specialties (13% of total cost)	Wired security, PA and surveillance systems. Video and network connections. Good day room and dispatch areas.	Surveillance and security system. Network connections throughout. Recreation facilities and good vehicle maintenance area.	Data network and AV connections throughout. Some cabinets and shelves. Plain day room and storage.	Minimum alarm system. Few cabinets and shelves. Few built-ins.
Plumbing (7% of total cost)	Top grade commercial fixtures. Copper supply, vent & drain pipe. Metal or synthetic stone toilet partitions.	Good commercial grade fixtures. Copper supply & drain pipe. Metal toilet partitions.	Standard grade fixtures. PEX or PVC supply, vent and drain pipe. Composition toilet partitions.	Minimum fixtures. Plastic supply, vent & drain pipe. Plastic-faced toilet partitions.
Lighting and Power (10% of total cost)	Decorative indirect fixtures in public rooms. Many recessed task lights on separate controls. Recessed LED fixtures in offices and hallways.	Indirect lighting fixtures in public rooms. Track lighting or recessed LED fixtures with some task lighting in offices and hallways.	Continuous 4 tube LED strips with decorative diffusers, 8' O.C. Some recessed ceiling fixtures.	Continuous exposed 2 tube fluorescent strip, fixtures 8' O.C.

Notes: Use the percent of total cost to help identify the correct quality classification. Includes vehicle and equipment bays, equipment storage and maintenance area, living accommodations, leisure time, administration and training facilities.

Square foot costs include the following components: Foundations as required for normal soil conditions. Floor, wall and roof structures. Interior ceiling, wall and floor finishes (including carpet). Exterior wall finish and roof cover. Interior partitions. Basic lighting and electrical systems. Rough and finish plumbing. Specialties as listed. Design and engineering fees. Typical utility hook-up. Contractor's mark-up.

Add the cost of: Canopies and canopy lighting. Installed PA and security systems beyond what appears in the quality classification above. Docks and ramps. Elevators. Draperies. Fire extinguishers and fire sprinklers. Heating and cooling systems. Exterior signs. Walks, paving and curbing. Yard improvements. See the section "Additional Costs for Commercial, Industrial and Public Structures" beginning on page 236.

Fire Stations – Masonry or Concrete

First Floor

Estimating Procedure

1. Use the tables in this section to estimate the cost of staffed fire stations including vehicle and equipment bays, equipment storage and maintenance areas, living accommodations, leisure time, administration and training facilities. Use the cost tables for Service Garages for volunteer fire stations with fire-fighting equipment bays but without living accommodations or administration facilities.
2. Establish the structure quality class by applying the information on page 68.
3. Calculate the area of the first floor. This should include all area within the building exterior walls and all inset areas outside the main walls but under the main building roof.
4. Find in the table below the square foot cost for the appropriate quality class and the nearest building area.
5. Use figures in the Wall Height Adjustment row to adjust that square foot cost for wall heights more or less than 14 feet in vehicle bays and 8 feet in other areas.
6. Multiply the adjusted square foot cost by the area of the first floor.
7. Multiply that total by the location factor on page 7.
8. Add costs from the section Additional Costs for Commercial, Industrial and Public Buildings beginning on page 236.
9. Using figures on the next page, add the cost of any second or higher floors or a basement.

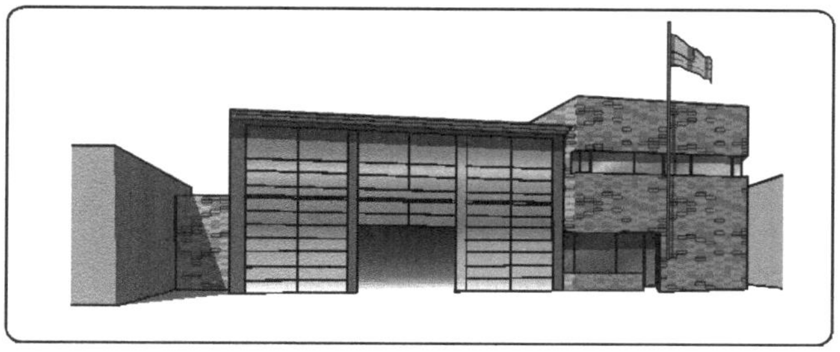

Fire Station, Class 2

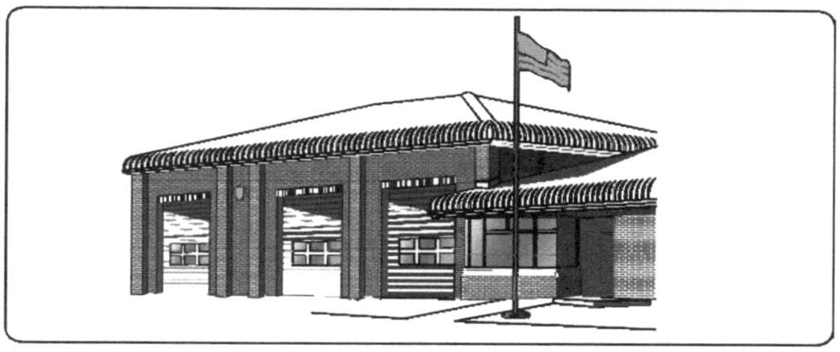

Fire Station, Class 3

First Floor – Square Foot Area

Quality Class	4,000	6,000	8,000	10,000	12,000	16,000	20,000	24,000	28,000	32,000	40,000
1, Best	420.19	401.60	386.84	374.84	364.68	355.93	348.44	336.03	325.84	317.60	304.59
1 & 2	376.81	360.17	346.99	336.17	326.98	319.13	312.49	301.23	292.26	284.94	273.11
2, Good	339.14	324.28	312.40	302.67	294.32	287.36	281.29	271.25	263.09	256.35	245.98
2 & 3	303.36	290.04	279.32	270.48	263.30	256.95	251.52	242.54	235.32	229.33	219.94
3, Average	272.03	260.06	250.58	242.79	236.02	230.55	225.66	217.49	211.01	205.65	197.26
3 & 4	237.41	226.96	218.66	211.88	206.06	201.24	196.96	189.85	184.09	179.56	172.17
4, Low	202.59	193.63	186.49	180.75	175.89	171.62	167.97	161.97	157.07	153.15	146.82
Wall Height Adjustment*	2.31	1.76	1.60	1.53	1.43	1.39	1.33	1.29	1.26	1.26	1.26

***Wall Height Adjustment:** Add or subtract the amount listed in this row to or from the square foot of floor cost for each foot of wall height more or less than 14 feet in equipment bays and 8 feet in other areas.

Fire Stations – Masonry or Concrete

Upper Floors and Basements

Estimating Procedure

1. Establish the quality class for second and higher floors and the basement. The quality class will usually be the same as the first floor of the building. Square foot costs for unfinished basements will be nearly the same regardless of the structure quality class.
2. Calculate the area of any second or higher floor or a basement.
3. Find in the tables below the square foot cost for the appropriate quality class and the nearest area.
4. Use figures in the Wall Height Adjustment row to adjust the square foot cost for wall heights more or less than 8 feet.
5. Multiply the adjusted square foot cost by the area.
6. Multiply that total by the location factor on page 7.
7. Add costs from the section Additional Costs for Commercial, Industrial and Public Buildings beginning on page 236.
8. Add totals from this page to the cost for the first floor to find the total building cost.

Second and Higher Floors – Square Foot Area

Quality Class	4,000	6,000	8,000	10,000	12,000	16,000	20,000	24,000	28,000	32,000	40,000
1, Best	369.02	352.66	339.72	329.19	320.29	312.57	306.02	295.07	286.17	278.91	267.49
1 & 2	330.93	316.33	304.72	295.16	287.13	280.26	274.42	264.51	256.63	250.23	239.88
2, Good	297.81	284.77	274.34	265.80	258.47	252.36	247.05	238.20	231.05	225.11	215.99
2 & 3	266.39	254.69	245.28	237.52	231.22	225.72	220.94	212.96	206.66	201.39	193.14
3, Average	238.90	228.37	219.99	213.23	207.33	202.46	198.15	191.03	185.30	180.62	173.21
3 & 4	208.48	199.29	192.01	186.03	180.98	176.70	172.97	166.73	161.66	157.68	151.19
4, Low	177.89	170.01	163.80	158.75	154.45	150.73	147.52	142.25	137.93	134.49	128.92
Wall Height Adjustment*	2.01	1.55	1.41	1.36	1.30	1.24	1.13	1.09	1.08	1.08	1.08

***Wall Height Adjustment:** Add or subtract the amount listed in this row to or from the square foot of floor cost for each foot of wall height more or less than 8 feet.

Finished Basement – Square Foot Area

Quality Class	4,000	6,000	8,000	10,000	12,000	16,000	20,000	24,000	28,000	32,000	40,000
1, Best	228.92	218.78	210.72	204.21	198.64	193.93	189.85	183.05	177.55	173.02	165.93
2, Good	184.75	176.66	170.16	164.87	160.36	156.55	153.24	147.79	143.32	139.63	134.01
3, Average	148.23	141.62	136.50	132.26	128.60	125.64	122.94	118.48	114.95	112.04	107.45
4, Low	110.32	105.50	101.62	98.48	95.80	93.51	91.49	88.28	85.55	83.40	79.98

Unfinished Basements

Area	4,000	6,000	8,000	10,000	12,000	16,000	20,000	24,000	28,000	32,000	40,000
Cost	76.55	73.15	70.44	68.30	66.45	64.87	63.49	61.19	59.33	57.88	55.48

Wall Height Adjustment: Add or subtract the amount listed in this table to or from the square foot of floor cost for each foot of basement wall height more or less than 8 feet.

Area	4,000	6,000	8,000	10,000	12,000	16,000	20,000	24,000	28,000	32,000	40,000
Finished	1.63	1.29	1.14	1.08	1.05	1.03	.94	.90	.88	.88	.88
Unfinished	1.48	1.14	1.05	.98	.96	.90	.82	.81	.80	.80	.80

Fire Stations – Wood or Steel Frame

Quality Classification

	Class 1 Best Quality	Class 2 Good Quality	Class 3 Average Quality	Class 4 Low Quality
Foundation (6% of total cost)	Reinforced concrete, depth to 8'.	Reinforced concrete, depth 6' to 8'.	Reinforced concrete, depth 4' to 6'.	Reinforced concrete, depth 4' or less.
Floor Structure (5% of total cost)	Concrete on steel beams and deck.	Sheathing on wood or steel floor trusses.	6" thickened edge concrete slab on 6" rock base.	4" reinforced slab on rock base.
Exterior Walls (15% of total cost)	Braced wood or steel studs. Decorative brick or stone veneer with ornate details.	Wood or steel studs. Good wood or composition siding. Some masonry veneer.	Wood studs. Stucco with integral color. Some brick trim.	Wood studs. Painted stucco or inexpensive wood panel siding.
Roof Structure & Cover (18% of total cost)	Glu-lams or steel trusses on steel intermediate columns. Panelized roof system with elastomeric or metal roof cover. Patio or roof deck. Engineered for earthquake or high wind zones.	Glu-lams or steel beams on steel intermediate columns. Panelized roof system with 5-ply, built-up or elastomeric roof cover. Good insulation.	Wood or metal trusses on intermediate columns. Panelized roof system with built-up or good composition roofing. Adequate insulation.	Beams or trusses on steel supports. OSB sheathing. Built-up or low cost membrane roofing. Foil insulation. Not designed for high wind or seismic zones.
Floor Finish (5% of total cost)	Stone or good terrazzo in public rooms. Carpeted leisure-time rooms. Tile in restrooms.	Terrazzo or hardwood in public rooms. Carpet in offices. Vinyl in restrooms.	Good sheet vinyl in meeting rooms. Carpet in offices. Composition tile in restrooms.	Low cost sheet vinyl in public rooms. Floor tile elsewhere.
Windows & Doors (6% of total cost)	Low-E glass in metal sash. Metal-frame laminated glass entrance doors. Metal interior doors. Institutional grade hardware. LEED certified.	Colored low-E glass. Decorative metal exterior doors. Solid core 8' high wood interior doors. Commercial grade hardware.	Insulated store front or glazed curtain wall. Metal exterior doors. Wood interior doors. Standard grade hardware.	Few standard grade metal windows. Low cost metal exterior doors. Low grade hardware.
Interior Wall Finish (13% of total cost)	Decorative plaster in public rooms. Plaster- or vinyl-covered wallboard in leisure time areas. Decorative metal or hardwood veneer wainscot and trim.	Plaster or good paneling in public rooms. Vinyl-covered wallboard in hallways. Textured or vinyl-covered wallboard with good trim elsewhere.	Textured and painted interior stucco or gypsum wallboard. Wood trim in public meeting rooms.	Painted gypsum wallboard.
Ceiling Finish (2% of total cost)	Suspended decorative textured plaster in administrative rooms. Acoustic ceiling tile elsewhere.	Suspended acoustical tile with concealed grid in meeting rooms. Ceiling tile elsewhere.	Suspended acoustical tile with exposed grid. Wallboard in vehicle bays.	Painted gypsum wallboard with trowel texture.
Specialties (13% of total cost)	Wired security, PA and surveillance systems. Video and network connections. Good day room and dispatch areas.	Surveillance and security system. Network connections throughout. Recreation facilities and good vehicle maintenance area.	Data network and AV connections throughout. Some cabinets and shelves. Plain day room and storage.	Minimum alarm system. Few cabinets and shelves. Few built-ins.
Plumbing (7% of total cost)	Top grade commercial fixtures. Copper supply, vent & drain pipe. Metal or synthetic stone toilet partitions.	Good commercial grade fixtures. Copper supply & drain pipe. Metal toilet partitions.	Standard grade fixtures. PEX or PVC supply, vent and drain pipe. Composition toilet partitions.	Minimum fixtures. Plastic supply, vent & drain pipe. Plastic-faced toilet partitions.
Lighting and Power (10% of total cost)	Decorative indirect fixtures in public rooms. Many recessed task lights on separate controls. Recessed LED fixtures in offices and hallways.	Indirect lighting fixtures in public rooms. Track lighting or recessed LED fixtures with some task lighting in offices and hallways.	Continuous 4 tube LED strips with decorative diffusers, 8' O.C. Some recessed ceiling fixtures.	Continuous exposed 2 tube fluorescent strip, fixtures 8' O.C.

Notes: Use the percent of total cost to help identify the correct quality classification. Includes vehicle and equipment bays, equipment storage and maintenance area, living accommodations, leisure time, administration and training facilities.

Square foot costs include the following components: Foundations as required for normal soil conditions. Floor, wall and roof structures. Interior ceiling, wall and floor finishes (including carpet). Exterior wall finish and roof cover. Interior partitions. Basic lighting and electrical systems. Rough and finish plumbing. Specialties as listed. Design and engineering fees. Typical utility hook-up. Contractor's mark-up.

Add the cost of: Canopies and canopy lighting. Installed PA and security systems beyond what appears in the quality classification above. Docks and ramps. Elevators. Draperies. Fire extinguishers and fire sprinklers. Heating and cooling systems. Exterior signs. Walks, paving and curbing. Yard improvements. See the section "Additional Costs for Commercial, Industrial and Public Structures" beginning on page 236.

Fire Stations – Wood or Steel Frame

First Floor

Estimating Procedure

1. Use the tables in this section to estimate the cost of staffed fire stations including vehicle and equipment bays, equipment storage and maintenance areas, living accommodations, leisure time, administration and training facilities. Use the cost tables for Service Garages for volunteer fire stations with fire-fighting equipment bays but without living accommodations or administration facilities.
2. Establish the structure quality class by applying the information on page 71.
3. Calculate the area of the first floor. This should include all area within the building exterior walls and all inset areas outside the main walls but under the main building roof.
4. Find in the table below the square foot cost for the appropriate quality class and the nearest building area.
5. Use figures in the Wall Height Adjustment row to adjust that square foot cost for wall heights more or less than 14 feet in equipment bays and 8 feet in other areas.
6. Multiply the adjusted square foot cost by the area of the first floor.
7. Multiply that total by the location factor on page 7.
8. Add costs from the section Additional Costs for Commercial, Industrial and Public Buildings beginning on page 236.
9. Using figures on the next page, add the cost of any second or higher floors or a basement.

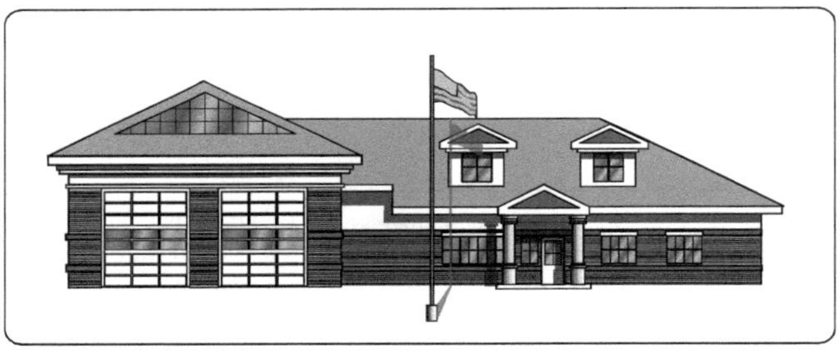

Fire Station, Class 3

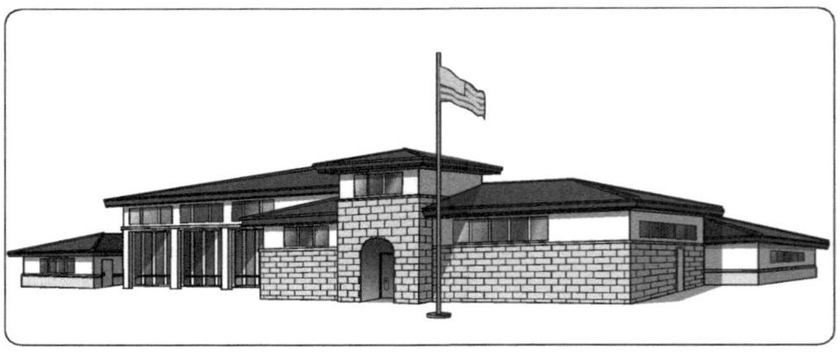

Fire Station, Class 2 & 3

Square Foot Area

Quality Class	4,000	6,000	8,000	10,000	12,000	16,000	20,000	24,000	28,000	32,000	40,000
1, Best	266.05	254.33	244.96	237.34	230.91	225.41	220.65	212.79	206.32	201.11	192.88
1 & 2	238.59	228.08	219.73	212.86	207.04	202.11	197.90	190.72	185.08	180.44	172.92
2, Good	214.73	205.32	197.83	191.66	186.39	181.94	178.11	171.75	166.59	162.30	155.75
2 & 3	192.10	183.67	176.88	171.29	166.70	162.74	159.30	153.58	149.03	145.24	139.30
3, Average	172.26	164.64	158.66	153.72	149.48	145.97	142.88	137.73	133.62	130.22	124.90
3 & 4	150.34	143.73	138.44	134.16	130.50	127.45	124.71	120.22	116.58	113.71	109.02
4, Low	128.28	122.60	118.08	114.48	111.33	108.69	106.34	102.57	99.44	96.99	92.96
Wall Height Adjustment*	1.46	1.11	1.01	.98	.91	.89	.81	.79	.77	.77	.77

***Wall Height Adjustment:** Add or subtract the amount listed in this table to or from the square foot of floor cost for each foot of wall height more or less than 14 feet in equipment bays and 8 feet in other areas.

Fire Stations – Wood or Steel Frame

Upper Floor and Basement

Estimating Procedure

1. Establish the quality class for second floor and the basement. The quality class will usually be the same as the first floor of the building. Square foot costs for unfinished basements will be nearly the same regardless of the structure quality class.
2. Calculate the area of any second floor or basement.
3. Find in the tables below the square foot cost for the appropriate quality class and the nearest area.
4. Use figures in the Wall Height Adjustment row to adjust the square foot cost for wall heights more or less than 8 feet.
5. Multiply the adjusted square foot cost by the area.
6. Multiply that total by the location factor on page 7.
7. Add costs from the section Additional Costs for Commercial, Industrial and Public Buildings beginning on page 236.
8. Add totals from this page to the cost for the first floor to find the total building cost.

Second Floor – Square Foot Area

Quality Class	4,000	6,000	8,000	10,000	12,000	16,000	20,000	24,000	28,000	32,000	40,000
1, Best	240.32	229.68	221.24	214.39	208.57	203.59	199.29	192.20	186.37	181.67	174.20
1 & 2	215.52	205.99	198.48	192.28	187.04	182.53	178.75	172.26	167.12	162.96	156.22
2, Good	193.95	185.45	178.66	173.10	168.37	164.35	160.89	155.18	150.46	146.62	140.68
2 & 3	173.52	165.90	159.73	154.70	150.58	147.01	143.88	138.70	134.62	131.15	125.83
3, Average	155.59	148.73	143.30	138.84	135.01	131.89	129.04	124.40	120.71	117.63	112.83
3 & 4	135.78	129.82	125.07	121.17	117.86	115.11	112.64	108.59	105.28	102.69	98.45
4, Low	115.87	110.73	106.67	103.42	100.57	98.17	96.06	92.63	89.83	87.59	83.98
Wall height Adjustment*	1.32	1.01	.91	.89	.82	.79	.73	.71	.70	.70	.70

***Wall Height Adjustment:** Add or subtract the amount listed in this table to or from the square foot of floor cost for each foot of second and higher floor wall height more or less than 8 feet.

Finished Basements – Square Foot Area

Quality Class	4,000	6,000	8,000	10,000	12,000	16,000	20,000	24,000	28,000	32,000	40,000
1, Best	164.70	157.41	151.61	146.91	142.94	139.52	136.59	131.71	127.70	124.49	119.39
2, Good	132.93	127.11	122.43	118.62	115.36	112.62	110.24	106.30	103.12	100.48	96.41
3, Average	106.64	101.90	98.20	95.17	92.52	90.38	88.43	85.25	82.72	80.62	77.32
4, Low	79.39	75.87	73.10	70.87	68.94	67.28	65.82	63.49	61.55	60.02	57.53

Unfinished Basements

Area	4,000	6,000	8,000	10,000	12,000	16,000	20,000	24,000	28,000	32,000	40,000
Cost	55.07	52.64	50.72	49.18	47.81	46.68	45.66	44.00	42.69	41.64	39.90

Wall Height Adjustment: Add or subtract the amount listed in this table to or from the square foot of floor cost for each foot of basement wall height more or less than 8 feet.

Area	4,000	6,000	8,000	10,000	12,000	16,000	20,000	24,000	28,000	32,000	40,000
Finished	1.15	.90	.82	.77	.73	.71	.65	.65	.64	.64	.64
Unfinished	1.05	.82	.73	.70	.66	.65	.61	.60	.59	.59	.59

Commercial Structures Section

Section Contents

Urban Stores

Urban store buildings are designed for retail sales and are usually found in strip or downtown commercial developments. Square foot costs in this section are representative of a building situation where construction activities are restricted to the immediate site. This restriction tends to make the cost slightly higher than suburban type stores where unlimited use of modern machinery and techniques is possible. Do not use these figures for department stores, discount houses or suburban stores. These building types are evaluated in later sections.

Costs are for shell-type buildings without permanent partitions and include all labor, material and equipment costs for the following:

1. Foundations as required for normal soil conditions.
2. Floor, rear wall, side wall and roof structures.
3. A front wall consisting of vertical support columns or pilasters and horizontal beams spanning the area between these members leaving an open space to receive a display front.
4. Interior floor, wall and ceiling finishes.
5. Exterior wall finish on the side and rear walls.
6. Roof cover.
7. Basic lighting and electrical systems.
8. Rough and finish plumbing.
9. Design and engineering fees.
10. Permits and fees.
11. Utility hook-up.
12. Contractor's contingency, overhead and profit.

The in-place costs of the following components should be added to the basic building cost to arrive at the total structure cost. See the section "Additional Costs for Commercial, Industrial and Public Structures" on page 236.

1. Heating and air conditioning systems.
2. Elevators and escalators.
3. Fire sprinklers and fire escapes.
4. All display front components.
5. Finish materials on the front wall.
6. Canopies, ramps and docks.
7. Interior partitions.
8. Exterior signs.
9. Mezzanines and basements.
10. Communication systems.

For valuation purposes, urban stores are divided into two building types: 1) masonry or concrete frame and, 2) wood or wood and steel frame. Masonry or concrete urban stores vary widely in cost. Consequently, 6 quality classifications are established. Wood or wood and steel frame urban stores are divided into 4 quality classes.

Urban Stores – Masonry or Concrete

Quality Classification

	Class 1 Best Quality	Class 2 Good Quality	Class 3 High Average Quality	Class 4 Low Average Quality	Class 5 Low Quality	Class 6 Minimum Quality
Foundation (15% of total cost)	Reinforced concrete.	Reinforced concrete.	Reinforced concrete.	Reinforced concrete.	Reinforced concrete.	Reinforced concrete.
Floor Structure (15% of total cost)	6" reinforced concrete on 6" rock fill or 2" x 12" joists 16" o.c.	4" to 6" reinforced concrete on 6" rock fill or 2" x 10" joists 16" o.c.	4" reinforced concrete on 6" rock fill or 2" x 10" joists 16" o.c.	4" reinforced concrete on 6" rock fill or 2" x 8" joists 16" o.c.	4" reinforced concrete on 6" rock fill or 2" x 6" joists 16" o.c.	4" reinforced concrete on 4" rock fill or 2" x 6" joists 16" o.c.
Wall Structure (15% of total cost)	Reinforced 8" concrete or 12" common brick or block.	Reinforced 8" concrete or 12" common brick or block.	Reinforced 8" concrete or 12" common brick or block.	Reinforced 8" concrete or 12" common brick or block.	Reinforced 8" concrete block or reinforced 6" concrete.	Reinforced 8" concrete block or reinforced 6" concrete or 8" clay tile or 8" brick.
Roof Covering (10% of total cost)	5 ply composition roof on 1" sheathing with insulation.	5 ply composition roof on 1" sheathing with insulation.	5 ply composition roof on 1" sheathing with insulation.	5 ply composition roof on 1" sheathing with insulation.	4 ply composition roof on 1" sheathing.	4 ply composition roof on 1" sheathing.
Floor Finish (5% of total cost)	Combination solid vinyl tile and terrazzo or very good carpet.	Combination solid vinyl tile and terrazzo or very good carpet.	Solid vinyl tile with some terrazzo or good carpet.	Vinyl tile with small areas of terrazzo, carpet or solid vinyl tile.	Resilient tile.	Composition tile.
Interior Wall Finish (5% of total cost)	Plaster on gypsum or metal lath or 2 layers of 5/8" gypsum wallboard with expensive wallpaper or vinyl wall cover.	Plaster on gypsum or metal lath or 2 layers of 5/8" gypsum wallboard with average wallpaper or vinyl wall cover.	Plaster with putty coat finish on gypsum or metal lath, or 5/8" gypsum wallboard taped, textured and painted, some vinyl wall covering.	Plaster with putty coat finish or gypsum or metal lath, or 5/8" gypsum wallboard taped, textured and painted or with wallpaper.	Lath, 2 coats plaster with putty coat finish or 1/2" gypsum wallboard taped, textured and painted.	Interior plaster on masonry. Colored finish.
Ceiling Finish (5% of total cost)	Acoustical plaster or suspended anodized acoustical metal panels.	Acoustical plaster or suspended anodized acoustical metal panels.	Plaster with putty coat finish and some acoustical plaster or suspended acoustical tile with gypsum wallboard backing.	Plaster with putty coat finish or suspended acoustical tile.	Gypsum wallboard taped and textured or lath, 2 coats of plaster and putty coat finish.	Ceiling tile or gypsum wallboard and paint.
Exterior Wall Finish (5% of total cost)	Waterproofed and painted finish with face brick on exposed walls.	Waterproofed and painted finish with face brick on exposed walls.	Waterproofed and painted finish, face brick on exposed walls.	Painted finish, face brick on exposed wall.	Painted finish.	Unfinished.
Lighting (10% of total cost)	Encased modular units and custom designed chandeliers. Many spotlights.	Encased modular units and stock design chandeliers. Many spotlights.	Encased modular units and stock chandeliers. Many spotlights.	Quad open strip fixtures or triple encased louvered strip fixtures. Average number of spotlights.	Triple open strip fixtures or double encased louvered strip fixtures. Some spotlights.	Double open strip fluorescent fixtures.
Plumbing *(Per each 5,000 S.F.)* (12% of total cost)	6 good fixtures, metal or marble toilet partitions.	6 good fixtures, metal or marble toilet partitions.	6 standard fixtures, metal or marble toilet partitions.	6 standard fixtures, metal toilet partitions.	4 standard commercial fixtures, metal toilet partitions.	4 standard commercial fixtures, wood toilet partitions.
Bath Wall Finish (3% of total cost)	Ceramic tile or marble or custom mosaic tile.	Ceramic tile or marble or custom mosaic tile.	Ceramic tile or marble or plain mosaic tile.	Gypsum wallboard and paint, some ceramic tile or plastic finish wallboard.	Gypsum wallboard and paint.	Gypsum wallboard and paint.

Note: Use the percent of total cost to help identify the correct quality classification

Urban Stores – Masonry or Concrete

First Floor, Length Less than Twice Width

Estimating Procedure

1. Establish the structure quality class by applying the information on page 76.
2. Compute the building ground floor area. This should include everything within the exterior walls and all insets outside the walls but under the main roof.
3. Add to or subtract from the cost below the appropriate amount from the Wall Height Adjustment Table on page 81 if the first floor wall height is more or less than 16 feet for large stores or 12 feet for small stores.
4. Multiply the adjusted square foot cost by the building area.
5. Deduct, if appropriate, for common walls, using the figures on page 81.
6. Multiply the total cost by the location factor on page 7 or 8.
7. Add the cost of heating and cooling equipment, elevators, escalators, fire escapes, fire sprinklers, display fronts, canopies, ramps, docks, interior partitions, mezzanines, basements, and communication systems from pages 236 to 248.
8. Add the cost of second and higher floors from page 80.

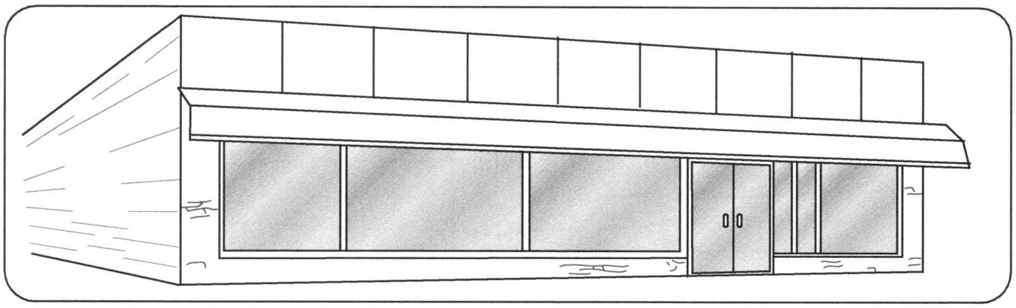

Urban Store, Class 4 & 5

Smaller Stores – Square Foot Area

Quality Class	500	600	700	800	900	1,000	1,200	1,500	1,700	2,000	2,500
4, Low Avg.	236.37	229.95	224.13	218.33	214.76	210.09	203.06	196.10	187.47	183.36	176.47
4 & 5	223.89	214.95	211.97	206.49	202.27	198.06	190.40	186.27	178.00	174.25	166.73
5, Low	210.93	204.27	198.46	195.63	190.69	187.77	180.66	174.83	168.03	163.45	156.40
5 & 6	199.25	192.56	188.49	184.80	178.97	174.15	166.35	165.40	159.83	155.58	147.66
6, Minimum	187.74	181.86	178.07	173.68	169.25	166.78	160.44	156.64	149.06	145.01	139.71

Larger Stores – Square Foot Area

Quality Class	3,000	3,500	4,000	4,500	5,000	6,000	7,500	10,000	15,000	20,000
1, Best	261.69	247.06	241.36	236.72	232.85	226.76	220.14	212.67	204.03	199.03
1 & 2	238.46	231.82	226.46	222.20	218.56	212.77	206.56	199.59	191.53	186.75
2, Good	224.58	218.37	213.34	209.27	205.87	200.40	194.55	187.94	180.34	175.99
2 & 3	214.35	208.43	203.57	196.80	196.43	191.26	185.66	179.34	172.14	167.77
3, Hi. Avg.	201.76	195.37	189.77	186.97	183.58	178.62	174.71	169.43	161.84	157.68
3 & 4	190.41	184.98	180.57	176.96	174.03	169.07	163.73	157.81	150.79	146.56
4, Low Avg.	174.48	170.16	166.10	162.79	160.04	155.62	150.76	145.23	138.81	134.95
4 & 5	165.99	161.20	157.38	154.18	151.64	147.38	142.76	137.52	131.32	127.72
5, Low	155.81	153.41	149.79	146.88	144.38	140.29	135.88	130.94	125.07	121.72
5 & 6	149.77	145.42	141.99	139.13	136.81	135.90	128.36	124.15	118.61	115.34
Minimum	138.44	137.75	134.46	131.76	129.59	125.91	121.97	117.52	112.38	109.22

Urban Stores – Masonry or Concrete

First Floor, Length Between 2 and 4 Times Width

Estimating Procedure

1. Establish the structure quality class by applying the information on page 76.
2. Compute the building ground floor area. This should include everything within the exterior walls and all insets outside the walls but under the main roof.
3. Add to or subtract from the cost below the appropriate amount from the Wall Height Adjustment Table on page 81 if the first floor wall height is more or less than 16 feet for large stores or 12 feet for small stores.
4. Multiply the adjusted square foot cost by the building area.
5. Deduct, if appropriate, for common walls, using the figures on page 81.
6. Multiply the total cost by the location factor on page 7 or 8.
7. Add the cost of heating and cooling equipment, elevators, escalators, fire escapes, fire sprinklers, display fronts, canopies, ramps, docks, interior partitions, mezzanines, basements, and communication systems from pages 236 to 248.
8. Add the cost of second and higher floors from page 80.

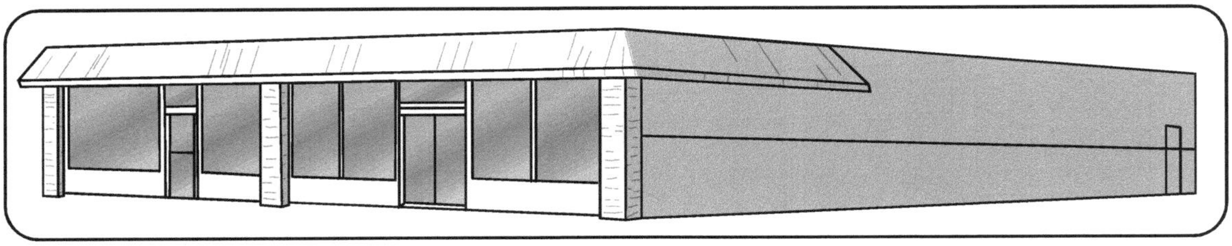

Urban Store, Class 4

Smaller Stores – Square Foot Area

Quality Class	500	600	700	800	900	1,000	1,250	1,500	1,700	2,000	2,500
4, Low Avg.	288.49	276.65	264.57	254.87	246.83	240.17	226.92	217.30	211.24	203.87	194.70
4 & 5	275.40	260.57	249.24	240.01	232.47	226.07	213.74	204.62	199.00	191.86	183.36
5, Low	260.40	246.35	235.56	226.88	219.78	213.75	202.09	193.41	188.01	181.47	173.31
5 & 6	248.92	235.46	225.17	216.90	210.07	204.36	193.11	184.93	179.72	173.55	165.70
6, Minimum	238.57	225.83	215.87	207.96	201.32	195.82	185.13	177.31	172.31	166.34	158.86

Larger Stores – Square Foot Area

Quality Class	3,000	3,500	4,000	4,500	5,000	6,000	7,500	10,000	15,000	20,000
1, Best	272.27	264.25	257.85	252.59	248.09	240.97	233.08	224.29	213.77	207.54
1 & 2	256.22	248.77	242.80	237.77	233.67	226.82	219.48	211.10	201.27	195.31
2, Good	240.94	233.88	228.25	223.51	219.64	213.30	206.35	198.38	189.28	183.69
2 & 3	229.97	223.25	217.83	213.41	209.63	203.58	197.03	189.49	180.66	175.40
3, Hi. Avg.	217.70	214.54	209.36	205.04	201.45	195.67	189.28	182.04	173.55	168.56
3 & 4	205.74	199.22	193.94	189.69	186.00	180.23	174.00	166.74	158.32	153.40
4, Low Avg.	189.70	183.05	178.31	174.29	171.02	165.66	159.77	153.21	145.45	141.02
4 & 5	180.10	174.35	169.72	166.01	162.78	157.74	152.15	145.92	138.61	134.23
5, Low	171.14	165.65	161.30	157.77	154.78	149.91	144.56	138.68	131.71	127.60
5 & 6	162.08	156.89	152.82	149.41	146.46	141.94	136.96	131.30	124.71	120.78
Minimum	153.68	148.88	144.98	141.70	139.04	134.67	130.00	124.63	118.35	113.71

Urban Stores – Masonry or Concrete

First Floor, Length More Than 4 Times Width

Estimating Procedure

1. Establish the structure quality class by applying the information on page 76.
2. Compute the building ground floor area. This should include everything within the exterior walls and all insets outside the walls but under the main roof.
3. Add to or subtract from the cost below the appropriate amount from the Wall Height Adjustment Table on page 81 if the first floor wall height is more or less than 16 feet for large stores or 12 feet for small stores.
4. Multiply the adjusted square foot cost by the building area.
5. Deduct, if appropriate, for common walls, using the figures on page 81.
6. Multiply the total cost by the location factor on page 7 or 8.
7. Add the cost of heating and cooling equipment, elevators, escalators, fire escapes, fire sprinklers, display fronts, canopies, ramps, docks, interior partitions, mezzanines, basements, and communication systems from pages 236 to 248.
8. Add the cost of second and higher floors from page 80.

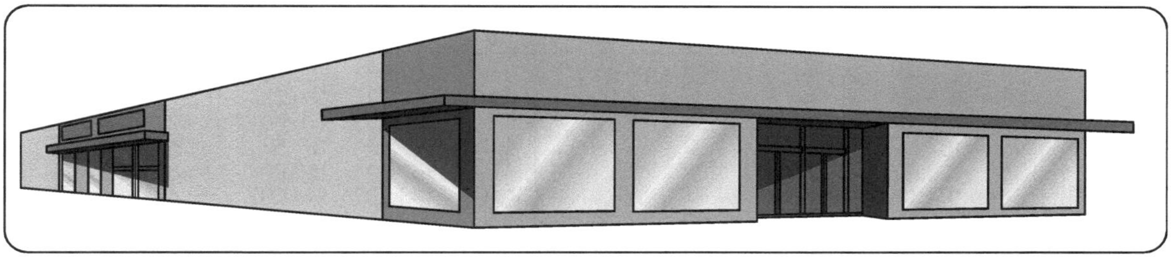

Urban Store, Class 3 & 4

Smaller Stores – Square Foot Area

Quality Class	500	600	700	800	900	1,000	1,200	1,500	1,700	2,000	2,500
4, Low Avg.	319.26	306.54	291.83	280.17	270.58	262.39	249.47	235.29	228.07	219.34	208.59
4 & 5	306.70	288.70	274.90	263.73	254.72	247.10	234.98	221.55	214.80	206.57	197.04
5, Low	289.94	272.99	259.83	249.47	240.89	233.74	222.18	209.49	203.10	195.37	188.01
5 & 6	278.14	261.76	249.28	239.25	231.03	224.14	213.06	200.99	194.74	187.28	178.15
6, Minimum	267.43	251.73	239.70	230.03	222.20	215.48	204.88	193.24	187.37	180.15	171.34

Larger Stores – Square Foot Area

Quality Class	3,000	3,500	4,000	4,500	5,000	6,000	7,500	10,000	15,000	20,000
1, Best	292.96	283.10	275.34	269.02	263.43	254.81	245.35	234.65	222.09	214.59
1 & 2	275.78	266.70	259.29	253.33	248.25	239.98	231.00	220.99	209.09	203.37
2, Good	259.39	250.81	243.80	238.16	233.42	225.75	217.23	207.78	196.58	190.07
2 & 3	238.47	230.57	224.17	218.93	214.54	207.45	199.73	191.02	180.74	174.72
3, Hi. Avg.	236.42	229.90	223.54	218.32	213.93	206.92	199.22	190.48	180.23	174.27
3 & 4	221.30	213.61	207.39	202.24	197.99	191.04	183.40	174.79	164.78	158.75
4, Low Avg.	204.44	197.21	191.58	186.79	182.83	176.49	169.38	161.57	152.15	146.62
4 & 5	194.23	187.43	181.93	177.48	173.65	167.59	160.84	153.47	144.49	139.36
5, Low	183.27	178.01	172.79	168.57	164.95	159.15	152.87	145.74	137.31	132.29
5 & 6	174.88	168.77	163.87	159.83	156.42	151.02	144.99	138.20	130.24	125.51
6, Minimum	165.99	160.17	155.14	151.69	146.79	140.92	135.64	129.47	123.62	119.10

Urban Stores – Masonry or Concrete

Second and Higher Floors

Estimating Procedure

1. Establish the structure quality class. The class for second and higher floors will usually be the same as the first floor.
2. Compute the square foot area of the second floor and each higher floor.
3. Add to or subtract from the square foot cost below the appropriate amount from the Wall Height Adjustment Table on page 81 if the wall height is more or less than 12 feet.
4. Multiply the adjusted square foot cost by the area of each floor.
5. Deduct, if appropriate, for common walls, using the figures on page 81.
6. Add 2% to the cost of each floor above the second floor. For example, the third floor cost would be 102% of the second floor cost and the fourth floor cost would be 104% of the second floor cost.
7. Multiply the total cost for each floor by the location factor on page 7 or 8.
8. Add the cost of heating and cooling equipment, escalators, fire escapes, fire sprinklers, and interior partitions from pages 236 to 248.

Length less than twice width – Square Foot Area

Quality Class	2,500	3,000	3,500	4,000	4,500	5,000	6,000	7,500	10,000	15,000	20,000
1, Best	234.99	228.29	223.07	218.93	215.51	212.64	207.96	202.78	196.95	190.01	185.93
1 & 2	219.26	213.00	208.14	204.36	201.06	198.38	194.09	189.23	183.79	177.26	173.55
2, Good	204.21	198.42	193.94	190.34	187.28	184.82	180.73	176.30	171.21	165.20	161.59
2 & 3	192.98	187.55	183.30	179.82	176.97	174.64	170.80	166.57	161.68	156.03	152.74
3, Hi. Avg.	183.88	178.64	174.58	171.32	168.70	166.34	162.70	158.68	154.06	148.72	145.42
3 & 4	169.66	164.41	160.26	157.01	154.27	151.94	148.20	144.09	139.49	133.99	130.72
4, Low Avg.	154.41	149.71	146.02	142.92	140.50	138.37	135.01	131.21	127.06	121.98	119.00
4 & 5	144.65	140.25	136.73	133.92	131.52	129.59	126.38	122.95	119.00	114.26	111.44
5, Low	135.05	130.89	127.60	124.97	122.71	120.95	118.03	114.72	110.97	106.66	103.96
5 & 6	127.53	123.52	120.49	117.93	115.92	114.14	111.36	108.33	104.80	100.62	98.15
6, Minimum	120.13	116.44	113.55	111.24	109.32	107.59	105.02	102.09	98.75	94.83	92.58

Length between 2 and 4 times width – Square Foot Area

Quality Class	2,500	3,000	3,500	4,000	4,500	5,000	6,000	7,500	10,000	15,000	20,000
1, Best	244.10	236.72	231.08	226.41	222.63	219.43	214.33	208.52	202.07	194.28	189.71
1 & 2	227.77	220.95	215.60	211.30	207.71	204.88	199.89	194.49	188.47	181.31	176.95
2, Good	211.99	205.65	200.72	196.75	193.41	190.59	186.11	181.09	175.50	168.87	164.79
2 & 3	200.65	194.63	189.93	186.19	183.10	180.38	176.07	171.41	166.02	159.69	155.86
3, Hi. Avg.	191.04	185.30	180.79	177.34	174.36	171.89	167.70	163.30	158.12	152.06	148.46
3 & 4	176.65	170.76	166.11	162.53	159.46	156.96	152.82	148.21	143.08	137.04	133.39
4, Low Avg.	160.75	155.42	151.37	147.87	145.21	142.90	139.09	135.01	130.25	124.66	121.43
4 & 5	150.80	145.77	141.93	138.80	136.16	134.01	130.41	126.57	122.21	116.88	113.88
5, Low	141.11	136.32	132.66	129.71	127.40	125.37	122.03	118.39	114.32	109.39	106.56
5 & 6	133.20	128.75	125.31	122.55	120.32	118.35	115.20	114.55	109.57	103.33	100.55
6, Minimum	125.49	121.40	118.09	115.53	113.35	111.53	108.58	105.34	101.72	97.45	94.76

Length over 4 times width – Square Foot Area

Quality Class	2,500	3,000	3,500	4,000	4,500	5,000	6,000	7,500	10,000	15,000	20,000
1, Best	253.84	247.26	241.73	237.15	233.15	229.74	224.09	217.43	209.56	199.79	193.72
1 & 2	236.72	230.60	225.57	221.22	217.53	214.34	208.99	202.81	195.49	186.38	180.68
2, Good	220.36	214.68	209.98	205.95	202.51	199.49	194.55	188.78	181.96	173.47	168.19
2 & 3	208.59	203.19	198.64	194.87	191.63	188.80	184.12	178.64	172.23	164.10	159.20
3, Hi. Avg.	198.49	193.38	189.21	185.48	182.40	179.71	175.31	170.13	163.95	156.31	151.60
3 & 4	184.67	179.19	174.68	170.86	167.73	164.95	160.27	155.05	148.88	141.20	136.50
4, Low Avg.	167.00	163.17	159.07	155.66	152.80	150.29	146.10	141.20	135.64	128.65	124.35
4 & 5	157.68	152.93	149.14	145.92	143.13	140.81	136.94	132.40	127.20	120.59	116.54
5, Low	147.79	143.32	139.69	136.72	134.08	132.00	128.27	124.03	119.10	112.93	109.16
5 & 6	139.46	135.31	131.98	129.09	126.63	124.55	121.17	117.12	112.48	106.72	103.05
6, Minimum	131.97	128.05	124.73	122.07	119.82	117.87	114.56	110.88	106.33	100.97	97.53

Urban Stores – Masonry or Concrete

Wall Height Adjustment

The square foot costs for urban stores are based on the wall heights listed on each page. The main or first floor height is the distance from the bottom of the floor slab or joists to the top of the roof slab or ceiling joists. Second and higher floors are measured from the top of the floor slab or floor joists to the top of the roof slab or ceiling joists. Add or subtract the amount listed to or from the square foot cost for each foot more or less than the standard wall height in the tables. For second and higher floors use only 75% of the wall height adjustment cost.

Area	500	600	700	800	900	1,000	1,250	1,500	1,750	2,000	2,500
Adjustment	7.54	6.91	6.37	6.01	5.72	5.36	4.95	4.31	4.04	3.87	3.46

Area	3,000	3,500	4,000	4,500	5,000	6,000	7,500	10,000	15,000	20,000
Adjustment	3.08	2.90	2.80	2.52	2.43	2.15	1.95	1.74	1.34	1.20

Perimeter (Common) Wall Adjustment

A common wall exists when two buildings share one wall. Adjust for common walls by deducting the linear foot costs below from the total structure cost. In some structures one or more walls are not owned at all. In this case, deduct the "No Ownership" cost per linear foot of wall not owned.

Small Urban Stores (to 2,500 S.F.)
Common wall, deduct $234 per linear foot
No ownership, deduct $467 per linear foot

Large Urban Stores (over 2,500 S.F.)
Common wall, deduct $465 per linear foot
No ownership, deduct $916 per linear foot

Urban Stores – Wood or Wood and Steel Frame

Quality Classification

	Class 1 Best Quality	Class 2 Good Quality	Class 3 Average Quality	Class 4 Low Quality
Foundation (15% of total cost)	Reinforced concrete.	Reinforced concrete.	Reinforced concrete.	Reinforced concrete.
Floor Structure (15% of total cost)	4" reinforced concrete on 6" rock fill or 2" x 10" joists 16" o.c.	4" reinforced concrete on 6" rock fill or 2" x 8" joists 16" o.c.	4" reinforced concrete on 6" rock fill or 2" x 6" joists 16" o.c.	4" reinforced concrete on 4" rock fill or 2" x 6" joists 16" o.c.
Wall Structure (15% of total cost)	2" x 6" studs 16" o.c.	2" x 4" or 2" x 6" studs 16" o.c.	2" x 4" studs 16" o.c. up to 14' high, 2" x 6" studs 16" o.c. over 14' high.	2" x 4" studs 16" o.c. up to 14' high, 2" x 6" studs 16" o.c. over 14' high.
Roof Covering (10% of total cost)	5 ply composition roof on O.S.B. sheathing with insulation.	5 ply composition roof on O.S.B. sheathing with insulation.	4 ply composition roof on O.S.B. sheathing.	4 ply composition roof on O.S.B. sheathing.
Floor Finish (5% of total cost)	Sheet vinyl with some terrazzo or good carpet.	Resilient tile with small areas of terrazzo, carpet or solid vinyl tile.	Composition tile.	Minimum grade tile.
Interior Wall Finish (5% of total cost)	Plaster with putty coat finish on gypsum or metal lath, or 5/8" gypsum wallboard taped, textured and painted or some vinyl wall cover.	Plaster with putty coat finish on gypsum or metal lath, or 5/8" gypsum wallboard taped, textured and painted or with wallpaper.	Lath, 2 coats plaster with putty coat finish or 1/2" gypsum wallboard taped, textured and painted.	1/2" gypsum wallboard taped, textured and painted.
Ceiling Finish (5% of total cost)	Plaster with putty coat finish and some acoustical plaster or suspended acoustical tile with gypsum wallboard backing.	Plaster with putty coat finish or suspended acoustical tile with exposed grid.	Gypsum wallboard taped and textured or lath, 2 coats of plaster and putty coat finish.	Ceiling tile or gypsum wallboard and paint.
Exterior Wall Finish (5% of total cost)	Good wood siding.	Average wood siding.	Stucco or average wood siding.	Stucco or inexpensive wood siding.
Lighting (10% of total cost)	Encased modular units and stock chandeliers. Many spotlights.	Quad open strip fixtures or triple encased louvered strip fixtures. Average number of spotlights.	Triple open strip fixtures or double encased louvered strip fixtures. Some spotlights.	Double open strip fixtures.
Plumbing (12% of total cost) *(Per 5,000 S.F.)*	6 standard fixtures, metal or marble toilet partitions.	6 standard fixtures, metal toilet partitions.	4 standard commercial fixtures, metal toilet partitions.	4 standard commercial fixtures, wood toilet partitions.
Bath Wall Finish (3% of total cost)	Ceramic tile, marble or plain mosaic tile.	Gypsum wallboard and paint, some ceramic tile or plastic finish wallboard.	Gypsum wallboard and paint.	Gypsum wallboard and paint.

Note: Use the percent of total cost to help identify the correct quality classification

Urban Stores – Wood or Wood and Steel Frame

First Floor, Length Less Than Twice Width

Estimating Procedure

1. Establish the structure quality class by applying the information on page 82.
2. Compute the building ground floor area. This should include everything within the exterior walls and all insets outside the walls but under the main roof.
3. Add to or subtract from the cost below the appropriate amount from the Wall Height Adjustment Table on page 87 if the first floor wall height is more or less than 16 feet for large stores or 12 feet for small stores.
4. Multiply the adjusted square foot cost by the building area.
5. Deduct, if appropriate, for common walls, using the figures on page 87.
6. Multiply the total cost by the location factor on page 7 or 8.
7. Add the cost of heating and cooling equipment, elevators, escalators, fire escapes, fire sprinklers, display fronts, canopies, ramps, docks, interior partitions, mezzanines, basements, and communication systems from pages 236 to 248.
8. Add the cost of second and higher floors from page 86.

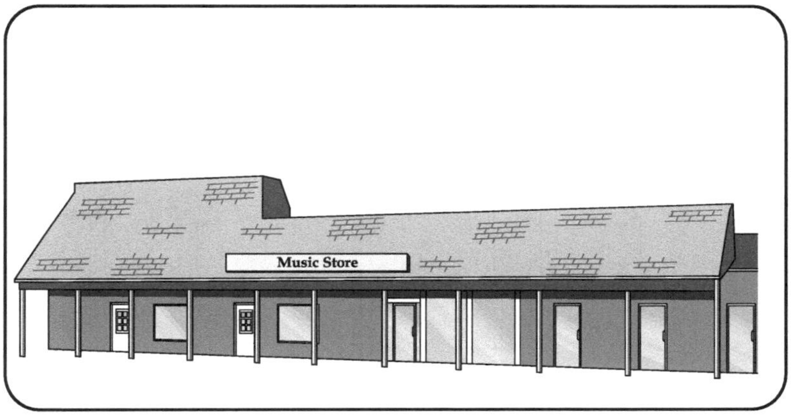

Urban Store, Class 3

Smaller Stores – Square Foot Area

Quality Class	500	600	700	800	900	1,000	1,250	1,500	1,750	2,000	2,500
2, Good	146.22	140.96	136.90	133.54	130.78	128.40	123.82	120.37	117.73	115.57	112.19
2 & 3	136.98	132.07	128.28	125.10	122.53	120.32	115.97	112.69	110.32	108.24	105.06
3, Average	128.62	124.00	120.39	117.42	115.08	112.97	108.87	105.93	103.47	101.62	98.71
3 & 4	118.04	113.78	110.47	107.82	105.57	103.68	99.93	97.17	94.98	93.26	90.59
4, Low	107.81	103.96	100.91	98.53	96.43	94.72	91.32	88.78	86.77	85.20	82.72

Larger Stores – Square Foot Area

Quality Class	3,000	3,500	4,000	4,500	5,000	6,000	7,500	10,000	15,000	20,000
1, Best	136.44	133.61	131.41	129.59	127.98	125.51	122.74	119.61	115.83	113.67
1 & 2	123.48	121.02	118.97	117.32	115.93	113.68	111.13	108.24	104.89	102.87
2, Good	111.81	109.52	107.66	106.19	104.84	102.83	100.53	98.01	94.98	93.12
2 & 3	104.70	102.57	100.81	99.44	98.24	96.36	94.23	91.78	88.91	87.22
3, Average	98.02	96.03	94.41	93.06	91.96	90.17	88.21	85.93	83.22	81.62
3 & 4	88.72	86.90	85.51	84.28	83.26	81.66	79.83	77.72	75.38	73.95
4, Low	81.24	79.62	78.26	77.22	76.29	74.75	73.13	71.19	69.02	67.68

Urban Stores – Wood or Wood and Steel Frame

First Floor, Length Between 2 and 4 Times Width

Estimating Procedure

1. Establish the structure quality class by applying the information on page 82.
2. Compute the building ground floor area. This should include everything within the exterior walls and all insets outside the walls but under the main roof.
3. Add to or subtract from the cost below the appropriate amount from the Wall Height Adjustment Table on page 87 if the first floor wall height is more or less than 16 feet for large stores or 12 feet for small stores.
4. Multiply the adjusted square foot cost by the building area.
5. Deduct, if appropriate, for common walls, using the figures on page 87.
6. Multiply the total cost by the location factor on page 7 or 8.
7. Add the cost of heating and cooling equipment, elevators, escalators, fire escapes, fire sprinklers, display fronts, canopies, ramps, docks, interior partitions, mezzanines, basements, and communication systems from pages 236 to 248.
8. Add the cost of second and higher floors from page 86.

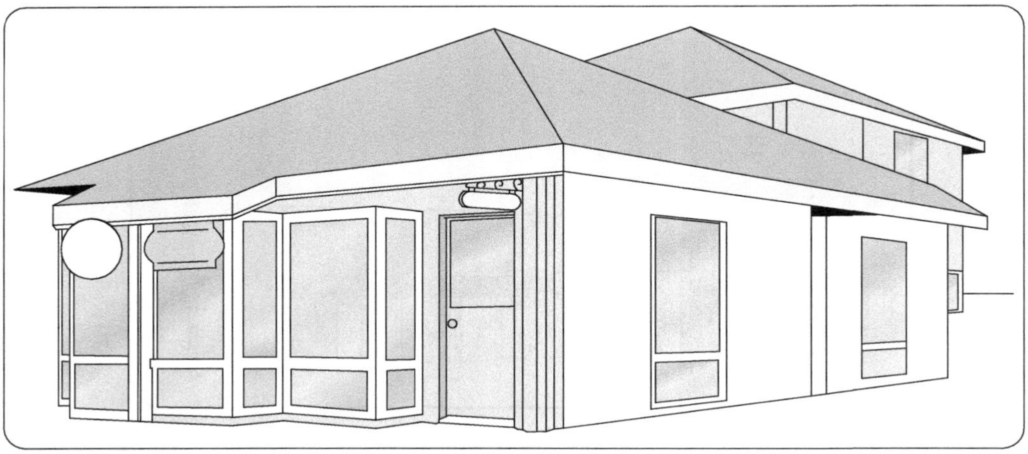

Urban Store, Class 3

Smaller Stores – Square Foot Area

Quality Class	500	600	700	800	900	1,000	1,250	1,500	1,750	2,000	2,500
2, Good	158.13	151.80	146.91	142.96	139.73	136.94	131.57	127.60	124.57	122.11	118.29
2 & 3	147.31	141.40	136.86	133.23	130.18	127.59	122.57	118.89	115.99	113.73	110.24
3, Average	138.88	133.49	129.14	125.65	122.83	120.44	115.66	112.19	109.52	107.31	103.96
3 & 4	127.59	122.53	118.56	115.41	112.74	110.54	106.21	103.00	100.53	98.53	95.49
4, Low	116.28	111.58	108.01	105.14	102.75	100.71	96.80	93.88	91.61	89.77	87.02

Larger Stores – Square Foot Area

Quality Class	3,000	3,500	4,000	4,500	5,000	6,000	7,500	10,000	15,000	20,000
1, Best	141.93	139.64	137.04	134.04	132.59	129.92	126.88	123.15	118.80	116.18
1 & 2	130.50	127.54	125.13	123.15	121.50	118.84	115.88	112.46	108.49	106.04
2, Good	116.71	114.08	111.96	110.22	108.71	106.27	103.66	100.67	97.00	94.90
2 & 3	108.93	106.48	104.54	102.87	101.43	99.16	96.68	93.89	90.59	88.53
3, Average	102.46	100.16	98.28	96.80	95.42	93.31	90.93	88.35	85.20	83.27
3 & 4	94.03	91.86	90.22	88.81	87.59	85.64	83.47	81.06	78.21	76.50
4, Low	86.09	84.15	82.56	81.27	80.16	78.36	76.40	74.21	71.52	69.98

Urban Stores – Wood or Wood and Steel Frame

First Floor, Length More Than 4 Times Width

Estimating Procedure

1. Establish the structure quality class by applying the information on page 82.
2. Compute the building ground floor area. This should include everything within the exterior walls and all insets outside the walls but under the main roof.
3. Add to or subtract from the cost below the appropriate amount from the Wall Height Adjustment Table on page 87 if the first floor wall height is more or less than 16 feet for large stores or 12 feet for small stores.
4. Multiply the adjusted square foot cost by the building area.
5. Deduct, if appropriate, for common walls, using the figures on page 87.
6. Multiply the total cost by the location factor on page 7 or 8.
7. Add the cost of heating and cooling equipment, elevators, escalators, fire escapes, fire sprinklers, display fronts, canopies, ramps, docks, interior partitions, mezzanines, basements, and communication systems from pages 236 to 248.
8. Add the cost of second and higher floors from page 86.

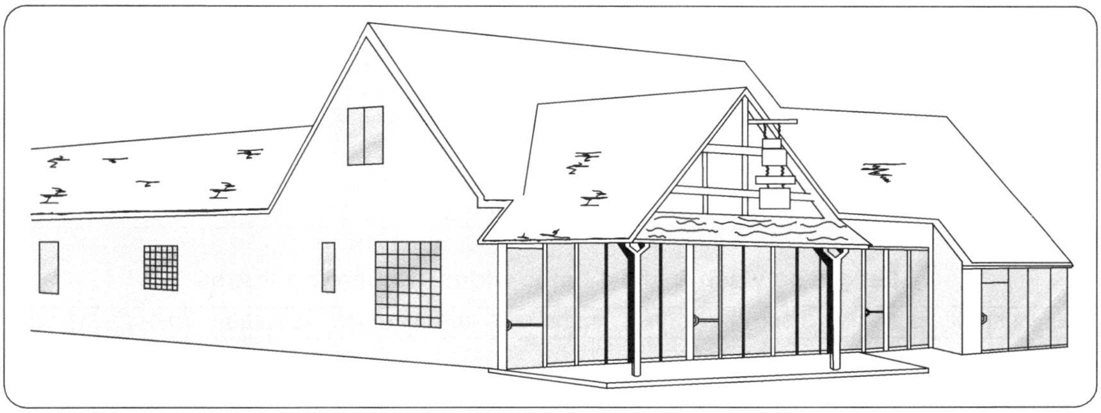

Urban Store, Class 3

Smaller Stores – Square Foot Area

Quality Class	500	600	700	800	900	1,000	1,250	1,500	1,750	2,000	2,500
2, Good	174.28	165.69	159.24	154.20	150.07	146.62	140.04	135.41	131.84	128.96	124.71
2 & 3	162.82	154.78	148.81	144.03	140.16	136.90	130.89	126.46	123.10	120.54	116.50
3, Average	152.68	145.14	139.49	135.04	131.43	128.49	122.73	118.56	115.44	112.99	109.32
3 & 4	139.64	132.80	127.60	123.56	120.30	117.51	112.35	108.50	105.63	103.30	99.93
4, Low	127.65	121.37	116.70	112.97	109.94	107.38	102.66	99.21	96.49	94.54	91.41

Larger Stores – Square Foot Area

Quality Class	3,000	3,500	4,000	4,500	5,000	6,000	7,500	10,000	15,000	20,000
1, Best	154.24	150.21	146.99	144.31	142.05	138.48	134.50	130.05	124.79	121.75
1 & 2	140.91	137.24	134.30	131.88	129.74	126.52	122.89	118.81	114.02	111.23
2, Good	128.96	125.60	122.89	120.63	118.74	115.79	112.45	108.73	104.39	101.81
2 & 3	114.41	111.43	109.03	107.07	105.34	102.71	99.81	96.46	92.62	90.40
3, Average	108.95	106.17	103.84	101.90	100.36	97.83	95.03	91.86	88.23	86.06
3 & 4	98.19	95.60	93.56	91.81	90.48	88.15	85.64	82.75	79.45	77.53
4, Low	88.87	86.54	84.65	83.13	81.84	79.79	77.53	74.95	71.91	70.14

Urban Stores – Wood or Wood and Steel Frame

Second and Higher Floors

Estimating Procedure

1. Establish the structure quality class. The class for second and higher floors will usually be the same as the first floor.
2. Compute the square foot area of the second floor and each higher floor.
3. Add to or subtract from the square foot cost below the appropriate amount from the Wall Height Adjustment Table on page 87 if the wall height is more or less than 12 feet.
4. Multiply the adjusted square foot cost by the area of each floor.
5. Deduct, if appropriate, for common walls, using the figures on page 87.
6. Add 2% to the cost for each floor above the second floor. For example, the third floor cost would be 102% of the second floor cost and the fourth floor cost would be 104% of the second floor cost.
7. Multiply the total cost by the location factor on page 7 or 8.
8. Add the cost of heating and cooling equipment, escalators, fire escapes, fire sprinklers, and interior partitions from pages 236 to 248.

Length less than twice width – Square Foot Area

Quality Class	2,500	3,000	3,500	4,000	4,500	5,000	6,000	7,500	10,000	15,000	20,000
1, Best	118.57	116.18	114.17	112.60	111.32	110.27	108.46	106.53	104.31	101.68	100.12
1 & 2	107.73	105.40	103.68	102.27	101.12	100.12	98.48	96.69	94.75	92.33	90.91
2, Good	97.73	95.69	94.15	92.80	91.68	90.85	89.32	87.77	85.94	83.74	82.46
2 & 3	90.61	88.74	87.26	86.08	85.08	84.23	82.91	81.37	79.65	77.68	76.51
3, Average	84.48	82.72	81.39	80.26	79.34	78.56	77.30	75.85	74.33	72.42	71.31
3 & 4	79.60	77.88	76.60	75.54	74.72	74.00	72.84	71.47	70.03	68.21	67.09
4, Low	74.53	73.05	71.74	70.81	70.05	69.28	68.20	66.97	65.55	63.92	62.91

Length between 2 and 4 times width – Square Foot Area

Quality Class	2,500	3,000	3,500	4,000	4,500	5,000	6,000	7,500	10,000	15,000	20,000
1, Best	122.10	119.31	117.19	115.43	113.98	112.75	110.79	108.66	106.19	103.28	101.44
1 & 2	110.78	108.24	106.28	104.73	103.46	102.41	100.54	98.63	96.36	93.74	92.16
2, Good	100.51	98.19	96.41	95.03	93.88	92.84	91.23	89.46	87.39	85.07	83.55
2 & 3	93.31	91.21	89.60	88.26	87.14	86.28	84.72	83.10	81.20	78.93	77.59
3, Average	86.99	85.07	83.50	82.33	81.24	80.44	78.96	77.46	75.64	73.58	72.38
3 & 4	81.84	80.01	78.59	77.44	76.41	75.60	74.33	72.89	71.19	69.23	68.10
4, Low	76.73	74.99	73.61	72.54	71.68	70.91	69.59	68.31	66.78	64.89	63.63

Length over 4 times width – Square Foot Area

Quality Class	2,500	3,000	3,500	4,000	4,500	5,000	6,000	7,500	10,000	15,000	20,000
1, Best	127.29	124.01	121.50	119.48	117.85	116.36	114.06	111.50	108.67	105.23	103.26
1 & 2	115.61	112.60	110.38	108.50	106.94	105.67	103.59	101.29	98.67	95.60	93.75
2, Good	104.84	102.20	100.11	98.45	97.09	95.92	93.99	91.86	89.54	86.66	85.10
2 & 3	97.49	94.98	93.06	91.55	90.18	89.08	87.33	85.41	83.22	80.62	79.07
3, Average	90.75	88.47	86.66	85.24	84.06	82.99	81.39	79.57	77.51	75.04	73.59
3 & 4	85.51	83.33	81.62	80.26	79.13	78.23	76.61	74.98	73.05	70.72	69.33
4, Low	79.93	77.87	76.33	75.04	74.01	73.09	71.68	70.08	68.23	66.16	64.83

Urban Stores – Wood or Wood and Steel Frame

Wall Height Adjustment

The square foot costs for urban stores are based on the wall heights listed on each page. The main or first floor height is the distance from the bottom of the floor slab or joists to the top of the roof slab or ceiling joists. Second and higher floors are measured from the top of the floor slab or floor joists to the top of the roof slab or ceiling joists. Add or subtract the amount listed to or from the square foot cost for each foot more or less than the standard wall height in the tables. For second and higher floors use only 75% of the wall height adjustment cost.

Area	**500**	**600**	**700**	**800**	**900**	**1,000**	**1,250**	**1,500**	**1,750**	**2,000**	**2,500**
Adjustment	2.36	2.14	2.00	1.85	1.75	1.73	1.53	1.39	1.27	1.23	1.09

Area	**3,000**	**3,500**	**4,000**	**4,500**	**5,000**	**6,000**	**7,500**	**10,000**	**15,000**	**20,000**
Adjustment	.92	.90	.89	.82	.77	.73	.71	.65	.64	.61

Perimeter (Common) Wall Adjustment

A common wall exists when two buildings share one wall. Adjust for common walls by deducting the linear foot costs below from the total structure cost. In some structures one or more walls are not owned at all. In this case, deduct the "No Ownership" cost per linear foot of wall not owned.

Small Urban Stores (2,500 S.F. or less)
Common wall, deduct $118 per linear foot
No ownership, deduct $236 per linear foot

Large Urban Stores (over 2,500 S.F.)
Common wall, deduct $174 per linear foot
No ownership, deduct $364 per linear foot

Suburban Stores

Suburban stores are usually built as part of shopping centers. They differ from urban stores in that they are built in open areas where modern construction techniques, equipment and more economical designs can be used. They are also subject to greater variations in size and shape than are urban stores. Do not use the figures in this section for department stores, discount houses or urban stores. These building types are evaluated in other sections.

Costs identified "building shell only" do not include permanent partitions, display fronts or finish materials on the front of the building. Costs for "multi-unit buildings" include partitions, display fronts and finish materials on the front of the building. All figures include the following costs:

1. Foundations as required for normal soil conditions.
2. Floor, rear wall, side wall and roof structures.
3. A front wall consisting of vertical support columns or pilasters and horizontal beams spanning the area between these members, leaving an open space to receive a display front.
4. Interior floor, wall and ceiling finishes.
5. Exterior wall finish on the side and rear walls.
6. Roof cover.
7. Basic lighting and electrical systems.
8. Rough and finish plumbing.
9. A usual or normal parapet wall.
10. Design and engineering fees.
11. Permits and fees.
12. Utility hook-ups.
13. Contractor's contingency, overhead and mark-up.

The in-place costs of these extra components should be added to the basic building cost to arrive at total structure cost. See the section "Additional Costs for Commercial, Industrial and Public Structures" on page 236.

1. Heating and air conditioning systems.
2. Fire sprinklers.
3. All display front components (shell-type buildings only).
4. Finish materials on the front wall of the building (shell-type building only).
5. Canopies.
6. Interior partitions (shell-type buildings only).
7. Exterior signs.
8. Mezzanines and basements.
9. Loading docks and ramps.
10. Miscellaneous yard improvements.
11. Communications systems.

For valuation purposes suburban stores are divided into two building types: masonry or concrete frame, or wood or wood and steel frame. Each building type is divided into four shape classes:

1. Buildings in which the depth is greater than the front.
2. Buildings in which the front is between one and two times the depth.
3. Buildings in which the front is between two and four times the depth.
4. Buildings in which the front is greater than four times the depth. Angular buildings should be classed by comparing the sum of the length of all wings to the width of the wings. All areas should be included, but no area should be included as part of two different wings. Note the example at the right.

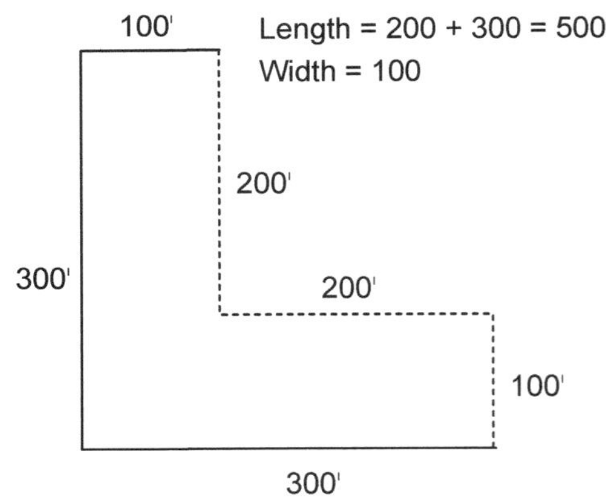

Suburban Stores – Masonry or Concrete

Quality Classification

	Class 1 Best Quality	Class 2 Good Quality	Class 3 Average Quality	Class 4 Low Quality
Foundation (15% of total cost)	Reinforced concrete.	Reinforced concrete.	Reinforced concrete.	Reinforced concrete.
Floor Structure (15% of total cost)	6" reinforced concrete on 6" rock base.	6" reinforced concrete on 6" rock base.	6" reinforced concrete on 6" rock base.	4" reinforced concrete on 6" rock base.
Wall Structure (15% of total cost)	8" reinforced decorative concrete block, 6" concrete tilt-up or 8" reinforced brick.	8" reinforced decorative concrete block, 6" concrete tilt-up or 8" reinforced brick.	8" reinforced concrete block, 6" concrete tilt-up or 8" reinforced common brick.	8" reinforced concrete block or 6" concrete tilt-up.
Roof (15% of total cost)	Glu-lam or steel beams on steel intermediate columns. Panelized roof system, 1/2" plywood sheathing, 5 ply built-up roof with insulation.	Glu-lam or steel beams on steel intermediate columns. Panelized roof system, 1/2" plywood sheathing, 5 ply built-up roof with insulation.	Glu-lam beams on steel intermediate columns. Panelized roof system, 1/2" plywood sheathing, 4 ply built-up roof.	Glu-lam beams on steel intermediate columns. Panelized roof system, 1/2" plywood sheathing, 4 ply built-up roof.
Floor Finish (5% of total cost)	Terrazzo, sheet vinyl or very good carpet.	Resilient tile with 50% solid vinyl tile, terrazzo, or good carpet.	Composition tile.	Minimum grade tile.
Interior Wall Finish (5% of total cost)	Inside of exterior walls furred out with gypsum wallboard or lath and plaster cover. Exterior walls and partitions finished with vinyl wall covers and hardwood veneers.	Interior stucco on inside of exterior walls, gypsum wallboard and texture or paper on partitions, some vinyl wall cover and plywood paneling.	Interior stucco on inside of exterior walls, gypsum wallboard and texture and paint on partitions.	Paint on inside of exterior walls, gypsum wallboard with texture and paint on partitions.
Ceiling Finish (5% of total cost)	Suspended good grade acoustical tile with gypsum wallboard backing.	Suspended acoustical tile with concealed grid system.	Suspended acoustical tile with exposed grid system.	Exposed beams with ceiling tile or paint.
Lighting (5% of total cost)	Recessed LED lighting in modular plastic panels.	Recessed LED strips with egg crate diffusers, 8' o.c.	Continuous 3 tube fluo-rescent strips with egg crate diffusers, 8' o.c.	Continuous exposed 2 tube fluorescent strips, 8' o.c.
Exterior (8% of total cost)	Face brick or stone veneer.	Exposed aggregate, some stone veneer.	Paint on exposed areas, some exposed aggregate.	Paint on exposed areas.
Plumbing (12% of total cost)	6 good fixtures per 5,000 S.F. of floor area, metal toilet partitions.	6 standard fixtures per 5,000 S.F. of floor area, metal toilet partitions.	4 standard fixtures per 5,000 S.F. of floor area, metal toilet partitions.	4 standard fixtures per 5,000 S.F. of floor area wood toilet partitions.

Note: Use the percent of total cost to help identify the correct quality classification.

The costs on pages 90 to 93, and 95 to 102 include display fronts. The quality of the display front will help establish the quality class of the building as a whole. Display fronts are classified as follows:

	Class 1 Best Quality	Class 2 Good Quality	Class 3 Average Quality	Class 4 Low Quality
Bulkhead *(0 to 4' high)*	Vitrolite, domestic marble or stainless steel.	Black flagstone, terrazzo or good ceramic tile.	Average ceramic tile, Roman brick or imitation flagstone.	Stucco, wood or common brick.
Window Frame	Bronze or stainless steel.	Heavy aluminum.	Aluminum.	Light aluminum with with stops.
Glass	1/4" float glass with mitered joints.	1/4" float glass, some mitered joints.	1/4" float glass.	Crystal or 1/4" float glass.
Sign Area *(4' high)*	Vitrolite, domestic marble or stainless steel.	Black flagstone, terrazzo, or good ceramic tile.	Average ceramic tile, Roman brick or imitation flagstone.	Stucco.
Pilasters	Vitrolite, domestic marble.	Black flagstone, terrazzo or good ceramic tile.	Average ceramic tile, Roman brick or imitation flagstone.	Stucco.

Suburban Stores – Masonry or Concrete

Building Shell Only

Estimating Procedure

1. Use these figures to estimate the cost of shell-type buildings without permanent partitions, display fronts or finish material on the front wall of the building.
2. Establish the structure quality class by applying the information on page 89.
3. Compute the building floor area. This should include everything within the exterior walls and all inset areas outside the main walls but under the main roof.
4. Add to or subtract from the cost below the appropriate amount from the Wall Height Adjustment Table (at the bottom of this page) if the wall height is more or less than 16 feet.
5. Multiply the adjusted square foot cost by the building area.
6. Deduct, if appropriate, for common walls, using the figures at the bottom of page 91.
7. Multiply the total cost by the location factor on page 7 or 8.
8. Add the cost of the appropriate additional components from page 236 to 248: heating and cooling equipment, fire sprinklers, display fronts, finish materials on the front wall, canopies, interior partitions, exterior signs, mezzanines and basements, loading docks and ramps, yard improvements, and communication systems.

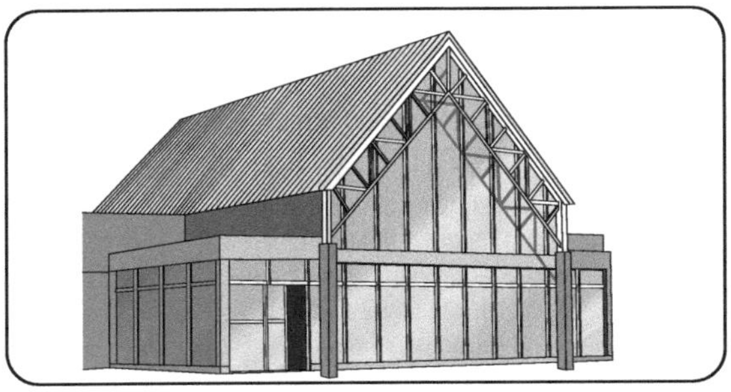

Suburban Store, Class 2

Depth greater than length of front – Square Foot Area

Quality Class	500	1,000	2,000	2,500	3,000	4,000	5,000	7,500	10,000	12,500	15,000
1, Best	297.12	247.81	213.00	204.20	197.67	188.53	182.27	172.52	166.72	162.81	159.83
1 & 2	275.03	229.44	197.26	189.02	182.97	174.45	168.70	159.72	154.27	150.67	147.98
2, Good	257.45	214.77	184.57	176.92	171.33	163.42	157.98	149.54	144.47	141.08	138.51
2 & 3	243.77	203.37	174.87	167.61	162.26	154.71	149.60	141.59	136.85	133.59	131.16
3, Average	234.27	195.46	167.97	160.97	155.89	148.65	143.71	136.04	131.51	128.45	126.07
3 & 4	220.80	184.19	158.44	151.75	146.87	140.04	135.39	128.22	123.94	120.96	118.80
4, Low	210.33	175.35	150.73	144.49	139.91	133.36	128.90	122.13	117.96	115.23	113.14

Length of front between 1 and 2 times depth – Square Foot Area

Quality Class	500	1,000	2,000	3,000	5,000	7,500	10,000	15,000	20,000	25,000	35,000
1, Best	283.88	238.43	206.54	192.40	178.35	169.38	164.00	157.74	154.07	151.41	148.07
1 & 2	262.32	220.43	190.77	177.82	164.66	156.52	151.60	145.78	142.29	139.94	136.74
2, Good	245.57	206.33	178.72	166.46	154.18	146.51	141.85	136.50	133.25	130.93	128.08
2 & 3	232.43	195.28	169.09	157.56	145.97	138.54	134.29	129.08	126.00	124.01	121.19
3, Average	223.73	187.96	162.81	151.68	140.43	133.55	129.27	124.31	121.35	119.40	116.65
3 & 4	211.98	178.12	154.23	143.63	133.07	126.55	122.53	117.74	115.00	113.13	110.59
4, Low	199.97	168.02	145.44	135.43	125.50	119.29	115.55	111.11	108.44	106.59	104.26

Wall Height Adjustment: Costs above are based on a 16' wall height, measured from the bottom of the floor slab or floor joists to the top of the roof cover. Add or subtract the amount listed to or from the square foot cost for each foot more or less than 16 feet.

Area	500	1,000	2,000	2,500	3,000	4,000	5,000	7,500	10,000	12,500
Cost	3.08	2.86	2.76	2.35	2.09	1.86	1.72	1.37	1.16	1.03

Area	15,000	20,000	25,000	30,000	35,000	50,000	70,000	75,000	100,000	150,000
Cost	.98	.96	.88	.85	.82	.72	.63	.60	.44	.39

Suburban Stores – Masonry or Concrete

Building Shell Only

Estimating Procedure

1. Use these figures to estimate the cost of shell-type buildings without permanent partitions, display fronts or finish material on the front wall of the building.
2. Establish the structure quality class by applying the information on page 89.
3. Compute the building floor area. This should include everything within the exterior walls and all inset areas outside the main walls but under the main building roof.
4. Add to or subtract from the cost below the appropriate amount from the Wall Height Adjustment Table on page 90 if the wall height is more or less than 16 feet.
5. Multiply the adjusted square foot cost by the building area.
6. Deduct, if appropriate, for common walls, using the figures at the bottom of this page.
7. Multiply the total cost by the location factor on page 7 or 8.
8. Add the cost of the appropriate additional components from page 236 to 248: heating and cooling equipment, fire sprinklers, display fronts, finish materials on the front wall, canopies, interior partitions, exterior signs, mezzanines and basements, loading docks and ramps, yard improvements, and communication systems.

Suburban Store, Class 1

Length between 2 and 4 times depth – Square Foot Area

Quality Class	1,000	2,000	3,000	5,000	7,500	10,000	15,000	20,000	30,000	50,000	70,000
1, Best	247.31	213.33	198.32	183.24	173.68	168.09	161.44	157.00	152.58	147.87	145.27
1 & 2	227.77	196.39	182.54	168.72	159.99	154.79	148.54	144.90	140.48	136.17	133.74
2, Good	212.05	182.91	169.99	157.07	148.91	144.16	138.32	134.90	130.81	126.75	124.60
2 & 3	200.86	173.25	160.98	148.82	141.11	136.50	131.03	127.81	124.00	120.07	117.96
3, Average	192.90	166.46	154.84	142.95	135.64	131.08	125.89	122.88	119.07	115.30	113.38
3 & 4	183.24	158.10	146.89	135.75	128.76	124.59	119.50	116.57	113.13	109.52	107.57
4, Low	173.00	149.27	138.66	128.18	121.52	117.59	112.91	110.14	106.69	103.46	101.73

Length greater than 4 times depth – Square Foot Area

Quality Class	2,000	3,000	5,000	10,000	15,000	20,000	30,000	50,000	75,000	100,000	150,000
1, Best	225.04	208.07	191.07	173.94	166.38	161.86	156.55	151.20	147.86	145.85	143.52
1 & 2	204.97	189.50	173.98	158.33	151.44	147.49	142.58	137.64	134.60	132.83	130.68
2, Good	190.74	176.30	161.84	147.39	141.02	137.20	132.66	128.08	125.23	123.51	121.48
2 & 3	180.51	166.87	153.22	139.49	133.45	129.84	125.55	121.32	118.55	116.89	115.07
3, Average	173.39	160.40	147.19	134.05	128.18	124.70	120.62	116.55	113.96	112.40	110.56
3 & 4	163.91	151.63	139.24	126.71	121.15	117.93	114.01	110.18	107.66	106.23	104.56
4, Low	155.17	143.53	131.72	119.97	114.68	111.65	107.94	104.26	101.94	100.59	98.85

Perimeter Wall Adjustment: A common wall exists when two buildings share one wall. Adjust for common walls by deducting $425 per linear foot of common wall from the total structure cost. In some structures one or more walls are not owned at all. In this case, deduct $816 per linear foot of wall not owned.

Suburban Stores – Masonry or Concrete

Multi-Unit Buildings

Estimating Procedure

1. Use these square foot costs to estimate the cost of stores designed for multiple occupancy. These costs include all components of shell buildings plus the cost of display fronts, finish materials on the front of the building and normal interior partitions.
2. Establish the structure quality class by applying the information on page 89. Evaluate the quality of the display front to help establish the correct quality class of the building as a whole. See also pages 242 to 245.
3. Compute the building floor area. This should include everything within the building exterior walls and all inset areas outside the main walls but under the main building roof.
4. Add to or subtract from the cost below the appropriate amount from the Wall Height Adjustment Table (at the bottom of this page) if the wall height is more or less than 16 feet.
5. Multiply the adjusted square foot cost by the building area.
6. Deduct, if appropriate, for common walls, using the figures at the bottom of page 93.
7. Multiply the total cost by the location factor on page 7 or 8.
8. Add the cost of the appropriate additional components from page 236 to 248: heating and cooling equipment, fire sprinklers, canopies, exterior signs, mezzanines and basements, loading docks and ramps, yard improvements, and communications systems.

Depth greater than length of front – Square Foot Area

Quality Class	500	1,000	2,000	2,500	3,000	4,000	5,000	7,500	10,000	12,500	15,000
1, Best	423.10	345.34	290.54	276.50	266.16	251.78	241.93	226.50	217.30	211.15	206.50
1 & 2	378.93	309.41	260.27	247.75	238.45	225.56	216.64	202.98	194.75	189.18	184.99
2, Good	340.58	277.94	233.87	222.57	214.33	202.65	194.67	182.35	174.96	169.96	166.19
2 & 3	322.23	255.39	214.78	204.50	196.80	186.13	178.82	167.38	160.72	156.05	152.71
3, Average	289.37	236.19	198.74	189.18	182.17	172.22	165.40	154.96	148.70	144.47	141.29
3 & 4	268.33	219.06	184.22	175.39	168.82	159.72	153.35	143.63	137.85	133.92	130.93
4, Low	247.18	201.77	169.74	161.59	155.63	147.00	141.37	132.38	127.00	123.35	120.69

Length of front between 1 and 2 times depth – Square Foot Area

Quality Class	500	1,000	2,000	3,000	5,000	7,500	10,000	15,000	20,000	25,000	35,000
1, Best	465.46	376.82	313.77	285.62	257.37	239.50	228.80	216.17	208.61	203.37	196.68
1 & 2	413.65	334.88	278.78	253.84	228.70	212.84	203.35	192.04	185.34	180.73	174.77
2, Good	367.83	297.87	247.96	225.75	203.33	189.23	180.78	170.86	164.84	160.77	155.39
2 & 3	335.69	271.82	226.28	206.03	185.61	172.72	165.06	155.90	150.42	146.76	141.76
3, Average	319.46	258.66	215.35	196.05	176.67	164.46	157.10	148.35	143.23	139.63	134.93
3 & 4	302.29	244.79	203.76	185.57	167.24	155.54	148.70	140.39	135.47	132.13	127.77
4, Low	275.78	223.33	185.91	169.29	152.50	141.94	135.65	128.08	123.63	120.58	116.57

Wall Height Adjustment: Costs above are based on a 16' wall height, measured from the bottom of the floor slab or floor joists to the top of the roof cover. Add or subtract the amount listed to or from the square foot cost for each foot more or less than 16 feet.

Square Foot Area

Class	2,000	3,000	5,000	10,000	15,000	25,000	35,000	50,000	75,000	100,000	150,000
1, Best	12.02	9.76	7.58	5.30	4.44	3.40	2.92	2.52	2.16	2.05	1.92
2, Good	8.95	7.34	5.63	3.92	3.39	2.52	2.16	1.95	1.64	1.54	1.37
3, Average	6.87	5.61	4.32	3.08	2.54	2.04	1.74	1.46	1.26	1.23	1.14
4, Low	5.41	4.39	3.40	2.35	2.05	1.55	1.29	1.19	1.06	.92	.89

Suburban Stores – Masonry or Concrete

Multi-Unit Buildings

Estimating Procedure

1. Use these square foot costs to estimate the cost of stores designed for multiple occupancy. These costs include all components of shell buildings plus the cost of display fronts, finish materials on the front of the building and normal interior partitions.
2. Establish the structure quality class by applying the information on page 89. Evaluate the quality of the display front to help establish the correct quality class of the building as a whole. See also pages 236 to 248.
3. Compute the building floor area. This should include everything within the building exterior walls and all inset areas outside the main walls but under the main building roof.
4. Add to or subtract from the cost below the appropriate amount from the Wall Height Adjustment Table (at the bottom of page 92) if the wall height is more or less than 16 feet.
5. Multiply the adjusted square foot cost by the building area.
6. Deduct, if appropriate, for common walls, using the figures at the bottom of this page.
7. Multiply the total cost by the location factor on page 7 or 8.
8. Add the cost of the appropriate additional components from page 236 to 248: heating and cooling equipment, fire sprinklers, canopies, exterior signs, mezzanines and basements, loading docks and ramps, yard improvements, and communications systems.

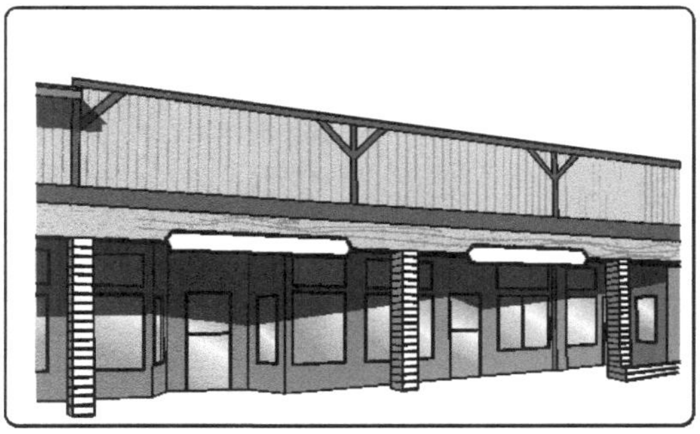

Suburban Store, Class 3

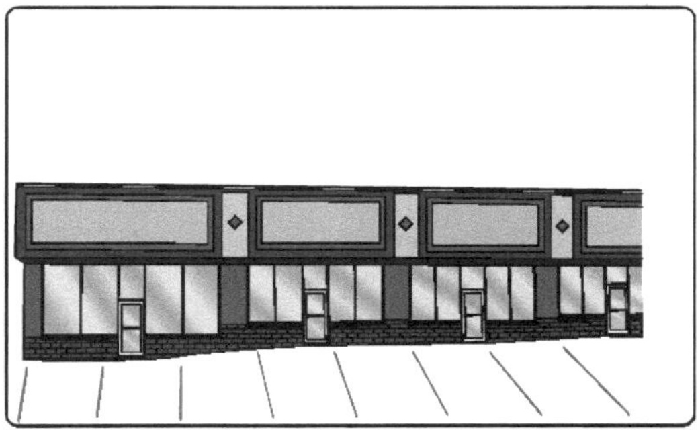

Suburban Store, Class 4

Length of front between 2 and 4 times depth – Square Foot Area

Quality Class	1,000	2,000	3,000	5,000	7,500	10,000	15,000	20,000	30,000	50,000	70,000
1, Best	425.00	348.59	314.57	280.24	258.53	245.55	230.29	220.99	210.15	199.12	193.28
1 & 2	378.87	310.83	280.51	249.93	230.54	218.96	205.22	197.08	187.29	177.56	172.30
2, Good	335.66	275.31	248.41	221.29	204.13	193.90	181.77	174.54	165.90	157.19	152.57
2 & 3	304.51	249.88	225.49	200.92	185.32	176.03	165.04	158.45	150.51	142.73	138.51
3, Average	286.85	228.28	206.03	183.62	169.23	160.80	150.71	144.71	137.61	130.39	126.56
3 & 4	256.71	210.64	189.96	169.35	156.16	148.40	139.09	133.57	126.90	120.27	116.83
4, Low	234.97	192.73	173.94	155.03	143.00	135.79	127.20	122.20	116.14	110.14	106.87

Length of front greater than 4 times depth – Square Foot Area

Quality Class	2,000	3,000	5,000	10,000	15,000	20,000	30,000	50,000	75,000	100,000	150,000
1, Best	389.65	347.73	305.85	264.14	245.68	234.79	221.87	208.89	200.76	195.88	190.23
1 & 2	344.64	307.53	270.48	233.53	217.30	207.69	196.28	184.82	177.57	173.26	168.23
2, Good	304.76	271.93	239.26	206.58	192.24	183.69	173.50	163.46	157.02	153.30	148.78
2 & 3	276.07	246.38	216.76	187.11	174.17	166.41	157.18	148.00	142.27	138.86	134.76
3, Average	252.75	225.53	198.35	171.30	159.41	152.26	143.87	135.47	130.16	127.03	123.35
3 & 4	233.34	208.21	183.12	158.18	147.15	140.59	132.83	125.08	120.24	117.23	113.87
4, Low	212.93	189.95	167.12	144.31	134.29	128.29	121.15	114.14	109.74	107.00	103.95

Perimeter Wall Adjustment: A common wall exists when two buildings share one wall. Adjust for common walls by deducting $427 per linear foot of common wall from the total structure cost. In some structures one or more walls are not owned at all. In this case, deduct $860 per linear foot of wall not owned.

Suburban Stores – Wood or Wood and Steel Frame

Quality Classification

	Class 1 Best Quality	Class 2 Good Quality	Class 3 Average Quality	Class 4 Low Quality
Foundation (15% of total cost)	Reinforced concrete.	Reinforced concrete.	Reinforced concrete.	Reinforced concrete.
Floor Structure (15% of total cost)	6" reinforced concrete on 6" rock base.	6" reinforced concrete on 6" rock base.	6" reinforced concrete on 6" rock base.	4" reinforced concrete on 6" rock base.
Wall Structure (15% of total cost)	2" x 6" - 16" o.c.	2" x 6" - 16" o.c.	2" x 6" - 16" o.c.	2" x 4" - 16" o.c.
Roof (15% of total cost)	Glu-lams or steel beams on steel intermediate columns. Panelized roof system, 1/2" plywood sheathing, 5 ply built-up roof with insulation.	Glu-lams or steel beams on steel intermediate columns. Panelized roof system, 1/2" plywood sheathing, 5 ply built-up roof with insulation.	Glu-lams on steel intermediate columns. Panelized roof system, 1/2" plywood sheathing, 4 ply built-up roof.	Glu-lams on steel intermediate columns. Panelized roof system, 1/2" plywood sheathing, 4 ply built-up roof.
Floor Finish (5% of total cost)	Terrazzo, sheet vinyl, or very good carpet.	Resilient tile with 50% solid vinyl tile, terrazzo, or good carpet.	Composition tile.	Minimum grade tile.
Interior Wall Finish (5% of total cost)	Gypsum wallboard or lath and plaster on exterior walls and partitions, finished with vinyl wall covers and hardwood veneers.	Gypsum wallboard, texture and paper on exterior walls and partitions, some vinyl wall cover and plywood paneling.	Gypsum wallboard, texture and paint on interior walls and partitions.	Gypsum wallboard, texture and paint on interior walls and partitions.
Ceiling Finish (5% of total cost)	Suspended good grade acoustical tile with gypsum board backing.	Suspended acoustical tile with concealed grid system.	Suspended acoustical tile with exposed grid system.	Exposed beams with ceiling tile or painted.
Lighting (5% of total cost)	Recessed LED lighting in modular plastic panels.	Continuous recessed 3 tube LED strips with egg crate diffusers, 8' o.c.	Continuous 3 tube fluorescent strips with egg crate diffusers, 8' o.c.	Continuous exposed 2 tube fluorescent strips, 8' o.c.
Exterior (8% of total cost)	Face brick or stone veneer.	Wood siding, some stone veneer.	Stucco on exposed areas, some brick trim.	Stucco on exposed areas.
Plumbing (12% of total cost)	6 good fixtures per 5,000 S.F. of floor area, metal toilet partitions.	6 standard fixtures per 5,000 S.F. of floor area, metal toilet partitions.	4 standard fixtures per 5,000 S.F. of floor area, metal toilet partitions.	4 standard fixtures per 5,000 S.F. of floor area wood toilet partitions.

Note: Use the percent of total cost to help identify the correct quality classification

Strip and Island Suburban Stores

For estimating purposes, wood frame suburban stores should be divided into strip type units or island type units. Strip type buildings have a front wall made up of display fronts. The side and rear walls, except for delivery or walk-through doors, are made up of solid, continuous wood frame walls. If there are any display areas in the sides or rear of these buildings, the cost of the display front must be added to the building cost and the cost of the wall that it replaces must be deducted from the building costs.

Island type suburban store buildings have display fronts on the major portion of all four sides. Stores may be arranged so that one store fronts on two sides or they may be partitioned in such a way that there are two separate stores fronting on each side of the building.

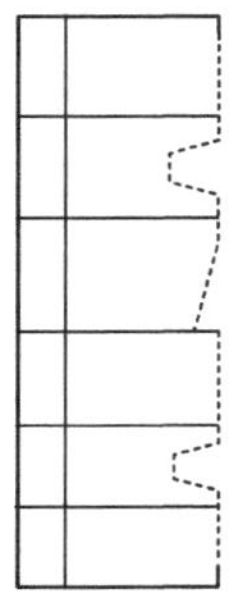

Strip Type

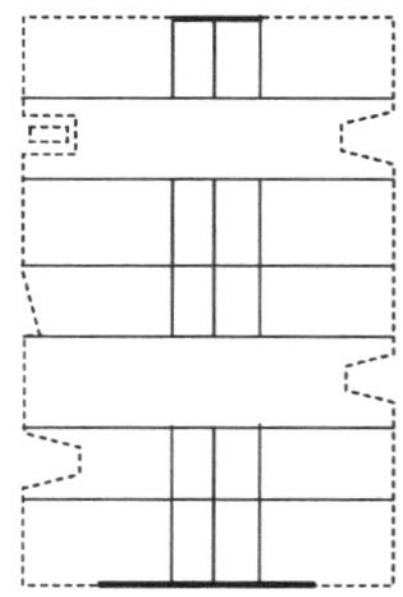

Island Type

Suburban Stores – Wood or Wood and Steel Frame

Building Shell Only, Island Type

Estimating Procedure

1. Use these figures to estimate the cost of shell-type buildings without permanent partitions, display fronts or finish material on the front wall of the building.
2. Establish the structure quality class by applying the information on page 94.
3. Compute the building floor area. This should include everything within the exterior walls and all inset areas outside the main walls but under the main roof.
4. Add to or subtract from the square foot cost below the appropriate amount from the Wall Height Adjustment Table (at the bottom of this page) if the wall height is more or less than 16 feet.
5. Multiply the adjusted square foot cost by the building area.
6. Deduct, if appropriate, for common walls, using the figures at the bottom of this page.
7. Multiply the total cost by the location factor on page 7 or 8.
8. Add the cost of the appropriate additional components from page 236 to 248: heating and cooling equipment, fire sprinklers, display fronts, finish materials on the front wall, canopies, interior partitions, exterior signs, mezzanines and basements, loading docks and ramps, yard improvements, and communication systems.

Length less than 1-1/2 times depth – Square Foot Area

Quality Class	3,500	5,000	7,500	10,000	12,500	15,000	20,000	30,000	40,000	50,000	75,000
1, Best	127.59	125.49	123.41	122.22	121.37	120.73	119.91	118.84	118.22	117.85	117.19
1 & 2	115.59	113.67	111.79	110.73	109.91	109.39	108.64	107.65	107.11	106.71	106.19
2, Good	105.34	103.66	101.92	100.95	100.27	99.77	99.07	98.19	97.70	97.37	96.79
2 & 3	101.75	98.29	95.06	94.17	93.52	93.08	92.42	91.57	91.06	90.78	90.31
3, Average	92.94	91.47	89.92	89.06	88.46	87.96	87.35	86.65	86.20	85.88	85.39
3 & 4	85.32	83.96	82.54	81.70	81.13	80.81	80.17	79.50	79.07	78.82	78.32
4, Low	77.56	76.23	75.00	74.31	73.71	73.41	72.86	72.26	71.86	71.65	71.17

Length between 1-1/2 and 2 times depth – Square Foot Area

Quality Class	4,500	5,000	7,500	10,000	15,000	20,000	30,000	40,000	50,000	75,000	100,000
1, Best	126.98	126.26	123.84	122.52	120.96	120.09	118.97	118.44	118.04	117.40	117.08
1 & 2	114.96	114.32	112.16	110.81	109.50	108.68	107.78	107.22	106.87	106.28	106.01
2, Good	104.87	104.30	102.30	101.20	99.84	99.13	98.28	97.78	97.49	96.97	96.68
2 & 3	97.71	97.15	95.26	94.25	93.08	92.39	91.57	91.11	90.85	90.41	90.10
3, Average	92.39	91.81	90.09	89.14	87.92	87.32	86.60	86.19	85.86	85.41	85.20
3 & 4	84.79	84.26	82.71	81.78	80.76	80.16	79.51	79.07	78.82	78.46	78.13
4, Low	76.89	76.44	75.00	74.21	73.31	72.74	72.08	71.74	71.44	71.11	70.91

Wall Height Adjustment: Add or subtract the amount listed to or from the square foot costs above for each foot of wall height more or less than 16 feet.

Area	3,500	5,000	7,500	10,000	12,500	15,000	Over 20,000
Cost	.30	.28	.23	.18	.17	.15	.13

Perimeter Wall Adjustment: For common wall deduct $72 per linear foot of common wall. For no wall ownership, deduct $146 per linear foot of wall.

Suburban Stores – Wood or Wood and Steel Frame

Building Shell Only, Island Type

Estimating Procedure

1. Use these figures to estimate the cost of shell-type buildings without permanent partitions, display fronts or finish material on the front wall of the building.
2. Establish the structure quality class by applying the information on page 94.
3. Compute the building floor area. This should include everything within the exterior walls and all inset areas outside the main walls but under the main roof.
4. Add to or subtract from the square foot cost below the appropriate amount from the Wall Height Adjustment Table (at the bottom of this page) if the wall height is more or less than 16 feet.
5. Multiply the adjusted square foot cost by the building area.
6. Deduct, if appropriate, for common walls, using the figures at the bottom of this page.
7. Multiply the total cost by the location factor on page 7 or 8.
8. Add the cost of the appropriate additional components from page 236 to 248: heating and cooling equipment, fire sprinklers, display fronts, finish materials on the front wall, canopies, interior partitions, exterior signs, mezzanines and basements, loading docks and ramps, yard improvements, and communication systems.

Length between 2 and 3 times depth – Square Foot Area

Quality Class	10,000	15,000	20,000	30,000	40,000	50,000	75,000	100,000	150,000	200,000	250,000
1, Best	124.35	122.53	121.44	120.19	119.45	118.91	118.07	117.72	117.17	116.83	116.57
1 & 2	112.49	110.81	109.88	108.73	108.14	107.63	106.93	106.52	106.04	105.70	105.55
2, Good	102.18	100.67	99.70	98.74	98.07	97.69	97.06	96.63	96.20	95.95	95.79
2 & 3	95.20	93.78	92.94	91.98	91.47	90.98	90.47	90.02	89.67	89.36	89.24
3, Average	89.53	88.16	87.35	86.42	85.93	85.51	85.00	84.64	84.32	84.06	83.87
3 & 4	82.07	80.84	80.14	79.27	78.82	78.46	77.97	77.58	77.34	77.01	76.85
4, Low	74.42	73.37	72.74	71.94	71.47	71.16	70.72	70.42	70.10	69.90	69.76

Length greater than 3 times depth – Square Foot Area

Quality Class	7,500	10,000	15,000	20,000	30,000	40,000	50,000	75,000	100,000	150,000	200,000
1, Best	124.01	123.24	121.78	120.94	119.78	119.05	118.54	117.77	117.21	116.56	116.22
1 & 2	115.99	115.25	113.92	109.50	108.49	107.81	107.33	106.60	106.19	105.56	105.21
2, Good	101.81	101.20	99.93	99.24	98.29	97.73	97.30	96.63	96.26	95.74	95.37
2 & 3	94.97	94.30	93.15	92.51	91.67	91.06	90.74	90.04	89.71	89.21	88.89
3, Average	89.28	88.74	87.71	87.02	86.29	85.69	85.38	84.72	84.39	83.96	83.68
3 & 4	81.84	81.37	80.44	79.83	79.07	78.57	78.30	77.68	77.44	76.94	76.73
4, Low	74.36	73.92	73.05	72.49	71.81	71.36	71.10	70.62	70.30	69.89	69.69

Wall Height Adjustment: Add or subtract the amount listed to or from the square foot costs above for each foot of wall height more or less than 16 feet.

Area	3,500	5,000	7,500	10,000	12,500	15,000	Over 20,000
Cost	.29	.28	.22	.18	.15	.14	.12

Perimeter Wall Adjustment: For common wall, deduct $70 per linear foot of common wall. For no wall ownership, deduct $143 per linear foot of wall.

Suburban Stores – Wood or Wood and Steel Frame

Building Shell Only, Strip Type

Estimating Procedure

1. Use these figures to estimate the cost of shell-type buildings without permanent partitions, display fronts or finish material on the front wall of the building.
2. Establish the structure quality class by applying the information on page 94.
3. Compute the building floor area. This should include everything within the exterior walls and all inset areas outside the main walls but under the main roof.
4. Add to or subtract from the square foot cost below the appropriate amount from the Wall Height Adjustment Table (at the bottom of this page) if the wall height is more or less than 16 feet.
5. Multiply the adjusted square foot cost by the building area.
6. Deduct, if appropriate, for common walls, using the figures at the bottom of this page.
7. Multiply the total cost by the location factor on page 7 or 8.
8. Add the cost of the appropriate additional components from page 236 to 248: heating and cooling equipment, fire sprinklers, display fronts, finish materials on the front wall, canopies, interior partitions, exterior signs, mezzanines and basements, loading docks and ramps, yard improvements, and communication systems.

Length less than 1-1/2 times depth – Square Foot Area

Quality Class	500	1,000	2,000	2,500	3,000	4,000	5,000	7,500	10,000	12,500	15,000
1, Best	199.36	172.24	153.05	148.16	144.53	139.49	136.02	130.66	127.48	125.34	123.66
1 & 2	182.16	157.41	139.86	135.41	132.07	127.51	124.29	119.39	116.49	114.50	113.02
2, Good	167.18	144.45	128.39	124.29	121.18	116.99	114.09	109.61	106.91	105.05	103.68
2 & 3	158.61	137.04	121.78	117.93	114.99	110.99	108.26	103.94	101.41	99.67	98.36
3, Average	152.01	131.34	116.71	112.97	110.27	106.35	103.78	99.64	97.17	95.52	94.25
3 & 4	142.43	122.99	109.32	105.84	103.26	99.64	97.16	93.29	91.06	89.52	88.28
4, Low	132.79	114.66	101.89	98.67	96.34	92.86	90.63	87.02	84.86	83.40	82.35

Length between 1-1/2 and 2 times depth – Square Foot Area

Quality Class	500	1,000	2,000	3,000	5,000	6,000	7,500	10,000	15,000	20,000	35,000
1, Best	192.20	166.76	148.66	140.57	132.46	130.01	127.37	124.27	120.61	118.44	114.99
1 & 2	175.67	152.40	135.79	128.47	121.04	118.81	116.37	113.55	110.22	108.19	105.15
2, Good	161.34	139.93	124.71	117.94	111.15	109.08	106.87	104.30	101.25	99.39	96.46
2 & 3	152.80	132.60	118.09	111.70	105.26	103.32	101.20	98.77	95.89	94.17	91.47
3, Average	146.27	126.88	113.05	106.94	100.74	98.87	96.84	94.52	91.77	90.09	87.51
3 & 4	137.46	119.24	106.25	100.51	94.77	92.89	91.06	88.85	86.29	84.65	82.24
4, Low	128.61	111.56	99.45	94.09	88.56	86.98	85.17	83.10	80.69	79.23	76.94

Wall Height Adjustment: Add or subtract the amount listed to or from the square foot cost above for each foot of wall height more or less than 16 feet.

Area	500	1,000	2,000	2,500	3,000	4,000	5,000	7,500	10,000	12,500
Cost	2.22	1.55	1.09	1.01	.90	.83	.73	.68	.62	.60

Area	15,000	20,000	25,000	30,000	35,000	50,000	70,000	75,000	100,000	150,000
Cost	.45	.33	.29	.27	.25	.22	.21	.19	.18	.17

Perimeter Wall Adjustment: For a common wall, deduct $241 per linear foot of common wall. For no wall ownership, deduct $492 per linear foot of wall.

Suburban Stores – Wood or Wood and Steel Frame

Building Shell Only, Strip Type

Estimating Procedure

1. Use these figures to estimate the cost of shell-type buildings without permanent partitions, display fronts or finish material on the front wall of the building.
2. Establish the structure quality class by applying the information on page 94.
3. Compute the building floor area. This should include everything within the exterior walls and all inset areas outside the main walls but under the main roof.
4. Add to or subtract from the square foot cost below the appropriate amount from the Wall Height Adjustment Table (at the bottom of this page) if the wall height is more or less than 16 feet.
5. Multiply the adjusted square foot cost by the building area.
6. Deduct, if appropriate, for common walls, using the figures at the bottom of this page.
7. Multiply the total cost by the location factor on page 7 or 8.
8. Add the cost of the appropriate additional components from page 236 to 248: heating and cooling equipment, fire sprinklers, display fronts, finish materials on the front wall, canopies, interior partitions, exterior signs, mezzanines and basements, loading docks and ramps, yard improvements, and communication systems.

Length between 2 and 3 times depth – Square Foot Area

Quality Class	1,000	2,000	3,000	5,000	7,500	10,000	15,000	20,000	30,000	50,000	70,000
1, Best	169.26	150.93	142.74	134.55	129.49	126.43	122.73	120.56	117.97	115.43	114.02
1 & 2	154.70	137.89	130.41	123.02	118.36	115.55	112.19	110.17	107.82	105.44	104.21
2, Good	142.01	126.53	119.74	112.86	108.59	106.00	102.92	101.12	98.97	96.83	95.61
2 & 3	134.74	120.21	113.71	107.16	103.09	100.54	97.70	95.95	93.90	91.93	90.81
3, Average	133.92	118.78	109.08	102.88	98.87	96.57	93.76	92.17	90.13	88.23	87.12
3 & 4	121.61	108.41	102.56	96.65	92.96	90.74	88.16	86.60	84.74	82.94	81.88
4, Low	113.71	101.29	95.82	90.31	86.83	84.79	82.42	80.94	79.17	77.49	76.54

Length greater than 3 times depth – Square Foot Area

Quality Class	2,000	3,000	5,000	10,000	15,000	20,000	30,000	50,000	75,000	100,000	150,000
1, Best	155.92	146.83	137.85	128.80	124.78	122.41	119.61	116.81	114.96	113.93	112.65
1 & 2	141.58	133.38	125.16	116.93	113.29	111.20	108.64	106.06	104.39	103.47	102.30
2, Good	129.97	122.40	114.91	107.33	104.04	102.05	99.69	97.31	95.92	94.99	93.90
2 & 3	123.25	116.09	108.95	101.78	98.65	96.69	94.52	92.28	90.89	90.02	89.08
3, Average	118.19	111.28	104.54	97.62	94.62	92.78	90.64	88.48	87.12	86.37	85.41
3 & 4	111.46	104.99	98.55	92.11	89.21	87.53	85.51	83.45	82.23	81.51	80.56
4, Low	104.55	98.48	92.39	86.37	83.68	82.07	80.18	78.32	77.13	76.40	75.52

Wall Height Adjustment: Add or subtract the amount listed to or from the square foot cost above for each foot of wall height more or less than 16 feet.

Area	500	1,000	2,000	2,500	3,000	4,000	5,000	7,500	10,000	12,500
Cost	2.26	1.61	1.11	1.02	0.95	0.88	0.78	0.72	0.65	0.61

Area	15,000	20,000	25,000	30,000	35,000	50,000	70,000	75,000	100,000	150,000
Cost	.44	.32	.30	.28	.26	.24	.22	.21	.19	.18

Perimeter Wall Adjustment: For a common wall, deduct $241 per linear foot of common wall. For no wall ownership, deduct $464 per linear foot of wall.

Suburban Stores – Wood or Wood and Steel Frame

Multi-Unit, Island Type

Estimating Procedure

1. Use these figures to estimate the cost of stores designed for multiple occupancy. These costs include all components of shell buildings plus the cost of display fronts, finish materials on the front of the building and normal interior partitions.
2. Establish the structure quality class by applying the information on page 94. Evaluate the quality of the display front to help establish the correct quality class of the building as a whole. The building classes have display fronts as classified on page 89. See also pages 242 to 245.
3. Compute the building floor area. This should include everything within the exterior walls and all inset areas outside the main walls but under the main roof.
4. Add to or subtract from the square foot cost below the appropriate amount from the Wall Height Adjustment Table (at the bottom of this page) if the wall height is more or less than 16 feet.
5. Multiply the adjusted square foot cost by the building area.
6. Deduct, if appropriate, for common walls, using the figures at the bottom of this page.
7. Multiply the total cost by the location factor on page 7 or 8.
8. Add the cost of the appropriate additional components from page 236 to 248: heating and cooling equipment, fire sprinklers, canopies, exterior signs, mezzanines and basements, loading docks and ramps, yard improvements, and communication systems.

Length less than 1-1/2 times depth – Square Foot Area

Quality Class	3,500	5,000	7,500	10,000	12,500	15,000	20,000	30,000	40,000	50,000	75,000
1, Best	283.86	258.40	235.15	219.06	210.00	202.80	191.21	179.49	172.63	162.87	158.61
1 & 2	248.13	226.25	206.43	192.52	184.53	178.61	168.69	167.32	158.60	148.69	142.01
2, Good	214.24	196.50	179.72	169.73	162.81	157.71	150.58	142.04	136.90	133.41	128.01
2 & 3	187.37	171.91	157.25	148.46	142.41	137.95	131.74	124.28	119.82	116.81	112.02
3, Average	166.25	152.48	139.51	131.71	126.37	122.38	116.83	110.27	106.25	103.63	99.33
3 & 4	148.91	136.56	124.89	117.94	113.14	109.56	104.63	98.67	95.17	92.78	89.01
4, Low	131.70	120.77	110.46	104.31	100.11	96.92	92.54	87.32	84.18	82.02	78.72

Length between 1-1/2 and 2 times depth – Square Foot Area

Quality Class	4,500	5,000	7,500	10,000	15,000	20,000	30,000	40,000	50,000	75,000	100,000
1, Best	270.33	262.67	237.34	222.79	205.99	196.26	184.93	178.35	173.89	167.10	163.10
1 & 2	238.25	231.52	209.15	196.38	181.60	173.03	162.98	157.19	153.33	147.33	143.78
2, Good	208.65	202.72	183.16	171.91	158.97	151.43	142.73	137.68	134.25	128.96	126.01
2 & 3	184.48	179.21	161.92	151.99	140.57	133.92	126.26	121.75	118.70	113.99	111.30
3, Average	162.20	157.59	142.44	133.73	123.66	117.84	111.09	107.04	104.39	100.28	97.87
3 & 4	146.51	142.38	128.62	120.68	111.68	106.28	100.28	96.65	94.25	90.59	88.40
4, Low	130.61	126.89	114.62	107.65	99.52	94.86	89.37	86.19	84.06	80.76	78.78

Wall Height Adjustment: Add or subtract the amount listed to or from the square foot cost above for each foot of wall height more or less than 16 feet.

Square Foot Area

Quality Class	3,500	5,000	7,500	10,000	12,500	15,000	20,000	30,000	40,000	50,000	75,000	100,000	150,000
1	13.59	11.42	9.32	3.89	5.08	3.19	5.85	4.80	4.19	3.80	3.08	2.71	2.30
2	9.33	7.92	6.48	2.83	3.48	2.34	4.08	3.32	2.93	2.63	2.19	1.86	1.56
3	6.48	5.46	4.44	6.29	2.58	4.46	2.79	2.30	2.00	1.78	1.50	1.35	1.06
4	4.76	4.01	3.28	4.37	5.46	3.17	2.06	1.71	1.43	1.32	1.10	0.91	0.81

Perimeter Wall Adjustment: For a common wall, deduct per linear foot: Class 1, $1,330, Class 2, $870, Class 3, $570, Class 4, $416. For no wall ownership, deduct per linear foot: Class 1, $2,550, Class 2, $1,740, Class 3, $1,190, Class 4, $830.

Suburban Stores – Wood or Wood and Steel Frame

Multi-Unit, Island Type

Estimating Procedure

1. Use these figures to estimate the cost of stores designed for multiple occupancy. These costs include all components of shell buildings plus the cost of display fronts, finish materials on the front of the building and normal interior partitions.
2. Establish the structure quality class by applying the information on page 94. Evaluate the quality of the display front to help establish the correct quality class of the building as a whole. The building classes have display fronts as classified on page 89. See also pages 242 to 245.
3. Compute the building floor area. This should include everything within the exterior walls and all inset areas outside the main walls but under the main roof.
4. Add to or subtract from the square foot cost below the appropriate amount from the Wall Height Adjustment Table (at the bottom of this page) if the wall height is more or less than 16 feet.
5. Multiply the adjusted square foot cost by the building area.
6. Deduct, if appropriate, for common walls, using the figures at the bottom of this page.
7. Multiply the total cost by the location factor on page 7 or 8.
8. Add the cost of the appropriate additional components from page 236 to 248: heating and cooling equipment, fire sprinklers, canopies, exterior signs, mezzanines and basements, loading docks and ramps, yard improvements, and communication systems.

Length between 2 and 3 times depth – Square Foot Area

Quality Class	7,500	10,000	15,000	20,000	30,000	40,000	50,000	75,000	100,000	150,000	200,000
1, Best	239.08	225.25	208.72	198.86	187.07	180.01	175.10	167.61	163.12	157.73	154.59
1 & 2	211.30	199.06	184.40	175.70	165.26	159.05	154.73	148.09	144.13	139.46	136.56
2, Good	185.71	174.96	162.06	154.41	145.25	139.80	136.00	130.20	126.65	122.47	120.04
2 & 3	164.84	155.28	143.81	137.04	128.89	124.03	120.71	115.58	112.41	108.72	106.52
3, Average	145.88	137.43	127.37	121.28	114.10	109.82	106.91	102.28	99.55	96.33	94.30
3 & 4	131.15	123.56	114.50	109.04	102.59	98.71	96.12	91.93	89.52	86.50	84.79
4, Low	116.23	109.53	101.46	96.63	90.91	87.46	85.16	81.40	79.24	76.72	75.17

Length greater than 3 times depth – Square Foot Area

Quality Class	10,000	15,000	20,000	30,000	40,000	50,000	75,000	100,000	150,000	200,000	250,000
1, Best	237.20	218.08	206.79	193.36	185.42	180.01	171.57	166.59	160.67	157.12	154.70
1 & 2	209.70	192.82	182.77	170.96	163.92	159.13	151.70	147.31	142.04	138.84	136.74
2, Good	184.17	169.38	160.61	150.20	144.01	139.81	133.29	129.38	124.72	122.03	120.13
2 & 3	163.81	150.66	142.75	133.59	128.03	124.29	118.50	115.08	110.91	108.56	106.90
3, Average	145.05	133.35	126.43	118.25	113.43	110.05	104.87	101.81	98.23	96.06	94.62
3 & 4	130.41	120.00	113.74	106.35	102.02	99.00	94.35	91.66	88.40	86.40	85.10
4, Low	115.62	106.35	100.80	94.28	90.44	87.77	83.68	81.22	78.36	76.61	75.49

Wall Height Adjustment: Add or subtract the amount listed to or from the square foot cost above for each foot of wall height more or less than 16 feet.

Square Foot Area

Quality Class	7,500	10,000	12,500	15,000	20,000	30,000	40,000	50,000	75,000	100,000	150,000	200,000	250,000
1	8.94	7.77	7.00	5.96	4.95	3.94	3.02	2.91	2.80	2.53	2.13	1.85	1.63
2	6.28	5.41	4.88	3.94	3.13	2.93	2.78	2.48	2.10	1.75	1.50	1.33	1.15
3	4.34	3.79	3.38	3.03	2.58	2.32	1.96	1.72	1.41	1.29	1.03	.91	.82
4	3.14	2.77	2.47	2.17	1.87	1.62	1.40	1.29	1.05	.87	.80	.70	.64

Perimeter Wall Adjustment: For a common wall, deduct per linear foot: Class 1, $1,350, Class 2, $860, Class 3, $630, Class 4, $414. For no wall ownership, deduct per linear foot: Class 1, $2,590, Class 2, $1,740, Class 3, $1,190, Class 4, $840.

Suburban Stores – Wood or Wood and Steel Frame

Multi-Unit, Strip Type

Estimating Procedure

1. Use these figures to estimate the cost of stores designed for multiple occupancy. These costs include all components of shell buildings plus the cost of display fronts, finish materials on the front of the building and normal interior partitions.
2. Establish the structure quality class by applying the information on page 94. Evaluate the quality of the display front to help establish the correct quality class of the building as a whole. The building classes have display fronts as classified on page 89. See also pages 242 to 245.
3. Compute the building floor area. This should include everything within the exterior walls and all inset areas outside the main walls but under the main roof.
4. Add to or subtract from the square foot cost below the appropriate amount from the Wall Height Adjustment Table (at the bottom of this page) if the wall height is more or less than 16 feet.
5. Multiply the adjusted square foot cost by the building area.
6. Deduct, if appropriate, for common walls, using the figures at the bottom of this page.
7. Multiply the total cost by the location factor on page 7 or 8.
8. Add the cost of the appropriate additional components from page 236 to 248: heating and cooling equipment, fire sprinklers, canopies, exterior signs, mezzanines and basements, loading docks and ramps, yard improvements, and communication systems.

Length less than 1-1/2 times depth – Square Foot Area

Quality Class	500	1,000	2,000	2,500	3,000	4,000	5,000	7,500	10,000	12,500	15,000
1, Best	292.06	243.49	209.33	200.59	194.19	185.14	179.02	169.51	163.80	159.88	157.03
1 & 2	258.75	215.77	185.42	177.73	172.01	163.99	158.55	150.11	145.05	141.61	139.08
2, Good	228.14	190.22	163.44	156.62	151.67	144.56	139.85	132.36	127.91	124.81	122.58
2 & 3	207.69	173.13	148.84	142.62	138.10	131.61	127.29	120.46	116.40	113.71	111.58
3, Average	190.22	158.61	136.36	130.72	126.45	120.62	116.61	110.40	106.66	104.13	102.27
3 & 4	175.26	146.09	125.59	124.79	116.50	111.10	107.38	101.66	98.23	95.92	94.23
4, Low	159.25	132.79	114.09	109.36	105.84	100.91	97.60	92.39	89.28	87.21	85.56

Length between 1-1/2 and 2 times depth – Square Foot Area

Quality Class	500	1,000	2,000	3,000	5,000	7,500	10,000	15,000	20,000	25,000	35,000
1, Best	326.21	268.49	227.57	209.42	191.22	179.78	172.93	164.83	159.98	156.67	152.33
1 & 2	287.79	236.87	200.79	184.81	168.74	158.60	152.54	145.41	141.17	138.19	134.42
2, Good	252.77	208.02	176.29	162.30	148.19	139.29	134.06	127.73	123.98	121.44	118.05
2 & 3	229.46	188.79	160.07	147.31	134.53	126.44	121.61	115.93	112.45	110.22	107.15
3, Average	208.96	171.97	145.76	134.11	122.44	115.13	110.78	106.19	102.44	100.38	97.60
3 & 4	191.17	157.35	133.34	122.74	112.04	105.33	101.38	96.49	93.76	91.81	89.28
4, Low	173.70	142.89	121.06	111.46	101.76	95.65	92.02	87.73	85.16	83.37	81.06

Wall Height Adjustment: Add or subtract the amount listed to or from the square foot of floor cost for each foot of wall height more or less than 16 feet.

Square Foot Area

Quality Class	500	1,000	2,000	3,000	5,000	7,500	10,000	15,000	20,000	25,000	35,000
1, Best	20.07	13.95	9.84	8.03	6.29	5.09	4.42	3.69	3.15	2.87	2.61
2, Good	13.15	9.16	6.45	5.28	4.04	3.34	2.92	2.39	2.14	1.85	1.74
3, Average	9.64	6.73	4.75	3.88	3.04	2.47	2.17	1.75	1.55	1.42	1.28
4, Low	7.03	4.84	3.40	2.80	2.18	1.76	1.55	1.28	1.05	.94	.91

Perimeter Wall Adjustment: For a common wall, deduct $248 per linear foot. For no wall ownership, deduct $494 per linear foot.

Suburban Stores – Wood or Wood and Steel Frame

Multi-Unit, Strip Type

Estimating Procedure

1. Use these figures to estimate the cost of stores designed for multiple occupancy. These costs include all components of shell buildings plus the cost of display fronts, finish materials on the front of the building and normal interior partitions.
2. Establish the structure quality class by applying the information on page 94. Evaluate the quality of the display front to help establish the correct quality class of the building as a whole. The building classes have display fronts as classified on page 89. See also pages 242 to 245.
3. Compute the building floor area. This should include everything within the exterior walls and all inset areas outside the main walls but under the main roof.
4. Add to or subtract from the square foot cost below the appropriate amount from the Wall Height Adjustment Table (at the bottom of this page) if the wall height is more or less than 16 feet.
5. Multiply the adjusted square foot cost by the building area.
6. Deduct, if appropriate, for common walls, using the figures at the bottom of this page.
7. Multiply the total cost by the location factor on page 7 or 8.
8. Add the cost of the appropriate additional components from page 236 to 248: heating and cooling equipment, fire sprinklers, canopies, exterior signs, mezzanines and basements, loading docks and ramps, yard improvements, and communication systems.

Length between 2 and 3 times depth – Square Foot Area

Quality Class	1,000	2,000	3,000	5,000	7,500	10,000	15,000	20,000	30,000	50,000	70,000
1, Best	309.87	257.31	234.11	210.83	196.21	187.44	177.04	170.89	163.55	156.32	152.42
1 & 2	273.06	226.78	206.32	185.83	172.89	165.25	156.03	150.64	144.21	137.76	134.28
2, Good	239.07	198.51	180.66	162.65	151.36	144.59	136.66	131.89	126.22	120.56	117.55
2 & 3	216.48	179.72	163.50	147.31	137.07	131.00	123.77	119.41	114.33	109.21	106.47
3, Average	196.21	162.93	148.20	133.47	124.19	118.74	112.15	108.19	103.59	98.95	96.49
3 & 4	179.53	149.06	135.62	122.08	113.60	108.60	102.53	99.03	94.79	90.54	88.24
4, Low	162.50	134.93	122.82	110.58	102.82	98.32	92.88	89.63	85.92	81.95	79.93

Length greater than 3 times depth – Square Foot Area

Quality Class	2,000	3,000	5,000	10,000	15,000	20,000	30,000	50,000	75,000	100,000	150,000
1, Best	282.77	254.48	226.46	198.71	186.59	179.35	170.76	162.29	156.95	153.79	149.99
1 & 2	249.01	224.11	199.35	174.92	164.26	157.89	150.40	142.92	138.23	135.40	132.10
2, Good	217.75	195.95	174.39	153.04	143.66	138.11	131.54	125.01	120.86	118.44	115.54
2 & 3	190.09	171.02	152.27	133.55	125.46	120.56	114.86	109.09	105.58	103.46	100.82
3, Average	178.15	160.32	142.66	125.16	117.49	113.02	107.60	102.27	98.87	96.89	94.54
3 & 4	163.96	147.55	131.34	115.20	108.14	103.94	99.07	94.10	91.01	89.16	87.04
4, Low	149.29	134.38	119.60	104.95	98.55	94.72	90.21	85.69	82.91	81.29	79.25

Wall Height Adjustment: Add or subtract the amount listed to or from the square foot cost above for each foot of wall height more or less than 16 feet.

Square Foot Area

Quality Class	2,000	3,000	5,000	10,000	15,000	20,000	30,000	50,000	75,000	100,000	150,000
1, Best	9.55	7.79	6.09	4.36	3.65	3.14	2.59	2.15	1.75	1.60	1.39
2, Good	6.14	5.05	3.88	2.80	2.29	2.03	1.72	1.35	1.12	1.03	.84
3, Average	4.51	3.74	2.88	2.10	1.73	1.50	1.26	.96	.84	.76	.65
4, Low	3.38	2.71	2.15	1.50	1.27	1.14	.90	.75	.65	.60	.47

Perimeter Wall Adjustment: For a common wall, deduct $239 per linear foot. For no wall ownership, deduct $481 per linear foot.

Supermarkets – Masonry or Concrete

Quality Classification

	Class 1 Best Quality	Class 2 Good Quality	Class 3 Average Quality	Class 4 Low Quality
Foundation (15% of total cost)	Reinforced concrete.	Reinforced concrete.	Reinforced concrete.	Reinforced concrete.
Floor Structure (12% of total cost)	4" reinforced concrete on 6" rock fill.	4" reinforced concrete on 6" rock fill.	4" reinforced concrete on 6" rock fill.	4" reinforced concrete on 6" rock fill.
Wall Structure (15% of total cost)	6" concrete tilt-up or ornamental block or brick.	6" concrete tilt-up, colored concrete block or brick.	6" concrete tilt-up or 8" concrete block.	6" concrete tilt-up or 8" concrete block.
Roof Structure (10% of total cost)	Glu-lams or steel "I" beams on steel intermediate columns, 2" x 12" purlins 16" o.c., 1/2" plywood sheathing.	Glu-lams or steel "I" beams on steel intermediate columns, 2" x 12" purlins 16" o.c., 1/2" plywood sheathing.	Glu-lams or steel "I" beams on steel intermediate columns, 2" x 12" purlins 16" o.c., 1/2" plywood sheathing.	Glu-lams or steel "I" beams on steel intermediate columns, 3" x 12" purlins 3' o.c., 1/2" plywood sheathing.
Floor Finish (5% of total cost)	Terrazzo in sales area. Sheet vinyl or carpet in cashiers' area.	Resilient tile in sales area. Terrazzo, solid vinyl tile or carpet in cashiers' area.	Composition tile in sales area.	Minimum grade tile in sales area.
Interior Wall Finish (5% of total cost)	Inside of exterior walls furred out with gypsum wallboard and paint or interior stucco, interior stucco or gypsum wallboard and vinyl wall cover on partitions.	Paint on inside of exterior walls, gypsum wallboard and paint or vinyl wall cover on partitions.	Paint on inside of exterior walls, wallboard and paint on partitions.	Paint on inside of exterior walls, wallboard and paint on partitions.
Ceiling Finish (5% of total cost)	Suspended acoustical tile, dropped ceiling over meat and produce departments.	Suspended acoustical tile or gypsum board and acoustical texture, dropped ceiling over meat and produce departments.	Ceiling tile on roof purlins, dropped ceiling over meat department.	Open.
Front (7% of total cost)	A large amount of float glass in good aluminum frames (18'-22' high for 3/4 of width), brick or stone veneer on remainder, 1 pair of good automatic doors per 7,000 S.F. of floor area, anodized aluminum sunshade over glass area, 8' canopy across front, 10'-12' raised walk across front.	A large amount of float glass in good aluminum frames (16'-18' high for 2/3 of width), brick or stone veneer on remainder, 1 pair of good automatic doors per 10,000 S.F. of floor area, 8' canopy across front, 10' raised walk across front.	A moderate amount of float glass in average quality aluminum frames (12'-16' high for 2/3 of width), exposed aggregate on remainder, 1 pair average automatic doors per 10,000 S.F. of floor area, 6' canopy across front, 8' raised walk across front.	Stucco or exposed aggregate with a small amount of float glass in an inexpensive aluminum frame (6'-10' high for 1/2 of width), 6' canopy across front, 6' ground level walk across front.
Exterior Finish (8% of total cost)	Large ornamental rock or brick veneer.	Large ornamental rock or brick veneer.	Paint, some exposed aggregate.	Paint.
Roof Cover (5% of total cost)	5 ply built-up roofing with large rock.	5 ply built-up roofing with tar and rock.	4 ply built-up roofing.	4 ply built-up roofing.
Plumbing (8% of total cost)	2 rest rooms with 3 fixtures, floor piping and drains to refrigerated cases, 2 double sinks with drain board.	2 rest rooms with 3 fixtures each, floor piping and drains to refrigerated cases, 2 double sinks with drain board.	2 rest rooms with 2 fixtures each, floor piping and drains to refrigerated cases, 2 double sinks.	1 rest room with 2 fixtures, floor piping and drains to refrigerated cases.
Electrical (5% of total cost)	Conduit wiring, recessed LED fixtures 8' o.c., 30-40 spotlights.	Conduit wiring, LED fixtures with diffusers 8' o.c., 30-40 spotlights.	Conduit wiring, 3 tube fluorescent fixtures, 8' o.c., 5 or 10 spotlights.	Conduit wiring, double tube fluorescent fixtures, 8' o.c.

Note: Use the percent of total cost to help identify the correct quality classification.

Square foot costs include the following components: Foundations as required for normal soil conditions. Floor, wall, and roof structures. Interior floor, wall and ceiling finishes. Exterior wall finish and roof cover. Display fronts. Interior partitions. Entry and delivery doors. A canopy and walk across the front of the building as described in the applicable building specifications. Basic lighting and electrical systems. Rough and finish plumbing. All plumbing, piping and wiring necessary to operate the usual refrigerated cases and vegetable cases. Design and engineering fees. Permits and hook-up fees. Contractor's mark-up.

Supermarkets – Masonry or Concrete

Estimating Procedure

1. Establish the structure quality class by using the information on page 103.
2. Compute the building floor area. This should include everything within the building exterior walls and all insets outside the main walls but under the main building roof.
3. Add to or subtract from the square foot cost below the appropriate amount from the Wall Height Adjustment Row (at the bottom of this page) if the wall height is more or less than 20 feet.
4. Multiply the adjusted square foot cost by the building area.
5. Deduct, if appropriate, for common walls, using the figures at the bottom of this page.
6. Multiply the total cost by the location factor listed on page 7 or 8.
7. Add the cost of heating and cooling equipment, fire sprinklers, exterior signs, yard improvements, loading docks, ramps and walk-in boxes if they are an integral part of the building. See pages 236 to 248.

Supermarket, Class 2

Square Foot Area

Quality Class	5,000	7,500	10,000	12,500	15,000	20,000	25,000	30,000	35,000	40,000	50,000
Exceptional	275.47	256.15	244.61	236.66	230.83	222.59	216.89	212.73	209.49	206.87	202.81
1, Best	262.51	244.13	233.09	225.57	219.93	212.11	206.72	202.70	199.66	197.16	193.25
1 & 2	243.40	226.28	216.13	209.05	203.87	196.58	191.62	187.92	185.00	182.71	179.28
2, Good	225.35	209.56	200.10	193.67	188.82	182.17	177.49	174.07	171.41	169.29	166.01
2 & 3	210.30	195.49	186.75	180.68	176.11	169.91	165.57	162.33	159.83	157.89	154.79
3, Average	197.64	183.86	175.49	169.77	165.57	159.75	155.58	152.62	150.28	148.37	145.44
3 & 4	179.29	166.62	159.10	154.00	150.21	144.86	141.13	138.43	136.24	134.63	132.01
4, Low	160.07	148.79	142.09	137.52	134.08	129.32	125.97	123.62	121.72	120.13	117.82
Wall Height Adjustment*	2.98	2.39	2.07	1.93	1.76	1.47	1.30	1.17	1.13	1.08	.94

***Wall Height Adjustment:** Add or subtract the amount listed in this row to or from the square foot of floor cost for each foot of wall height more or less than 20 feet.

Perimeter Wall Adjustment: A common wall exists when two buildings share one wall. Adjust for common walls by deducting the linear foot costs below from the total structure cost. In some structures one or more walls are not owned at all. In this case, deduct the "No Ownership" cost per linear foot of wall not owned. For common wall, deduct $524 per linear foot. For no wall ownership, deduct $1,080 per linear foot.

Supermarkets – Wood and Steel Frame

Quality Classification

	Class 1 Best Quality	Class 2 Good Quality	Class 3 Average Quality	Class 4 Low Quality
Foundation (15% of total cost)	Reinforced concrete.	Reinforced concrete.	Reinforced concrete.	Reinforced concrete.
Floor Structure (12% of total cost)	4" reinforced concrete on 6" rock fill.	4" reinforced concrete on 6" rock fill.	4" reinforced concrete on 6" rock fill.	4" reinforced concrete on 6" rock fill.
Wall Structure (10% of total cost)	2" x 6" - 16" o.c.	2" x 6" - 16" o.c.	2" x 4" - 16" o.c.	2" x 4" - 16" o.c.
Roof Structure (10% of total cost)	Glu-lams or steel "I" beams on steel intermediate columns, 2" x 12" purlins 16" o.c., 1/2" plywood sheathing.	Glu-lams or steel "I" beams on steel intermediate columns, 2" x 12" purlins 16" o.c., 1/2" plywood sheathing.	Glu-lams or steel "I" beams on steel intermediate columns, 2" x 12" purlins 16" o.c., 1/2" plywood sheathing.	Glu-lams or steel "I" beams on steel intermediate columns, 3" x 12" purlins 3' o.c., 1/2" plywood sheathing.
Floor Finish (5% of total cost)	Terrazzo in sales area. Sheet vinyl or carpet in cashiers' area.	Resilient tile in sales area. Terrazzo, solid vinyl tile or carpet in cashiers' area.	Composition tile in sales area.	Minimum tile or inexpensive composition tile in sales area.
Interior Wall Finish (7% of total cost)	Gypsum wallboard and vinyl wall cover or interior stucco on inside of exterior walls and on partitions.	Gypsum wallboard, texture and paint or vinyl wall cover on inside of exterior walls, and on partitions.	Gypsum wallboard, texture and paint on inside of exterior walls, and on partitions.	Gypsum wallboard and paint on inside of exterior walls, and on partitions.
Ceiling Finish (5% of total cost)	Suspended acoustical tile, dropped ceiling over meat and produce departments.	Suspended acoustical tile or gypsum board and acoustical texture, dropped ceiling over meat and produce departments.	Ceiling tile on roof purlins, dropped ceiling over meat department.	Open.
Front (10% of total cost)	A large amount of float glass in good aluminum frames (18'-22' high for 3/4 of width), brick or stone veneer on remainder, 1 pair good automatic doors per 7,000 S.F. of floor area, anodized aluminum sunshade over glass area, 8' canopy across front, 10'-12' raised walk across front.	A large amount of float glass in good aluminum frames (16'-18' high for 2/3 of width), brick or stone veneer on remainder, 1 pair good automatic doors per 10,000 S.F. of floor area, 9' canopy across front, 10' raised walk across front.	Moderate amount of float glass in average quality aluminum frames (12'-16' high for 2/3 of width), wood siding on remainder, 1 pair average automatic doors per 10,000 S.F. of floor area, 6' canopy across front, 8' raised walk across front.	Stucco with small amount of float glass in an inexpensive aluminum frame (6'-10' high for 1/2 of width), 6' canopy across front, 6' ground level walk across front.
Exterior Finish (8% of total cost)	Large ornamental rock or brick veneer.	Good wood siding, some masonry veneer.	Stucco or wood siding.	Stucco.
Roof Cover (5% of total cost)	5 ply built-up roofing with large rock.	5 ply built-up roofing with tar and rock.	4 ply built-up roofing.	4 ply built-up roofing.
Plumbing (8% of total cost)	2 rest rooms with 3 fixtures each, floor piping and drains to refrigerated cases, 2 double sinks with drain board.	2 rest rooms with 3 fixtures each, floor piping and drains to refrigerated cases, 2 double sinks with drain board.	2 rest rooms with 2 fixtures each, floor piping and drains to refrigerated cases, 2 double sinks.	1 rest room with 2 fixtures, floor piping and drains to refrigerated cases.
Electrical (5% of total cost)	Conduit wiring, LED fixtures, 8' o.c., 30 to 40 spotlights.	Conduit wiring, LED fixtures with diffusers, 8' o.c., 30 to 40 spotlights.	Conduit wiring, 3 tube fluorescent fixtures, 8' o.c., 5 or 10 spotlights.	Conduit wiring, double tube fluorescent fixtures, 8' o.c.

Note: Use the percent of total cost to help identify the correct quality classification.

Square foot costs include the following components: Foundations as required for normal soil conditions. Floor, wall, and roof structures. Interior floor, wall and ceiling finishes. Exterior wall finish and roof cover. Display fronts. Interior partitions. Entry and delivery doors. A canopy and walk across the front of the building as described in the applicable building specifications. Basic lighting and electrical systems. Rough and finish plumbing. All plumbing, piping and wiring necessary to operate the usual refrigerated cases and vegetable cases. Design and engineering fees. Permits and hook-up fees. Contractor's mark-up.

Supermarkets – Wood and Steel Frame

Estimating Procedure

1. Establish the structure quality class by using the information on page 105.
2. Compute the building floor area. This should include everything within the building exterior walls and all insets outside the main walls but under the main building roof.
3. Add to or subtract from the square foot cost below the appropriate amount from the Wall Height Adjustment Row (at the bottom of this page) if the wall height is more or less than 20 feet.
4. Multiply the adjusted square foot cost by the building area.
5. Deduct, if appropriate, for common walls, using the figures at the bottom of this page.
6. Multiply the total cost by the location factor listed on page 7 or 8.
7. Add the cost of heating and cooling equipment, fire sprinklers, exterior signs, yard improvements, loading docks, ramps and walk-in boxes if they are an integral part of the building. See pages 236 to 248.

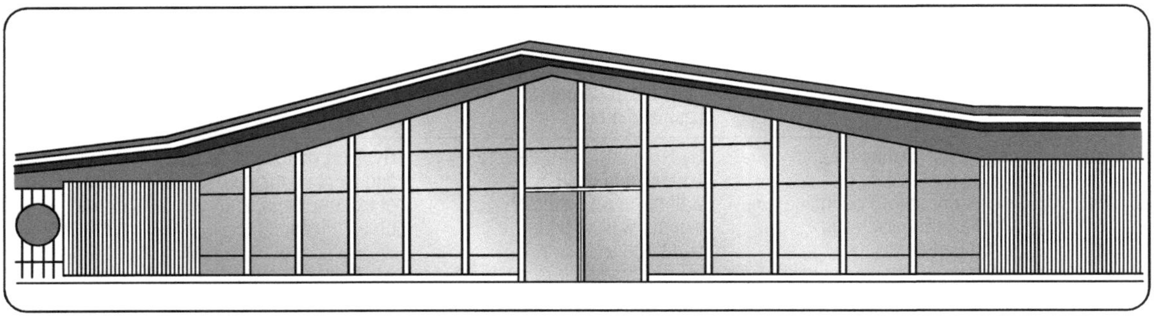

Supermarket, Class 3

Square Foot Area

Quality Class	5,000	7,500	10,000	12,500	15,000	20,000	25,000	30,000	35,000	40,000	50,000
Exceptional	185.51	174.81	168.37	163.95	160.69	156.12	152.98	150.63	148.89	147.45	145.14
1, Best	176.97	166.76	160.60	156.45	153.35	148.91	145.96	143.76	142.01	140.63	138.50
1 & 2	162.60	153.28	147.54	143.76	140.85	136.83	134.11	132.10	130.48	129.17	127.24
2, Good	149.22	140.48	135.37	131.78	129.28	125.53	123.03	121.10	119.71	118.52	116.71
2 & 3	139.37	131.35	126.53	123.24	120.83	117.34	115.01	113.25	111.85	110.79	109.14
3, Average	131.25	123.70	119.12	115.96	113.67	110.46	108.18	106.56	105.31	104.21	102.66
3 & 4	120.22	113.29	109.10	106.25	104.13	101.18	99.13	97.65	96.46	95.54	94.07
4, Low	108.30	102.06	98.29	95.66	93.78	91.14	89.28	87.90	86.88	86.01	84.72
Wall Height Adjustment*	1.15	.98	.87	.79	.72	.66	.65	.64	.60	.59	.52

***Wall Height Adjustment:** Add or subtract the amount listed in this row to or from the square foot of floor cost for each foot of wall height more or less than 20 feet.

Perimeter Wall Adjustment: A common wall exists when two buildings share one wall. Adjust for common walls by deducting the linear foot costs below from the total structure cost. In some structures one or more walls are not owned at all. In this case, deduct the "No Ownership" cost per linear foot of wall not owned. For common wall, deduct $263 per linear foot. For no wall ownership, deduct $531 per linear foot.

Small Food Stores – Masonry Construction

Quality Classification

	Class 1 Best Quality	Class 2 Good Quality	Class 3 Average Quality	Class 4 Low Quality
Foundation (15% of total cost)	Reinforced concrete.	Reinforced concrete.	Reinforced concrete.	Reinforced concrete.
Floor Structure (12% of total cost)	4" reinforced concrete on 6" rock fill.	4" reinforced concrete on 6" rock fill.	4" reinforced concrete on 6" rock fill.	4" reinforced concrete on 6" rock fill.
Wall Structure (10% of total cost)	6" concrete tilt-up or ornamental block or brick.	6" concrete tilt-up, colored concrete block or brick.	6" concrete tilt-up or 8" concrete block.	6" concrete tilt-up or 8" concrete block.
Roof Structure (10% of total cost)	Glu-lams or steel "I" beams on steel interme-diate columns, 2" x 12" purlins 16" o.c., 1/2" plywood sheathing.	Glu-lams or steel "I" beams, 2" x 12" purlins 16" o.c., 1/2" plywood sheathing.	Glu-lams, 3" x 12" purlins 3' o.c., 1/2" plywood sheathing.	Glu-lams, 3" x 12" purlins 3' o.c., 1/2" plywood sheathing.
Floor Finish (5% of total cost)	Resilient tile in sales area.	Composition tile in sales area.	Minimum grade tile in sales area.	Concrete.
Interior Wall Finish (7% of total cost)	Paint on inside of exterior walls, gypsum wallboard, texture and paint or vinyl wall cover on partitions.	Paint on inside of exterior walls, gypsum wallboard, texture and paint on partitions.	Paint on inside of exterior walls, gypsum wallboard and paint on partitions.	Paint on inside of exterior walls, gypsum wallboard and paint on partitions.
Ceiling Finish (5% of total cost)	Suspended acoustical tile or gypsum board and acoustical texture.	Ceiling tile on roof purlins.	Open.	Open.
Front (10% of total cost)	A large amount of float glass in good aluminum frames (10'-12' high for 2/3 of width), brick or stone veneer on remainder, 1 pair good aluminum and glass doors per 3,000 S.F. of floor area, 8' canopy across front, 10' raised walk across front.	A moderate amount of float glass in average aluminum frames (8' to 10' high for 2/3 of width), exposed aggregate on remainder, 1 pair average aluminum and glass doors per 3,000 S.F. of floor area, 6' canopy across front, 8' raised walk across front.	Painted concrete block with a small amount of float glass in an inex-pensive aluminum frame (6' to 8' high for 1/2 of width), 6' canopy across front, 6' ground level walk across front.	Painted concrete block with small amount of crystal glass in wood frames, wood and glass doors, small canopy over entrance, 6' ground level walk at entrances.
Exterior Finish (8% of total cost)	Colored block.	Paint.	Paint.	Paint.
Roof Cover (5% of total cost)	5 ply built-up roofing with tar and rock.	4 ply built-up roofing.	4 ply built-up roofing.	4 ply built-up roofing.
Plumbing (8% of total cost)	2 rest rooms with 3 fixtures each, floor piping and drains to refrigerated cases.	1 rest room with 3 fixtures, floor piping and drains to refrigerated cases.	1 rest room with 2 fixtures, floor piping and drains to refrigerated cases.	1 rest room with 2 fixtures, floor piping and drains to refrigerated cases.
Electrical (5% of total cost)	Conduit wiring, LED fixtures with diffusers, 8' o.c., 5 spotlights.	Conduit wiring, LED fixtures, 8' o.c.	Conduit wiring, double tube fluorescent fixtures, 8' o.c.	Conduit wiring, incan-descent fixtures, 10' o.c. or single tube fluorescent fixtures, 8' o.c.

Note: Use the percent of total cost to help identify the correct quality classification.

Square foot costs include the following components: Foundations as required for normal soil conditions. Floor, wall, and roof structures. Interior floor, wall and ceiling finishes. Exterior wall finish and roof cover. Display fronts. Interior partitions. Entry and delivery doors. A canopy and walk across the front of the building as described in the applicable building specifications. Basic lighting and electrical systems. Rough and finish plumbing. All plumbing, piping and wiring necessary to operate the usual refrigerated cases and vegetable cases. Design and engineering fees. Permits and hook-up fees. Contractor's mark-up.

Small Food Stores – Masonry Construction

Estimating Procedure

1. Establish the structure quality class by using the information on page 107.
2. Compute the building floor area. This should include everything within the building exterior walls and all insets outside the main walls but under the main building roof.
3. Add to or subtract from the square foot cost below the appropriate amount from the Wall Height Adjustment Row (at the bottom of this page) if the wall height is more or less than 12 feet.
4. Multiply the adjusted square foot cost by the building area.
5. Deduct, if appropriate, for common walls, using the figures at the bottom of this page.
6. Multiply the total cost by the location factor listed on page 7 or 8.
7. Add the cost of heating and cooling equipment, fire sprinklers, exterior signs, yard improvements, loading docks, ramps and walk-in boxes if they are an integral part of the building. See pages 236 to 248.

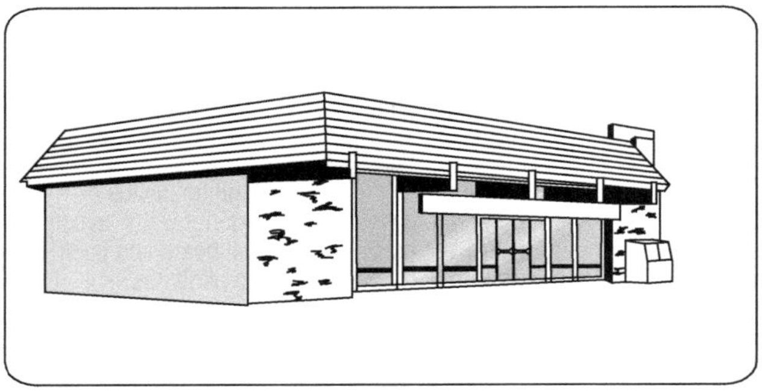

Small Food Store, Class 1 & 2

Small Food Store, Class 2

Square Foot Area

Quality Class	500	1,000	1,500	2,000	2,500	3,000	3,500	4,000	4,500	5,000	6,000
1, Best	406.03	307.65	269.00	247.30	233.13	223.03	215.51	209.36	204.44	200.29	193.79
1 & 2	373.96	283.97	248.62	228.80	215.84	206.56	199.55	194.09	189.46	185.71	179.70
2, Good	344.75	262.54	230.22	212.01	200.13	191.63	185.20	180.07	175.82	172.41	166.91
2 & 3	325.70	249.77	218.89	201.32	189.70	181.25	174.76	169.66	165.46	161.92	156.25
3, Average	307.70	235.43	206.34	189.85	178.95	171.14	165.20	160.55	156.58	153.31	148.11
3 & 4	254.81	214.68	192.69	178.45	168.25	160.48	154.08	149.14	144.77	141.09	134.95
4, Low	237.38	198.60	178.01	164.59	155.07	147.83	141.99	137.31	133.23	129.86	124.25
Wall Height Adjustment*	9.33	6.58	5.36	4.65	4.07	3.86	3.53	3.26	3.08	2.98	2.80

***Wall Height Adjustment:** Add or subtract the amount listed in this row to or from the square foot of floor cost for each foot of wall height more or less than 12 feet.

Perimeter Wall Adjustment: For common wall, deduct $270 per linear foot. For no wall ownership, deduct $545 per linear foot.

Small Food Stores – Wood Frame Construction

Quality Classification

	Class 1 Best Quality	Class 2 Good Quality	Class 3 Average Quality	Class 4 Low Quality
Foundation (15% of total cost)	Reinforced concrete.	Reinforced concrete.	Reinforced concrete.	Reinforced concrete.
Floor Structure (12% of total cost)	4" reinforced concrete on 6" rock fill.	4" reinforced concrete on 6" rock fill.	4" reinforced concrete on 6" rock fill.	4" reinforced concrete on 6" rock fill.
Wall Structure (10% of total cost)	2" x 6" - 16" o.c.	2" x 6" - 16" o.c.	2" x 4" - 16" o.c.	2" x 4" - 16" o.c.
Roof Structure (10% of total cost)	Glu-lams or steel "I" beams on steel intermediate columns, 2" x 12" purlins 16" o.c., 1/2" plywood sheathing.	Glu-lams or steel "I" beams on steel intermediate columns, 2" x 12" purlins 16" o.c., 1/2" plywood sheathing.	Glu-lams, or steel "I" beams, 3" x 12" purlins 3' o.c., 1/2" plywood sheathing.	Glu-lams, 3" x 12" purlins 3' o.c. 1/2" plywood sheathing.
Floor Finish (5% of total cost)	Resilient tile in sales area.	Resilient tile in sales area.	Inexpensive composition tile in sales area.	Concrete.
Interior Wall Finish (7% of total cost)	Gypsum wallboard, texture and paint or vinyl wall cover on inside of exterior walls, and on partitions.	Gypsum wallboard, texture and paint on inside of exterior walls, and on partitions.	Gypsum wallboard and paint on inside of exterior walls, and on partitions.	Gypsum wallboard and paint on inside of exterior walls and on partitions.
Ceiling Finish (5% of total cost)	Suspended acoustical tile or gypsum board and acoustical texture.	Ceiling tile on roof purlins.	Open.	Open.
Front (10% of total cost)	A large amount of float glass in good aluminum frames (10'-12' high for 2/3 of width), brick or stone veneer on remainder, 1 pair good aluminum and glass doors per 3,000 S.F. of floor area, 8' canopy across front, 10' raised walk across front.	A moderate amount of float glass in average quality aluminum frames (8' to 10' high for 2/3 of width), ornamental concrete block on remainder, 1 pair average aluminum and glass doors per 3,000 S.F. of floor area, 6' canopy across front, 8' raised walk across front.	Painted stucco with a small amount of float glass in an inexpensive aluminum frame (6' to 8' high for 1/2 of width) 6' canopy across front. 6' ground level walk across front.	Stucco with small amount of crystal glass in wood frames, wood and glass doors, small canopy over entrance, 6' ground level walk at entrances.
Exterior Finish (8% of total cost)	Stucco and paint or wood siding.	Stucco and paint.	Stucco.	Stucco.
Roof Cover (5% of total cost)	5 ply built-up roofing with tar and rock.	4 ply built-up roofing.	4 ply built-up roofing.	4 ply built-up roofing.
Plumbing (8% of total cost)	2 rest rooms with 3 fixtures each, floor piping and drains to refrigerated cases.	1 rest room with 3 fixtures, floor piping and drains to refrigerated cases.	1 rest room with 2 fixtures, floor piping and drains to refrigerated cases.	1 rest room with 2 fixtures, floor piping and drains to refrigerated cases.
Electrical (5% of total cost)	Conduit wiring, LED fixtures with diffusers, 8' o.c., 5 spotlights.	Conduit wiring, LED fixtures, 8' o.c.	Conduit wiring, double tube fluorescent fixtures, 8' o.c.	Conduit wiring, incandescent fixtures, 10' o.c. or single tube fluorescent fixtures, 8' o.c.

Note: Use the percent of total cost to help identify the correct quality classification.

Square foot costs include the following components: Foundations as required for normal soil conditions. Floor, wall, and roof structures. Interior floor, wall and ceiling finishes. Exterior wall finish and roof cover. Display fronts. Interior partitions. Entry and delivery doors. A canopy and walk across the front of the building as described in the applicable building specifications. Basic lighting and electrical systems. Rough and finish plumbing. All plumbing, piping and wiring necessary to operate the usual refrigerated cases and vegetable cases. Design and engineering fees. Permits and hook-up fees. Contractor's mark-up.

Small Food Stores – Wood Frame Construction

Estimating Procedure

1. Establish the structure quality class by using the information on page 109.
2. Compute the building floor area. This should include everything within the building exterior walls and all insets outside the main walls but under the main building roof.
3. Add to or subtract from the square foot cost below the appropriate amount from the Wall Height Adjustment Row (at the bottom of this page) if the wall height is more or less than 12 feet.
4. Multiply the adjusted square foot cost by the building area.
5. Deduct, if appropriate, for common walls, using the figures at the bottom of this page.
6. Multiply the total cost by the location factor listed on page 7 or 8.
7. Add the cost of heating and cooling equipment, fire sprinklers, exterior signs, yard improvements, loading docks, ramps and walk-in boxes if they are an integral part of the building. See pages 236 to 248.

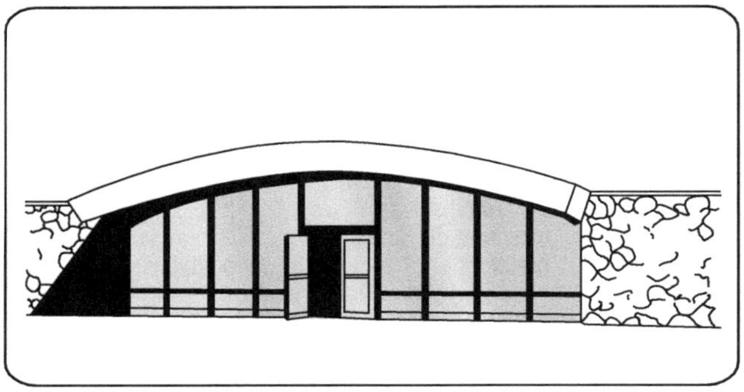

Small Food Store, Class 1

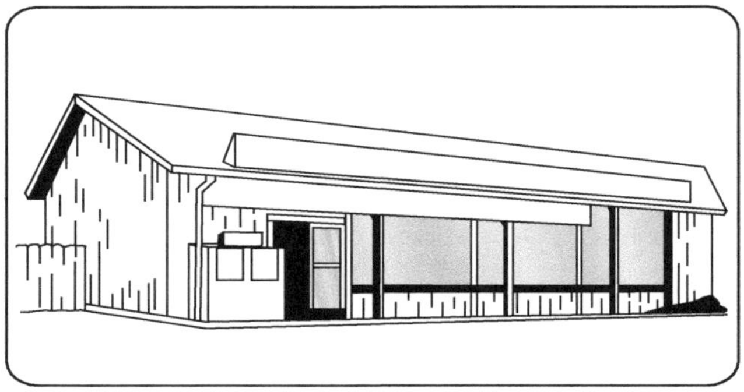

Small Food Store, Class 3

Square Foot Area

Quality Class	500	1,000	1,500	2,000	2,500	3,000	3,500	4,000	4,500	5,000	6,000
1, Best	263.66	200.91	177.23	164.25	155.81	149.94	145.54	142.07	139.25	136.94	133.26
1 & 2	243.71	186.15	164.33	152.40	144.79	139.32	135.28	132.12	129.49	127.37	124.01
2, Good	226.78	173.63	153.55	142.50	135.37	130.38	126.57	123.68	121.22	119.22	116.10
2 & 3	206.58	158.56	140.44	130.51	124.14	119.64	116.21	113.49	111.31	109.53	106.81
3, Average	188.82	145.49	129.04	120.10	114.30	110.22	107.13	104.79	102.88	101.20	98.65
3 & 4	171.15	132.14	117.30	109.26	104.02	100.39	97.62	95.43	93.73	92.23	90.00
4, Low	153.90	117.88	104.64	97.63	93.14	90.02	87.73	85.96	84.50	83.27	81.41
Wall Height Adjustment*	4.11	2.99	2.42	2.13	1.91	1.69	1.64	1.50	1.42	1.36	1.26

***Wall Height Adjustment:** Add or subtract the amount listed in this row to or from the square foot of floor cost for each foot of wall height more or less than 12 feet.

Perimeter Wall Adjustment: For common wall, deduct $135 per linear foot. For no wall ownership, deduct $263 per linear foot.

Discount Houses – Masonry or Concrete

Quality Classification

	Class 1 Best Quality	Class 2 Good Quality	Class 3 Average Quality	Class 4 Low Quality
Foundation (15% of total cost)	Reinforced concrete.	Reinforced concrete.	Reinforced concrete.	Reinforced concrete.
Floor Structure (12% of total cost)	4" reinforced concrete on 6" rock fill.	4" reinforced concrete on 6" rock fill.	4" reinforced concrete on 6" rock fill.	4" reinforced concrete on 6" rock fill.
Wall Structure (10% of total cost)	6" concrete tilt-up or ornamental block or brick.	6" concrete tilt-up, colored concrete block or brick.	6" concrete tilt-up or 8" concrete block.	6" concrete tilt-up or 8" concrete block.
Roof Structure (10% of total cost)	Glu-lams or steel "I" beams on steel intermediate columns, plywood sheathing.	Glu-lams or steel "I" beams on steel intermediate columns, plywood sheathing.	Glu-lams, or steel "I" beams on steel intermediate columns, plywood sheathing.	Glu-lams, or steel "I" beams on steel intermediate columns, plywood sheathing.
Floor Finish (5% of total cost)	Terrazzo, sheet vinyl or good carpet in sales area.	Resilient tile in sales area with some terrazzo, solid vinyl tile, or carpet.	Composition tile in sales area.	Inexpensive composition tile in sales area.
Interior Wall Finish (7% of total cost)	Inside of exterior walls furred out with gypsum wallboard and paint or interior stucco or gypsum wallboard and vinyl wall cover on partitions.	Paint on inside of exterior walls, gypsum wallboard and paint or vinyl wall cover on partitions.	Paint on inside of exterior walls, gypsum wallboard and paint on partitions.	Paint on inside of exterior walls, gypsum wallboard and paint on partitions.
Ceiling Finish (5% of total cost)	Suspended acoustical tile, dropped ceilings in some areas.	Suspended acoustical tile or gypsum board and acoustical texture, dropped ceilings in some areas.	Ceiling tile on roof purlins, dropped ceiling over some areas.	Open.
Front (10% of total cost)	A large amount of float glass in good aluminum frames (18'-22' high for 3/4 of width), brick or stone veneer on remainder, 1 pair good automatic doors per 7,000 S.F. of floor area, anodized aluminum sunshade over glass area, 8' canopy across front, 10'-12' raised walk across front.	A large amount of float glass in good aluminum frames (16'-18' high for 2/3 of width), brick or stone veneer on remainder, 1 pair good automatic doors per 10,000 S.F. of floor area, 8' canopy across front, 10' raised walk across front.	A moderate amount of float glass in average quality aluminum frames (12'-16' high extending 20' on each side of entrances), exposed aggregate on remainder, 1 pair average automatic doors per 20,000 S.F. of floor area, 6' canopy over glass area.	Stucco or exposed aggregate with a small amount of float glass in inexpensive aluminum frames (6'-10' high for 1/2 of width), 6' canopy at entrances, 6' ground level walk across front.
Exterior Finish (8% of total cost)	Large ornamental rock or brick veneer on exposed walls.	Exposed aggregate on exposed walls.	Paint, some exposed aggregate.	Paint.
Roof Cover (5% of total cost)	5 ply built-up roofing with large rock.	5 ply built-up roofing with tar and rock.	4 ply built-up roofing.	4 ply built-up roofing.
Plumbing (8% of total cost)	2 rest rooms with 8 fixtures each, floor piping and drains to refrigerated cases.	2 rest rooms with 6 fixtures each, floor piping and drains to refrigerated cases.	2 rest rooms with 4 fixtures each, floor piping and drains to refrigerated cases.	2 rest rooms with 2 fixtures each, floor piping and drains to refrigerated cases.
Electrical (5% of total cost)	Conduit wiring, LED fixtures 8' o.c., 60-80 spotlights.	Conduit wiring, LED fixtures with diffusers 8' o.c., 60-80 spotlights.	Conduit wiring, 3 tube fluorescent fixtures 8' o.c., 20 to 40 spotlights.	Conduit wiring, double tube fluorescent fixtures, 8' o.c.

Note: Use the percent of total cost to help identify the correct quality classification.

Square foot costs include the following components: Foundations as required for normal soil conditions. Floor, wall, and roof structures. Interior floor, wall and ceiling finishes. Exterior wall finish and roof cover. Display fronts. Interior partitions. Entry and delivery doors. A canopy and walk across the front of the building as described in the applicable building specifications. Basic lighting and electrical systems. Rough and finish plumbing. Design and engineering fees. Permits and hook-up fees. Contractor's mark-up.

Discount Houses – Masonry or Concrete

Estimating Procedure

1. Establish the structure quality class by using the information on page 111.
2. Compute the building floor area. This should include everything within the building exterior walls and all insets outside the main walls but under the main building roof.
3. Add to or subtract from the square foot cost below the appropriate amount from the Wall Height Adjustment Row (at the bottom of this page) if the wall height is more or less than 20 feet.
4. Multiply the adjusted square foot cost by the building area.
5. Deduct, if appropriate, for common walls, using the figures at the bottom of this page.
6. Multiply the total cost by the location factor listed on page 7 or 8.
7. Add the cost of heating and cooling equipment, fire sprinklers, exterior signs, yard improvements, loading docks, ramps and walk-in boxes if they are an integral part of the building. See pages 236 to 248.

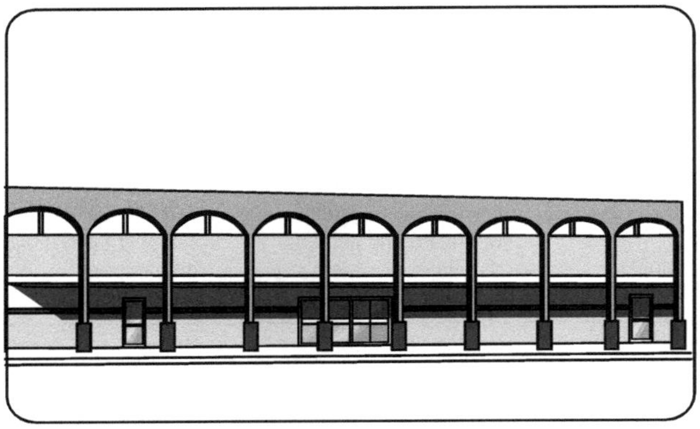

Discount House, Class 3

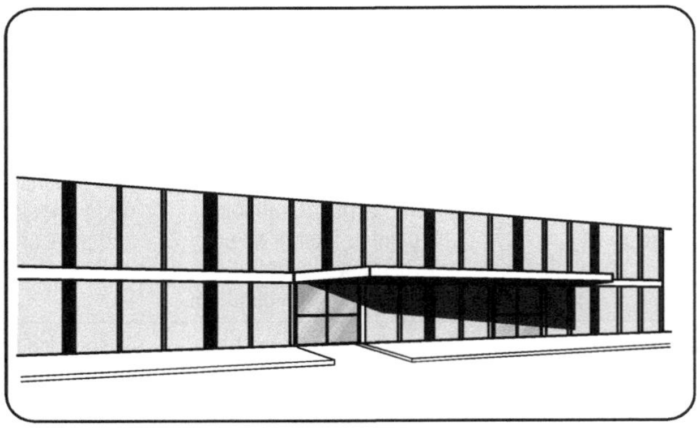

Discount House, Class 4

Square Foot Area

Quality Class	15,000	20,000	25,000	30,000	35,000	40,000	50,000	75,000	100,000	150,000	200,000
1, Best	169.07	163.29	159.35	156.45	154.08	152.23	149.42	144.94	142.36	139.06	137.07
1 & 2	158.87	153.57	149.81	147.03	144.86	143.13	140.49	136.17	133.70	130.76	128.94
2, Good	151.87	146.66	143.10	140.50	138.46	136.74	134.22	130.22	127.79	124.88	123.29
2 & 3	141.20	136.32	133.14	130.68	128.78	127.20	124.82	121.06	118.82	116.21	114.54
3, Average	132.89	128.45	125.32	122.96	121.21	119.73	117.50	113.97	111.89	109.37	107.76
3 & 4	121.48	117.33	114.54	112.40	110.75	109.39	107.29	104.13	102.23	99.96	98.54
4, Low	109.78	106.04	103.48	101.56	100.10	98.91	97.06	94.11	92.34	90.35	89.02
Wall Height Adjustment*	1.74	1.57	1.29	1.14	1.08	1.06	.98	.81	.78	.69	.68

***Wall Height Adjustment:** Add or subtract the amount listed in this row to or from the square foot of floor cost for each foot of wall height more or less than 20 feet.

Perimeter Wall Adjustment: A common wall exists when two buildings share one wall. Adjust for common walls by deducting the linear foot costs below from the total structure cost. In some structures one or more walls are not owned at all. In this case, deduct the "No Ownership" cost per linear foot of wall not owned. For common wall, deduct $431 per linear foot. For no wall ownership, deduct $872 per linear foot.

Discount Houses – Wood or Wood and Steel Frame

Quality Classification

	Class 1 Best Quality	Class 2 Good Quality	Class 3 Average Quality	Class 4 Low Quality
Foundation (15% of total cost)	Reinforced concrete.	Reinforced concrete.	Reinforced concrete.	Reinforced concrete.
Floor Structure (12% of total cost)	4" reinforced concrete on 6" rock fill.	4" reinforced concrete on 6" rock fill.	4" reinforced concrete on 6" rock fill.	4" reinforced concrete on 6" rock fill.
Wall Structure (10% of total cost)	2" x 6" - 16" o.c.	2" x 6" - 16" o.c.	2" x 6" - 16" o.c.	2" x 4" - 16" o.c.
Roof Structure (10% of total cost)	Glu-lams or steel "I" beams on steel intermediate columns, plywood sheathing.	Glu-lams or steel "I" beams on steel intermediate columns, plywood sheathing.	Glu-lams, or steel "I" beams on steel intermediate columns, plywood sheathing.	Glu-lams, or steel "I" beams on steel intermediate columns, plywood sheathing.
Floor Finish (5% of total cost)	Terrazzo, sheet vinyl or good carpet in sales area.	Resilient tile in sales area with some terrazzo, solid vinyl tile or carpet.	Resilient tile in sales area.	Inexpensive composition tile in sales area.
Interior Wall Finish (5% of total cost)	Gypsum wallboard and vinyl wall cover or interior stucco on inside of walls and on partitions.	Gypsum wallboard, texture and paint or vinyl wall cover on inside of exterior walls and on partitions.	Gypsum wallboard, texture and paint on inside of exterior walls and on partitions.	Gypsum wallboard and paint on inside of exterior walls and on partitions.
Ceiling Finish (5% of total cost)	Suspended acoustical tile, dropped ceilings over some areas.	Suspended acoustical tile or gypsum board and acoustical texture, dropped ceilings in some areas.	Ceiling tile on roof purlins, dropped ceiling over some areas.	Open.
Front (10% of total cost)	A large amount of float glass in good aluminum frames (18'-22' high for 3/4 of width), brick or stone veneer on remainder, 1 pair good automatic doors 1 per 7,000 S.F. of floor area, anodized aluminum sunshade over glass area, 8' canopy across front, 10'-12' raised walk across front.	A large amount of float glass in good aluminum frames (16'-18' high for 2/3 of width), brick or stone veneer on remainder, 1 pair good automatic doors per 10,000 S.F. of floor area, 8' canopy across front, 10' raised walk across front.	A moderate amount of float glass in average quality aluminum frames (12'-16' high extending 20' on each side of entrances), exposed aggregate on remainder, 1 pair average automatic doors per 20,000 S.F. of floor area, 6' canopy over glass areas.	Stucco or exposed aggregate with a small amount of float glass in inexpensive aluminum frames (6'-10' high for 1/2 of width), 6' canopy at entrances, 6' ground level walk across front.
Exterior Finish (7% of total cost)	Good wood siding or masonry veneer.	Wood siding, some masonry veneer.	Stucco, some masonry trim.	Stucco.
Roof Cover (5% of total cost)	5 ply built-up roofing with large rock.	5 ply built-up roofing with tar and rock.	4 ply built-up roofing.	4 ply built-up roofing.
Plumbing (10% of total cost)	2 rest rooms with 8 fixtures each, floor piping and drains to refrigerated cases.	2 rest rooms with 6 fixtures each, floor piping and drains to refrigerated cases.	2 rest rooms with 4 fixtures each, floor piping and drains to refrigerated cases.	2 rest rooms with 2 fixtures each, floor piping and drains to refrigerated cases.
Electrical (6% of total cost)	Conduit wiring, recessed LED fixtures 8' o.c., 60-80 spotlights.	Conduit wiring, LED fixtures with diffusers 8' o.c., 60-80 spotlights.	Conduit wiring, 3 tube fluorescent fixtures 8' o.c., 20 to 40 spotlights.	Conduit wiring, double tube fluorescent fixtures, 8' o.c.

Note: Use the percent of total cost to help identify the correct quality classification.

Square foot costs include the following components: Foundations as required for normal soil conditions. Floor, wall, and roof structures. Interior floor, wall and ceiling finishes. Exterior wall finish and roof cover. Display fronts. Interior partitions. Entry and delivery doors. A canopy and walk across the front of the building as described in the applicable building specifications. Basic lighting and electrical systems. Rough and finish plumbing. Design and engineering fees. Permits and hook-up fees. Contractor's mark-up.

Discount Houses – Wood or Wood and Steel Frame

Estimating Procedure

1. Establish the structure quality class by using the information on page 113.
2. Compute the building floor area. This should include everything within the building exterior walls and all insets outside the main walls but under the main building roof.
3. Add to or subtract from the square foot cost below the appropriate amount from the Wall Height Adjustment Row at the bottom of this page) if the wall height is more or less than 20 feet.
4. Multiply the adjusted square foot cost by the building area.
5. Deduct, if appropriate, for common walls, using the figures at the bottom of this page.
6. Multiply the total cost by the location factor listed on page 7 or 8.
7. Add the cost of heating and cooling equipment, fire sprinklers, exterior signs, yard improvements, loading docks, ramps and walk-in boxes if they are an integral part of the building. See pages 236 to 248.

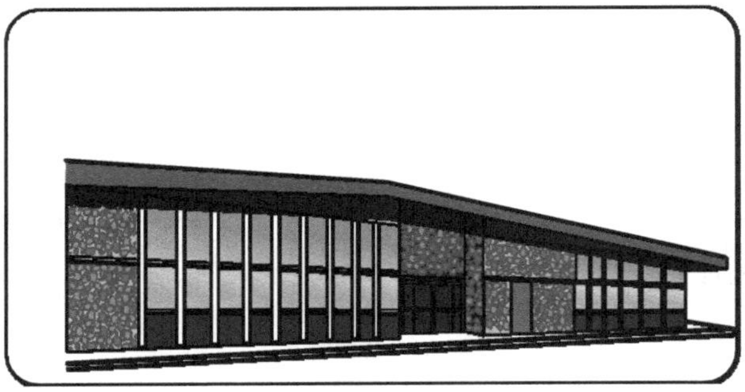

Discount House, Class 2

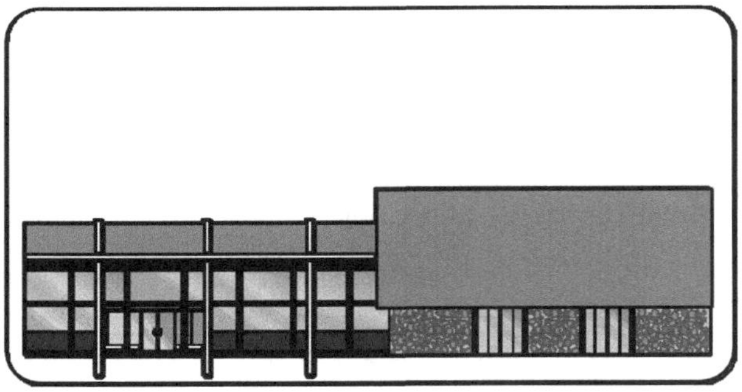

Discount House, Class 2 & 3

Square Foot Area

Quality Class	15,000	20,000	25,000	30,000	35,000	40,000	50,000	75,000	100,000	150,000	200,000
1, Best	121.03	117.98	115.96	114.48	113.28	112.34	110.88	108.60	107.23	105.63	104.63
1 & 2	114.07	111.22	109.31	107.90	106.81	105.84	104.50	102.34	101.12	99.57	98.67
2, Good	108.83	106.09	104.30	102.90	101.86	101.03	99.74	97.69	96.43	95.00	94.12
2 & 3	101.25	98.75	97.06	95.79	94.86	93.99	92.75	90.89	89.73	88.40	87.59
3, Average	95.17	92.86	91.33	90.04	89.10	88.40	87.24	85.43	84.37	83.12	82.36
3 & 4	87.02	84.84	83.39	82.34	81.40	80.78	79.72	78.08	77.12	75.96	75.30
4, Low	78.30	76.38	75.06	74.11	73.32	72.77	71.81	70.33	69.39	68.34	67.69
Wall Height Adjustment*	.71	.68	.64	.62	.59	.58	.56	.44	.41	.29	.27

***Wall Height Adjustment:** Add or subtract the amount listed in this row to or from the square foot of floor cost for each foot of wall height more or less than 20 feet.

Perimeter Wall Adjustment: A common wall exists when two buildings share one wall. Adjust for common walls by deducting the linear foot costs below from the total structure cost. In some structures one or more walls are not owned at all. In this case, deduct the "No Ownership" cost per linear foot of wall not owned. For common wall, deduct $232 per linear foot. For no wall ownership, deduct $455 per linear foot.

Banks and Savings Offices – Masonry or Concrete

Quality Classification

	Class 1 Best Quality	Class 2 Good Quality	Class 3 High Average Quality	Class 4 Low Average Quality	Class 5 Low Quality
Foundation (11% of total cost)	Reinforced concrete.	Reinforced concrete.	Reinforced concrete.	Reinforced concrete.	Reinforced concrete.
Floor Structure (10% of total cost)	Reinforced concrete.	Reinforced concrete.	Reinforced concrete.	Reinforced concrete.	Reinforced concrete.
Wall Structure (10% of total cost)	12" reinforced brick, 8" reinforced concrete or concrete columns with noncombustible filler walls.	12" reinforced brick, 8" reinforced concrete or concrete columns with noncombustible filler walls.	6" concrete tilt-up, 12" reinforced brick, 8" reinforced con. or concrete columns with noncombustible filler walls.	8" decorative concrete block, 6" concrete tilt-up, 8" reinforced brick or concrete columns with noncombustible filler walls.	8" concrete block, 6" reinforced concrete, 6" concrete tilt-up or 8" reinforced brick.
Roof & Cover (7% of total cost)	5 ply built-up roof, some portions copper or slate.	5 ply built-up roof, some portions heavy shake or mission tile.	4 ply built-up roof, some portions heavy shake or mission tile.	4 ply built-up roof, some portions shake or composition shingles.	4 ply built-up roof.
Exterior Wall Finish (9% of total cost)	Expensive veneers.	Ornamental brick or stone veneer.	Brick, or ornamental block veneer.	Ornamental rock imbedded in tilt-up panels or stucco with masonry trim.	Exposed aggregate or stucco.
Interior Wall Finish (7% of total cost)	Ornamental plaster painted with murals or designs. Ornamental moldings, marble and similar expensive finishes.	Ornamental plaster furred out from walls. Ornamental moldings, some hardwood panels or marble finishes.	Hard plaster furred out from walls. Ornamental cove at ceiling. Vinyl wall covering. Some hardwood veneer.	Plaster on lath with putty coat finish. Gypsum wallboard, texture and paint. Some vinyl wall cover.	Plaster on lath with putty coat finish or gypsum wallboard, texture and paint.
Glass (5% of total cost)	Tinted float glass in customized frames, (50% of exterior wall area.)	Tinted float glass in bronze frames, (50% of exterior wall area).	Tinted float glass in heavy anodized aluminum frames or custom wood frames, (50% of exterior wall area).	Moderate amount of float glass in heavy frames, (25% of exterior wall area).	Small to moderate amount of float glass in average aluminum frames, (5 to 10% of exterior wall area).
Overhang (4% of total cost)	4' closed overhang with copper gutters.	4' closed overhang with copper gutters.	4' closed overhang with painted gutters.	3' closed overhang with painted gutters.	None.
Floor Finish					
Public Area: (7% of total cost)	Terrazzo and marble.	Detailed terrazzo.	Detailed terrazzo.	Plain terrazzo.	Sheet vinyl or vinyl tile.
Officers, Area: (3% of total cost)	Excellent carpet.	Very good carpet.	Very good carpet.	Good carpet.	Average quality carpet.
Work area: (2% of total cost)	Good sheet vinyl or carpet.	Average sheet vinyl, tile or carpet.	Resilient tile.	Resilient tile.	Composition tile.
Ceiling Finish (8% of total cost)	Ornamental plaster with exposed ornamental beams.	Ornamental plaster.	Suspended acoustical tile with gypsum wallboard backing or acoustical plaster.	Suspended acoustical tile with exposed grid system.	Acoustical tile on wood furring.
Lighting (5% of total cost)	Recessed panelized LED lighting ,custom light fixtures or chandeliers, and many spotlights.	Recessed panelized LED lighting. Many spotlights.	Recessed panelized LED lighting. Many spotlights.	Recessed panelized fluorescent lighting. Some spotlights.	Nonrecessed panelized lighting.
Plumbing (12% of total cost)	2 or more rest rooms with special fixtures, good ceramic tile, marble or terrazzo wainscot walls. Hard plaster with putty coat and enamel paint. Marble toilet screens.	2 or more rest rooms with many good fixtures, good ceramic tile, marble or terrazzo wainscot walls. Hard plaster with putty coat and enamel paint. Marble toilet screens.	2 rest rooms with 4 or more fixtures, ceramic tile or terrazzo floors. Ceramic tile or terrazzo wainscot. Hard plaster walls with putty coat and enamel paint. Marble or metal toilet screens.	2 rest rooms with 4 or more fixtures each. Ceramic tile or terrazzo floor. Ceramic tile or terrazzo wainscot plaster walls with putty coat. Enamel paint. Good metal toilet partitions.	2 rest rooms with 4 fixtures each. Ceramic tile or vinyl asbestos tile floors. Plaster walls with enamel paint. Metal toilet screens.

Note: Use the percent of total cost to help identify the correct quality classification.

Square foot costs include the following components: Foundations as required for normal soil conditions. Floor, wall, and roof structures. Interior floor, wall and ceiling finishes. Exterior wall finish and roof cover. All glass and glazing. Interior partitions. Roof overhang as described above. Basic electrical systems and lighting fixtures. Rough and finish plumbing. Typical bank vault. Alarm systems. Night depository. Typical record vault and fire doors. Fire exits. Design and engineering fees. Permits and hook-up fees. Contractor's mark-up.

Banks and Savings Offices – Masonry or Concrete

Length Less Than Twice Width

Estimating Procedure

1. Establish the structure quality class by using the information on page 115.
2. Compute the building floor area. This should include everything within the building exterior walls and all insets outside the main walls but under the main building roof.
3. Add to or subtract from the square foot cost below the appropriate amount from the Wall Height Adjustment Table on page 119 if the wall height is more or less than 16 feet for the first floor or 12 feet for higher floors.
4. Multiply the adjusted square foot cost by the building area.
5. Deduct, if appropriate, for common walls, using the figures on page 119.
6. Multiply the total cost by the location factor listed on page 7 or 8.
7. Add the cost of heating and cooling equipment, fire sprinklers, exterior signs, elevators, and yard improvements from pages 236 to 248. Add the cost of mezzanines, bank fixtures, external windows, safe deposit boxes and vault doors from page 125.

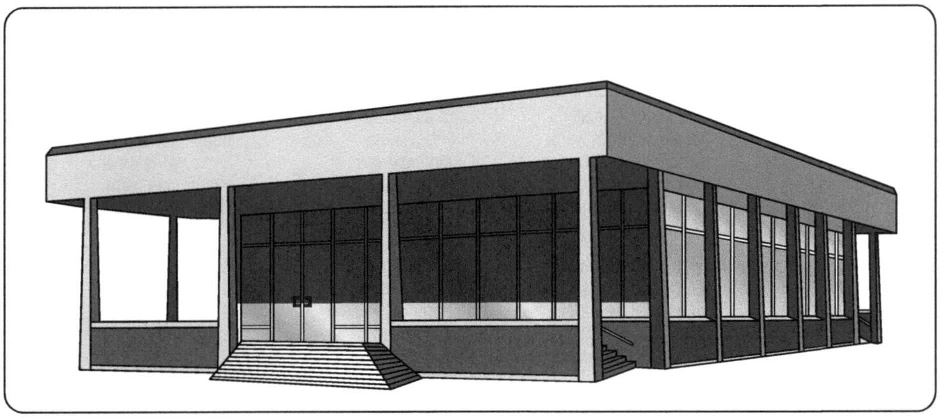

Savings Office, Class 3

First Story – Square Foot Area

Quality Class	2,500	3,000	3,500	4,000	4,500	5,000	6,000	7,500	10,000	15,000	20,000
1, Best	650.91	630.93	613.40	597.96	584.29	572.22	551.59	526.99	497.06	458.56	434.13
1 & 2	612.68	593.79	577.36	562.73	549.97	538.61	519.11	496.00	467.80	431.52	408.56
2, Very Good	582.33	564.45	548.73	534.94	522.80	511.95	493.42	471.50	444.62	410.27	388.41
2 & 3	545.04	528.19	513.61	500.68	489.30	479.19	461.88	441.29	416.13	384.01	363.47
3, Good	517.26	501.38	487.53	475.35	464.42	454.85	438.48	419.10	395.04	364.36	345.00
3 & 4	487.79	472.82	459.74	448.22	437.99	428.97	413.35	395.06	372.53	343.69	325.39
4, Average	464.43	450.16	437.61	426.67	416.98	408.29	393.64	376.00	354.57	327.26	309.73
4 & 5	428.90	415.71	404.19	393.99	385.00	376.96	363.37	347.18	327.48	302.15	285.99
5, Low	390.78	378.72	368.22	358.97	350.88	343.55	331.20	316.40	298.52	275.34	260.66

Second and Higher Stories – Square Foot Area

Quality Class	500	750	1,000	1,500	2,000	3,000	4,000	5,000	7,500	10,000	20,000
1, Best	530.68	475.39	442.70	404.36	381.54	354.62	338.67	327.93	311.05	301.10	282.29
1 & 2	501.80	449.45	418.56	382.16	360.71	335.26	320.25	310.03	294.12	284.62	266.90
2, Very Good	481.78	431.53	401.87	366.99	346.41	322.02	307.62	297.71	282.43	273.37	256.35
2 & 3	453.82	406.51	378.60	345.76	326.31	303.27	289.66	280.42	266.08	257.54	241.39
3, Good	434.44	389.18	362.38	330.94	312.35	290.31	277.31	268.49	254.69	246.48	231.09
3 & 4	409.74	367.05	341.82	312.16	294.49	273.79	261.54	253.16	240.19	232.53	217.89
4, Average	390.67	349.95	325.92	297.63	280.85	261.10	249.35	241.37	228.93	221.70	207.77
4 & 5	368.57	330.26	307.46	280.85	265.08	246.29	235.26	227.70	215.96	209.10	196.07
5, Low	344.17	308.34	287.24	262.21	247.50	230.00	219.65	212.67	201.73	195.28	183.05

Banks and Savings Offices – Masonry or Concrete

Length Between 2 and 4 Times Width

Estimating Procedure

1. Establish the structure quality class by using the information on page 115.
2. Compute the building floor area. This should include everything within the building exterior walls and all insets outside the main walls but under the main building roof.
3. Add to or subtract from the square foot cost below the appropriate amount from the Wall Height Adjustment Table on page 119 if the wall height is more or less than 16 feet for the first floor or 12 feet for higher floors.
4. Multiply the adjusted square foot cost by the building area.
5. Deduct, if appropriate, for common walls, using the figures on page 119.
6. Multiply the total cost by the location factor listed on page 7 or 8.
7. Add the cost of heating and cooling equipment, fire sprinklers, exterior signs, elevators, and yard improvements from pages 236 to 248. Add the cost of mezzanines, bank fixtures, external windows, safe deposit boxes and vault doors from page 125.

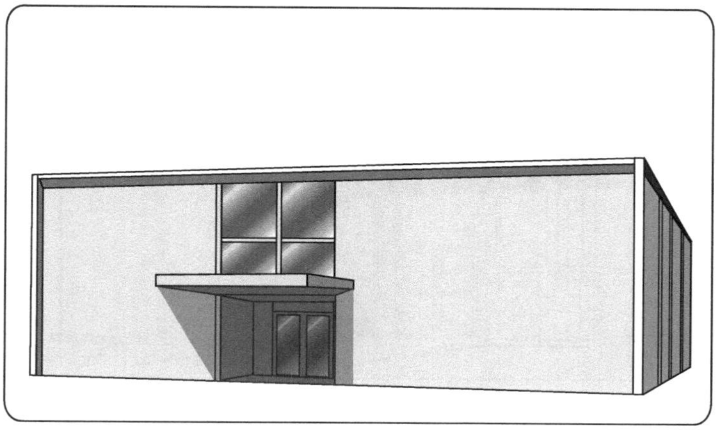

Savings Office, Class 4

First Story – Square Foot Area At 16 Foot Wall Height

Quality Class	2,500	3,000	3,500	4,000	4,500	5,000	6,000	7,500	10,000	15,000	20,000
1, Best	693.57	657.77	630.93	610.06	593.04	579.11	557.16	533.83	508.38	479.76	463.51
1 & 2	654.98	620.64	594.97	574.87	558.70	545.12	524.18	501.94	477.51	449.97	434.34
2, Very Good	625.30	591.87	566.84	547.29	531.48	518.42	498.04	476.32	452.72	426.10	410.98
2 & 3	587.22	554.73	530.50	511.74	496.66	484.22	464.86	444.34	422.36	397.82	384.14
3, Good	558.87	528.24	505.39	487.53	473.14	461.17	442.53	422.55	400.89	376.63	362.85
3 & 4	526.07	497.11	475.61	458.77	445.24	434.02	416.48	397.84	377.79	355.20	342.32
4, Average	499.97	472.33	451.71	435.74	422.80	412.24	395.76	378.20	359.27	338.17	326.31
4 & 5	460.25	434.79	415.85	401.18	389.49	379.75	364.70	348.74	331.58	312.63	302.04
5, Low	417.67	394.66	377.66	364.45	353.86	345.20	331.67	317.49	302.26	285.43	276.07

Second and Higher Stories – Square Foot Area

Quality Class	500	750	1,000	1,500	2,000	3,000	4,000	5,000	7,500	10,000	20,000
1, Best	567.71	507.35	471.20	428.05	402.35	371.73	353.40	340.90	321.42	309.75	287.69
1 & 2	537.80	480.63	446.34	405.60	381.17	352.03	334.68	322.91	304.49	293.41	272.55
2, Very Good	516.20	461.26	428.34	389.24	365.82	337.95	321.34	309.84	292.22	281.62	261.63
2 & 3	485.40	433.76	402.84	366.01	343.98	317.79	302.15	291.39	274.75	264.85	245.96
3, Good	461.32	412.29	382.93	347.91	327.00	302.05	287.25	277.01	261.21	251.73	233.80
3 & 4	436.57	390.06	362.29	329.24	309.44	285.79	271.71	262.16	247.10	238.16	221.21
4, Average	416.13	371.96	345.46	313.84	294.96	272.49	259.11	249.94	235.63	227.12	210.95
4 & 5	392.03	350.35	325.35	295.58	277.80	256.63	244.01	235.32	222.05	213.84	198.68
5, Low	365.82	326.81	303.62	275.89	259.24	239.44	227.70	219.64	207.16	199.59	185.40

Banks and Savings Offices – Masonry or Concrete

Length More Than 4 Times Width

Estimating Procedure

1. Establish the structure quality class by using the information on page 115.
2. Compute the building floor area. This should include everything within the building exterior walls and all insets outside the main walls but under the main building roof.
3. Add to or subtract from the square foot cost below the appropriate amount from the Wall Height Adjustment Table on page 119 if the wall height is more or less than 16 feet for the first floor or 12 feet for higher floors.
4. Multiply the adjusted square foot cost by the building area.
5. Deduct, if appropriate, for common walls, using the figures on page 119.
6. Multiply the total cost by the location factor listed on page 7 or 8.
7. Add the cost of heating and cooling equipment, fire sprinklers, exterior signs, elevators, and yard improvements from pages 236 to 248. Add the cost of mezzanines, bank fixtures, external windows, safe deposit boxes and vault doors from page 125.

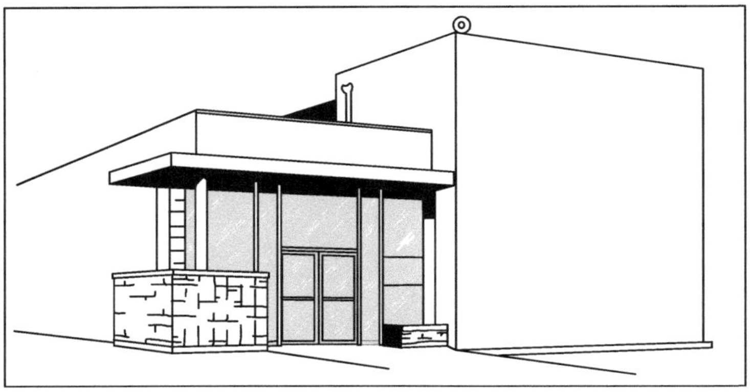

Bank, Class 5

First Story – Square Foot Area

Quality Class	2,500	3,000	3,500	4,000	4,500	5,000	6,000	7,500	10,000	15,000	20,000
1, Best	740.89	697.75	665.79	641.04	621.09	604.61	579.19	552.29	523.33	491.27	473.14
1 & 2	697.66	657.04	627.00	603.73	584.95	569.43	545.39	520.11	492.84	462.58	445.58
2, Very Good	663.07	624.53	595.92	573.67	555.93	541.21	518.36	494.30	468.40	439.66	423.50
2 & 3	621.88	585.69	558.87	538.09	521.32	507.57	486.14	463.58	439.23	412.29	397.18
3, Good	590.75	556.44	530.87	511.16	495.31	482.16	461.88	440.39	417.30	391.60	377.30
3 & 4	555.46	523.10	499.19	480.60	465.62	453.33	434.27	414.08	392.30	368.28	354.70
4, Average	526.69	496.14	473.45	455.78	441.61	429.98	411.78	392.61	372.05	349.17	336.36
4 & 5	484.67	456.48	435.63	419.39	406.31	395.62	378.91	361.26	342.32	321.34	309.58
5, Low	439.12	413.62	394.68	380.05	368.19	358.47	343.39	327.42	310.26	291.19	280.54

Second and Higher Stories – Square Foot Area

Quality Class	500	750	1,000	1,500	2,000	3,000	4,000	5,000	7,500	10,000	20,000
1, Best	628.32	556.04	513.10	462.24	431.90	396.03	374.67	360.12	337.44	323.97	298.47
1 & 2	594.78	526.31	485.73	437.57	408.81	374.92	354.64	340.87	319.43	306.68	282.48
2, Very Good	569.11	503.61	464.80	418.65	391.28	358.77	339.36	326.25	305.63	293.41	270.35
2 & 3	534.44	472.92	436.38	393.15	367.42	336.83	318.71	306.33	287.02	275.49	253.81
3, Good	509.39	450.85	415.95	374.68	350.14	321.12	303.75	291.92	273.50	262.60	241.87
3 & 4	478.97	423.81	391.09	352.40	329.30	301.86	285.65	274.49	257.25	246.93	227.58
4, Average	454.32	402.07	371.01	334.25	312.23	286.36	270.97	260.38	244.07	234.29	215.78
4 & 5	428.05	378.78	349.46	314.81	294.19	269.77	255.14	245.27	229.85	220.69	203.26
5, Low	399.20	353.29	325.95	293.66	274.42	251.65	237.96	228.80	214.35	205.85	189.69

Banks and Savings Offices – Masonry or Concrete

Wall Adjustments

Wall Height Adjustment

Add or subtract the amount listed in this table to the square foot of floor cost for each foot of wall height more or less than 16 feet, if adjusting for a first floor, and 12 feet if adjusting for upper floors.

Square Foot Area

Quality Class	500	750	1,000	1,500	2,000	2,500	3,000	3,500
1, Best	25.66	21.04	18.33	14.94	18.58	11.69	10.62	9.86
2, Very Good	23.12	18.98	16.52	13.61	11.72	10.51	9.58	8.91
3, Good	20.74	17.02	14.50	12.10	10.50	9.40	8.59	7.91
4, Average	17.35	14.21	12.34	10.16	8.77	7.85	7.22	6.66
5, Low	13.71	11.25	9.75	7.96	6.96	6.30	5.99	5.25

Quality Class	4,000	4,500	5,000	6,000	7,500	10,000	15,000	20,000
1, Best	9.24	8.67	8.28	7.55	6.72	5.79	4.66	4.00
2, Very Good	8.28	7.80	7.41	6.74	6.01	5.26	4.22	3.63
3, Good	7.53	6.98	6.68	6.09	5.41	4.65	3.84	3.27
4, Average	6.30	5.77	5.55	5.04	4.54	3.90	3.17	2.78
5, Low	4.94	4.59	4.30	3.98	3.53	3.03	2.42	2.10

Perimeter (Common) Wall Adjustment

First Story

Class	For a Common Wall, Deduct Per L.F.	For No Wall Ownership, Deduct Per L.F.	For Lack of Exterior Finish, Deduct Per L.F.
1	$1,470	$2,945	$998
2	1,233	2,444	806
3	1,014	2,026	555
4	611	1,260	376
5	500	1,022	107

Second and Higher Stories

Class	For a Common Wall, Deduct Per L.F.	For No Wall Ownership, Deduct Per L.F.	For Lack of Exterior Finish, Deduct per L.F.
1	$1,522	$3,048	$1,033
2	739	1,431	371
3	585	1,176	238
4	498	998	164
5	371	705	114

Note: First floor costs include the cost of overhang as described on page 115. Second floor costs do not include any allowance for overhang.

Banks and Savings Offices – Wood Frame

Quality Classification

	Class 1 Best Quality	Class 2 Good Quality	Class 3 High Average Quality	Class 4 Low Average Quality	Class 5 Low Quality
Foundation (10% of total cost)	Reinforced concrete.	Reinforced concrete.	Reinforced concrete.	Reinforced concrete.	Reinforced concrete.
Floor Structure (10% of total cost)	Reinforced concrete.	Reinforced concrete.	Reinforced concrete.	Reinforced concrete.	Reinforced concrete.
Wall Structure (10% of total cost)	2" x 6" - 16" o.c. brick or concrete columns with combustible filler walls.	2" x 6" - 16" o.c. brick or concrete columns with combustible filler walls.	2" x 6" - 16" o.c.	2" x 6" - 16" o.c.	2" x 6" - 16" o.c.
Roof & Cover (10% of total cost)	Copper or slate.	Mission tile or heavy shakes.	5 ply built-up roof. Some portions heavy shake or mission tile.	4 ply built-up roof. Some portions shingle or composition shingles.	4 ply built-up roof.
Exterior Wall Finish (7% of total cost)	Expensive veneers.	Expensive ornamental veneer filler walls.	Wood siding combined with brick or stone veneers.	Good wood siding.	Good stucco or wood siding.
Interior Wall Finish (7% of total cost)	Ornamental plaster painted with murals or designs, ornamental moldings, marble and similar expensive finishes.	Ornamental plaster ornamental moldings, some hardwood panels or marble finishes.	Hard plaster, ornamental cove at ceiling, vinyl wall covering, some hardwood veneer.	Plaster on lath with putty coat finish. Gypsum wallboard, texture and paint. Some vinyl wall cover.	Plaster on lath with putty coat finish. Gypsum wallboard, texture and paint.
Glass (5% of total cost)	Tinted float glass in customized frames (50% of exterior wall area).	Tinted float glass in bronze frames (50% of exterior wall area).	Tinted float glass in heavy anodized aluminum frames or custom wood frames (50% of exterior wall area).	Moderate amount of float glass in heavy frames (25% of exterior wall area).	Small to moderate amount of float glass in average aluminum frames (5 to 10% of exterior wall area).
Overhang (4% of total cost)	4' closed overhang with copper gutters.	4' closed overhang with copper gutters.	4' closed overhang with painted gutters.	3' closed overhang with painted gutters.	None.
Floor Finish					
Public Area: (7% of total cost)	Terrazzo and marble.	Detailed terrazzo.	Detailed terrazzo.	Plain terrazzo.	Sheet vinyl or resilient tile.
Officers Area: (3% of total cost)	Excellent carpet.	Very good carpet.	Very good carpet.	Good carpet.	Average quality carpet.
Work Area: (2% of total cost)	Good sheet vinyl or carpet.	Average sheet vinyl or carpet.	Resilient tile.	Composition tile.	Composition tile.
Ceiling Finish (8% of total cost)	Ornamental plaster with exposed ornamental beams.	Ornamental plaster.	Suspended acoustical tile with gypsum wallboard backing or plain acoustical plaster.	Suspended acoustical tile with exposed grid system.	Acoustical tile on wood strips.
Lighting (5% of total cost)	Recessed panelized LED light fixtures, custom light fixtures or chandeliers. Many spotlights.	Recessed panelized LED lighting. Custom chandeliers. Many spotlights.	Recessed panelized LED lighting. Many spotlights.	Recessed panelized fluorescent lighting. Some spotlights.	Nonrecessed panelized lighting.
Plumbing (12% of total cost)	2 or more rest rooms with special fixtures. Good ceramic tile, marble or terrazzo wainscot walls. Hard plaster with putty coat and enamel paint. Marble toilet screens. toilet screens.	2 or more rest rooms with many good fixtures. Good ceramic tile, marble or terrazzo wainscot walls. Hard plaster with putty coat and enamel paint. Marble screens.	2 rest rooms with 4 or more fixtures. Ceramic tile or terrazzo wainscot, hard plaster walls with putty coat and enamel paint. Marble or metal toilet screens.	2 rest rooms with 4 or more fixtures each. Ceramic tile or terrazzo floor. Ceramic tile or terrazzo wainscot plaster walls with putty coat. Enamel paint. Good metal toilet partitions.	2 rest rooms with 4 fixtures each. Metal toilet screens. Ceramic tile or vinyl asbestos tile floors. Plaster walls with enamel paint.

Note: Use the percent of total cost to help identify the correct quality classification.

Square foot costs include the following components: Foundations as required for normal soil conditions. Floor, wall, and roof structures. Interior floor, wall and ceiling finishes. Exterior wall finish and roof cover. All glass and glazing. Interior partitions. Roof overhang as described above. Basic electrical systems and lighting fixtures. Rough and finish plumbing. Typical bank vault. Alarm systems. Night depository. Typical record vault and fire doors. Fire exits. Design and engineering fees, permits and hook-up fees, contractor's mark-up.

Length Less Than Twice Width

Estimating Procedure

1. Establish the structure quality class by using the information on page 120.
2. Compute the building floor area. This should include everything within the building exterior walls and all insets outside the main walls but under the main building roof.
3. Add to or subtract from the square foot cost below the appropriate amount from the Wall Height Adjustment Table on page 124 if the wall height is more or less than 16 feet for the first floor or 12 feet for higher floors.
4. Multiply the adjusted square foot cost by the building area.
5. Deduct, if appropriate, for common walls, using the figures on page 124.
6. Multiply the total cost by the location factor listed on page 7 or 8.
7. Add the cost of heating and cooling equipment, fire sprinklers, exterior signs, elevators, and yard improvements from pages 236 to 248. Add the cost of mezzanines, bank fixtures, external windows, safe deposit boxes and vault doors from page 125.

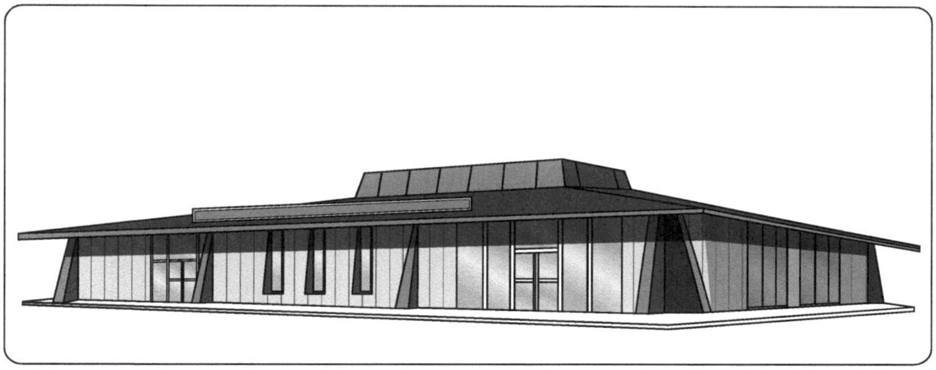

Savings Office, Class 5

First Story – Square Foot Area

Quality Class	2,500	3,000	3,500	4,000	4,500	5,000	6,000	7,500	10,000	15,000	20,000
1, Best	473.67	447.90	429.26	415.08	403.83	394.79	380.89	366.73	352.03	336.56	328.27
1 & 2	446.96	422.69	405.05	391.72	381.14	372.54	359.49	346.08	332.23	317.52	309.78
2, Very Good	421.95	398.92	382.35	369.64	359.67	351.56	339.22	326.63	313.50	299.68	292.38
2 & 3	377.89	357.38	342.47	331.18	322.17	314.99	303.96	292.62	280.81	268.51	261.92
3, Good	372.83	352.62	337.94	326.76	317.92	310.75	299.87	288.71	277.08	264.92	258.38
3 & 4	347.38	328.51	314.79	304.39	296.13	289.49	279.39	269.01	258.15	246.82	240.76
4, Average	326.76	309.01	296.11	286.41	278.64	272.46	262.86	253.04	242.83	232.11	226.51
4 & 5	299.24	283.06	271.27	262.28	255.21	249.44	240.70	231.73	222.39	212.62	207.43
5, Low	270.66	255.95	245.30	237.19	230.80	225.58	217.68	209.54	201.19	192.29	187.57

Second and Higher Stories – Square Foot Area

Quality Class	500	750	1,000	1,500	2,000	3,000	4,000	5,000	7,500	10,000	20,000
1, Best	326.69	296.46	278.41	257.05	244.21	229.13	220.10	213.99	204.34	198.60	187.83
1 & 2	304.62	277.43	261.28	242.22	230.84	217.39	209.37	203.94	195.42	190.51	180.89
2, Very Good	287.94	263.21	248.46	231.02	220.70	208.34	201.07	196.05	188.32	183.70	174.99
2 & 3	266.02	244.08	231.03	215.66	206.52	195.68	189.22	184.82	177.94	173.90	166.20
3, Good	247.47	228.27	216.86	203.35	195.27	185.67	180.01	176.20	170.09	166.50	159.70
3 & 4	228.03	210.96	200.87	188.98	181.90	173.49	168.48	165.16	159.92	156.78	150.84
4, Average	211.04	196.42	187.62	177.30	171.15	163.81	159.38	156.46	151.86	149.11	143.87
4 & 5	195.29	182.53	174.99	165.89	160.53	154.22	150.43	147.90	143.86	141.50	137.07
5, Low	182.98	169.86	162.49	154.22	149.49	144.15	141.05	138.99	135.80	133.99	130.63

Banks and Savings Offices – Wood Frame

Length Between 2 and 4 Times Width

Estimating Procedure

1. Establish the structure quality class by using the information on page 120.
2. Compute the building floor area. This should include everything within the building exterior walls and all insets outside the main walls but under the main building roof.
3. Add to or subtract from the square foot cost below the appropriate amount from the Wall Height Adjustment Table on page 124 if the wall height is more or less than 16 feet for the first floor or 12 feet for higher floors.
4. Multiply the adjusted square foot cost by the building area.
5. Deduct, if appropriate, for common walls, using the figures on page 124.
6. Multiply the total cost by the location factor listed on page 7 or 8.
7. Add the cost of heating and cooling equipment, fire sprinklers, exterior signs, elevators, and yard improvements from pages 236 to 248. Add the cost of mezzanines, bank fixtures, external windows, safe deposit boxes and vault doors from page 125.

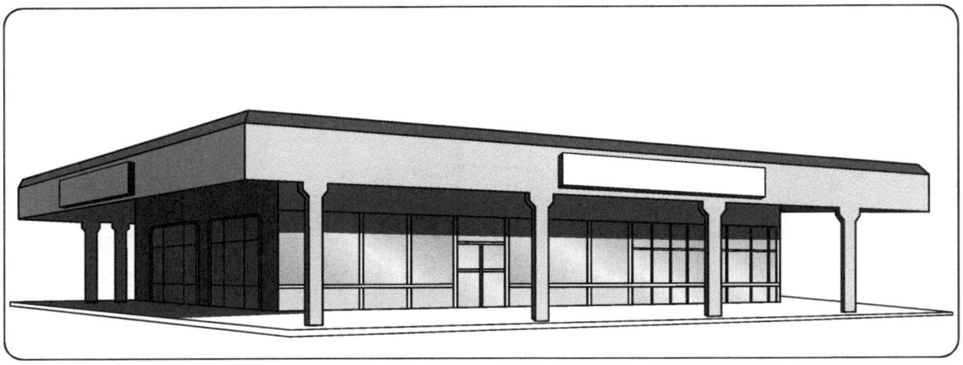

Bank, Class 4

First Story – Square Foot Area

Quality Class	2,500	3,000	3,500	4,000	4,500	5,000	6,000	7,500	10,000	15,000	20,000
1, Best	485.87	460.79	442.21	427.83	416.32	406.84	392.17	376.60	359.95	341.70	331.50
1 & 2	458.60	435.03	417.44	403.87	392.94	384.07	370.19	355.56	339.84	322.60	312.82
2, Very Good	435.04	412.59	395.98	383.02	372.75	364.30	351.16	337.25	322.39	305.93	296.74
2 & 3	405.88	384.86	369.42	357.42	347.80	339.82	327.56	314.66	300.75	285.44	276.95
3, Good	382.75	363.04	348.41	337.05	327.98	320.52	308.91	296.74	283.65	269.21	261.14
3 & 4	355.13	336.72	323.16	312.70	304.22	297.35	286.58	275.23	263.15	249.73	242.27
4, Average	333.47	316.22	303.54	293.55	285.70	279.20	269.08	258.51	247.04	234.55	227.49
4 & 5	304.47	288.75	277.13	268.13	260.84	254.97	245.74	236.06	225.51	214.10	207.73
5, Low	275.04	260.80	250.25	242.12	235.63	230.28	221.98	213.15	203.75	193.38	187.57

Second and Higher Stories – Square Foot Area

Quality Class	500	750	1,000	1,500	2,000	3,000	4,000	5,000	7,500	10,000	20,000
1, Best	330.01	302.16	285.63	266.00	254.25	240.38	232.18	226.53	217.82	212.57	202.68
1 & 2	312.57	286.17	270.48	251.90	240.84	227.74	219.89	214.56	206.32	201.31	191.99
2, Very Good	297.52	272.53	257.48	239.82	229.28	216.75	209.31	204.23	196.38	191.65	182.71
2 & 3	278.05	254.66	240.61	224.15	214.36	202.58	195.68	190.91	183.56	179.10	170.81
3, Good	262.30	240.22	227.04	211.44	202.18	191.13	184.53	180.02	173.13	168.99	161.11
3 & 4	243.89	223.29	211.06	196.66	188.04	177.74	171.60	167.44	160.95	157.08	149.78
4, Average	229.15	209.86	198.36	184.81	176.56	167.03	161.24	157.33	151.28	147.64	140.74
4 & 5	214.83	196.74	185.93	173.15	165.57	156.58	151.17	147.52	141.75	138.40	131.98
5, Low	200.21	183.32	173.28	161.38	154.31	145.91	140.87	137.42	132.17	128.96	123.00

Banks and Savings Offices – Wood Frame

Length More Than 4 Times Width

Estimating Procedure

1. Establish the structure quality class by using the information on page 120.
2. Compute the building floor area. This should include everything within the building exterior walls and all insets outside the main walls but under the main building roof.
3. Add to or subtract from the square foot cost below the appropriate amount from the Wall Height Adjustment Table on page 124 if the wall height is more or less than 16 feet for the first floor or 12 feet for higher floors.
4. Multiply the adjusted square foot cost by the building area.
5. Deduct, if appropriate, for common walls, using the figures on page 124.
6. Multiply the total cost by the location factor listed on page 7 or 8.
7. Add the cost of heating and cooling equipment, fire sprinklers, exterior signs, elevators, and yard improvements from pages 236 to 248. Add the cost of mezzanines, bank fixtures, external windows, safe deposit boxes and vault doors from page 125.

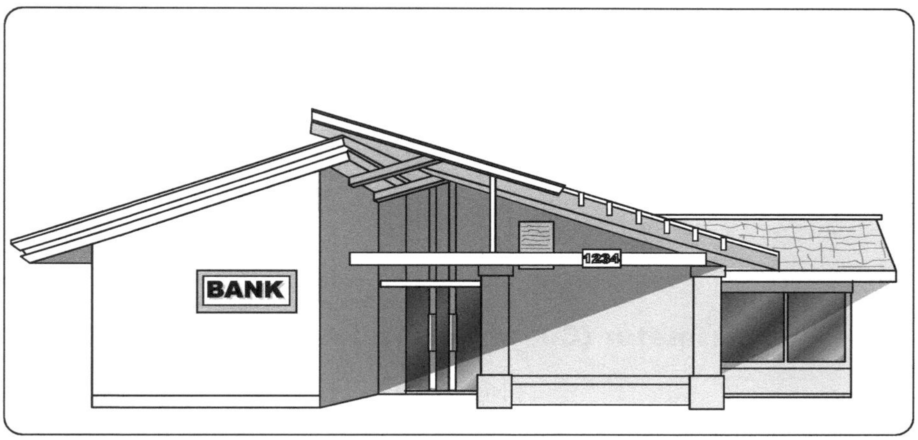

Bank, Class 4 & 5

First Story – Square Foot Area

Quality Class	2,500	3,000	3,500	4,000	4,500	5,000	6,000	7,500	10,000	15,000	20,000
1, Best	510.42	482.43	461.71	445.64	432.97	422.45	406.19	389.12	370.99	351.07	339.99
1 & 2	481.94	455.44	435.84	420.74	408.72	398.81	383.49	367.42	350.30	331.42	320.98
2, Very Good	456.80	431.78	413.19	398.81	387.41	378.10	363.49	348.25	332.00	314.13	304.21
2 & 3	427.12	403.59	386.33	372.90	362.20	353.48	339.82	325.61	310.43	293.74	284.54
3, Good	402.50	380.33	364.05	351.41	341.38	333.08	320.30	306.84	292.51	276.76	268.06
3 & 4	373.09	352.57	337.48	325.78	316.41	308.77	296.86	284.44	271.17	256.52	248.40
4, Average	348.44	329.32	315.19	304.21	295.51	288.32	277.28	265.64	253.25	239.63	232.06
4 & 5	316.95	299.54	286.68	276.75	268.80	262.29	252.22	241.65	230.35	218.04	211.06
5, Low	284.91	269.24	257.71	248.76	241.60	235.76	226.70	217.18	207.00	195.90	189.75

Second and Higher Stories – Square Foot Area

Quality Class	500	750	1,000	1,500	2,000	3,000	4,000	5,000	7,500	10,000	20,000
1, Best	379.14	339.08	315.20	286.92	270.12	250.22	238.38	230.30	217.75	210.24	196.07
1 & 2	353.56	317.26	295.70	270.13	254.97	236.95	226.24	218.90	207.51	200.80	188.04
2, Very Good	330.60	297.91	278.52	255.51	241.85	225.64	215.99	209.40	199.24	193.20	181.70
2 & 3	303.67	274.87	257.71	237.42	225.29	211.04	202.48	196.74	187.68	182.31	172.22
3, Good	280.49	255.20	240.17	222.37	211.73	199.19	191.70	186.52	178.63	173.91	164.97
3 & 4	256.94	234.70	221.39	205.73	196.50	185.39	178.82	174.30	167.37	163.24	155.41
4, Average	236.51	216.99	205.43	191.75	183.56	173.96	168.22	164.33	158.30	154.67	147.94
4 & 5	217.22	200.33	190.36	178.43	171.42	163.13	158.19	154.71	149.51	146.40	140.50
5, Low	197.87	183.62	175.11	165.08	159.12	152.13	147.98	145.11	140.68	138.06	133.10

Wall Adjustments

Wall Height Adjustment

Add or subtract the amount listed in this table to or from the square foot of floor cost for each foot of wall height more or less than 16 feet, if adjusting for first floor, and 12 feet if adjusting for upper floors.

Square Foot Area

Quality Class	500	750	1,000	1,500	2,000	2,500	3,000	3,500
1, Best	15.88	13.04	11.32	9.31	8.03	7.22	6.58	6.13
2, Very Good	14.30	11.79	10.23	8.37	7.30	6.55	5.94	5.58
3, Good	12.73	10.45	9.06	7.39	6.42	5.74	5.33	4.86
4, Average	8.66	7.14	6.21	5.10	4.41	3.96	3.64	3.30
5, Low	5.71	4.64	4.11	3.33	2.97	2.59	2.37	2.18

Square Foot Area

Quality Class	4,000	4,500	5,000	6,000	7,500	10,000	15,000	20,000
1, Best	5.98	5.55	5.16	4.61	4.11	3.51	3.00	2.65
2, Very Good	5.39	4.95	4.65	4.12	3.64	3.18	2.59	2.36
3, Good	4.71	4.40	4.07	3.69	3.22	2.80	2.35	2.09
4, Average	3.23	3.05	2.85	2.54	2.21	1.93	1.66	1.47
5, Low	2.13	2.03	1.95	1.72	1.50	1.27	1.07	0.92

Perimeter (Common) Wall Adjustment

First Story

Class	For a Common Wall, Deduct Per L.F.	For No Wall Ownership, Deduct Per L.F.	For Lack of Exterior Finish, Deduct Per L.F.
1	$765	$1,479	$729
2	648	1,258	629
3	520	1,064	423
4	381	758	305
5	175	331	106

Second and Higher Stories

Class	For a Common Wall, Deduct Per L.F.	For No Wall Ownership, Deduct Per L.F.	For Lack of Exterior Finish, Deduct Per L.F.
1	$456	$936	$413
2	387	758	306
3	322	629	200
4	200	399	140
5	140	282	712

Note: First floor costs include the cost of overhang as described on page 120. Second floor costs do not include any allowance for overhang.

Banks and Savings Offices Additional Costs

Mezzanines Without Partitions

Quality Class	1, Best	2, Good	3, High Avg.	4, Low Avg.	5, Low
S.F. Cost	$115	$110	$95.60	$87.40	$79.60

With Partitions

Quality Class	1, Best	2, Good	3, High Avg.	4, Low Avg.	5, Low
S.F. Cost	$149	$136	$121	$114	$98.50

Mezzanine costs include: Floor system, floor finish, typical stairway, lighting and structural support costs.

Fixtures, cost per square foot of floor

Shell-type counters, plastic finish, no counter screens or drawers.	$26.10 to $29.30
Counters with drawers, good hardwood, plain counter screens.	30.50 to 41.60
Counters with drawers, good hardwood, plastic countertops, average counter screens.	41.60 to 60.80
Counters with drawers, terrazzo or marble finish, marble counter tops, fancy counter screens.	60.80 to 102.80

Costs include counters, screens, and partitions. The square-foot cost should be applied only to the first floor area. Office areas used for purposes other than conducting bank business related to the immediate site should be excluded.

External Access Facilities, cost each unit

Drive-up teller, flush window	$13,290 to $18,100
Drive-up teller, projected window	15,900 to 23,960
Walk-up teller, flush window	8,340 to 9,790
Automatic teller, cash dispensing, with phone	65,850 to 78,700
Night deposit vault whole chest	14,880 to 16,000

Safe Deposit Boxes, cost per box

Box Sizes	Modular Unit	Custom Built
3" x 5"	$122 to $143	$194 to $207
5" x 5"	154 to 160	200 to 260
3" x 10"	133 to 172	228 to 267
5" x 10"	176 to 205	284 to 338
10" x 10"	250 to 286	373 to 480

Bank fixtures and safe deposit boxes are part of the structure cost only if they are fixed to the building.

Vault Doors Record Storage

Description	3' x 7', 2 hour	3' x 7', 4 hour	4' x 7', 2 hour	4' x 7', 4 hour
In Place Cost	$4,880	$5,630	$6,410	$6,920

Prices include frames, time locks, and architrave.

Department Stores – Reinforced Concrete

Quality Classification

	Class 1 Best Quality	Class 2 Good Quality	Class 3 Average Quality	Class 4 Low Quality
Foundation (17% of total cost)	Reinforced concrete.	Reinforced concrete.	Reinforced concrete.	Reinforced concrete.
Ground Floor Structure (10% of total cost)	6" reinforced concrete on 6" rock base.	6" reinforced concrete on 6" rock base.	6" reinforced concrete on 6" rock base.	4" reinforced concrete on 6" rock base.
Wall Structure (10% of total cost)	Reinforced concrete.	Reinforced concrete.	Reinforced concrete.	Reinforced concrete.
Upper Floor Structure (12% of total cost)	Reinforced concrete.	Reinforced concrete.	Reinforced concrete.	Reinforced concrete.
Roof & Cover (10% of total cost)	Reinforced concrete with 5 ply built-up roofing and insulation.	Reinforced concrete with 5 ply built-up roofing and insulation.	Reinforced concrete with 4 ply built-up roofing.	Reinforced concrete with 4 ply built-up roofing.
Floor Finish (5% of total cost)	Terrazzo and very good carpet.	Resilient tile with 50% vinyl tile, terrazzo or good carpet.	Composition tile.	Minimum grade tile.
Interior Wall Finish (5% of total cost)	Gypsum wallboard or lath and plaster finished with good paper or vinyl wall cover on hardwood veneer paneling.	Gypsum wallboard and texture or paper. Some vinyl wall cover or hardwood veneer paneling.	Interior stucco or gypsum wallboard, texture and paint.	Gypsum wallboard, texture and paint.
Ceiling Finish (5% of total cost)	Suspended good grade acoustical tile with gypsum wallboard backing.	Suspended acoustical tile with concealed grid system.	Suspended acoustical tile with exposed grid system.	Painted.
Lighting (6% of total cost)	Recessed LED lighting in modular plastic panels. Many spotlights.	Continuous recessed LED light strips with egg crate diffusers 8' o.c. Average number of spotlights.	Continuous LED light strips with egg crate diffusers, 8' o.c. Some spotlights.	Continuous exposed 2 tube fluorescent strips, 8' o.c.
Display Fronts (6% of total cost)	Best quality front as described on page 77. 15 to 25% of the first floor exterior wall is made up of display fronts.	Good quality front as described on page 77. 15 to 25% of the first floor exterior wall is made up of display fronts.	Average quality front as described on page 77. 10 to 20% of the first floor exterior wall is made up of display fronts.	Low quality flat type as described on page 77. 10 to 20% of the first floor exterior wall is made up of display fronts.
Exterior Wall Finish (8% of total cost)	Ornamental block or brick, some marble veneer.	Decorative block, some stone veneer.	Paint.	Paint.
Plumbing (6% of total cost)	6 good fixtures per 20,000 square feet of floor area. Metal toilet partitions.	6 standard fixtures per 20,000 square feet of floor area. Metal toilet partitions.	4 standard fixtures per 20,000 square feet of floor area. Metal toilet partitions.	4 standard fixtures per 20,000 square feet of floor area. Wood toilet partitions.

Note: Use the percent of total cost to help identify the correct quality classification.

Square foot costs include the following components: Foundations as required for normal soil conditions. Floor, wall and roof structures. Interior ceiling, wall and floor finishes (including carpet). Exterior wall finish and roof cover. Display fronts. Interior partitions (including perimeter wall partitions). Entry and delivery doors. Basic lighting and electrical systems. Rough and finish plumbing. Design and engineering fees. Permits and hook-up fees. Contractor's mark-up.

The in-place cost of these extra components should be added to the basic building cost to arrive at the total structure cost. See the section "Additional Costs for Commercial Structures" on page 236 to 248. Heating and air conditioning systems. Elevators and escalators. Fire sprinklers. Exterior signs. Canopies and walks. Paving and curbing. Loading docks or ramps. Miscellaneous yard improvements. Mezzanines.

Department Stores – Reinforced Concrete

First Floor

Estimating Procedure

1. Use these square foot costs to estimate the cost of retail stores designed to sell a wider variety of goods. These buildings differ from discount houses in that they have more interior partitions and more elaborate interior and exterior finishes.
2. Establish the structure quality class by using the information on page 126.
3. Compute the building floor area. This should include everything within the building exterior walls and all inset areas outside the main walls but under the main building roof.
4. Add to or subtract from the square foot cost below the appropriate amount from the Wall Height Adjustment Table shown below if the wall height is more or less than 20 feet.
5. Multiply the adjusted square foot cost by the building area.
6. Deduct, if appropriate, for common walls, using the figures at the bottom of this page.
7. Multiply the total cost by the location factor listed on page 7 or 8.
8. Add the cost of appropriate additional components from page 236 to 248: heating and air conditioning systems, elevators and escalators, fire sprinklers, exterior signs, canopies and walks, paving and curbing, loading docks and ramps, miscellaneous yard improvements, and mezzanines.
9. Add the cost of second and higher floors and basements from page 128.

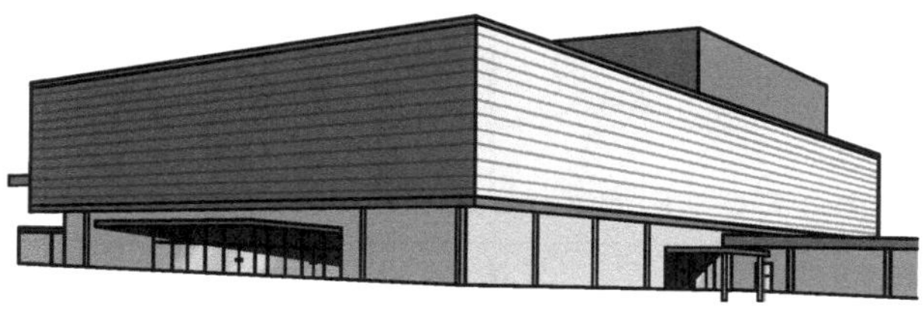

Department Store, Class 1

First Floor – Square Foot Area

Quality Class	20,000	25,000	30,000	35,000	40,000	45,000	50,000	60,000	70,000	80,000	100,000
1, Best	398.63	383.04	370.34	359.85	350.94	343.19	336.52	325.35	316.30	308.78	297.02
1 & 2	370.81	356.99	345.84	336.48	328.61	321.89	315.94	306.02	298.01	291.38	281.00
2, Good	349.17	337.00	327.01	318.82	311.89	305.92	300.63	291.92	284.88	279.12	269.97
2 & 3	322.98	312.46	304.01	296.99	291.03	285.91	281.56	274.08	268.10	263.21	255.34
3, Average	302.13	293.24	285.95	280.03	275.13	270.80	267.16	261.15	256.13	252.04	245.73
3 & 4	273.23	266.08	260.36	255.56	251.69	248.36	245.43	240.54	236.65	233.48	228.30
4, Low	247.87	242.64	238.47	235.14	232.10	229.66	227.31	223.69	220.71	218.17	214.31

Perimeter (Common) Wall Adjustment – First Floor

Class	For a Common Wall, Deduct Per L.F.	For No Wall Ownership, Deduct Per L.F.	For Lack of Exterior Finish, Deduct Per L.F.
1	$1,370	$2,748	$901
2	1,049	2,061	534
3	768	1,492	238
4	519	1,036	147

Wall Height Adjustment: Add or subtract the amount listed to or from the square foot of floor cost for each foot of first and upper story wall height more or less than 20 feet.

Square Foot Area

Quality Class	20,000	25,000	30,000	35,000	40,000	45,000	50,000	60,000	70,000	80,000	100,000
1, Best	4.07	3.46	3.08	2.88	2.63	2.44	2.39	2.15	2.09	2.05	2.01
2, Good	2.90	2.40	2.10	2.00	1.81	1.76	1.69	1.57	1.53	1.36	1.34
3, Average	2.39	2.01	1.81	1.69	1.47	1.36	1.33	1.20	1.19	1.17	1.14
4, Low	1.94	1.74	1.47	1.26	1.19	1.17	1.13	1.08	1.06	1.03	.94

Department Stores – Reinforced Concrete

Estimating Procedure

1. Establish the basement and upper floor quality class. The quality class will usually be the same as the first floor of the building. Square foot costs for unfinished basements will be nearly the same regardless of the structure quality class.
2. Compute the floor area.
3. Add to or subtract from the square foot cost below the appropriate amount from the Wall Height Adjustment Table on page 127 for second and higher floors and from the bottom of this page for basements for each foot of wall height more or less than 20 feet for second and higher floors and 16 feet for basements.
4. Multiply the adjusted square foot cost from one of the three tables below by the floor area.
5. Deduct, if appropriate, for common or unfinished upper floor walls, using the costs in the table below titled "Second and Higher Floor Perimeter Wall Adjustments."
6. Multiply the total cost by the location factor listed on page 7 or 8.
7. Add the cost of appropriate additional components from page 236 to 248: heating and air conditioning systems, elevators and escalators, fire sprinklers, and mezzanines.
8. Add the total from this page to the total from page 127 to find the building cost.

Second and Higher Stories – Square Foot Area

Quality Class	20,000	25,000	30,000	35,000	40,000	45,000	50,000	60,000	70,000	80,000	100,000
1, Best	357.30	342.67	330.68	320.70	312.17	304.86	298.47	287.75	279.09	271.77	260.40
1 & 2	333.29	320.05	309.44	300.46	292.97	286.47	280.67	271.16	263.40	257.09	247.02
2, Good	314.57	302.73	293.17	285.31	278.54	272.77	267.76	259.39	252.61	247.01	238.15
2 & 3	291.83	281.78	273.56	266.75	261.15	256.15	251.95	244.69	238.97	234.09	226.70
3, Average	274.16	265.46	258.49	252.79	247.99	243.77	240.13	234.11	229.28	225.22	218.93
3 & 4	250.28	243.12	237.49	232.83	228.89	225.62	222.78	217.89	214.20	210.95	205.96
4, Low	224.77	219.26	214.87	211.44	208.52	206.11	203.97	200.48	197.65	195.38	191.82

Second and Higher Floor Perimeter (Common) Wall Adjustment

Class	For a Common Wall, Deduct Per L.F.	For No Wall Ownership, Deduct Per L.F.	For Lack of Exterior Finish, Deduct Per L.F.
1	$1,055	$2,088	$720
2	779	1,534	428
3	561	1,119	204
4	494	944	148

Finished Basements – Square Foot Area

Quality Class	20,000	25,000	30,000	35,000	40,000	45,000	50,000	60,000	70,000	80,000	100,000
1, Best	304.52	294.84	286.83	280.02	274.21	269.16	264.67	257.36	251.25	246.11	238.09
1 & 2	285.18	276.55	269.41	263.37	258.30	253.84	249.99	243.48	238.22	233.81	226.81
2, Good	275.25	267.20	260.66	255.05	250.30	246.29	242.70	236.66	231.84	227.80	221.31
2 & 3	256.86	249.99	244.32	239.69	235.73	232.22	229.24	224.25	220.25	216.80	211.47
3, Average	246.82	241.47	236.89	232.84	228.98	226.31	223.62	219.07	215.31	212.11	207.09
3 & 4	228.09	223.06	218.99	215.71	212.98	210.62	208.60	205.17	202.48	200.12	196.58
4, Low	207.36	203.31	200.24	197.73	195.76	194.19	192.70	190.33	188.49	186.90	184.67

Unfinished Basements

Area	20,000	25,000	30,000	35,000	40,000	45,000	50,000	60,000	70,000	80,000	100,000
Cost	141.57	139.04	136.93	135.44	134.03	133.23	132.13	130.72	129.64	128.91	127.42

Basement Wall Height Adjustment: Add or subtract the amount listed to or from the square foot of floor cost for each foot of basement wall height more or less than 16 feet.

Area	20,000	25,000	30,000	35,000	40,000	45,000	50,000	60,000	70,000	80,000	100,000
Finished	1.95	1.76	1.53	1.33	1.26	1.20	1.19	1.17	1.14	1.13	1.13
Unfinished	1.77	1.53	1.26	1.20	1.17	1.13	1.09	1.08	1.06	1.06	1.03

Department Stores – Masonry or Concrete

Quality Classification

	Class 1 Best Quality	Class 2 Good Quality	Class 3 Average Quality	Class 4 Low Quality
Foundation (20% of total cost)	Reinforced concrete.	Reinforced concrete.	Reinforced concrete.	Reinforced concrete.
Floor Structure (15% of total cost)	6" reinforced concrete on 6" rock base.	6" reinforced concrete on 6" rock base.	6" reinforced concrete on 6" rock base.	4" reinforced concrete on 6" rock base.
Wall Structure (12% of total cost)	8" reinforced decorative concrete block, 6" concrete tilt-up or 8" reinforced brick.	8" reinforced decorative concrete block, 6" concrete tilt-up or 8" reinforced brick.	8" reinforced concrete. block, 6" concrete tilt-up or 8" reinforced common brick.	8" reinforced concrete block or 6" concrete tilt-up.
Roof & Cover (12% of total cost)	Glu-lams or steel beams on steel intermediate columns. Panelized roof system, 1/2" plywood sheathing, 5 ply built-up roof with insulation.	Glu-lams or steel beams on intermediate columns. Panelized roof system, 1/2" plywood sheathing, 5 ply built-up roof with insulation.	Glu-lams on steel intermediate columns. Panelized roof system, 1/2" plywood sheathing, 4 ply built-up roof.	Glu-lams on steel intermediate columns. Panelized roof system, 1/2" plywood sheath-ing, 4 ply built-up roof.
Floor Finish (5% of total cost)	Terrazzo and very good carpet.	Resilient tile with 50% sheet vinyl, terrazzo or good carpet.	Composition tile.	Minimum grade tile.
Interior Wall Finish (5% of total cost)	Gypsum wallboard or lath and plaster, finished with good paper or vinyl wall covers, or hardwood veneer paneling.	Gypsum wallboard and texture or paper, some vinyl wall cover or hardwood veneer paneling.	Interior stucco or gypsum wallboard, texture and paint.	Gypsum wallboard, texture and paint.
Ceiling Finish (5% of total cost)	Suspended good grade acoustical tile with gypsum wallboard backing.	Suspended acoustical tile with concealed grid system.	Suspended acoustical tile with exposed grid system.	Painted.
Lighting (6% of total cost)	Recessed LED lighting in modular plastic panels. Many spotlights.	Continuous recessed LED strips with egg crate diffusers, 8' o.c. Average number of spotlights.	Continuous LED strips with egg crate diffusers, 8' o.c. Some spotlights.	Continuous exposed 2 tube fluorescent strips, 8' o.c.
Display Fronts (6% of total cost)	Best quality front as described on page 77. 15 to 25% of the first floor exterior wall is made up of display fronts.	Good quality front as described on page 77. 15 to 25% of the first floor exterior wall is made up of display fronts.	Average quality front as described on page 77. 10 to 20% of the first floor exterior wall is made up of display fronts.	Low quality flat type as described on page 77. 10 to 20% of the first floor exterior wall is made up of display fronts.
Exterior Wall Finish (8% of total cost)	Ornamental block or large rock imbedded in tilt-up panels, some stone veneer.	Exposed aggregate or decorative block. Some stone veneer.	Paint, some exposed aggregate.	Paint.
Plumbing (6% of total cost)	6 good fixtures per 20,000 square feet of floor area. Metal toilet partitions.	6 standard fixtures per 20,000 square feet of floor area. Metal toilet partitions.	4 standard fixtures per 20,000 square feet of floor area. Metal toilet partitions.	4 standard fixtures per 20,000 square feet of floor area. Wood toilet partitions.

Note: Use the percent of total cost to help identify the correct quality classification.

Square foot costs include the following components: Foundations as required for normal soil conditions. Floor, wall and roof structures. Interior ceiling, wall and floor finishes (including carpet). Exterior wall finish and roof cover. Display fronts. Interior partitions (including perimeter wall partitions). Entry and delivery doors. Basic lighting and electrical systems. Rough and finish plumbing. Design and engineering fees. Permits and hook-up fees. Contractor's mark-up.

The in-place cost of these extra components should be added to the basic building cost to arrive at the total structure cost. See the section "Additional Costs for Commercial Structures" on page 236 to 248. Heating and air conditioning systems. Elevators and escalators. Fire sprinklers. Exterior signs. Canopies and walks. Paving and curbing. Loading docks or ramps. Miscellaneous yard improvements. Mezzanines.

Department Stores – Masonry or Concrete

First Floor

Estimating Procedure

1. Use these square foot costs to estimate the cost of retail stores designed to sell a wider variety of goods. These buildings differ from discount houses in that they have more interior partitions and more elaborate interior and exterior finishes.
2. Establish the structure quality class by using the information on page 129.
3. Compute the building floor area. This should include everything within the building exterior walls and all inset areas outside the main walls but under the main building roof.
4. Add to or subtract from the square foot cost below the appropriate amount from the Wall Height Adjustment Row (near the bottom of this page) if the wall height is more or less than 16 feet.
5. Multiply the adjusted square foot cost by the building area.
6. Deduct, if appropriate, for common or unfinished walls, using the figures at the bottom of this page.
7. Multiply the total cost by the location factor listed on page 7 or 8.
8. Add the cost of appropriate additional components from page 236 to 248: heating and air conditioning systems, elevators and escalators, fire sprinklers, exterior signs, canopies and walks, paving and curbing, loading docks and ramps, miscellaneous yard improvements, and mezzanines.
9. Add the cost of second and higher floors and basements from page 131.

Department Store, Class 1 & 2

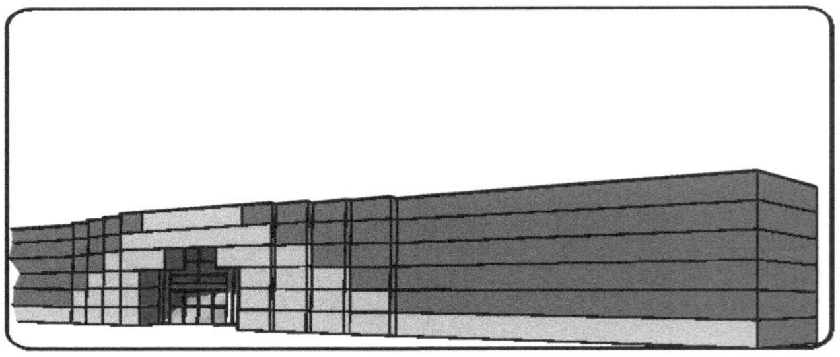

Department Store, Class 2 & 3

First Floor – Square Foot Area

Quality Class	20,000	25,000	30,000	35,000	40,000	45,000	50,000	60,000	70,000	80,000	100,000
1, Best	302.05	288.70	278.09	269.46	262.14	255.91	250.56	241.55	234.22	228.29	218.95
1 & 2	270.87	258.94	249.47	241.65	235.04	229.46	224.62	216.48	210.08	204.88	196.33
2, Good	243.80	233.09	224.57	217.58	211.60	206.58	202.15	195.00	189.14	184.27	176.79
2 & 3	218.10	208.49	200.75	194.45	189.27	184.74	180.79	174.35	169.15	164.86	158.10
3, Average	195.58	186.96	180.12	174.55	169.67	165.70	162.21	156.39	151.73	147.87	141.78
3 & 4	170.68	163.17	157.19	152.27	148.16	144.65	141.57	136.50	132.31	129.09	123.74
4, Low	145.62	139.18	134.07	130.00	126.39	123.37	120.74	116.42	112.91	110.09	105.55
Wall Height Adjustment*	1.64	1.30	1.17	1.09	1.06	1.03	.96	.94	.90	.88	.86

***Wall Height Adjustment:** Add or subtract the amount listed in this row to or from the square foot of floor cost for each foot of wall height more or less than 16 feet.

Perimeter Wall Adjustment: For common wall, deduct $442 per linear foot. For no wall ownership, deduct $870 per linear foot.

Department Stores – Masonry or Concrete

Upper Floors and Basements

Estimating Procedure

1. Establish the basement and upper floor quality class. The quality class will usually be the same as the first floor of the building. Square foot costs for unfinished basements will be nearly the same regardless of the structure quality class.
2. Compute the floor area.
3. Add to or subtract from the square foot cost below the appropriate amount from the Wall Height Adjustment Row for each foot of wall height more or less than 12 feet for second and higher floors and 12 feet for basements.
4. Multiply the adjusted square foot cost by the floor area.
5. Deduct, if appropriate, for common or unfinished upper floor walls, using the Perimeter Wall Adjustment costs listed below.
6. Multiply the total cost by the location factor listed on page 7 or 8.
7. Add the cost of the appropriate additional components from page 236 to 248: heating and air conditioning systems, elevators and escalators, fire sprinklers, and mezzanines.
8. Add the total from this page to the total from page 130 to find the building cost.

Second and Higher Floors – Square Foot Area

Quality Class	20,000	25,000	30,000	35,000	40,000	45,000	50,000	60,000	70,000	80,000	100,000
1, Best	264.42	254.71	245.79	237.79	230.73	224.43	218.89	209.21	201.29	194.55	183.63
1 & 2	233.26	224.72	216.82	209.87	203.57	198.10	193.14	184.71	177.59	171.62	161.92
2, Good	206.15	198.49	191.62	185.42	179.89	174.95	170.68	163.13	156.96	151.69	143.11
2 & 3	183.01	176.27	170.13	164.58	159.74	155.35	151.48	144.93	139.36	134.63	127.13
3, Average	161.39	155.35	149.91	145.05	140.82	136.94	133.43	127.61	122.79	118.72	112.04
3 & 4	132.92	128.03	123.47	119.41	115.99	112.84	110.03	105.12	101.23	97.79	92.32
4, Low	107.31	103.78	101.02	98.61	96.61	94.82	93.47	90.88	88.91	87.21	84.57
Wall Height Adjustment*	1.17	1.07	.94	.88	.86	.82	.81	.80	.80	.78	.78

***Wall Height Adjustment:** Add or subtract the amount listed in this row to or from the square foot of floor cost for each foot of wall height more or less than 12 feet.

Perimeter Wall Adjustment: For common wall, deduct $237 per linear foot. For no wall ownership, deduct $478 per linear foot.

Finished Basement – Square Foot Area

Quality Class	20,000	25,000	30,000	35,000	40,000	45,000	50,000	60,000	70,000	80,000	100,000
1, Best	234.99	225.55	217.89	211.60	206.34	201.90	197.84	191.29	185.97	181.54	174.68
2, Good	188.30	180.72	174.58	169.55	165.38	161.67	158.55	153.27	149.06	145.49	139.92
3, Average	152.87	146.63	141.74	137.75	134.23	131.23	128.68	124.42	120.95	118.14	113.64
4, Low	109.46	105.09	101.54	98.60	96.08	94.05	92.14	89.10	86.71	84.62	81.38

Unfinished Basements

Area	20,000	25,000	30,000	35,000	40,000	45,000	50,000	60,000	70,000	80,000	100,000
Cost	64.69	61.61	59.44	57.82	56.41	55.33	54.34	52.97	51.77	50.98	49.43

Wall Height Adjustment: Add or subtract the amount listed in this table to or from the square foot of floor cost for each foot of basement wall height more or less than 12 feet.

Area	20,000	25,000	30,000	35,000	40,000	45,000	50,000	60,000	70,000	80,000	100,000
Finished	1.33	1.24	1.20	1.17	1.14	1.14	1.13	1.09	1.09	1.08	1.06
Unfinished	1.26	1.19	1.17	1.14	1.13	1.09	1.09	1.08	1.06	1.06	1.03

Department Stores – Wood Frame

Quality Classification

	Class 1 Best Quality	Class 2 Good Quality	Class 3 Average Quality	Class 4 Low Quality
Foundation (20% of total cost)	Reinforced concrete.	Reinforced concrete.	Reinforced concrete.	Reinforced concrete.
Floor Structure (10% of total cost)	6" reinforced concrete on 6" rock base.	6" reinforced concrete on 6" rock base.	6" reinforced concrete on 6" rock base.	4" reinforced concrete on 6" rock base.
Wall Structure (9% of total cost)	2" x 6", 16" o.c.	2" x 6", 16" o.c.	2" x 6", 16" o.c.	2" x 4", 16" o.c.
Roof & Cover (12% of total cost)	Glu-lams or steel beams on steel intermediate columns. Panelized roof system, 1/2" plywood sheathing, 5 ply built-up roof with insulation.	Glu-lams or steel beams on steel intermediate columns. Panelized roof system, 1/2" plywood sheathing, 5 ply built-up roof with insulation.	Glu-lams on steel intermediate columns. Panelized roof system, 1/2" plywood sheathing, 4 ply built-up roof.	Glu-lams on steel intermediate columns. Panelized roof system, 1/2" plywood sheathing, 4 ply built-up roof.
Floor Finish (6% of total cost)	Terrazzo and very good carpet.	Resilient tile with 50% sheet vinyl, terrazzo or good carpet.	Composition tile.	Minimum grade tile.
Interior Wall Finish (5% of total cost)	Gypsum wallboard or lath and plaster finished with good paper or vinyl wall covers, or hardwood veneer paneling.	Gypsum wallboard, texture or paper, some vinyl wall cover or hardwood veneer paneling.	Interior stucco or gypsum wallboard. Texture and paint.	Gypsum wallboard. Texture and paint.
Ceiling Finish (5% of total cost)	Suspended good grade acoustical tile with gypsum wallboard backing.	Suspended acoustical tile with concealed grid system.	Suspended acoustical tile with exposed grid system.	Painted.
Lighting (8% of total cost)	Recessed LED lighting in modular plastic panels. Many spotlights.	Continuous recessed LED strips with egg crate diffusers, 8' o.c. Average number of spotlights.	Continuous LED strips with egg crate diffusers, 8' o.c. Some spotlights.	Continuous exposed 2 tube fluorescent strips, 8' o.c.
Display Fronts (8% of total cost)	Best quality fronts as described on page 77. 15 to 25% of the first floor exterior wall is made up of display fronts.	Good quality front as described on page 77. 15 to 25% of the first floor exterior wall is made up of display fronts.	Average quality front as described on page 77. 10 to 20% of the first floor exterior wall is made up of display fronts.	Low quality flat type as described on page 77. 10 to 20% of the first floor exterior wall is made up of display fronts.
Exterior Wall Finish (10% of total cost)	Good wood siding. Extensive stone veneer.	Wood siding. Some brick or stone veneer.	Stucco. Some brick trim.	Stucco.
Plumbing (7% of total cost)	6 good fixtures per 20,000 square feet of floor area. Metal toilet partitions.	6 standard fixtures per 20,000 square feet of floor area. Metal toilet partitions.	4 standard fixtures per 20,000 square feet of floor area. Metal toilet partitions.	4 standard fixtures per 20,000 square feet of floor area. Wood toilet partitions.

Note: Use the percent of total cost to help identify the correct quality classification.

Square foot costs include the following components: Foundations as required for normal soil conditions. Floor, wall and roof structures. Interior ceiling, wall and floor finishes (including carpet). Exterior wall finish and roof cover. Display fronts. Interior partitions (including perimeter wall partitions). Entry and delivery doors. Basic lighting and electrical systems. Rough and finish plumbing. Design and engineering fees. Permits and hook-up fees. Contractor's mark-up.

The in-place cost of these extra components should be added to the basic building cost to arrive at the total structure cost. See the section "Additional Costs for Commercial Structures" on page 236 to 248. Heating and air conditioning systems. Elevators and escalators. Fire sprinklers. Exterior signs. Canopies and walks. Paving and curbing. Loading docks or ramps. Miscellaneous yard improvements. Mezzanines.

Department Stores – Wood Frame

First Floor

Estimating Procedure

1. Use these square foot costs to estimate the cost of retail stores designed to sell a wider variety of goods. These buildings differ from discount houses in that they have more interior partitions and more elaborate interior and exterior finishes.
2. Establish the structure quality class by applying the information on page 132.
3. Compute the floor area. This should include everything within the building exterior walls and all inset areas outside the main walls but under the main building roof.
4. Add to or subtract from the square foot cost below the appropriate amount from the Wall Height Adjustment Row (near the bottom of this page) if the wall height is more or less than 16 feet.
5. Multiply the adjusted square foot cost by the building area.
6. Deduct, if appropriate, for common walls or no wall ownership, using the figures at the bottom of this page.
7. Multiply the total cost by the location factor on page 7 or 8.
8. Add the cost of appropriate additional components from page 236 to 248: heating and air conditioning systems, elevators and escalators, fire sprinklers, exterior signs, canopies and walks, paving and curbing, loading docks and ramps, miscellaneous yard improvements, and mezzanines.
9. Add the cost of second and higher floors and basements from page 134.

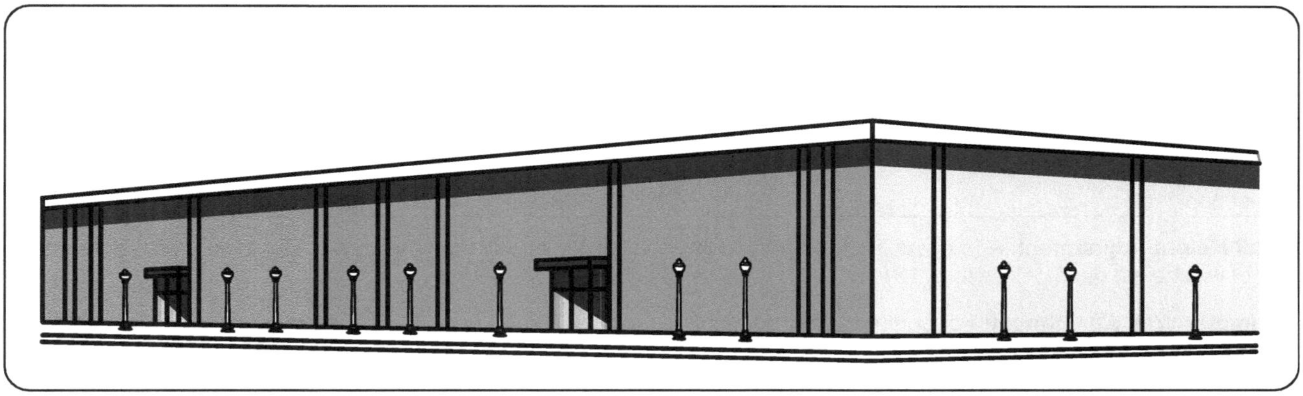

Department Store, Class 2

Square Foot Area

Quality Class	20,000	25,000	30,000	35,000	40,000	45,000	50,000	60,000	70,000	80,000	100,000
1, Best	216.29	207.47	200.38	194.52	189.59	185.24	181.55	175.29	170.24	166.07	159.45
1 & 2	194.28	186.43	180.02	174.77	170.36	166.50	163.18	157.53	152.98	149.29	143.31
2, Good	173.92	166.85	161.18	156.45	152.50	149.02	146.04	141.02	136.92	133.57	128.32
2 & 3	155.28	149.02	143.86	139.64	136.09	133.02	130.29	125.85	122.23	119.22	114.53
3, Average	139.48	133.81	129.18	125.40	122.23	119.42	117.00	112.97	109.74	107.11	102.85
3 & 4	121.09	116.18	112.21	108.93	106.13	103.74	101.62	98.13	95.37	93.01	89.28
4, Low	102.66	98.55	95.22	92.35	89.97	87.98	86.19	83.23	80.85	78.87	75.72
Wall Height Adjustment*	.61	.60	.59	.52	.49	.49	.47	.47	.44	.40	.27

***Wall Height Adjustment:** Add or subtract the amount listed in this table to or from the square foot of floor cost for each foot of wall height more or less than 16 feet.

Perimeter Wall Adjustment: For common wall, deduct $65 per linear foot. For no wall ownership, deduct $130 per linear foot.

Department Stores – Wood Frame

Upper Floors and Basements

Estimating Procedure

1. Establish the basement and upper floor quality class. The quality class will usually be the same as the first floor of the building. Square foot costs for unfinished basements will be nearly the same regardless of the structure quality class.
2. Compute the floor area.
3. Add to or subtract from the square foot cost below the appropriate amount from the Wall Height Adjustment Row for each foot of wall height more or less than 12 feet.
4. Multiply the adjusted square foot cost by the floor area.
5. Deduct, if appropriate, for common upper floor walls and walls not owned, using the Perimeter Wall Adjustment costs.
6. Multiply the total cost by the location factor on page 7 or 8.
7. Add the cost of appropriate additional components from page 236 to 248: heating and air conditioning systems, elevators and escalators, fire sprinklers, and mezzanines.
8. Add the total from this page to the total from page 133 to find the building cost.

Second and Higher Floors – Square Foot Area

Quality Class	20,000	25,000	30,000	35,000	40,000	45,000	50,000	60,000	70,000	80,000	100,000
1, Best	179.29	172.97	167.60	162.96	159.09	155.62	152.53	147.43	143.23	139.73	134.17
1 & 2	158.81	153.12	148.43	144.32	140.82	137.85	135.14	130.63	126.86	123.77	118.91
2, Good	140.91	136.00	131.72	128.18	125.08	122.34	119.92	115.94	112.65	109.87	105.47
2 & 3	125.21	120.82	117.00	113.87	111.11	108.69	106.59	103.02	100.04	97.62	93.74
3, Average	111.21	107.23	103.93	101.05	98.70	96.49	94.62	91.48	88.85	86.68	83.25
3 & 4	94.31	90.93	88.07	85.68	83.61	81.85	80.21	77.58	75.39	73.52	70.61
4, Low	77.13	74.36	72.05	70.16	68.44	66.95	65.71	63.50	61.66	60.16	57.73
Wall Height Adjustment*	.59	.49	.47	.40	.37	.29	.27	.22	.21	.20	.14

***Wall Height Adjustment:** Add or subtract the amount listed in this table to or from the square foot of floor cost for each foot of wall height more or less than 12 feet.

Perimeter Wall Adjustment: For common wall, deduct $161 per linear foot. For no wall ownership, deduct $322 per linear foot.

Finished Basements – Square Foot Area

Quality Class	20,000	25,000	30,000	35,000	40,000	45,000	50,000	60,000	70,000	80,000	100,000
1, Best	175.29	169.72	164.36	159.45	155.16	151.27	147.73	141.62	136.55	132.22	125.16
2, Good	140.57	136.09	131.74	127.88	124.38	121.22	118.46	113.54	109.44	105.96	100.38
3, Average	114.92	111.28	107.79	104.63	101.78	99.16	96.87	92.88	89.55	86.66	82.08
4, Low	79.51	76.95	74.51	72.38	70.37	68.60	66.97	64.27	61.94	59.94	56.82

Unfinished Basements

Area	20,000	25,000	30,000	35,000	40,000	45,000	50,000	60,000	70,000	80,000	100,000
Cost	44.91	43.14	41.74	40.58	39.51	38.71	38.02	36.75	35.79	34.97	33.64

Wall Height Adjustment: Add or subtract the amount listed in this table to or from the square foot of floor cost for each foot of basement wall height more or less than 12 feet.

Area	20,000	25,000	30,000	35,000	40,000	45,000	50,000	60,000	70,000	80,000	100,000
Finished	1.01	.91	.90	.89	.87	.83	.82	.77	.75	.75	.73
Unfinished	.91	.90	.87	.83	.82	.77	.77	.75	.73	.72	.70

General Office Buildings – Masonry or Concrete

Quality Classification

	Class 1 Best Quality	Class 2 Good Quality	Class 3 Average Quality	Class 4 Low Quality
Foundation (12% of total cost)	Reinforced concrete.	Reinforced concrete.	Reinforced concrete.	Reinforced concrete.
First Floor Structure (8% of total cost)	Reinforced concrete slab on grade or standard wood frame.	Reinforced concrete slab on grade or standard wood frame.	Reinforced concrete slab on grade or 4" x 6" girders with plywood sheathing.	Reinforced concrete slab on grade.
Upper Floor Structures (9% of total cost)	Standard wood frame, plywood and 1-1/2" light weight concrete sub floor.	Standard wood frame, plywood and 1-1/2" lightweight concrete sub floor.	Standard wood frame, 5/8" plywood sub floor.	Standard wood frame, 5/8" plywood sub floor.
Walls (10% of total cost)	8" decorative concrete block or 6" concrete tilt-up.	8" decorative concrete block or 6" concrete tilt-up.	8" reinforced concrete block or 8" reinforced brick or 8"" clay tile.	8" reinforced concrete block or 8" clay tile.
Roof Structure (6% of total cost)	Standard wood frame, flat or low pitch.	Standard wood frame, flat or low pitch.	Standard wood frame, flat or low pitch.	Standard wood frame, flat or low pitch.
Exterior Wall Finish (8% of total cost)	Decorative block or large rock imbedded in tilt-up panels with 10 - 20% brick or stone veneer.	Decorative block or exposed aggregate and 10 - 20% brick or stone veneer.	Stucco or colored concrete block.	Painted concrete block or tile.
Windows (5% of total cost)	Average number in good aluminum frame. Fixed float glass in good frame on front side.	Average number in good aluminum frame. Some fixed float glass in front.	Average number of average aluminum sliding type.	Average number of low cost aluminum sliding type.
Roof Cover (5% of total cost)	5 ply built-up roofing on flat roofs. Heavy shake or tile on sloping roofs.	5 ply built-up roofing on flat roofs. Average shake or composition, tar and large rock on sloping roofs.	4 ply built-up roofing on flat roofs. Wood shingle or composition, tar and pea gravel on sloping roofs.	3 ply built-up roofing on flat roofs. Composition shingle on sloping roofs.
Overhang (3% of total cost)	3' closed overhang, fully guttered.	2' closed overhang, fully guttered.	None on flat roofs. 18" open on sloping roofs. Fully guttered.	None on flat roofs. 12" to 16" open on sloping roofs. Gutters over entrances.
Floor Finishes:				
Offices (3% of total cost)	Very good carpet.	Good carpet.	Average grade carpet.	Minimum grade tile.
Corridors (2% of total cost)	Solid vinyl tile or carpet.	Resilient tile.	Composition tile.	Minimum grade tile.
Bathrooms (1% of total cost)	Sheet vinyl or ceramic tile.	Sheet vinyl or ceramic tile.	Vinyl asbestos tile.	Minimum grade tile.
Interior Wall Finishes:				
Offices (6% of total cost)	Good hardwood veneer.	Hardwood veneer paneling or vinyl wall cover.	Gypsum wallboard, texture and paint.	Gypsum wallboard, texture and paint.
Corridors (4% of total cost)	Good hardwood veneer.	Gypsum wallboard and vinyl wall cover	Gypsum wallboard, texture and paint	Gypsum wallboard, texture and paint.
Bathrooms (2% of total cost)	Gypsum wallboard and enamel with ceramic tile wainscot.	Gypsum wallboard and enamel or vinyl wall covering	Gypsum wallboard and enamel.	Gypsum wallboard and texture and paint.
Ceiling Finish (4% of total cost)	Suspended "T" bar and acoustical tile.	Gypsum wallboard and acoustical tile.	Gypsum wallboard and acoustical texture.	Gypsum wallboard and paint.
Plumbing (6% of total cost)	Copper tubing and top quality fixtures.	Copper tubing and good fixtures.	Copper tubing and standard fixtures.	Copper tubing and economy fixtures.
Lighting (6% of total cost)	Conduit wiring, good fixtures.	Conduit wiring, good fixtures.	Romex or conduit wiring, average fixtures.	Romex wiring, economy fixtures.

Note: Use the percent of total cost to help identify the correct quality classification.

Square foot costs include the following components: Foundations as required for normal soil conditions. Floor, wall and roof structures. Interior ceiling, wall and floor finishes. Exterior wall finish and roof cover. Interior partitions. Cabinets, doors and windows. Basic electrical systems and lighting fixtures. Rough plumbing and fixtures. Permits and fees. Contractor's mark-up.

General Office Buildings – Masonry or Concrete

Exterior Suite Entrances, Length Less Than Twice Width

Estimating Procedure

1. Use these figures to estimate general office buildings in which access to each suite is through an exterior entrance. Medical and dental offices have smaller rooms and more plumbing fixtures than general offices and should be estimated with figures from the Medical and Dental Buildings Section. See page 151.
2. Establish the building quality class by applying the information on page 135.
3. Compute the first floor area. This should include everything within the exterior walls and all insets outside the main walls but under the main roof.
4. If the first floor wall height is more or less than 10 feet, add to or subtract from the first floor square foot cost below the appropriate amount from the Wall Height Adjustment Table on page 142.
5. Multiply the adjusted square foot cost by the first floor area.
6. Deduct, if appropriate, for common walls or no wall finish, Use the figures on page 142.
7. If there are second or higher floors, compute the square foot area on each floor. Locate the appropriate square foot cost from the table at the bottom of this page. Adjust this figure for a wall height more or less than 9 feet, using the figures on page 142. Multiply the adjusted cost by the square foot area on each floor. Use the figures on page 142 to deduct for common walls or no wall finish. Add the result to the cost from step 6 above.
8. Multiply the total cost by the location factor on page 7 or 8.
9. Add the cost of heating and air conditioning systems, elevators, fire sprinklers, exterior signs, paving and curbing, miscellaneous yard improvements. See pages 236 to 248.

First Story – Square Foot Area

Quality Class	1,000	1,500	2,000	2,500	3,000	4,000	5,000	7,500	10,000	15,000	20,000
Exceptional	419.58	385.08	365.21	351.99	342.41	329.13	320.28	306.64	298.72	289.42	284.02
1, Best	387.12	355.23	336.98	324.68	315.84	303.62	295.45	282.91	275.60	266.95	262.00
1 & 2	351.36	322.45	305.94	294.78	286.78	275.74	268.25	256.93	250.29	242.45	237.90
2, Good	325.04	298.29	282.89	272.56	265.27	254.95	248.05	237.56	231.38	224.21	219.97
2 & 3	296.60	272.27	258.15	248.88	242.05	232.75	226.42	216.81	211.25	204.60	200.75
3, Average	274.32	251.75	238.67	230.08	223.89	215.24	209.38	200.48	195.28	189.25	185.67
3 & 4	248.18	227.73	215.96	208.19	202.47	194.65	189.36	181.37	176.67	171.19	167.98
4, Low	219.73	201.62	191.26	184.31	179.31	172.31	167.75	160.63	156.45	151.54	148.75

Second and Higher Stories – Square Foot Area

Quality Class	1,000	1,500	2,000	2,500	3,000	4,000	5,000	7,500	10,000	15,000	20,000
Exceptional	380.26	351.14	334.14	322.71	314.32	302.67	294.80	282.69	275.48	267.09	262.12
1, Best	357.44	330.08	314.09	303.28	295.45	284.43	277.13	265.68	258.93	251.07	246.28
1 & 2	326.52	301.50	286.97	277.13	269.89	259.86	253.09	242.73	236.52	229.27	225.09
2, Good	304.11	280.88	267.28	258.07	251.38	242.02	235.78	225.98	220.33	213.61	209.54
2 & 3	276.73	255.69	243.20	234.87	228.76	220.31	214.53	205.81	200.48	194.39	190.76
3, Average	253.74	234.38	220.51	215.31	209.74	201.97	196.61	188.60	183.77	178.26	174.91
3 & 4	227.18	209.82	199.63	192.72	187.73	180.75	176.19	168.83	164.63	159.61	156.53
4, Low	202.41	186.90	177.84	171.71	167.27	160.97	156.85	150.40	146.55	142.07	139.43

Exterior Suite Entrances, Length Between 2 and 4 Times Width

Estimating Procedure

1. Use these figures to estimate general office buildings in which access to each suite is through an exterior entrance. Medical and dental offices have smaller rooms and more plumbing fixtures than general offices and should be estimated with figures from the Medical and Dental Buildings Section. See page 151.
2. Establish the building quality class by applying the information on page 135.
3. Compute the first floor area. This should include everything within the exterior walls and all insets outside the main walls but under the main roof.
4. If the first floor wall height is more or less than 10 feet, add to or subtract from the first floor square foot cost below the appropriate amount from the Wall Height Adjustment Table on page 142.
5. Multiply the adjusted square foot cost by the first floor area.
6. Deduct, if appropriate, for common walls or no wall finish. Use the figures on page 142.
7. If there are second or higher floors, compute the square foot area on each floor. Locate the appropriate square foot cost from the table at the bottom of this page. Adjust this figure for a wall height more or less than 9 feet, using the figures on page 142. Multiply the adjusted cost by the square foot area on each floor. Use the figures on page 142 to deduct for common walls or no wall finish. Add the result to the cost from step 6 above.
8. Multiply the total cost by the location factor on page 7 or 8.
9. Add the cost of heating and air conditioning systems, elevators, fire sprinklers, exterior signs, paving and curbing, miscellaneous yard improvements. See pages 236 to 248.

First Story – Square Foot Area

Quality Class	1,000	1,500	2,000	2,500	3,000	4,000	5,000	7,500	10,000	15,000	20,000
Exceptional	455.74	411.52	386.42	370.04	358.16	342.01	331.31	315.10	305.69	294.89	288.57
1, Best	420.42	379.63	356.63	341.43	330.47	315.56	305.68	290.74	282.07	272.02	266.29
1 & 2	382.20	345.13	324.20	310.26	300.35	286.93	277.86	264.36	256.49	247.38	242.02
2, Good	353.33	319.01	299.63	286.89	277.73	265.12	256.85	244.27	237.01	228.65	223.79
2 & 3	322.87	291.53	273.77	262.12	253.74	242.32	234.72	223.27	216.62	208.89	204.43
3, Average	298.27	269.55	253.14	242.33	234.65	224.07	216.93	206.33	200.24	193.13	189.03
3 & 4	270.82	244.54	229.70	219.87	212.86	203.19	196.86	187.29	181.73	175.22	171.53
4, Low	240.54	217.24	203.95	195.28	189.03	180.53	174.91	166.32	161.41	155.66	152.35

Second and Higher Stories – Square Foot Area

Quality Class	1,000	1,500	2,000	2,500	3,000	4,000	5,000	7,500	10,000	15,000	20,000
Exceptional	420.96	381.02	358.53	343.73	333.05	318.55	309.01	294.37	285.97	276.29	270.58
1, Best	387.92	351.22	330.41	316.79	306.99	293.58	284.68	271.26	263.57	254.63	249.39
1 & 2	355.51	321.83	302.73	290.28	281.24	268.98	260.83	248.62	241.48	233.24	228.63
2, Good	330.55	299.10	281.61	269.89	261.53	250.12	242.56	231.09	224.50	216.90	212.41
2 & 3	300.61	272.04	256.02	245.43	237.80	227.34	220.53	210.13	204.16	197.23	193.26
3, Average	274.56	248.57	233.88	224.23	217.31	207.83	201.58	192.08	186.56	180.26	176.52
3 & 4	249.47	225.93	212.51	203.65	197.36	188.74	183.02	174.43	169.48	163.69	160.37
4, Low	220.80	199.88	184.73	180.35	174.68	167.09	162.02	154.36	150.07	144.91	141.86

General Office Buildings – Masonry or Concrete

Exterior Suite Entrances, Length More Than 4 Times Width

Estimating Procedure

1. Use these figures to estimate general office buildings in which access to each suite is through an exterior entrance. Medical and dental offices have smaller rooms and more plumbing fixtures than general offices and should be estimated with figures from the Medical and Dental Buildings Section. See page 151.
2. Establish the building quality class by applying the information on page 135.
3. Compute the first floor area. This should include everything within the exterior walls and all insets outside the main walls but under the main roof.
4. If the first floor wall height is more or less than 10 feet, add to or subtract from the first floor square foot cost below the appropriate amount from the Wall Height Adjustment Table on page 142.
5. Multiply the adjusted square foot cost by the first floor area.
6. Deduct, if appropriate, for common walls or no wall finish. Use the figures on page 142.
7. If there are second or higher floors, compute the square foot area on each floor. Locate the appropriate square foot cost from the table at the bottom of this page. Adjust this figure for a wall height more or less than 9 feet, using the figures on page 142. Multiply the adjusted cost by the square foot area on each floor. Use the figures on page 142 to deduct for common walls or no wall finish. Add the result to the cost from step 6 above.
8. Multiply the total cost by the location factor on page 7 or 8.
9. Add the cost of heating and air conditioning systems, elevators, fire sprinklers, exterior signs, paving and curbing, miscellaneous yard improvements. See pages 236 to 248.

First Story – Square Foot Area

Quality Class	1,000	1,500	2,000	2,500	3,000	4,000	5,000	7,500	10,000	15,000	20,000
Exceptional	497.95	444.27	413.99	394.17	379.83	360.48	347.64	328.24	317.16	304.19	296.77
1, Best	458.14	408.65	380.89	362.55	349.47	331.70	319.82	301.97	291.72	279.84	272.99
1 & 2	417.60	372.51	347.12	330.50	318.55	302.32	291.53	275.25	265.93	255.07	248.80
2, Good	384.90	343.31	319.97	304.63	293.61	278.60	268.71	253.74	245.13	235.08	229.32
2 & 3	352.31	314.33	292.87	278.81	268.75	254.97	245.87	232.19	224.31	215.29	209.96
3, Average	327.39	292.10	272.26	259.09	249.82	237.04	228.65	215.89	208.53	200.01	195.11
3 & 4	296.27	264.35	246.28	234.49	226.07	214.51	206.87	195.28	188.69	181.03	176.52
4, Low	263.59	235.10	219.18	208.61	201.14	190.83	184.09	173.82	167.88	160.97	157.03

Second and Higher Stories – Square Foot Area

Quality Class	1,000	1,500	2,000	2,500	3,000	4,000	5,000	7,500	10,000	15,000	20,000
Exceptional	457.24	408.34	380.92	362.88	350.07	332.43	320.80	303.39	293.25	281.67	274.87
1, Best	421.80	376.65	351.40	334.68	322.81	306.64	295.81	279.76	270.54	259.76	253.55
1 & 2	386.52	345.16	322.02	306.72	295.88	281.04	271.20	256.49	247.93	238.15	232.33
2, Good	360.68	322.18	300.50	286.30	276.15	262.37	253.09	239.37	231.36	222.17	210.51
2 & 3	329.45	294.28	274.44	261.52	252.20	239.52	231.10	218.57	211.32	202.96	198.04
3, Average	304.12	271.72	253.41	241.48	232.84	221.13	213.41	201.83	195.08	187.32	182.87
3 & 4	275.88	246.44	229.86	219.05	211.25	200.57	193.56	183.01	176.98	169.96	165.82
4, Low	246.15	219.77	204.95	195.28	188.38	178.89	172.72	163.29	157.89	151.59	147.97

General Office Buildings – Masonry or Concrete

Interior Suite Entrances, Length Less Than Twice Width

Estimating Procedure

1. Use these figures to estimate general office buildings in which access to each suite is through an interior corridor. Medical and dental offices have smaller rooms and more plumbing fixtures than general offices and should be estimated with figures from the Medical and Dental Buildings Section. See page 151.
2. Establish the building quality class by applying the information on page 135.
3. Compute the first floor area. This should include everything within the exterior walls and all insets outside the main walls but under the main roof.
4. If the first floor wall height is more or less than 10 feet, add to or subtract from the first floor square foot cost below the appropriate amount from the Wall Height Adjustment Table on page 142.
5. Multiply the adjusted square foot cost by the first floor area.
6. Deduct, if appropriate, for common walls or no wall finish. Use the figures on page 142.
7. If there are second or higher floors, compute the square foot area on each floor. Locate the appropriate square foot cost from the table at the bottom of this page. Adjust this figure for a wall height more or less than 9 feet, using the figures on page 142. Multiply the adjusted cost by the square foot area on each floor. Use the figures on page 142 to deduct for common walls or no wall finish. Add the result to the cost from step 6 above.
8. Multiply the total cost by the location factor on page 7 or 8.
9. Add the cost of heating and air conditioning systems, elevators, fire sprinklers, exterior signs, paving and curbing, miscellaneous yard improvements. See pages 236 to 248.

First Story – Square Foot Area

Quality Class	2,000	2,500	3,000	4,000	5,000	7,500	10,000	15,000	20,000	30,000	40,000
Exceptional	354.23	341.27	331.82	318.80	310.07	296.80	288.99	279.85	274.54	268.36	264.61
1, Best	326.63	314.70	305.95	294.01	285.85	273.59	266.49	258.11	253.19	247.44	244.05
1 & 2	299.56	288.59	280.62	269.64	262.15	251.04	244.45	236.70	232.18	226.98	223.88
2, Good	279.84	269.64	262.15	251.92	244.99	234.47	228.29	221.20	216.90	212.04	209.07
2 & 3	255.78	246.40	239.52	230.09	223.88	214.20	208.61	202.15	198.25	193.78	191.13
3, Average	236.38	227.76	221.40	212.77	206.87	198.05	192.79	186.83	183.25	179.11	176.56
3 & 4	213.51	205.78	200.03	192.19	186.84	178.82	174.23	168.74	165.54	161.75	159.52
4, Low	188.69	181.80	176.80	169.79	165.22	158.07	153.97	149.05	146.27	143.03	140.94

Second and Higher Stories – Square Foot Area

Quality Class	2,000	2,500	3,000	4,000	5,000	7,500	10,000	15,000	20,000	30,000	40,000
Exceptional	330.85	319.11	310.53	298.76	290.93	278.82	271.73	263.43	258.60	252.79	249.39
1, Best	305.17	294.39	286.60	275.65	268.40	257.16	250.67	243.06	238.48	233.14	230.01
1 & 2	280.73	270.85	263.63	253.71	246.92	236.65	230.69	223.61	219.44	214.51	211.63
2, Good	262.64	251.73	246.66	237.32	230.96	221.40	215.70	209.10	205.30	200.69	197.99
2 & 3	239.37	230.88	224.65	216.18	210.44	201.72	196.55	190.57	187.05	182.89	181.23
3, Average	219.87	212.10	206.50	198.59	193.33	185.30	180.56	175.10	171.80	168.02	165.73
3 & 4	197.97	190.93	185.83	178.79	174.01	166.82	162.54	157.66	154.69	151.21	149.15
4, Low	174.42	168.25	163.77	157.60	153.44	146.98	143.29	138.93	136.37	133.25	131.51

General Office Buildings – Masonry or Concrete

Interior Suite Entrances, Length Between 2 and 4 Times Width

Estimating Procedure

1. Use these figures to estimate general office buildings in which access to each suite is through an interior corridor. Medical and dental offices have smaller rooms and more plumbing fixtures than general offices and should be estimated with figures from the Medical and Dental Buildings Section. See page 151.
2. Establish the building quality class by applying the information on page 135.
3. Compute the first floor area. This should include everything within the exterior walls and all insets outside the main walls but under the main roof.
4. If the first floor wall height is more or less than 10 feet, add to or subtract from the first floor square foot cost below the appropriate amount from the Wall Height Adjustment Table on page 142.
5. Multiply the adjusted square foot cost by the first floor area.
6. Deduct, if appropriate, for common walls or no wall finish. Use the figures on page 142.
7. If there are second or higher floors, compute the square foot area on each floor. Locate the appropriate square foot cost from the table at the bottom of this page. Adjust this figure for a wall height more or less than 9 feet, using the figures on page 142. Multiply the adjusted cost by the square foot area on each floor. Use the figures on page 142 to deduct for common walls or no wall finish. Add the result to the cost from step 6 above.
8. Multiply the total cost by the location factor on page 7 or 8.
9. Add the cost of heating and air conditioning systems, elevators, fire sprinklers, exterior signs, paving and curbing, miscellaneous yard improvements. See pages 236 to 248.

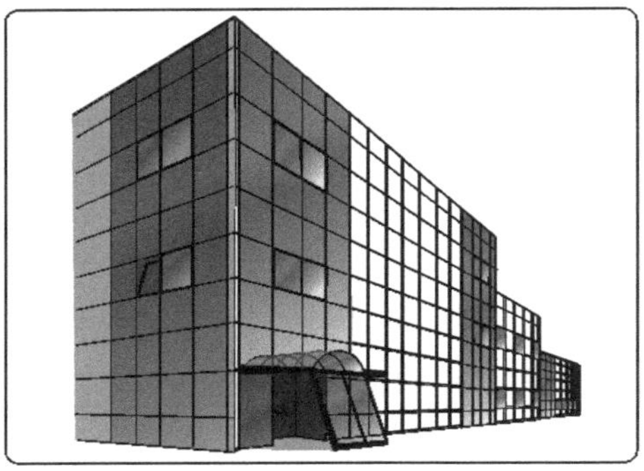

General Offices, Class 1 & 2

First Story – Square Foot Area

Quality Class	2,000	2,500	3,000	4,000	5,000	7,500	10,000	15,000	20,000	30,000	40,000
Exceptional	373.80	357.80	346.36	330.51	320.04	304.03	294.84	284.09	277.84	270.47	266.16
1, Best	344.74	330.04	319.43	304.90	295.20	280.54	272.01	262.13	256.22	249.51	245.54
1 & 2	315.82	302.67	292.96	279.60	270.72	257.17	249.45	240.27	235.00	228.77	225.07
2, Good	277.08	265.50	257.28	245.95	238.39	226.92	220.25	212.46	207.94	202.68	199.49
2 & 3	270.05	258.56	250.23	238.84	231.19	219.71	213.05	205.22	200.74	195.42	192.34
3, Average	249.81	239.15	231.48	220.91	213.92	203.26	197.08	189.85	185.71	180.74	177.89
3 & 4	226.02	216.37	209.38	199.83	193.62	183.92	178.30	171.89	167.98	163.56	160.89
4, Low	200.01	191.51	185.21	176.81	171.26	162.67	157.78	151.97	148.59	144.74	142.42

Second and Higher Stories – Square Foot Area

Quality Class	2,000	2,500	3,000	4,000	5,000	7,500	10,000	15,000	20,000	30,000	40,000
Exceptional	347.71	333.39	322.90	308.71	299.16	284.78	276.44	266.69	260.94	254.28	250.30
1, Best	320.51	307.14	297.67	284.48	275.76	262.50	254.79	245.82	240.58	234.42	230.83
1 & 2	295.17	282.94	274.11	262.01	253.97	241.74	234.68	226.33	221.55	215.85	212.64
2, Good	277.08	265.50	257.28	245.95	238.39	226.92	220.25	212.46	207.94	202.68	199.49
2 & 3	252.21	241.74	234.22	223.97	217.02	206.56	200.48	193.38	189.33	184.47	181.66
3, Average	231.85	222.40	215.56	206.03	199.69	190.08	184.52	178.01	174.27	169.77	167.24
3 & 4	209.59	200.94	194.65	186.03	180.30	171.65	166.58	160.71	157.34	153.27	150.91
4, Low	185.21	177.57	172.10	164.49	159.39	151.71	147.21	142.05	139.03	135.54	133.41

Interior Suite Entrances, Length More Than 4 Times Width

Estimating Procedure

1. Use these figures to estimate general office buildings in which access to each suite is through an interior corridor. Medical and dental offices have smaller rooms and more plumbing fixtures than general offices and should be estimated with figures from the Medical and Dental Buildings Section. See page 151.
2. Establish the building quality class by applying the information on page 135.
3. Compute the first floor area. This should include everything within the exterior walls and all insets outside the main walls but under the main roof.
4. If the first floor wall height is more or less than 10 feet, add to or subtract from the first floor square foot cost below the appropriate amount from the Wall Height Adjustment Table on page 142.
5. Multiply the adjusted square foot cost by the first floor area.
6. Deduct, if appropriate, for common walls or no wall finish. Use the figures on page 142.
7. If there are second or higher floors, compute the square foot area on each floor. Locate the appropriate square foot cost from the table at the bottom of this page. Adjust this figure for a wall height more or less than 9 feet, using the figures on page 142. Multiply the adjusted cost by the square foot area on each floor. Use the figures on page 142 to deduct for common walls or no wall finish. Add the result to the cost from step 6 above.
8. Multiply the total cost by the location factor on page 7 or 8.
9. Add the cost of heating and air conditioning systems, elevators, fire sprinklers, exterior signs, paving and curbing, miscellaneous yard improvements. See pages 236 to 248.

First Story – Square Foot Area

Quality Class	2,000	2,500	3,000	4,000	5,000	7,500	10,000	15,000	20,000	30,000	40,000
Exceptional	400.39	380.51	366.37	347.08	334.29	315.11	304.08	291.39	284.04	275.48	270.51
1, Best	369.43	351.11	338.03	320.13	308.45	290.73	280.55	268.79	262.04	254.23	249.56
1 & 2	338.67	321.92	309.84	293.52	282.74	266.57	257.16	246.48	240.24	233.01	228.80
2, Good	316.33	300.63	289.37	274.18	264.12	248.92	240.22	230.13	224.37	217.69	213.74
2 & 3	289.16	274.74	264.48	250.56	241.36	227.50	219.52	210.38	205.04	198.98	195.31
3, Average	268.10	254.79	245.34	232.31	223.83	210.99	203.57	195.17	190.13	184.47	181.09
3 & 4	242.93	230.91	222.29	210.59	202.81	191.26	184.55	176.80	172.35	167.16	164.05
4, Low	215.73	205.07	197.38	186.96	180.17	169.72	163.87	157.01	153.11	148.38	145.79

Second and Higher Stories – Square Foot Area

Quality Class	2,000	2,500	3,000	4,000	5,000	7,500	10,000	15,000	20,000	30,000	40,000
Exceptional	367.63	349.65	337.01	320.11	309.26	293.50	284.68	274.75	269.13	262.84	259.24
1, Best	336.36	319.86	308.26	292.97	282.99	268.63	260.51	251.41	246.29	240.54	237.23
1 & 2	311.66	296.44	285.72	271.42	262.29	248.88	241.42	233.01	228.17	222.90	219.75
2, Good	293.44	279.08	269.00	255.51	246.88	234.29	227.24	219.36	214.87	209.87	206.91
2 & 3	267.08	253.97	244.75	232.59	224.75	213.23	206.88	199.67	195.56	191.02	188.40
3, Average	247.06	234.99	226.39	215.12	207.90	197.21	191.35	184.67	180.94	176.66	174.26
3 & 4	243.33	231.30	223.03	211.88	204.63	194.25	188.45	181.82	178.14	173.90	171.49
4, Low	198.27	188.55	181.81	172.73	166.84	158.32	153.62	148.24	145.25	141.78	139.82

General Office Buildings – Masonry or Concrete

Wall Height Adjustments

The square foot costs for general offices are based on the wall heights of 10 feet for first floors and 9 feet for higher floors. The main or first floor height is the distance from the bottom of the floor slab or joists to the top of the roof slab or ceiling joists. Second and higher floors are measured from the top of the floor slab or floor joists to the top of the roof slab or ceiling joists. Add or subtract the amount listed in this table to or from the square foot of floor cost for each foot of wall height more or less than 10 feet, if adjusting for a first floor, and 9 feet, if adjusting for upper floors.

Square Foot Area

Quality Class	1,000	1,500	2,000	3,000	4,000	5,000	7,500	10,000	15,000	20,000	40,000
1, Best	11.17	8.77	7.41	5.85	5.04	4.46	3.58	3.06	2.50	2.09	1.57
2, Good	9.71	7.58	6.40	5.07	4.40	3.87	3.12	2.65	2.10	1.91	1.33
3, Average	8.53	6.70	5.64	4.54	3.87	3.41	2.65	2.32	1.94	1.64	1.17
4, Low	7.58	5.84	5.06	3.99	3.42	3.02	2.40	2.09	1.74	1.36	1.08

Perimeter (Common) Wall Adjustment

A common wall exists when two buildings share one wall. Adjust for common walls by deducting the linear foot cost below from the total structure cost. In some structures one or more walls are not owned at all. In this case, deduct the "No Ownership" cost per linear foot of wall not owned. Where a perimeter wall remains unfinished, deduct the "lack of exterior finish" cost.

First Story

Class	For a Common Wall, Deduct Per L.F.	For No Wall Ownership, Deduct Per L.F.	For Lack of Exterior Finish, Deduct Per L.F.
1	$565	$1,142	$440
2	474	946	285
3	360	692	179
4	274	547	107

Second and Higher Stories

Class	For a Common Wall, Deduct Per L.F.	For No Wall Ownership, Deduct Per L.F.	For Lack of Exterior Finish, Deduct Per L.F.
1	$547	$1,089	$445
2	470	916	273
3	360	694	179
4	339	669	111

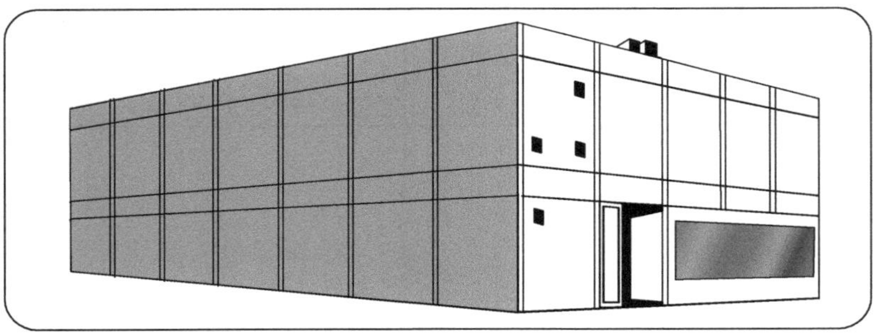

General Office, Class 3

General Office Buildings – Wood Frame

Quality Classification

	Class 1 Best Quality	Class 2 Good Quality	Class 3 Average Quality	Class 4 Low Quality
Foundation (12% of total cost)	Reinforced concrete.	Reinforced concrete.	Reinforced concrete.	Reinforced concrete.
First Floor Structure (8% of total cost)	Reinforced concrete slab on grade or standard wood frame.	Reinforced concrete slab on grade or standard wood frame.	Reinforced concrete slab on grade or 4" x 6" girders with 2" T&G subfloor.	Reinforced concrete slab on grade.
Upper Floor Structure (10% of total cost)	Standard wood frame. Plywood and 1-1/2" light-weight concrete subfloor.	Standard wood frame. Plywood and 1-1/2" light- weight concrete subfloor.	Standard wood frame. 5/8" plywood subfloor.	Standard wood frame. 5/8" plywood subfloor.
Walls (9% of total cost)	Standard wood frame.	Standard wood frame.	Standard wood frame.	Standard wood frame.
Roof Structure (6% of total cost)	Standard wood frame, flat or low pitch.	Standard wood frame, flat or low pitch.	Standard wood frame, flat or low pitch.	Standard wood frame, flat or low pitch.
Exterior Wall Finish (8% of total cost)	Good wood siding with 10 to 20% brick or stone veneer.	Average wood siding or stucco and 10 to 20% brick or stone veneer.	Stucco with some wood trim or cheap wood siding.	Stucco.
Windows (5% of total cost)	Average number in good aluminum frame. Fixed float glass in good frame on front side.	Average number in good aluminum frame. Some fixed float glass in front.	Average number of average aluminum sliding type.	Average number of low cost aluminum sliding type.
Roof Cover (5% of total cost)	5 ply built-up roofing on flat roofs. Heavy shake or tile on sloping roofs.	5 ply built-up roofing on flat roofs. Average shake or composition, tar and large rock on sloping roofs. Single ply membrane.	4 ply built-up roofing on flat roofs. Wood shingle or composition, tar and pea gravel on sloping roofs.	3 ply built-up roofing on flat roofs. Composition shingle on sloping roofs.
Overhang (3% of total cost)	3' closed overhang, fully guttered.	2' closed overhang, fully guttered.	None on flat roofs. 18" on sloping roofs, fully guttered.	None on flat roofs. 12" to 16" open on sloping roofs, gutters over entrances.
Floor Finishes:				
Offices (3% of total cost)	Very good carpet.	Good carpet.	Average grade carpet.	Minimum grade tile.
Corridors (2% of total cost)	Solid vinyl tile or carpet.	Resilient tile.	Composition tile.	Minimum grade tile.
Bathrooms (1% of total cost)	Sheet vinyl or ceramic tile.	Sheet vinyl or ceramic tile.	Composition tile.	Minimum grade tile.
Interior Wall Finish:				
Offices (6% of total cost)	Good hardwood veneer paneling.	Hardwood veneer paneling or vinyl wall cover.	Gypsum wallboard, texture and paint.	Gypsum wallboard, texture and paint.
Corridors (4% of total cost)	Good hardwood veneer paneling.	Gypsum wallboard and vinyl wall cover.	Gypsum wallboard, texture and paint.	Gypsum wallboard, texture and paint.
Bathrooms (2% of total cost)	Gypsum wallboard and enamel with ceramic tile wainscot.	Gypsum wallboard and enamel or vinyl wall covering.	Gypsum wallboard and enamel.	Gypsum wallboard, texture and paint.
Ceiling Finish (4% of total cost)	Suspended "T" bar and acoustical tile.	Gypsum wallboard and acoustical tile.	Gypsum wallboard and acoustical texture.	Gypsum wallboard and paint.
Plumbing (6% of total cost)	Copper tubing, top quality fixtures.	Copper tubing, good fixtures.	Copper tubing, standard fixtures.	Copper tubing, economy fixtures.
Lighting (6% of total cost)	Conduit wiring, good fixtures.	Conduit wiring, good fixtures.	Romex or conduit wiring, average fixtures.	Romex wiring economy fixtures.

Note: Use the percent of total cost to help identify the correct quality classification.

Square foot costs include the following components: Foundations as required for normal soil conditions. Floor, wall and roof structures. Interior floor, wall and ceiling finishes. Exterior wall finish and roof cover. Interior partitions. Cabinets, doors and windows. Basic electrical systems and lighting fixtures. Rough plumbing and fixtures. Permits and fees. Contractor's mark-up.

General Office Buildings – Wood Frame

Exterior Suite Entrances, Length Less Than Twice Width

Estimating Procedure

1. Use these figures to estimate general office buildings in which access to each suite is through an exterior entrance. Medical and dental offices have smaller rooms and more plumbing fixtures than general offices and should be estimated with figures from the Medical and Dental Buildings Section. See page 151.
2. Establish the building quality class by applying the information on page 143.
3. Compute the first floor area. This should include everything within the exterior walls and all insets outside the main walls but under the main roof.
4. If the first floor wall height is more or less than 10 feet, add to or subtract from the first floor square foot cost below the appropriate amount from the Wall Height Adjustment Table on page 150.
5. Multiply the adjusted square foot cost by the first floor area.
6. Deduct, if appropriate, for common walls or no wall finish. Use the figures on page 150.
7. If there are second or higher floors, compute the square foot area on each floor. Locate the appropriate square foot cost from the table at the bottom of this page. Adjust this figure for a wall height more or less than 9 feet, using the figures on page 150. Multiply the adjusted cost by the square foot area on each floor. Use the figures on page 150 to deduct for common walls or no wall finish. Add the result to the cost from step 6 above.
8. Multiply the total cost by the location factor on page 7 or 8.
9. Add the cost of heating and air conditioning systems, elevators, fire sprinklers, exterior signs, paving and curbing, miscellaneous yard improvements. See pages 236 to 248.

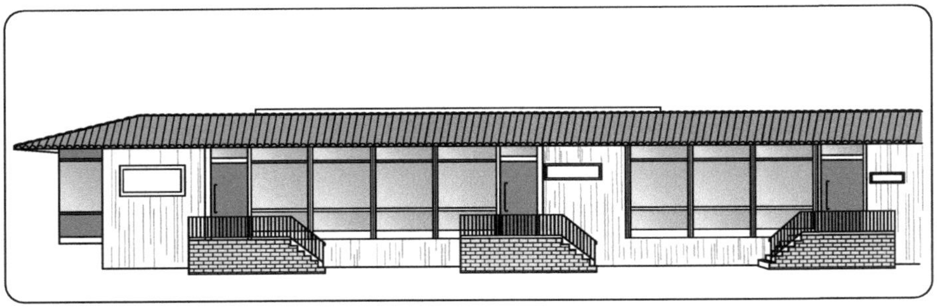

General Offices, Class 1 & 2

First Story – Square Foot Area

Quality Class	1,000	1,500	2,000	2,500	3,000	4,000	5,000	7,500	10,000	15,000	20,000
Exceptional	256.85	242.86	234.91	229.51	225.63	220.30	216.75	211.30	208.19	204.43	202.26
1, Best	237.03	224.09	216.72	211.81	208.20	203.30	199.97	195.01	192.04	188.64	186.53
1 & 2	214.67	202.94	196.21	191.81	188.53	184.10	181.13	176.56	173.91	170.87	169.01
2, Good	195.57	184.94	178.85	174.73	171.91	167.74	165.04	160.94	158.46	155.69	154.04
2 & 3	178.49	168.79	163.23	159.46	156.78	153.08	150.56	146.84	144.66	142.06	140.56
3, Average	163.30	154.41	149.28	145.91	143.44	140.01	137.74	134.30	132.27	129.96	128.57
3 & 4	145.95	138.01	133.37	130.38	128.15	125.16	123.15	120.02	118.28	116.10	114.87
4, Low	127.68	120.82	116.81	114.09	112.22	109.53	107.78	105.10	103.51	101.72	100.54

Second and Higher Stories – Square Foot Area

Quality Class	1,000	1,500	2,000	2,500	3,000	4,000	5,000	7,500	10,000	15,000	20,000
Exceptional	224.40	215.45	210.26	206.84	204.40	201.01	198.68	195.25	193.23	190.85	189.49
1, Best	206.75	198.44	193.66	190.63	188.32	185.14	183.05	179.84	177.97	175.77	174.54
1 & 2	188.71	181.18	176.88	173.96	171.91	169.01	167.09	164.22	162.47	160.50	159.32
2, Good	172.37	165.50	161.58	158.95	156.99	154.36	152.58	150.07	148.43	146.53	145.56
2 & 3	155.51	149.28	145.72	143.32	141.62	139.31	137.73	135.30	133.89	132.23	131.31
3, Average	139.78	134.17	131.03	128.84	127.34	125.23	123.80	121.61	120.40	118.91	118.05
3 & 4	123.68	118.70	115.93	114.02	112.64	110.81	109.52	107.63	106.48	105.21	104.44
4, Low	107.65	103.34	100.91	99.21	98.02	96.40	95.27	93.62	92.64	91.56	90.91

General Office Buildings – Wood Frame

Exterior Suite Entrances, Length Between 2 and 4 Times Width

Estimating Procedure

1. Use these figures to estimate general office buildings in which access to each suite is through an exterior entrance. Medical and dental offices have smaller rooms and more plumbing fixtures than general offices and should be estimated with figures from the Medical and Dental Buildings Section. See page 151.
2. Establish the building quality class by applying the information on page 153.
3. Compute the first floor area. This should include everything within the exterior walls and all insets outside the main walls but under the main roof.
4. If the first floor wall height is more or less than 10 feet, add to or subtract from the first floor square foot cost below the appropriate amount from the Wall Height Adjustment Table on page 150.
5. Multiply the adjusted square foot cost by the first floor area.
6. Deduct, if appropriate, for common walls or no wall finish. Use the figures on page 150.
7. If there are second or higher floors, compute the square foot area on each floor. Locate the appropriate square foot cost from the table at the bottom of this page. Adjust this figure for a wall height more or less than 9 feet, using the figures on page 150. Multiply the adjusted cost by the square foot area on each floor. Use the figures on page 150 to deduct for common walls or no wall finish. Add the result to the cost from step 6 above.
8. Multiply the total cost by the location factor on page 7 or 8.
9. Add the cost of heating and air conditioning systems, elevators, fire sprinklers, exterior signs, paving and curbing, miscellaneous yard improvements. See pages 236 to 248.

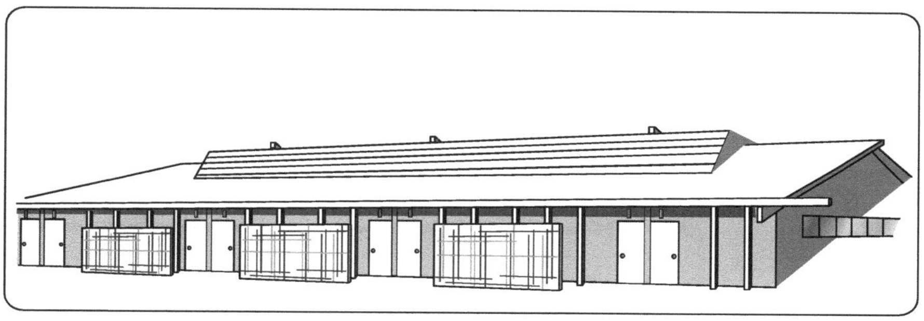

General Offices, Class 3

First Story – Square Foot Area

Quality Class	1,000	1,500	2,000	2,500	3,000	4,000	5,000	7,500	10,000	15,000	20,000
Exceptional	271.36	253.62	243.53	236.95	232.13	225.63	221.35	214.85	211.06	206.74	204.18
1, Best	250.25	233.92	224.62	218.57	214.14	208.14	204.21	198.11	194.69	190.70	188.35
1 & 2	226.53	211.65	203.31	197.79	193.83	188.41	184.82	179.38	176.21	172.53	170.40
2, Good	206.49	192.97	185.24	180.23	176.64	171.70	168.41	163.46	160.55	157.33	155.35
2 & 3	188.64	176.27	169.29	164.64	161.44	156.90	153.86	149.39	146.77	143.76	141.95
3, Average	172.57	161.24	154.85	150.63	147.65	143.55	140.74	136.59	134.21	131.43	129.85
3 & 4	154.53	144.32	138.63	134.87	132.15	128.50	126.06	122.32	120.21	117.70	116.25
4, Low	135.46	126.49	121.47	118.32	115.90	112.62	110.47	107.22	105.38	103.16	101.88

Second and Higher Stories – Square Foot Area

Quality Class	1,000	1,500	2,000	2,500	3,000	4,000	5,000	7,500	10,000	15,000	20,000
Exceptional	233.63	221.81	215.30	210.97	208.00	203.82	201.16	197.11	194.84	192.13	190.61
1, Best	215.51	204.62	198.59	194.65	191.81	188.04	185.51	181.86	179.72	177.32	175.87
1 & 2	195.93	186.22	180.69	177.13	174.58	171.16	168.90	165.51	163.55	161.36	160.00
2, Good	178.77	169.72	164.64	161.44	159.12	156.03	153.86	150.83	149.09	147.04	145.85
2 & 3	162.27	154.04	149.48	146.49	144.40	141.56	139.64	136.84	135.29	133.41	132.35
3, Average	146.15	138.72	134.66	131.98	130.05	127.46	125.79	123.26	121.86	120.25	119.25
3 & 4	128.92	122.44	118.81	116.49	114.81	112.55	111.08	108.81	107.55	106.02	105.23
4, Low	112.87	107.18	103.96	101.92	100.46	98.54	97.17	95.24	94.15	92.83	92.16

General Office Buildings – Wood Frame

Exterior Suite Entrances, Length More Than 4 Times Width

Estimating Procedure

1. Use these figures to estimate general office buildings in which access to each suite is through an exterior entrance. Medical and dental offices have smaller rooms and more plumbing fixtures than general offices and should be estimated with figures from the Medical and Dental Buildings Section. See page 151.
2. Establish the building quality class by applying the information on page 153.
3. Compute the first floor area. This should include everything within the exterior walls and all insets outside the main walls but under the main roof.
4. If the first floor wall height is more or less than 10 feet, add to or subtract from the first floor square foot cost below the appropriate amount from the Wall Height Adjustment Table on page 150.
5. Multiply the adjusted square foot cost by the first floor area.
6. Deduct, if appropriate, for common walls or no wall finish. Use the figures on page 150.
7. If there are second or higher floors, compute the square foot area on each floor. Locate the appropriate square foot cost from the table at the bottom of this page. Adjust this figure for a wall height more or less than 9 feet, using the figures on page 150. Multiply the adjusted cost by the square foot area on each floor. Use the figures on page 150 to deduct for common walls or no wall finish. Add the result to the cost from step 6 above.
8. Multiply the total cost by the location factor on page 7 or 8.
9. Add the cost of heating and air conditioning systems, elevators, fire sprinklers, exterior signs, paving and curbing, miscellaneous yard improvements. See pages 236 to 248.

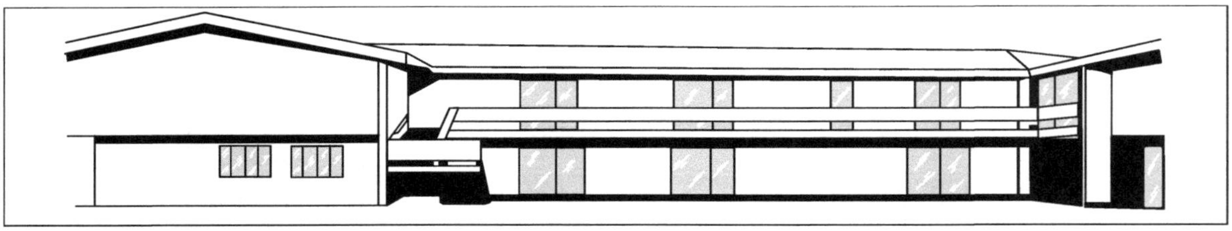

General Offices, Class 2

First Story – Square Foot Area

Quality Class	1,000	1,500	2,000	2,500	3,000	4,000	5,000	7,500	10,000	15,000	20,000
Exceptional	288.70	266.64	254.25	246.16	240.30	232.49	227.18	219.40	214.85	209.69	206.68
1, Best	266.30	245.97	234.54	227.03	221.68	214.40	209.58	202.37	198.11	193.42	190.63
1 & 2	241.20	222.83	212.40	205.68	200.85	194.18	189.84	183.27	179.57	175.18	172.57
2, Good	219.58	202.80	193.33	187.18	182.75	176.70	172.84	166.83	163.43	159.43	157.11
2 & 3	200.80	185.42	176.88	171.18	167.11	161.61	158.00	152.53	149.46	145.84	143.76
3, Average	184.13	170.07	162.14	156.99	153.35	148.24	144.88	139.90	137.08	133.71	131.77
3 & 4	164.82	152.24	145.14	140.56	137.24	132.77	129.71	125.28	122.64	119.78	117.94
4, Low	144.70	133.61	127.46	123.35	120.42	116.50	113.83	109.96	107.66	105.10	103.54

Second and Higher Stories – Square Foot Area

Quality Class	1,000	1,500	2,000	2,500	3,000	4,000	5,000	7,500	10,000	15,000	20,000
Exceptional	243.73	229.73	221.87	216.72	213.01	208.03	204.64	199.61	196.80	193.45	191.48
1, Best	224.84	211.98	204.72	199.95	196.54	191.88	188.81	184.17	181.52	178.49	176.66
1 & 2	204.86	193.09	186.43	182.13	179.02	174.76	172.02	167.75	165.36	162.55	160.95
2, Good	187.59	176.71	170.72	166.76	163.93	160.07	157.47	153.62	151.43	148.83	147.33
2 & 3	169.67	159.89	154.44	150.83	148.24	144.77	142.39	138.95	136.92	134.64	133.29
3, Average	153.08	144.32	139.48	136.14	133.84	130.66	128.61	125.43	123.63	121.55	120.35
3 & 4	136.49	128.66	124.26	121.37	119.25	116.50	114.62	111.79	110.19	108.34	107.20
4, Low	118.71	111.91	108.08	105.56	103.83	101.30	99.70	97.27	95.81	94.18	93.31

General Office Buildings – Wood Frame

Interior Suite Entrances, Length Less Than Twice Width

Estimating Procedure

1. Use these figures to estimate general office buildings in which access to each suite is through an interior corridor. Medical and dental offices have smaller rooms and more plumbing fixtures than general offices and should be estimated with figures from the Medical and Dental Buildings Section. See page 151.
2. Establish the building quality class by applying the information on page 153.
3. Compute the first floor area. This should include everything within the exterior walls and all insets outside the main walls but under the main roof.
4. If the first floor wall height is more or less than 10 feet, add to or subtract from the first floor square foot cost below the appropriate amount from the Wall Height Adjustment Table on page 150.
5. Multiply the adjusted square foot cost by the first floor area.
6. Deduct, if appropriate, for common walls or no wall finish. Use the figures on page 150.
7. If there are second or higher floors, compute the square foot area on each floor. Locate the appropriate square foot cost from the table at the bottom of this page. Adjust this figure for a wall height more or less than 9 feet, using the figures on page 150. Multiply the adjusted cost by the square foot area on each floor. Use the figures on page 150 to deduct for common walls or no wall finish. Add the result to the cost from step 6 above.
8. Multiply the total cost by the location factor on page 7 or 8.
9. Add the cost of heating and air conditioning systems, elevators, fire sprinklers, exterior signs, paving and curbing, miscellaneous yard improvements. See page 236 to 248.

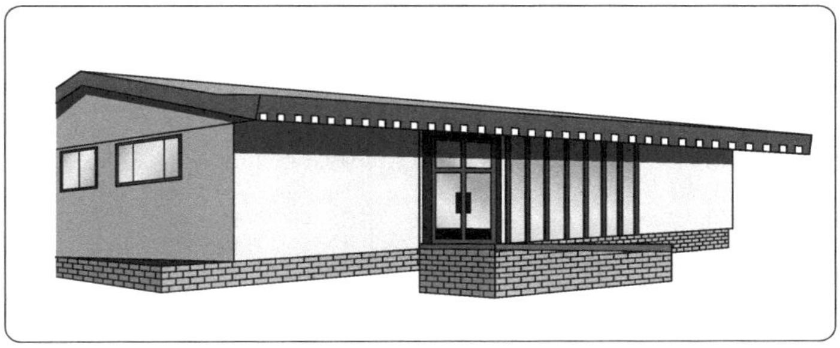

General Offices, Class 3 & 4

First Story – Square Foot Area

Quality Class	2,000	2,500	3,000	4,000	5,000	7,500	10,000	15,000	20,000	30,000	40,000
Exceptional	226.65	221.44	217.73	212.53	209.00	203.74	200.61	196.91	194.78	192.16	190.64
1, Best	208.80	204.07	200.61	195.84	192.59	187.62	184.85	181.42	179.43	177.12	175.67
1 & 2	200.49	186.84	183.65	179.22	176.35	171.89	169.15	166.11	164.28	162.14	160.83
2, Good	176.46	172.45	169.58	165.51	162.74	158.60	156.22	153.36	151.69	149.69	148.50
2 & 3	161.11	157.41	154.77	151.09	148.56	144.81	142.58	139.95	138.42	136.61	135.53
3, Average	147.33	144.02	141.58	138.19	135.91	132.43	130.42	128.03	126.57	124.97	124.03
3 & 4	132.10	129.09	126.87	123.83	121.78	118.71	116.84	114.81	113.51	111.99	111.13
4, Low	115.57	112.92	111.08	108.35	106.62	103.88	102.28	100.45	99.31	98.00	97.27

Second and Higher Stories – Square Foot Area

Quality Class	2,000	2,500	3,000	4,000	5,000	7,500	10,000	15,000	20,000	30,000	40,000
Exceptional	202.22	198.95	196.54	193.22	190.95	187.56	185.56	183.17	181.88	180.21	179.30
1, Best	186.52	183.56	181.27	178.23	176.20	173.05	171.18	169.02	167.74	166.25	165.41
1 & 2	171.79	168.99	166.87	164.13	162.28	159.32	157.66	155.62	154.49	153.07	152.28
2, Good	158.19	155.56	153.63	151.07	149.33	146.56	145.08	143.29	142.23	140.95	140.20
2 & 3	147.20	144.88	138.78	136.54	134.90	132.49	131.12	129.49	128.51	127.38	126.73
3, Average	128.96	126.84	125.35	123.20	121.75	119.60	118.34	116.83	115.94	114.91	114.33
3 & 4	114.74	112.85	111.48	109.62	108.35	106.39	105.26	103.90	103.16	102.24	101.75
4, Low	99.64	97.99	96.76	95.22	94.05	92.37	91.41	90.22	89.60	88.77	88.36

General Office Buildings – Wood Frame

Interior Suite Entrances, Length Between 2 and 4 Times Width

Estimating Procedure

1. Use these figures to estimate general office buildings in which access to each suite is through an interior corridor. Medical and dental offices have smaller rooms and more plumbing fixtures than general offices and should be estimated with figures from the Medical and Dental Buildings Section. See page 151.
2. Establish the building quality class by applying the information on page 153.
3. Compute the first floor area. This should include everything within the exterior walls and all insets outside the main walls but under the main roof.
4. If the first floor wall height is more or less than 10 feet, add to or subtract from the first floor square foot cost below the appropriate amount from the Wall Height Adjustment Table on page 150.
5. Multiply the adjusted square foot cost by the first floor area.
6. Deduct, if appropriate, for common walls or no wall finish. Use the figures on page 150.
7. If there are second or higher floors, compute the square foot area on each floor. Locate the appropriate square foot cost from the table at the bottom of this page. Adjust this figure for a wall height more or less than 9 feet, using the figures on page 150. Multiply the adjusted cost by the square foot area on each floor. Use the figures on page 150 to deduct for common walls or no wall finish. Add the result to the cost from step 6 above.
8. Multiply the total cost by the location factor on page 7 or 8.
9. Add the cost of heating and air conditioning systems, elevators, fire sprinklers, exterior signs, paving and curbing, miscellaneous yard improvements. See page 236 to 248.

First Story – Square Foot Area

Quality Class	2,000	2,500	3,000	4,000	5,000	7,500	10,000	15,000	20,000	30,000	40,000
Exceptional	234.60	228.54	224.15	218.04	213.87	207.43	203.62	199.19	196.54	193.44	191.55
1, Best	217.91	212.32	208.22	202.47	198.65	192.67	189.16	185.04	182.55	179.65	177.94
1 & 2	197.88	192.79	189.13	183.86	180.43	174.99	171.76	167.98	165.70	163.13	161.58
2, Good	182.54	177.92	174.49	169.67	166.42	161.44	158.46	154.99	152.92	150.51	149.09
2 & 3	166.83	162.54	159.41	155.01	152.04	147.52	144.81	141.62	139.73	137.47	136.22
3, Average	152.78	148.83	145.96	141.98	139.25	135.01	132.55	129.70	127.95	125.90	124.72
3 & 4	137.04	133.40	130.89	127.27	124.84	121.09	118.91	116.29	114.81	112.93	111.82
4, Low	119.91	116.83	114.57	111.41	109.34	106.01	104.02	101.83	100.46	98.86	97.87

Second and Higher Stories – Square Foot Area

Quality Class	2,000	2,500	3,000	4,000	5,000	7,500	10,000	15,000	20,000	30,000	40,000
Exceptional	207.43	203.30	200.30	196.22	193.59	189.57	187.20	184.48	182.97	181.06	180.00
1, Best	191.45	187.56	184.87	181.09	178.63	174.90	172.71	170.24	168.74	167.07	166.07
1 & 2	175.89	172.38	169.95	166.51	164.24	160.73	158.81	156.49	155.16	153.59	152.69
2, Good	162.30	159.05	156.72	153.57	151.47	148.27	146.46	144.40	143.09	141.70	140.81
2 & 3	146.71	143.77	141.64	138.78	136.92	134.05	132.35	130.46	129.36	128.04	127.32
3, Average	132.55	129.96	128.03	125.46	123.75	121.13	119.64	117.94	116.92	115.74	115.10
3 & 4	118.08	115.79	114.08	111.72	110.22	107.94	106.64	105.06	104.17	103.08	102.57
4, Low	102.18	100.25	98.75	96.74	95.43	93.40	92.30	90.92	90.14	89.24	88.74

General Office Buildings – Wood Frame

Interior Suite Entrances, Length More Than 4 Times Width

Estimating Procedure

1. Use these figures to estimate general office buildings in which access to each suite is through an interior corridor. Medical and dental offices have smaller rooms and more plumbing fixtures than general offices and should be estimated with figures from the Medical and Dental Buildings Section. See page 151.
2. Establish the building quality class by applying the information on page 153.
3. Compute the first floor area. This should include everything within the exterior walls and all insets outside the main walls but under the main roof.
4. If the first floor wall height is more or less than 10 feet, add to or subtract from the first floor square foot cost below the appropriate amount from the Wall Height Adjustment Table on page 150.
5. Multiply the adjusted square foot cost by the first floor area.
6. Deduct, if appropriate, for common walls or no wall finish. Use the figures on page 150.
7. If there are second or higher floors, compute the square foot area on each floor. Locate the appropriate square foot cost from the table at the bottom of this page. Adjust this figure for a wall height more or less than 9 feet, using the figures on page 150. Multiply the adjusted cost by the square foot area on each floor. Use the figures on page 150 to deduct for common walls or no wall finish. Add the result to the cost from step 6 above.
8. Multiply the total cost by the location factor on page 7 or 8.
9. Add the cost of heating and air conditioning systems, elevators, fire sprinklers, exterior signs, paving and curbing, miscellaneous yard improvements. See page 236 to 248.

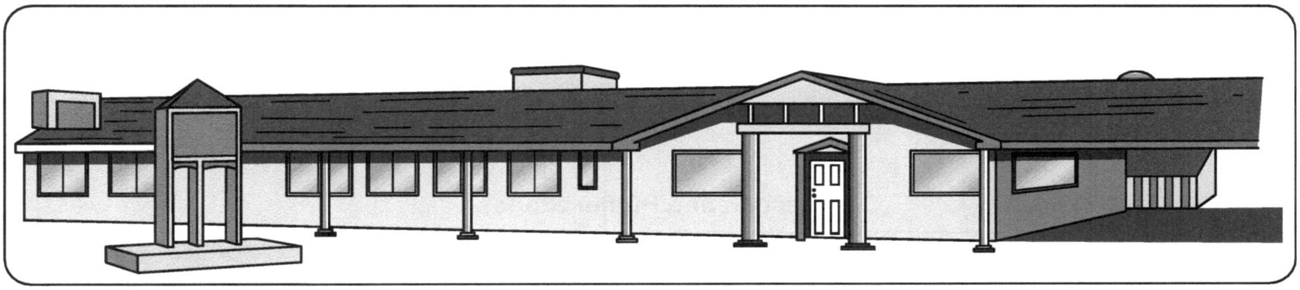

General Offices, Class 2

First Story – Square Foot Area

Quality Class	2,000	2,500	3,000	4,000	5,000	7,500	10,000	15,000	20,000	30,000	40,000
Exceptional	245.91	237.96	232.18	224.43	219.28	211.58	207.15	202.04	199.13	195.59	193.57
1, Best	226.82	219.44	214.18	207.00	202.34	195.04	191.04	186.31	183.65	180.43	178.57
1 & 2	207.35	200.66	195.84	189.31	184.93	178.41	174.65	170.60	167.79	164.98	163.25
2, Good	191.22	185.02	180.55	174.52	170.52	164.47	161.06	157.07	154.77	152.09	150.51
2 & 3	174.86	169.12	165.08	159.66	155.91	150.40	147.24	143.76	141.58	139.11	137.72
3, Average	160.33	155.18	151.43	146.38	142.98	138.00	135.02	131.72	129.75	127.48	126.26
3 & 4	143.73	139.04	135.60	131.13	128.14	123.68	121.06	118.04	116.30	114.29	113.10
4, Low	126.15	122.09	119.12	115.12	112.52	108.51	106.27	103.66	102.06	100.33	99.31

Second and Higher Stories – Square Foot Area

Quality Class	2,000	2,500	3,000	4,000	5,000	7,500	10,000	15,000	20,000	30,000	40,000
Exceptional	213.48	208.46	204.86	199.95	196.67	191.78	188.96	185.72	183.79	181.59	180.34
1, Best	197.44	192.79	189.46	184.90	181.88	177.37	174.74	171.72	170.00	167.96	166.74
1 & 2	181.86	177.56	174.50	170.24	167.50	163.33	160.95	158.19	156.57	154.69	153.59
2, Good	167.58	163.55	160.73	156.82	154.31	150.47	148.20	145.68	144.22	142.51	141.42
2 & 3	151.64	148.06	145.45	141.98	139.64	136.22	134.18	131.90	130.51	128.93	128.09
3, Average	137.24	134.04	131.63	128.54	126.42	123.24	121.39	119.33	118.17	116.71	115.87
3 & 4	122.40	119.42	117.42	114.58	112.68	109.95	108.30	106.40	105.38	104.09	103.37
4, Low	106.48	103.95	102.16	99.74	98.06	95.63	94.25	92.62	91.68	90.57	89.90

General Office Buildings – Wood Frame

Wall Height Adjustment

The square foot costs for general offices are based on the wall heights of 10 feet for first floors and 9 feet for higher floors. Add to or subtract from the amount listed in this table the square foot of floor cost for each foot of wall height more or less than 10 feet, if adjusting for a first floor, and 9 feet, if adjusting for upper floors.

Square Foot Area

Quality Class	1,000	1,500	2,000	3,000	4,000	5,000	7,500	10,000	15,000	20,000	40,000
1, Best	2.90	2.47	2.14	1.83	1.55	1.41	1.10	.90	.73	.64	.27
2, Good	2.55	2.11	1.86	1.60	1.41	1.13	.92	.87	.65	.59	.21
3, Average	2.19	1.86	1.69	1.38	1.13	1.05	.87	.73	.61	.49	.21
4, Low	2.04	1.69	1.50	1.24	1.04	.91	.77	.65	.52	.44	.20

Perimeter (Common) Wall Adjustment

A common wall exists when two buildings share one wall. Adjust for common walls by deducting the linear foot cost below from the total structure cost. In some structures one or more walls are not owned at all. In this case, deduct the "No Ownership" cost per linear foot of wall not owned. Where a perimeter wall remains unfinished, deduct the "Lack of Exterior Finish" cost.

First Story

Class	For a Common Wall, Deduc8t Per L.F.	For No Wall Ownership, Deduct Per L.F.	For Lack of Exterior Finish, Deduct Per L.F.
1	$483	$974	$429
2	416	810	374
3	384	744	298
4	286	549	219

Second and Higher Stories

Class	For a Common Wall, Deduct Per L.F.	For No Wall Ownership, Deduct Per L.F.	For Lack of Exterior Finish, Deduct Per L.F.
1	$327	$625	$428
2	273	518	374
3	261	494	298
4	242	472	219

Medical-Dental Buildings – Masonry or Concrete

	Class 1 Best Quality	Class 2 Good Quality	Class 3 Average Quality	Class 4 Low Quality
First Floor Structure (10% of total cost)	Reinforced concrete slab on grade or standard wood frame.	Reinforced concrete slab on grade or standard wood frame.	Reinforced concrete slab on grade.	Reinforced concrete slab on grade.
Upper Floor Structure (8% of total cost)	Standard wood frame, plywood and 1-1/2" lightweight concrete subfloor.	Standard wood frame, plywood and 1-1/2" lightweight concrete subfloor.	Standard wood frame. 5/8" plywood subfloor.	Standard wood frame 5/8" plywood subfloor.
Walls (9% of total cost)	8" decorative concrete block or 6" concrete tilt-up.	8" decorative concrete block or 6" concrete tilt-up.	8" reinforced concrete block or 8" reinforced brick.	8" reinforced concrete block or clay tile.
Roof Structure (6% of total cost)	Standard wood frame, flat or low pitch.	Standard wood frame, flat or low pitch.	Standard wood frame, flat or low pitch.	Standard wood frame, flat or low pitch.
Exterior Wall Finish (8% of total cost)	Decorative block or large rock imbedded in tilt-up panels with 10 to 20% brick or stone veneer.	Decorative block or exposed aggregate and 10 to 20% brick or stone veneer.	Stucco or colored concrete block.	Painted.
Windows (5% of total cost)	Average number in good aluminum frame. Fixed float glass in good frame on front.	Average number in good aluminum frame. Some fixed float glass in front.	Average number of average aluminum sliding type.	Average number of low cost aluminum sliding type.
Roof Cover (5% of total cost)	5 ply built-up roofing on flat roofs. Heavy shake or tile on sloping roofs.	5 ply built-up roofing on flat roofs. Average shake or composition, tar and large rock on sloping roofs.	4 ply built-up roofing on flat roofs. Wood shingle or composition, tar and pea gravel on sloping roofs.	3 ply built-up roofing on flat roofs. Composition shingle on sloping roofs.
Overhang (3% of total cost)	3' closed overhang, fully guttered.	2' closed overhang, fully guttered.	None on flat roofs. 18" open on sloping roofs, fully guttered.	None on flat roofs. 12" to 16" open on sloping roofs, gutters over entrances.
Business Offices (3% of total cost)	Good hardwood veneer paneling. Solid vinyl or carpet.	Hardwood paneling or vinyl wall cover. Resilient tile or carpet.	Gypsum wallboard, texture and paint. composition tile.	Gypsum wallboard, texture and paint. Minimum grade tile.
Corridors (6% of total cost)	Good hardwood veneer paneling. Solid vinyl or carpet.	Gypsum wallboard and vinyl wall cover. Resilient tile.	Gypsum wallboard, texture and paint. Composition tile.	Gypsum wallboard, texture and paint. Minimum grade tile.
Waiting Rooms (7% of total cost)	Good hardwood veneer paneling. Carpet.	Hardwood paneling. Carpet.	Gypsum wallboard, some paneling. Resilient tile or carpet.	Gypsum wallboard, texture and paint. Minimum grade tile.
Private Offices (2% of total cost)	Good hardwood veneer paneling. Carpet.	Textured wall cover and hardwood paneling. Carpet.	Gypsum wallboard and paper, wood paneling. Resilient tile or carpet.	Gypsum wallboard, texture and paint. Minimum grade tile.
Treatment Rooms (5% of total cost)	Gypsum wallboard and vinyl wall covering. Sheet vinyl or carpet.	Gypsum wallboard and enamel. Sheet vinyl.	Gypsum wallboard and enamel. Resilient tile.	Gypsum wallboard, texture and paint. Minimum grade tile.
Bathrooms (3% of total cost)	Gypsum wallboard and enamel with ceramic tile wainscot. Sheet vinyl or ceramic tile.	Gypsum wallboard and enamel or vinyl wall covering. Sheet vinyl or ceramic tile.	Gypsum wallboard and enamel. Resilient tile.	Gypsum wallboard, texture and paint. Minimum grade tile.
Ceiling Finish (4% of total cost)	Suspended "T" bar and acoustical tile.	Gypsum wallboard and acoustical tile.	Gypsum wallboard and acoustical tile.	Gypsum wallboard and paint.
Utilities				
Plumbing (6% of total cost)	Copper tubing, good economy fixtures.	Copper tubing, good fixtures.	Copper tubing, average fixtures.	Copper tubing, fixtures.
Lighting (6% of total cost)	Conduit wiring, good fixtures.	Conduit wiring, good fixtures.	Romex or conduit wiring, average fixtures.	Romex wiring, economy fixtures.
Cabinets (4% of total cost)	Formica faced with formica tops.	Good grade of hardwood with formica tops.	Average amount of painted wood or low grade hardwood with formica top.	Minimum amount of painted wood with formica top.

Note: Use the percent of total cost to help identify the correct quality classification.

Square foot costs include the following components: Foundations as required for normal soil conditions. Floor, wall and roof structures. Interior floor, wall and ceiling finishes. Exterior wall finish and roof cover. Interior partitions. Cabinets, doors and windows. Basic electrical systems and lighting fixtures. Rough plumbing and fixtures. Permits and fees. Contractor's mark-up. In addition to the above components, costs for buildings with more than 10,000 feet include the cost of lead shielding for typical x-ray rooms.

Medical-Dental Buildings – Masonry or Concrete

Exterior Suite Entrances, Length Less Than Twice Width

Estimating Procedure

1. Use these figures to estimate medical, dental, psychiatric, optometry and similar professional buildings in which access to each office suite is through an exterior entrance. Buildings in this section have more plumbing fixtures per square foot of floor and smaller room sizes than general office buildings. Note also that buildings with more than 10,000 square feet are assumed to have lead shielded x-ray rooms.
2. Establish the building quality class by applying the information on page 151.
3. Compute the first floor area. This should include everything within the exterior walls and all insets outside the main walls but under the main roof.
4. If the first floor wall height is more or less than 10 feet, add to or subtract from the first floor square foot cost below the appropriate amount from the Wall Height Adjustment Table on page 158.
5. Multiply the adjusted square foot cost by the first floor area.
6. Deduct, if appropriate, for common walls or no wall finish. Use the figures on page 158.
7. If there are second or higher floors, compute the square foot area on each floor. Locate the appropriate square foot cost from the table at the bottom of this page. Adjust this figure for a wall height more or less than 9 feet, using the figures on page 158. Multiply the adjusted cost by the square foot area on each floor. Use the figures on page 158 to deduct for common walls or no wall finish. Add the result to the cost from step 6 above.
8. Multiply the total cost by the location factor listed on page 7 or 8.
9. Add the cost of heating and air conditioning systems, elevators, fire sprinklers, exterior signs, paving and curbing, miscellaneous yard improvements. See page 236 to 248.

Medical-Dental Building, Class 2

First Story – Square Foot Area

Quality Class	1,000	1,500	2,000	2,500	3,000	4,000	5,000	7,500	10,000	15,000	20,000
Exceptional	470.19	442.44	426.30	415.55	407.52	396.64	389.24	378.01	371.29	363.46	358.87
1, Best	435.19	409.50	394.56	384.53	377.23	367.14	360.30	349.85	343.65	336.43	332.20
1 & 2	412.63	388.19	374.13	364.64	357.68	348.07	341.64	331.73	325.90	318.99	314.94
2, Good	394.02	370.84	357.28	348.31	341.64	332.50	326.31	316.79	311.26	304.72	300.82
2 & 3	364.27	342.71	330.27	321.89	315.76	307.23	301.61	292.81	287.66	281.60	278.04
3, Average	341.61	321.74	309.95	302.13	296.32	285.49	282.99	274.78	269.97	264.27	260.92
3 & 4	322.04	303.02	291.99	284.49	279.17	271.69	266.61	258.88	254.30	248.93	245.79
4, Low	299.87	282.29	272.00	265.11	259.97	253.03	248.37	241.13	236.89	231.85	228.88

Second and Higher Stories – Square Foot Area

Quality Class	1,000	1,500	2,000	2,500	3,000	4,000	5,000	7,500	10,000	15,000	20,000
Exceptional	420.97	392.81	377.57	367.81	361.01	351.92	346.04	337.58	332.71	327.35	324.22
1, Best	411.04	383.59	368.72	359.22	352.59	343.73	337.99	329.58	324.84	319.63	316.61
1 & 2	390.49	364.35	350.35	341.26	334.97	326.50	321.01	313.11	308.66	303.63	300.83
2, Good	376.84	351.72	338.11	329.40	323.28	315.10	309.88	302.18	297.75	293.07	290.30
2 & 3	347.95	324.76	312.16	304.03	298.45	291.03	286.05	278.98	275.06	270.65	268.01
3, Average	326.68	304.86	293.07	285.46	280.30	273.15	268.65	261.94	258.23	253.98	251.65
3 & 4	307.65	287.02	275.93	268.77	263.77	257.25	252.84	246.64	242.62	239.15	236.89
4, Low	286.73	267.58	257.17	250.61	245.96	239.71	235.73	229.90	226.56	223.01	220.91

Medical-Dental Buildings – Masonry or Concrete

Exterior Suite Entrances, Length Between 2 and 4 Times Width

Estimating Procedure

1. Use these figures to estimate medical, dental, psychiatric, optometry and similar professional buildings in which access to each office suite is through an exterior entrance. Buildings in this section have more plumbing fixtures per square foot of floor and smaller room sizes than general office buildings. Note also that buildings with more than 10,000 square feet are assumed to have lead shielded x-ray rooms.
2. Establish the building quality class by applying the information on page 151.
3. Compute the first floor area. This should include everything within the exterior walls and all insets outside the main walls but under the main roof.
4. If the first floor wall height is more or less than 10 feet, add to or subtract from the first floor square foot cost below the appropriate amount from the Wall Height Adjustment Table on page 158.
5. Multiply the adjusted square foot cost by the first floor area.
6. Deduct, if appropriate, for common walls or no wall finish. Use the figures on page 158.
7. If there are second or higher floors, compute the square foot area on each floor. Locate the appropriate square foot cost from the table at the bottom of this page. Adjust this figure for a wall height more or less than 9 feet, using the figures on page 158. Multiply the adjusted cost by the square foot area on each floor. Use the figures on page 158 to deduct for common walls or no wall finish. Add the result to the cost from step 6 above.
8. Multiply the total cost by the location factor listed on page 7 or 8.
9. Add the cost of heating and air conditioning systems, elevators, fire sprinklers, exterior signs, paving and curbing, miscellaneous yard improvements. See page 236 to 248.

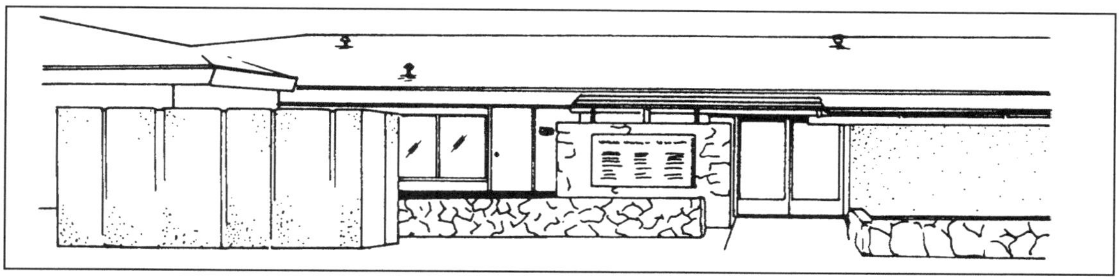

Medical-Dental Building, Class 3 & 4

First Story – Square Foot Area

Quality Class	1,000	1,500	2,000	2,500	3,000	4,000	5,000	7,500	10,000	15,000	20,000
Exceptional	506.79	469.28	448.18	434.20	424.18	410.47	401.44	387.80	379.76	370.62	365.23
1, Best	469.92	435.14	415.45	402.52	393.34	380.56	372.17	359.47	352.17	343.65	338.67
1 & 2	441.87	409.31	390.83	378.66	369.97	358.06	350.07	338.17	331.25	323.23	318.61
2, Good	421.72	390.58	372.87	361.31	352.99	341.61	334.14	322.72	316.10	308.47	304.01
2 & 3	389.74	360.91	344.62	333.91	326.25	315.65	308.71	298.19	292.15	285.00	280.92
3, Average	365.95	338.82	323.59	313.54	306.33	296.43	289.88	279.95	274.21	267.55	263.73
3 & 4	344.52	318.99	304.58	295.10	288.30	279.08	272.80	263.59	258.16	251.96	248.35
4, Low	320.83	297.12	283.72	274.92	268.57	259.90	254.04	245.43	240.49	234.62	231.19

Second and Higher Stories – Square Foot Area

Quality Class	1,000	1,500	2,000	2,500	3,000	4,000	5,000	7,500	10,000	15,000	20,000
Exceptional	467.08	434.27	415.79	403.52	394.70	382.79	374.84	362.85	355.80	347.78	343.09
1, Best	433.09	402.97	385.47	374.04	365.95	354.75	347.48	336.31	329.83	322.35	318.04
1 & 2	411.33	382.35	366.01	355.32	347.60	337.00	329.94	319.39	313.33	306.27	301.96
2, Good	396.80	368.77	353.14	342.71	335.30	325.06	318.34	308.15	302.18	295.45	291.37
2 & 3	366.26	340.58	325.98	316.43	309.58	300.15	293.86	284.46	279.06	272.75	269.03
3, Average	343.78	319.63	306.02	297.02	290.46	281.73	275.93	267.04	261.90	256.00	252.59
3 & 4	323.89	301.14	288.28	279.89	273.78	265.36	259.90	251.56	246.81	241.10	237.80
4, Low	302.09	280.90	268.82	260.96	255.33	247.56	242.40	234.62	230.03	224.94	221.94

Medical-Dental Buildings – Masonry or Concrete

Exterior Suite Entrances, Length More Than 4 Times Width

Estimating Procedure

1. Use these figures to estimate medical, dental, psychiatric, optometry and similar professional buildings in which access to each office suite is through an exterior entrance. Buildings in this section have more plumbing fixtures per square foot of floor and smaller room sizes than general office buildings. Note also that buildings with more than 10,000 square feet are assumed to have lead shielded x-ray rooms.
2. Establish the building quality class by applying the information on page 151.
3. Compute the first floor area. This should include everything within the exterior walls and all insets outside the main walls but under the main roof.
4. If the first floor wall height is more or less than 10 feet, add to or subtract from the first floor square foot cost below the appropriate amount from the Wall Height Adjustment Table on page 158.
5. Multiply the adjusted square foot cost by the first floor area.
6. Deduct, if appropriate, for common walls or no wall finish. Use the figures on page 158.
7. If there are second or higher floors, compute the square foot area on each floor. Locate the appropriate square foot cost from the table at the bottom of this page. Adjust this figure for a wall height more or less than 9 feet, using the figures on page 158. Multiply the adjusted cost by the square foot area on each floor. Use the figures on page 158 to deduct for common walls or no wall finish. Add the result to the cost from step 6 above.
8. Multiply the total cost by the location factor listed on page 7 or 8.
9. Add the cost of heating and air conditioning systems, elevators, fire sprinklers, exterior signs, paving and curbing, miscellaneous yard improvements. See page 236 to 248.

Medical-Dental Building, Class 2 & 3

First Story – Square Foot Area

Quality Class	1,000	1,500	2,000	2,500	3,000	4,000	5,000	7,500	10,000	15,000	20,000
Exceptional	546.19	498.91	472.50	455.24	442.84	426.17	415.14	398.67	389.12	378.25	371.92
1, Best	506.40	462.58	438.03	422.00	410.62	395.08	384.86	369.53	360.77	350.63	344.74
1 & 2	475.54	434.24	411.18	396.18	385.51	370.85	361.23	346.86	338.61	329.24	323.66
2, Good	455.01	415.57	393.61	379.17	368.91	354.99	345.84	331.95	324.06	315.04	309.83
2 & 3	418.68	382.44	362.18	348.93	339.47	326.60	318.25	305.53	298.23	289.92	285.00
3, Average	392.68	358.66	339.58	327.26	318.34	306.35	298.40	286.48	279.76	271.82	267.30
3 & 4	369.63	337.48	319.66	307.95	299.67	288.30	280.90	270.72	263.21	255.89	251.58
4, Low	344.11	314.27	297.63	286.74	278.96	268.49	261.43	251.07	245.05	238.22	234.15

Second and Higher Stories – Square Foot Area

Quality Class	1,000	1,500	2,000	2,500	3,000	4,000	5,000	7,500	10,000	15,000	20,000
Exceptional	499.61	459.29	436.52	421.65	410.82	396.18	386.36	371.73	363.12	353.29	347.60
1, Best	461.26	424.01	403.09	389.19	379.29	365.73	356.73	343.16	335.30	326.25	320.96
1 & 2	439.67	404.24	384.22	371.02	361.51	348.63	340.06	327.03	319.62	310.93	305.84
2, Good	423.95	389.83	370.50	357.72	348.69	336.22	327.94	315.42	308.23	299.82	295.00
2 & 3	391.51	360.02	342.10	330.36	321.93	310.37	302.73	291.29	284.49	276.85	272.31
3, Average	367.36	337.77	320.99	310.10	302.15	291.37	284.12	273.30	267.07	259.81	255.56
3 & 4	346.60	318.54	302.79	292.39	284.97	274.75	268.03	257.79	251.95	245.02	241.02
4, Low	322.75	296.70	282.01	272.34	265.36	255.89	249.55	240.15	234.62	228.28	224.43

Interior Suite Entrances, Length Less Than Twice Width

Estimating Procedure

1. Use these figures to estimate medical, dental, psychiatric, optometry and similar professional buildings in which access to each office suite is through an interior corridor. Buildings in this section have more plumbing fixtures per square foot of floor and smaller room sizes than general office buildings. Note also that buildings with more than 10,000 square feet are assumed to have lead shielded x-ray rooms.
2. Establish the building quality class by applying the information on page 151.
3. Compute the first floor area. This should include everything within the exterior walls and all insets outside the main walls but under the main roof.
4. If the first floor wall height is more or less than 10 feet, add to or subtract from the first floor square foot cost below the appropriate amount from the Wall Height Adjustment Table on page 158.
5. Multiply the adjusted square foot cost by the first floor area.
6. Deduct, if appropriate, for common walls or no wall finish. Use the figures on page 158.
7. If there are second or higher floors, compute the square foot area on each floor. Locate the appropriate square foot cost from the table at the bottom of this page. Adjust this figure for a wall height more or less than 9 feet, using the figures on page 158. Multiply the adjusted cost by the square foot area on each floor. Use the figures on page 158 to deduct for common walls or no wall finish. Add the result to the cost from step 6 above.
8. Multiply the total cost by the location factor listed on page 7 or 8.
9. Add the cost of heating and air conditioning systems, elevators, fire sprinklers, exterior signs, paving and curbing, miscellaneous yard improvements. See page 236 to 248.

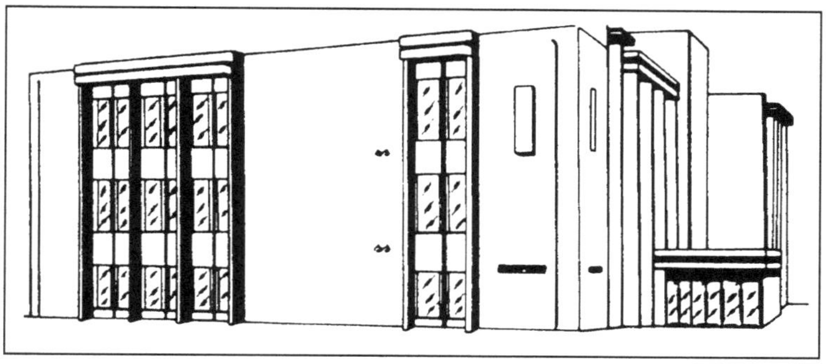

Medical-Dental Building, Class 3 & 4

First Story – Square Foot Area

Quality Class	2,000	2,500	3,000	4,000	5,000	7,500	10,000	15,000	20,000	30,000	40,000
Exceptional	397.43	386.71	379.17	369.87	360.90	349.92	343.49	335.90	331.53	326.38	323.31
1, Best	378.71	368.52	361.01	350.88	343.95	333.49	327.38	320.13	316.01	310.99	308.15
1 & 2	357.28	347.64	340.53	330.93	324.49	314.52	308.77	301.92	298.08	293.41	290.73
2, Good	341.96	332.70	325.95	316.78	310.52	301.03	295.56	289.08	285.31	280.76	278.15
2 & 3	316.93	308.40	302.13	293.58	287.80	279.06	273.99	267.92	264.41	260.27	257.81
3, Average	295.70	287.69	281.89	273.96	268.57	260.37	255.50	249.97	246.73	242.89	240.54
3 & 4	279.89	272.28	266.76	259.24	254.04	246.34	241.83	236.58	233.48	229.82	227.69
4, Low	262.35	255.15	250.02	242.93	238.22	230.96	226.72	221.78	218.84	215.48	213.41

Second and Higher Stories – Square Foot Area

Quality Class	2,000	2,500	3,000	4,000	5,000	7,500	10,000	15,000	20,000	30,000	40,000
Exceptional	375.77	366.13	359.19	349.60	343.16	333.36	327.62	320.98	317.11	312.42	309.77
1, Best	358.35	349.16	342.61	333.41	327.35	317.97	312.47	306.28	302.47	298.01	295.56
1 & 2	340.07	331.45	325.06	316.46	310.68	301.70	296.60	290.63	287.02	282.82	280.41
2, Good	327.65	319.33	313.28	304.86	299.23	290.76	285.77	279.95	276.50	272.55	270.10
2 & 3	302.96	295.17	289.54	281.96	276.64	268.79	264.25	258.82	255.67	251.96	249.77
3, Average	284.35	277.12	271.91	264.54	259.72	252.34	248.03	243.01	240.04	236.50	234.46
3 & 4	267.49	260.58	255.70	248.90	244.37	237.49	233.26	228.66	225.83	222.47	220.64
4, Low	251.93	245.43	240.78	234.32	230.01	223.45	219.71	215.25	212.64	209.49	207.69

Interior Suite Entrances, Length Between 2 and 4 Times Width

Estimating Procedure

1. Use these figures to estimate medical, dental, psychiatric, optometry and similar professional buildings in which access to each office suite is through an interior corridor. Buildings in this section have more plumbing fixtures per square foot of floor and smaller room sizes than general office buildings. Note also that buildings with more than 10,000 square feet are assumed to have lead shielded x-ray rooms.
2. Establish the building quality class by applying the information on page 151.
3. Compute the first floor area. This should include everything within the exterior walls and all insets outside the main walls but under the main roof.
4. If the first floor wall height is more or less than 10 feet, add to or subtract from the first floor square foot cost below the appropriate amount from the Wall Height Adjustment Table on page 158.
5. Multiply the adjusted square foot cost by the first floor area.
6. Deduct, if appropriate, for common walls or no wall finish. Use the figures on page 158.
7. If there are second or higher floors, compute the square foot area on each floor. Locate the appropriate square foot cost from the table at the bottom of this page. Adjust this figure for a wall height more or less than 9 feet, using the figures on page 158. Multiply the adjusted cost by the square foot area on each floor. Use the figures on page 158 to deduct for common walls or no wall finish. Add the result to the cost from step 6 above.
8. Multiply the total cost by the location factor listed on page 7 or 8.
9. Add the cost of heating and air conditioning systems, elevators, fire sprinklers, exterior signs, paving and curbing, miscellaneous yard improvements. See page 236 to 248.

Medical-Dental Building, Class 3

First Story – Square Foot Area

Quality Class	2,000	2,500	3,000	4,000	5,000	7,500	10,000	15,000	20,000	30,000	40,000
Exceptional	416.03	402.81	393.23	380.11	371.38	358.14	350.45	341.48	336.24	330.18	326.49
1, Best	396.54	383.96	374.84	362.31	353.99	341.29	334.08	325.46	320.44	314.59	311.17
1 & 2	373.72	361.89	353.27	341.44	333.58	321.70	314.76	306.80	301.96	296.58	293.36
2, Good	357.28	346.02	337.84	326.49	319.02	307.70	301.03	293.36	288.90	283.63	280.51
2 & 3	330.95	320.44	312.73	302.48	295.54	284.84	278.66	271.66	267.51	262.60	259.80
3, Average	308.54	298.72	291.62	281.89	275.40	265.61	259.83	253.21	249.33	244.84	242.09
3 & 4	292.27	282.94	276.34	267.07	260.87	251.62	246.15	239.75	236.30	231.88	229.50
4, Low	273.56	264.87	258.58	249.96	244.17	235.53	230.54	224.43	221.12	217.04	214.68

Second and Higher Stories – Square Foot Area

Quality Class	2,000	2,500	3,000	4,000	5,000	7,500	10,000	15,000	20,000	30,000	40,000
Exceptional	392.15	380.30	371.83	360.14	352.34	340.53	333.69	325.74	321.01	315.73	312.38
1, Best	373.77	362.42	354.35	343.23	335.69	324.67	317.99	310.43	305.95	300.89	297.73
1 & 2	354.55	343.90	336.19	325.64	318.61	307.87	301.65	294.50	290.28	285.41	282.47
2, Good	341.72	331.45	323.91	313.70	306.99	296.80	290.79	283.91	279.79	275.05	272.27
2 & 3	315.45	305.95	299.09	289.70	283.41	274.03	268.48	262.01	258.31	253.84	251.41
3, Average	295.96	287.08	280.51	271.72	265.88	256.94	251.73	245.79	242.31	238.20	235.73
3 & 4	280.24	271.72	265.64	257.37	251.73	243.33	238.45	232.75	229.37	225.44	223.21
4, Low	262.57	254.68	249.00	241.19	235.96	228.08	223.40	218.17	215.01	211.43	209.20

Medical-Dental Buildings – Masonry or Concrete

Interior Suite Entrances, Length More Than 4 Times Width

Estimating Procedure

1. Establish the building quality class by applying the information on page 151.
3. Compute the first floor area. This should include everything within the exterior walls and all insets outside the main walls but under the main roof.
4. If the first floor wall height is more or less than 10 feet, add to or subtract from the first floor square foot cost below the appropriate amount from the Wall Height Adjustment Table on page 158.
5. Multiply the adjusted square foot cost by the first floor area.
6. Deduct, if appropriate, for common walls or no wall finish. Use the figures on page 158.
7. If there are second or higher floors, compute the square foot area on each floor. Locate the appropriate square foot cost from the table at the bottom of this page. Adjust this figure for a wall height more or less than 9 feet, using the figures on page 158. Multiply the adjusted cost by the square foot area on each floor. Use the figures on page 158 to deduct for common walls or no wall finish. Add the result to the cost from step 6 above.
8. Multiply the total cost by the location factor listed on page 7 or 8.
9. Add the cost of heating and air conditioning systems, elevators, fire sprinklers, exterior signs, paving and curbing, miscellaneous yard improvements. See page 236 to 248.

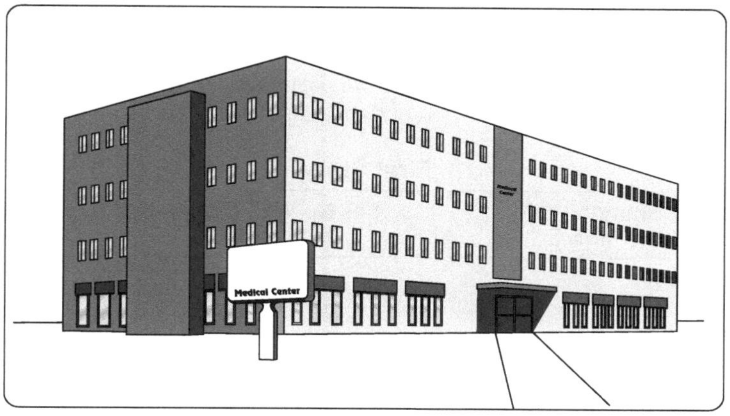

Medical-Dental Building, Class 2 & 3

First Story – Square Foot Area

Quality Class	2,000	2,500	3,000	4,000	5,000	7,500	10,000	15,000	20,000	30,000	40,000
Exceptional	439.91	423.29	411.39	395.25	384.60	368.57	359.31	348.69	342.43	335.36	331.24
1, Best	419.30	403.39	392.14	376.66	366.45	351.25	342.42	332.33	326.38	319.63	315.56
1 & 2	394.83	379.95	369.24	354.70	345.15	330.78	322.56	313.02	307.44	301.02	297.23
2, Good	377.23	363.05	352.88	338.98	329.79	316.01	308.15	298.98	293.70	287.57	283.96
2 & 3	348.99	335.76	326.38	313.56	305.09	292.32	285.09	276.60	271.71	266.10	262.75
3, Average	328.07	315.77	306.89	294.79	286.80	274.90	268.00	260.07	255.47	250.05	247.02
3 & 4	310.43	298.72	290.28	278.96	271.36	259.99	253.56	246.07	241.73	236.62	233.77
4, Low	290.75	279.77	271.96	261.26	254.18	243.58	237.41	230.54	226.33	221.68	218.84

Second and Higher Stories – Square Foot Area

Quality Class	2,000	2,500	3,000	4,000	5,000	7,500	10,000	15,000	20,000	30,000	40,000
Exceptional	412.50	397.72	387.08	372.69	363.23	348.99	340.79	331.46	325.95	319.66	316.01
1, Best	393.39	379.24	369.12	355.41	346.42	332.74	324.99	316.05	310.77	304.76	301.23
1 & 2	373.02	359.62	350.00	337.05	328.47	315.56	308.23	299.67	294.78	289.05	285.68
2, Good	359.31	346.44	337.15	324.71	316.43	304.02	296.92	288.70	283.95	278.41	275.17
2 & 3	331.44	319.57	310.96	299.46	291.83	280.47	273.93	266.36	261.94	256.74	253.82
3, Average	311.32	300.12	292.18	281.33	274.15	263.40	257.28	250.05	246.00	241.25	238.39
3 & 4	294.72	284.07	276.50	266.28	259.42	249.35	243.44	236.72	232.85	228.29	225.70
4, Low	276.15	266.31	259.11	249.51	243.12	233.74	228.15	221.94	218.21	213.95	211.46

Medical-Dental Buildings – Masonry or Concrete

Wall Height Adjustment

The square foot costs for medical-dental buildings are based on the wall heights of 10 feet for first floors and 9 feet for higher floors. The main or first floor height is the distance from the bottom of the floor slab or joists to the top of the roof slab or ceiling joists. Second and higher floors are measured from the top of the floor slab or floor joists to the top of the roof slab or ceiling joists. Add or subtract the amount listed in this table to or from the square foot of floor cost for each foot of wall height more or less than 10 feet, if adjusting for a first floor, and 9 feet, if adjusting for upper floors.

Square Foot Area

Quality Class	1,000	1,500	2,000	3,000	4,000	5,000	7,500	10,000	15,000	20,000	40,000
1, Best	11.93	9.29	7.81	6.14	5.22	4.63	3.65	3.13	2.54	2.09	1.55
2, Good	10.27	8.02	6.73	5.36	4.54	4.04	3.21	2.75	2.15	1.93	1.30
3, Average	8.99	6.99	5.81	4.66	3.98	3.53	2.88	2.39	1.95	1.69	1.17
4, Low	7.88	6.30	5.26	4.05	3.51	3.06	2.43	2.10	1.74	1.36	1.06

Perimeter (Common) Wall Adjustment

A common wall exists when two buildings share one wall. Adjust for common walls by deducting the linear foot costs below from the total structure cost. In some structures, one or more walls are not owned at all. In this case, deduct the "No Ownership" cost per linear foot of wall not owned. If a wall has no exterior finish, deduct the "Lack of Exterior Finish" cost.

First Story

Class	For a Common Wall, Deduct Per L.F.	For No Wall Ownership, Deduct Per L.F.	For Lack of Exterior Finish, Deduct Per L.F.
1	$677	$1,372	$522
2	533	927	341
3	380	765	185
4	290	584	120

Second and Higher Stories

Class	For a Common Wall, Deduct Per L.F.	For No Wall Ownership, Deduct Per L.F.	For Lack of Exterior Finish, Deduct Per L.F.
1	$635	$1,238	$519
2	522	1,000	338
3	382	766	184
4	343	676	116

Medical-Dental Buildings – Wood Frame

	Class 1 Best Quality	Class 2 Good Quality	Class 3 Average Quality	Class 4 Low Quality
Foundation (9% of total cost)	Reinforced concrete.	Reinforced concrete.	Reinforced concrete.	Reinforced concrete.
First Floor Structure (4% of total cost)	Reinforced concrete slab on grade or standard wood frame.	Reinforced concrete slab on grade or standard wood frame.	Reinforced concrete slab on grade or 4" x 6" girders with 2" T&G subfloor.	Reinforced concrete slab on grade.
Upper Floor Structure (6% of total cost)	Standard wood frame, plywood and 1-1/2" light-weight concrete subfloor.	Standard wood frame, plywood and 1-1/2" light-weight concrete subfloor.	Standard wood frame, 5/8" plywood subfloor.	Standard wood frame, 5/8" plywood subfloor.
Walls (9% of total cost)	Standard wood frame.	Standard wood frame.	Standard wood frame.	Standard wood frame.
Roof Structure (6% of total cost)	Standard wood frame, flat or low pitch.	Standard wood frame, flat or low pitch.	Standard wood frame, flat or low pitch.	Standard wood frame, flat or low pitch.
Exterior Finishes:				
Walls (8% of total cost)	Good wood siding with 10 to 20% brick or stone veneer.	Average wood siding or stucco and 10 to 20% brick or stone veneer.	Stucco with some wood trim or cheap wood siding.	Stucco.
Windows (5% of total cost)	Average number in good aluminum frame. Fixed float glass in good frame on front.	Average number in good aluminum frame. Some fixed float glass in front.	Average amount of average aluminum sliding type.	Average number of low cost aluminum sliding type.
Roof Cover (5% of total cost)	5 ply built-up roofing on flat roofs. Heavy shake or tile on sloping roofs.	5 ply built-up roofing on flat roofs. Avg. shake or composition, tar and large rock on sloping roofs.	4 ply built-up roofing on flat roofs. Wood shingle or composition, tar and pea gravel on sloping roofs.	3 ply built-up roofing on flat roofs. Composition shingle on sloping roofs.
Overhang (3% of total cost)	3' sealed overhang, fully guttered.	2' sealed overhang, fully guttered.	None on flat roofs. 18" unsealed on sloping roofs, fully guttered.	None on flat roofs. 12" to 16" unsealed on sloping roofs, gutters over entrances
Floor Finishes:				
Business Offices (3% of total cost)	Solid vinyl tile or carpet.	Resilient tile or carpet.	Composition tile.	Minimum grade tile.
Corridors (2% of total cost)	Solid vinyl tile or carpet.	Resilient tile or carpet.	Composition tile.	Minimum grade tile.
Waiting Rooms (1% of total cost)	Carpet.	Carpet.	Composition tile or carpet.	Minimum grade tile.
Private Offices (2% of total cost)	Carpet.	Carpet.	Composition tile or carpet.	Minimum grade tile.
Treatment Rooms (2% of total cost)	Sheet vinyl or carpet.	Sheet vinyl.	Composition tile.	Minimum grade tile.
Bathrooms (1% of total cost)	Sheet vinyl or ceramic tile.	Sheet vinyl or ceramic tile.	Composition tile.	Minimum grade tile.
Interior Wall Finishes:				
Business Offices (3% of total cost)	Good hardwood veneer paneling.	Hardwood paneling or vinyl wall cover.	Gypsum wallboard, texture and paint.	Gypsum wallboard, texture and paint.
Corridors (2% of total cost)	Good hardwood veneer paneling.	Gypsum wallboard and vinyl wall cover.	Gypsum wallboard, texture and paint.	Gypsum wallboard, texture and paint.
Waiting Rooms (1% of total cost)	Good hardwood veneer paneling.	Hardwood paneling.	Gypsum wallboard and paper, some wood paneling.	Gypsum wallboard, texture and paint.
Treatment Rooms (3% of total cost)	Gypsum wallboard and vinyl wall covering.	Gypsum wallboard and enamel.	Gypsum wallboard and enamel.	Gypsum wallboard, texture and paint.
Bathrooms (2% of total cost)	Gypsum wallboard and enamel with ceramic tile wainscot.	Gypsum wallboard and enamel or vinyl wall covering.	Gypsum wallboard and enamel.	Gypsum wallboard, texture and paint.
Ceiling Finish (4% of total cost)	Suspended "T" bar and acoustical tile.	Gypsum wallboard and acoustical tile.	Gypsum wallboard and acoustical tile.	Gypsum wallboard and paint.
Plumbing (6% of total cost)	Copper tubing, good fixtures.	Copper tubing, good fixtures.	Copper tubing, average fixtures.	Copper tubing, economy fixtures.
Lighting (6% of total cost)	Conduit wiring, good fixtures.	Conduit wiring, good fixtures.	Romex or conduit wiring, average fixtures.	Romex wiring, economy fixtures.
Cabinets (7% of total cost)	Formica faced with formica tops.	Good grade of hardwood with formica tops.	Average amount of painted wood or low grade hardwood with formica top.	Minimum amount of painted wood with formica top.

Note: Use the percent of total cost to help identify the correct quality classification.

Square foot costs include the following components: Foundations as required for normal soil conditions. Floor, wall and roof structures. Interior floor, wall and ceiling finishes. Exterior wall finish and roof cover. Interior partitions. Cabinets, doors and windows. Basic electrical systems and lighting fixtures. Rough plumbing and fixtures. Permits and fees. Contractor's mark-up. In addition to the above components, costs for buildings with more than 10,000 feet include the cost of lead shielding for typical x-ray rooms.

Medical-Dental Buildings – Wood Frame

Exterior Suite Entrances, Length Less Than Twice Width

Estimating Procedure

1. Use these figures to estimate medical, dental, psychiatric, optometry and similar professional buildings in which access to each office suite is through an exterior entrance. Buildings in this section have more plumbing fixtures per square foot of floor and smaller room sizes than general office buildings. Note also that buildings with more than 10,000 square feet are assumed to have lead shielded x-ray rooms.
2. Establish the building quality class by applying the information on page 159.
3. Compute the first floor area. This should include everything within the exterior walls and all insets outside the main walls but under the main roof.
4. If the first floor wall height is more or less than 10 feet, add to or subtract from the first floor square foot cost below the appropriate amount from the Wall Height Adjustment Table on page 166.
5. Multiply the adjusted square foot cost by the first floor area.
6. Deduct, if appropriate, for common walls or no wall finish. Use the figures on page 166.
7. If there are second or higher floors, compute the square foot area on each floor. Locate the appropriate square foot cost from the table at the bottom of this page. Adjust this figure for a wall height more or less than 9 feet, using the figures on page 166. Multiply the adjusted cost by the square foot area on each floor. Use the figures on page 166 to deduct for common walls or no wall finish. Add the result to the cost from step 6 above.
8. Multiply the total cost by the location factor listed on page 7 or 8.
9. Add the cost of heating and air conditioning systems, elevators, fire sprinklers, exterior signs, paving and curbing, miscellaneous yard improvements. See page 236 to 248.

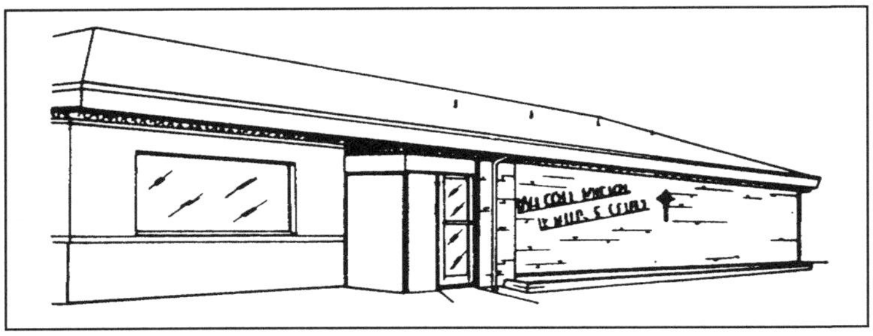

Medical-Dental Building, Class 2 & 3

First Story – Square Foot Area

Quality Class	1,000	1,500	2,000	2,500	3,000	4,000	5,000	7,500	10,000	15,000	20,000
Exceptional	308.48	296.52	289.64	285.06	281.79	277.25	274.24	269.72	267.06	263.88	262.11
1, Best	283.21	272.14	265.95	261.73	258.76	254.57	251.84	247.64	245.13	242.33	240.59
1 & 2	267.03	256.62	250.75	246.80	243.95	240.04	237.40	233.45	231.04	228.38	226.82
2, Good	252.58	242.71	237.15	233.43	230.76	227.02	224.55	220.81	218.68	216.08	214.58
2 & 3	234.55	225.42	220.19	216.76	214.26	210.78	208.57	204.98	203.05	200.58	199.21
3, Average	219.52	211.01	206.11	202.85	200.56	197.31	195.18	191.95	190.03	187.82	186.50
3 & 4	205.19	197.22	192.65	189.62	187.44	184.39	182.48	179.42	177.60	175.56	174.28
4, Low	191.58	184.16	179.97	177.11	175.09	172.28	170.38	167.61	165.90	163.95	162.81

Second and Higher Stories – Square Foot Area

Quality Class	1,000	1,500	2,000	2,500	3,000	4,000	5,000	7,500	10,000	15,000	20,000
Exceptional	275.04	268.05	264.00	261.24	259.21	256.40	254.47	251.48	249.75	247.69	246.47
1 Best	252.82	246.38	242.67	240.17	238.33	235.76	234.00	231.20	229.65	227.74	226.61
1 & 2	236.44	230.52	227.02	224.66	222.91	220.49	218.80	216.35	214.79	212.96	211.98
2, Good	230.32	224.54	221.18	218.82	217.17	214.82	213.21	210.74	209.13	207.50	206.49
2 & 3	212.87	207.45	204.34	202.22	200.58	198.44	196.97	194.70	193.29	191.75	190.80
3, Average	198.11	193.87	190.19	188.19	186.76	184.77	183.32	181.18	179.97	178.52	177.60
3 & 4	185.14	180.49	177.76	175.89	174.54	172.63	171.38	169.44	168.22	166.77	166.01
4, Low	173.26	168.82	166.33	164.55	163.28	161.57	160.23	158.44	157.41	156.07	155.24

Medical-Dental Buildings – Wood Frame

Exterior Suite Entrances, Length Between 2 and 4 Times Width

Estimating Procedure

1. Use these figures to estimate medical, dental, psychiatric, optometry and similar professional buildings in which access to each office suite is through an exterior entrance. Buildings in this section have more plumbing fixtures per square foot of floor and smaller room sizes than general office buildings. Note also that buildings with more than 10,000 square feet are assumed to have lead shielded x-ray rooms.
2. Establish the building quality class by applying the information on page 159.
3. Compute the first floor area. This should include everything within the exterior walls and all insets outside the main walls but under the main roof.
4. If the first floor wall height is more or less than 10 feet, add to or subtract from the first floor square foot cost below the appropriate amount from the Wall Height Adjustment Table on page 166.
5. Multiply the adjusted square foot cost by the first floor area.
6. Deduct, if appropriate, for common walls or no wall finish. Use the figures on page 166.
7. If there are second or higher floors, compute the square foot area on each floor. Locate the appropriate square foot cost from the table at the bottom of this page. Adjust this figure for a wall height more or less than 9 feet, using the figures on page 166. Multiply the adjusted cost by the square foot area on each floor. Use the figures on page 166 to deduct for common walls or no wall finish. Add the result to the cost from step 6 above.
8. Multiply the total cost by the location factor listed on page 7 or 8.
9. Add the cost of heating and air conditioning systems, elevators, fire sprinklers, exterior signs, paving and curbing, miscellaneous yard improvements. See page 236 to 248.

Medical-Dental Building, Class 1 & 2

First Story – Square Foot Area

Quality Class	1,000	1,500	2,000	2,500	3,000	4,000	5,000	7,500	10,000	15,000	20,000
Exceptional	315.11	300.25	291.84	286.35	282.47	277.01	273.42	268.00	264.91	261.24	259.11
1, Best	289.27	275.63	267.93	262.90	259.24	254.28	251.05	246.07	243.19	239.83	237.92
1 & 2	272.31	259.54	252.30	247.50	244.12	239.42	236.32	231.68	228.91	225.88	225.19
2, Good	257.20	245.42	238.44	234.00	230.81	226.35	223.48	219.00	216.45	213.46	211.81
2 & 3	238.99	227.77	221.34	217.21	214.21	210.06	207.37	203.31	200.88	198.13	196.55
3, Average	223.56	213.12	207.18	203.30	200.39	196.55	194.02	190.19	187.95	185.30	183.89
3 & 4	208.85	199.07	193.49	189.87	187.26	183.66	181.26	177.66	175.60	173.26	171.80
4, Low	195.06	185.87	180.67	177.29	174.77	171.49	169.27	165.87	163.95	161.70	160.36

Second and Higher Stories – Square Foot Area

Quality Class	1,000	1,500	2,000	2,500	3,000	4,000	5,000	7,500	10,000	15,000	20,000
Exceptional	278.34	269.51	264.40	261.04	258.55	255.21	252.89	249.44	247.48	245.11	243.73
1, Best	255.55	247.37	242.70	239.61	237.40	234.24	232.21	229.01	227.18	225.01	223.80
1 & 2	242.81	235.14	230.68	227.69	225.55	222.66	220.67	217.62	215.89	213.88	212.62
2, Good	233.08	225.63	221.39	218.59	216.51	213.60	211.82	208.91	207.21	205.18	204.08
2 & 3	215.51	208.72	204.81	202.13	200.19	197.56	195.89	193.12	191.58	189.77	188.69
3, Average	199.67	193.36	189.68	187.30	185.55	183.10	181.52	178.97	177.51	175.87	174.84
3 & 4	187.68	181.80	178.29	175.97	174.31	172.08	170.54	168.22	166.88	165.31	164.36
4, Low	174.18	168.69	165.57	163.35	161.85	159.72	158.31	156.16	154.92	153.41	152.53

Medical-Dental Buildings – Wood Frame

Exterior Suite Entrances, Length More Than 4 Times Width

Estimating Procedure

1. Use these figures to estimate medical, dental, psychiatric, optometry and similar professional buildings in which access to each office suite is through an exterior entrance. Buildings in this section have more plumbing fixtures per square foot of floor and smaller room sizes than general office buildings. Note also that buildings with more than 10,000 square feet are assumed to have lead shielded x-ray rooms.
2. Establish the building quality class by applying the information on page 159.
3. Compute the first floor area. This should include everything within the exterior walls and all insets outside the main walls but under the main roof.
4. If the first floor wall height is more or less than 10 feet, add to or subtract from the first floor square foot cost below the appropriate amount from the Wall Height Adjustment Table on page 166.
5. Multiply the adjusted square foot cost by the first floor area.
6. Deduct, if appropriate, for common walls or no wall finish. Use the figures on page 166.
7. If there are second or higher floors, compute the square foot area on each floor. Locate the appropriate square foot cost from the table at the bottom of this page. Adjust this figure for a wall height more or less than 9 feet, using the figures on page 166. Multiply the adjusted cost by the square foot area on each floor. Use the figures on page 166 to deduct for common walls or no wall finish. Add the result to the cost from step 6 above.
8. Multiply the total cost by the location factor listed on page 7 or 8.
9. Add the cost of heating and air conditioning systems, elevators, fire sprinklers, exterior signs, paving and curbing, miscellaneous yard improvements. See page 236 to 248.

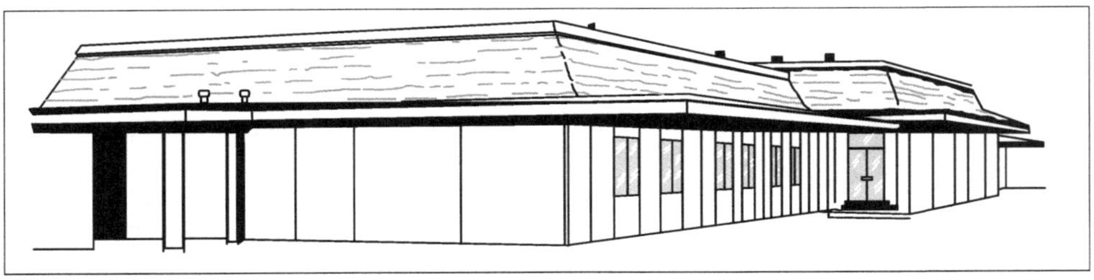

Medical-Dental Building, Class 1

First Story – Square Foot Area

Quality Class	1,000	1,500	2,000	2,500	3,000	4,000	5,000	7,500	10,000	15,000	20,000
Exceptional	329.44	311.13	300.96	294.22	289.42	282.87	278.63	272.04	268.34	264.00	261.51
1, Best	303.08	286.29	276.93	270.76	266.30	260.28	256.37	250.34	246.90	242.86	240.58
1 & 2	285.01	269.24	260.38	254.56	250.35	244.77	241.01	235.44	232.13	228.40	226.27
2, Good	269.06	254.23	245.85	240.38	236.43	231.12	227.64	222.26	219.16	215.67	213.58
2 & 3	249.72	235.82	228.12	223.01	219.42	214.40	211.19	206.18	203.31	200.08	198.21
3, Average	233.51	220.65	213.28	208.57	205.15	200.55	197.43	192.82	190.19	187.12	185.30
3 & 4	217.99	205.95	199.19	194.75	191.52	187.19	184.33	180.03	177.54	174.73	173.12
4, Low	203.47	192.20	185.91	181.77	178.76	174.69	172.06	168.00	165.69	163.12	161.57

Second and Higher Stories – Square Foot Area

Quality Class	1,000	1,500	2,000	2,500	3,000	4,000	5,000	7,500	10,000	15,000	20,000
Exceptional	287.69	276.34	269.91	265.77	262.83	258.70	255.98	251.95	249.57	246.88	245.26
1, Best	264.27	253.83	248.01	244.13	241.41	237.67	235.16	231.39	229.27	226.73	225.31
1 & 2	250.83	240.90	235.29	231.73	229.13	225.47	223.13	219.59	217.51	215.21	213.91
2, Good	239.95	230.39	225.15	221.56	219.10	215.71	213.41	210.06	208.09	205.87	204.51
2 & 3	221.77	212.96	208.05	204.86	202.47	199.36	197.27	194.13	192.28	190.27	189.01
3, Average	206.11	198.05	193.46	190.47	188.29	185.39	183.44	180.49	178.82	176.92	175.77
3 & 4	192.85	185.17	181.03	178.17	176.19	173.43	171.60	168.87	167.33	165.56	164.45
4, Low	180.05	173.04	168.94	166.41	164.50	161.91	160.21	157.73	156.22	154.53	153.57

Medical-Dental Buildings – Wood Frame

Interior Suite Entrances, Length Less Than Twice Width

Estimating Procedure

1. Use these figures to estimate medical, dental, psychiatric, optometry and similar professional buildings in which access to each office suite is through an interior corridor. Buildings in this section have more plumbing fixtures per square foot or floor and smaller room sizes than general office buildings. Note also that buildings with more than 10,000 square feet are assumed to have lead shielded x-ray rooms.
2. Establish the building quality class by applying the information on page 159.
3. Compute the first floor area. This should include everything within the exterior walls and all insets outside the main walls but under the main roof.
4. If the first floor wall height is more or less than 10 feet, add or subtract from the first floor square foot cost below the appropriate amount from the Wall Height Adjustment Table on page 166.
5. Multiply the adjusted square foot cost by the first floor area.
6. Deduct, if appropriate, for common walls or no wall finish. Use the figures on page 166.
7. If there are second or higher floors, compute the square foot area on each floor. Locate the appropriate square foot cost from the table at the bottom of this page. Adjust this figure for a wall height more or less than 9 feet, using the figures on page 166. Multiply the adjusted cost by the square foot area on each floor. Use the figures on page 166 to deduct for common walls or no wall finish. Add the result to the cost from step 6 above.
8. Multiply the total cost by the location factor listed on page 7 or 8.
9. Add the cost of heating and air conditioning systems, elevators, fire sprinklers, exterior signs, paving and curbing, miscellaneous yard improvements. See page 236 to 248.

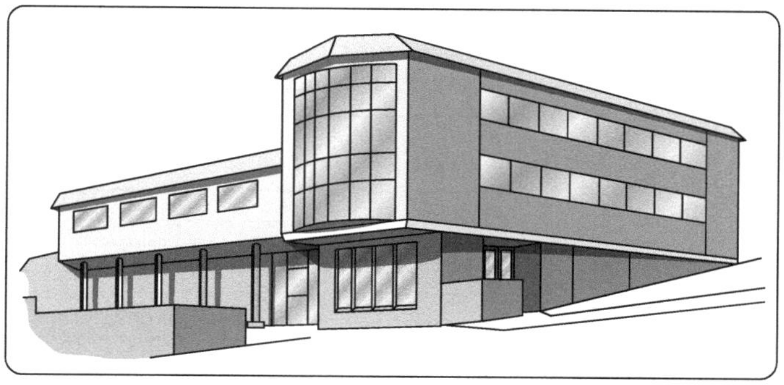

Medical-Dental Building, Class 3

First Story – Square Foot Area

Quality Class	2,000	2,500	3,000	4,000	5,000	7,500	10,000	15,000	20,000	30,000	40,000
Exceptional	261.26	257.03	253.87	249.63	246.82	242.48	239.95	236.95	235.18	233.09	231.98
1, Best	247.36	243.35	240.33	236.35	233.59	229.51	227.13	224.34	222.68	220.74	219.58
1 & 2	233.63	229.84	227.03	223.25	220.73	216.78	214.58	211.89	210.26	208.44	207.43
2, Good	221.63	218.09	215.42	211.82	209.33	205.68	203.48	201.04	199.53	197.84	196.78
2 & 3	205.87	202.47	200.06	196.74	194.46	191.04	188.98	186.53	185.30	183.70	182.81
3, Average	193.09	189.92	187.61	184.43	182.38	179.12	177.29	175.10	173.80	172.36	171.42
3 & 4	181.60	178.60	176.46	173.46	171.53	168.47	166.69	164.64	163.44	162.03	161.19
4, Low	170.78	167.98	165.97	163.19	161.29	158.46	156.80	154.91	153.63	152.44	151.64

Second and Higher Stories – Square Foot Area

Quality Class	2,000	2,500	3,000	4,000	5,000	7,500	10,000	15,000	20,000	30,000	40,000
Exceptional	239.55	236.95	235.06	232.42	230.67	227.93	226.31	224.43	223.31	222.10	221.33
1, Best	227.16	224.65	222.90	220.34	218.72	216.12	214.60	212.87	211.83	210.59	209.83
1 & 2	215.66	213.29	211.51	209.21	207.61	205.14	203.76	202.11	201.07	199.95	198.55
2, Good	205.97	203.66	202.04	199.79	198.27	195.93	194.65	192.97	192.04	190.92	190.23
2 & 3	190.28	188.18	186.71	184.59	183.15	181.04	179.78	178.30	177.43	176.41	175.80
3, Average	177.02	175.10	173.63	171.70	170.38	168.41	167.22	165.87	165.04	164.10	163.55
3 & 4	166.41	164.56	163.28	161.61	160.21	158.34	157.19	155.92	155.16	154.25	153.74
4, Low	156.28	154.57	153.36	151.64	150.47	148.69	147.69	146.48	145.69	144.89	144.40

Medical-Dental Buildings – Wood Frame

Interior Suite Entrances, Length Between 2 and 4 Times Width

Estimating Procedure

1. Use these figures to estimate medical, dental, psychiatric, optometry and similar professional buildings in which access to each office suite is through an interior corridor. Buildings in this section have more plumbing fixtures per square foot of floor and smaller room sizes than general office buildings. Note also that buildings with more than 10,000 square feet are assumed to have lead shielded x-ray rooms.
2. Establish the building quality class by applying the information on page 159.
3. Compute the first floor area. This should include everything within the exterior walls and all insets outside the main walls but under the main roof.
4. If the first floor wall height is more or less than 10 feet, add to or subtract from the first floor square foot cost below the appropriate amount from the Wall Height Adjustment Table on page 166.
5. Multiply the adjusted square foot cost by the first floor area.
6. Deduct, if appropriate, for common walls or no wall finish. Use the figures on page 166.
7. If there are second or higher floors, compute the square foot area on each floor. Locate the appropriate square foot cost from the table at the bottom of this page. Adjust this figure for a wall height more or less than 9 feet, using the figures on page 166. Multiply the adjusted cost by the square foot area on each floor. Use the figures on page 166 to deduct for common walls or no wall finish. Add the result to the cost from step 6 above.
8. Multiply the total cost by the location factor listed on page 7 or 8.
9. Add the cost of heating and air conditioning systems, elevators, fire sprinklers, exterior signs, paving and curbing, miscellaneous yard improvements. See page 236 to 248.

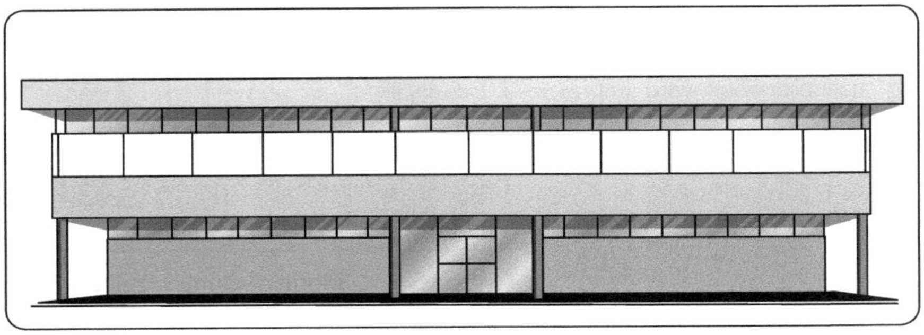

Medical-Dental Building, Class 1 & 2

First Story – Square Foot Area

Quality Class	2,000	2,500	3,000	4,000	5,000	7,500	10,000	15,000	20,000	30,000	40,000
Exceptional	267.79	262.59	258.89	253.74	250.34	245.17	242.21	238.60	236.65	234.22	232.87
1, Best	254.46	249.49	245.97	241.11	237.83	232.90	230.04	226.71	224.84	222.51	221.17
1 & 2	239.96	235.27	231.97	227.34	224.26	219.70	216.97	213.87	212.03	209.92	208.57
2, Good	227.41	223.01	219.86	215.49	212.59	208.20	205.65	202.68	201.00	198.94	197.79
2 & 3	211.30	207.20	204.29	200.21	197.48	193.44	191.06	188.29	186.71	184.82	183.75
3, Average	197.98	194.12	191.32	187.56	185.11	181.18	178.97	176.38	174.97	173.14	172.08
3 & 4	186.84	183.21	180.55	177.01	174.65	171.02	168.90	166.50	165.01	163.40	162.42
4, Low	175.05	171.61	169.23	165.72	163.58	160.21	158.24	156.00	154.62	153.08	152.13

Second and Higher Stories – Square Foot Area

Quality Class	2,000	2,500	3,000	4,000	5,000	7,500	10,000	15,000	20,000	30,000	40,000
Exceptional	244.75	241.16	238.58	235.16	232.89	229.53	227.67	225.47	224.26	222.87	221.98
1, Best	233.19	229.80	227.39	224.12	221.93	218.76	216.97	214.85	213.71	212.27	211.50
1 & 2	220.30	217.02	214.73	211.62	209.61	206.68	204.94	203.01	201.86	200.58	199.83
2, Good	210.26	207.20	204.98	202.05	200.09	197.22	195.59	193.77	192.67	191.48	190.79
2 & 3	195.04	192.16	190.17	187.38	185.59	182.98	181.39	179.67	178.68	177.54	176.90
3, Average	180.79	178.09	176.27	173.68	172.06	169.59	168.19	166.55	165.64	164.56	163.97
3 & 4	170.78	168.23	166.47	164.10	162.49	160.19	158.84	157.38	156.45	155.46	154.83
4, Low	159.67	157.33	155.63	153.41	151.95	149.71	148.53	147.08	146.37	145.41	144.82

Medical-Dental Buildings – Wood Frame

Interior Suite Entrances, Length More Than 4 Times Width

Estimating Procedure

1. Use these figures to estimate medical, dental, psychiatric, optometry and similar professional buildings in which access to each office suite is through an interior corridor. Buildings in this section have more plumbing fixtures per square foot of floor and smaller room sizes than general office buildings. Note also that buildings with more than 10,000 square feet are assumed to have lead shielded x-ray rooms.
2. Establish the building quality class by applying the information on page 159.
3. Compute the first floor area. This should include everything within the exterior walls and all insets outside the main walls but under the main roof.
4. If the first floor wall height is more or less than 10 feet, add to or subtract from the first floor square foot cost below the appropriate amount from the Wall Height Adjustment Table on page 166.
5. Multiply the adjusted square foot cost by the first floor area.
6. Deduct, if appropriate, for common walls or no wall finish. Use the figures on page 166.
7. If there are second or higher floors, compute the square foot area on each floor. Locate the appropriate square foot cost from the table at the bottom of this page. Adjust this figure for a wall height more or less than 9 feet, using the figures on page 166. Multiply the adjusted cost by the square foot area on each floor. Use the figures on page 166 to deduct for common walls or no wall finish. Add the result to the cost from step 6 above.
8. Multiply the total cost by the location factor listed on page 7 or 8.
9. Add the cost of heating and air conditioning systems, elevators, fire sprinklers, exterior signs, paving and curbing, miscellaneous yard improvements. See page 236 to 248.

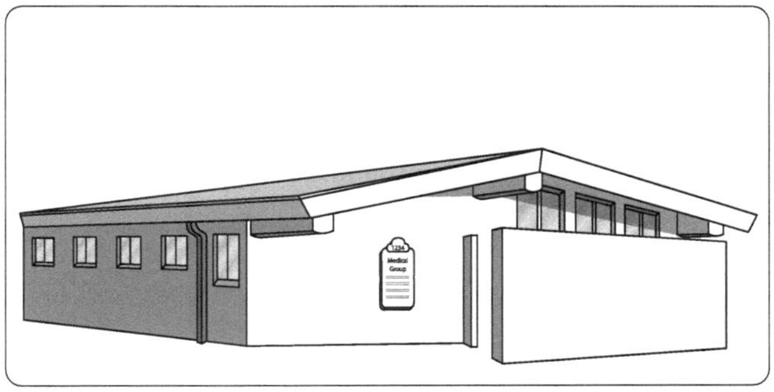

Medical-Dental Building, Class 2

First Story – Square Foot Area

Quality Class	2,000	2,500	3,000	4,000	5,000	7,500	10,000	15,000	20,000	30,000	40,000
Exceptional	278.22	271.99	267.53	261.15	256.85	250.19	246.24	241.63	238.78	235.49	233.58
1, Best	263.78	257.96	253.67	247.64	243.57	237.20	233.51	229.03	226.39	223.28	221.44
1 & 2	248.61	243.07	239.04	233.39	229.50	223.56	219.99	215.88	213.35	210.47	208.75
2, Good	235.35	230.18	226.31	220.96	217.30	211.71	208.32	204.42	202.04	199.27	197.60
2 & 3	219.69	214.78	211.26	206.14	202.84	197.56	194.41	190.79	188.56	185.93	184.45
3, Average	204.74	200.22	196.90	192.21	189.04	184.16	181.22	177.85	175.70	173.32	171.91
3 & 4	198.50	194.14	190.88	186.39	183.27	178.57	175.76	172.38	170.39	168.12	166.67
4, Low	180.71	176.70	173.80	169.68	166.88	162.55	160.00	156.99	155.16	152.99	151.74

Second and Higher Stories – Square Foot Area

Quality Class	2,000	2,500	3,000	4,000	5,000	7,500	10,000	15,000	20,000	30,000	40,000
Exceptional	249.78	245.74	242.83	238.89	236.27	232.23	230.06	227.41	225.94	224.12	223.11
1, Best	236.86	233.02	230.28	226.53	224.04	220.30	218.10	215.66	214.18	212.53	211.48
1 & 2	224.80	221.12	218.51	214.90	212.57	209.00	206.96	204.62	203.26	201.61	200.67
2, Good	214.51	211.06	208.57	205.14	202.88	199.53	197.55	195.29	193.94	192.53	191.56
2 & 3	199.24	195.98	193.66	190.52	188.44	185.25	183.44	181.36	180.12	178.77	177.94
3, Average	184.43	181.52	179.38	176.38	174.47	171.53	169.89	167.96	166.80	165.51	164.70
3 & 4	174.17	171.35	169.33	166.52	164.72	161.96	160.36	158.56	157.28	156.28	155.54
4, Low	162.72	160.14	158.30	155.63	153.94	151.43	149.94	148.17	147.16	146.08	145.41

Medical-Dental Buildings – Wood Frame

Wall Height Adjustment

The square foot costs for medical and dental buildings are based on the wall heights of 10 feet for first floors and 9 feet for higher floors. The first floor height is the distance from the bottom of the floor slab or joists to the top of the roof slab or ceiling joists. Second and higher floors are measured from the top of the floor slab or floor joists to the top of the roof slab or ceiling joists. Add or subtract the amount listed in this table to or from the square foot of floor cost for each foot of wall height more or less than 10 feet, if adjusting for first floor, and 9 feet, if adjusting for upper floors.

Square Foot Area

Quality Class	1,000	1,500	2,000	3,000	4,000	5,000	7,500	10,000	15,000	20,000	40,000
1, Best	2.90	2.32	1.96	1.58	1.38	1.15	.91	.83	.70	.64	.44
2, Good	2.56	1.95	1.72	1.39	1.13	1.04	.83	.73	.64	.59	.40
3, Average	2.27	1.73	1.49	1.22	1.03	.90	.75	.65	.60	.47	.22
4, Low	1.98	1.55	1.33	1.04	.90	.83	.65	.61	.47	.40	.22

Perimeter (Common) Wall Adjustment

A common wall exists when two buildings share one wall. Adjust for common walls by deducting the linear foot costs below from the total structure cost. In some structures, one or more walls are not owned at all. In this case, deduct the "No Ownership" cost per linear foot of wall not owned. If a wall has no exterior finish, deduct the "Lack of Exterior Finish" cost.

First Story

Class	For a Common Wall, Deduct Per L.F.	For No Wall Ownership, Deduct Per L.F.	For Lack of Exterior Finish, Deduct Per L.F.
1	$372	$716	$314
2	297	575	249
3	240	461	172
4	184	382	134

Second and Higher Stories

Class	For a Common Wall, Deduct Per L.F.	For No Wall Ownership, Deduct Per L.F.	For Lack of Exterior Finish, Deduct Per L.F.
1	$249	$478	$312
2	186	372	247
3	172	362	171
4	166	336	132

Convalescent Hospitals – Masonry or Concrete

Quality Classification

	Class 1 Best Quality	Class 2 Good Quality	Class 3 Average Quality	Class 4 Low Quality
Foundation (18% of total cost)	Reinforced concrete.	Reinforced concrete.	Reinforced concrete.	Reinforced concrete.
Floor Structure (10% of total cost)	Reinforced concrete.	Reinforced concrete.	Reinforced concrete.	Reinforced concrete.
Roof & Cover (7% of total cost)	Standard wood frame and composition roofing with large colored rock or tile. 4' to 6' sealed overhang.	Standard wood frame and composition roofing with large colored rock or asbestos shingles. 3' to 4' sealed overhang.	Standard wood frame and composition roofing small colored rock. 4' exposed or 3' sealed overhang.	Standard wood frame and composition roofing pea gravel or composition shingles. 2' exposed overhang.
Exterior Finishes				
Walls (12% of total cost)	8" colored concrete block.	8" colored concrete block.	8" colored concrete block.	8" colored concrete block.
Front (5% of total cost)	Brick or stone veneer and float glass in aluminum frame.	Stone or brick veneer. Small amount of float glass in aluminum frame.	Rustic siding or some brick or stone veneer.	Stucco and small amount siding.
Windows (5% of total cost)	Good aluminum sliding glass doors or anodized aluminum casement windows.	Good aluminum sliding glass doors with aluminum jalousie window for vent.	Average aluminum sliding type. 6' sliding glass door in bedrooms.	Low cost aluminum sliding type.
Doors (5% of total cost)	Anodized aluminum entry doors.	Aluminum and glass entry doors.	Good hardwood slab door with side lights or aluminum and glass entry door.	Wood slab door.
Interior Finishes				
Floors (5% of total cost)	Resilient tile. Good carpet in entry and day room. Quarry tile or epoxy type floor in kitchen.	Resilient tile. Good carpet in day room. May have coved sheet vinyl in kitchen.	Composition tile. Day room may have average grade carpet.	Low cost tile. Composition tile in kitchen. Concrete in storage areas.
Walls (10% of total cost)	Gypsum wallboard and enamel. Large amount of plastic wall cover. Some wood paneling or glass walls in entry and day rooms. 4' ceramic tile wainscot in kitchen. Gypsum wallboard and paint in storage areas.	Gypsum wallboard and enamel. Gypsum wallboard and some plastic wall cover or wood paneling in entry, bedrooms, dining room, day room, and kitchen. Gypsum wallboard and paint in storage areas.	Gypsum wallboard and enamel. Some plastic wall cover or wood paneling in entry and day room. Gypsum wallboard and paint in storage areas.	Gypsum wallboard and texture. Gypsum wallboard and enamel in kitchen. Unfinished gypsum wallboard in storage areas.
Ceilings (8% of total cost)	Acoustical tile. Washable acoustical tile in kitchen. Gypsum wallboard and paint in storage areas.	Gypsum wallboard and enamel. Acoustical tile in corridor. Open beam ceilings with wood decking in entry and day room. Gypsum wallboard and paint in storage areas.	Gypsum wallboard & enamel or acoustical texture. Open beams with wood decking. Gypsum wallboard and paint in storage areas.	Gypsum wallboard and texture. Gypsum wallboard and acoustical texture in entry. Gypsum wallboard and enamel in kitchen. Unfinished gypsum in storage areas.
Bath Finish Detail				
Main Bath & Showers (10% of total cost)	Ceramic tile floors. Ceramic tile walls. Gypsum wallboard and enamel ceilings.	Ceramic tile floors. 6' ceramic tile wainscot. Gypsum wallboard and enamel walls and ceilings.	Ceramic tile floors. 4' ceramic tile wainscot. Gypsum wallboard and enamel walls and ceilings.	Resilient tile floors. Gypsum wallboard and enamel walls and ceilings.
Toilet Rooms (5% of total cost)	Sheet vinyl or ceramic tile floors. 4' ceramic tile wainscot. Gypsum wallboard and enamel walls and ceilings.	Sheet vinyl floors. 4' plastic or ceramic tile wainscot.	Resilient tile or linoleum floors. 4' plastic or marlite wainscot.	Composition tile floors. Gypsum wallboard and enamel walls and ceilings.

Note: Use the percent of total cost to help identify the correct quality classification.

Square foot costs include the following components: Foundations as required for normal soil conditions. Floor, wall and roof structures. Interior floor, wall and ceiling finishes. Exterior wall finish and roofing. Interior partitions. Cabinets, doors and windows. Basic electrical systems and all plumbing. Permits and fees. Contractor's mark-up. Cost of nurses' built-in station desks and cabinets. Nurses' call system. Emergency lighting system. Cubicle curtain tracks. Kitchen range hood.

The in-place cost of these extra components should be added to the basic building cost to arrive at the total structure cost. See the section "Additional Costs for Commercial Buildings" on page 236 to 248. Heating and air conditioning systems. Elevators. Fire sprinklers. Dumbwaiters. Fireplaces. *Square foot costs do not include the following items:* Drapes. Cubicle curtains. Incinerators. Laundry or kitchen equipment, other than the range hood.

Convalescent Hospitals – Masonry or Concrete

Estimating Procedure

1. Use these figures to estimate institutions such as nursing or rest homes which have facilities limited to bed care for people unable to care for themselves.
2. Establish the building quality class by applying the information on page 167.
3. Compute the floor area. This should include everything within the exterior walls and all insets outside the main walls but under the main roof. Areas such as storage rooms which have a substantially inferior finish or any area with a slightly interior finish and which does not perform a function related to the hospital operation should be estimated at 1/2 to 3/4 of the square foot costs listed.
4. Estimate second stories and basements at 100% of the first floor cost if they are of the same general quality. A storage area of the same or slightly inferior quality located in the basement, second floor or separate building should be estimated at 100% of the building square foot cost.
5. Multiply the appropriate square foot cost below by the floor area.
6. Multiply the total cost by the location factor on page 7 or 8.
7. Add the cost of heating and air conditioning systems, fire sprinklers, drapes, incinerators, elevators, dumbwaiters, and fireplaces. See the section beginning on page 236.

Convalescent Hospital, Class 2

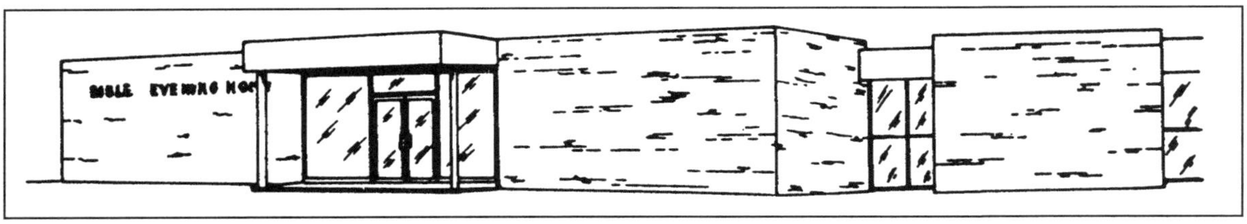

Convalescent Hospital, Class 2

Square Foot Area

Quality Class	5,000	6,000	7,000	8,000	9,000	10,000	12,500	15,000	17,500	20,000	25,000	30,000
Exceptional	360.58	349.24	341.07	334.97	330.33	326.51	319.83	315.37	312.18	309.82	306.59	304.12
1, Best	340.37	329.59	322.02	316.17	311.79	308.23	301.88	297.68	294.72	292.39	289.37	287.32
1 & 2	320.85	310.69	303.47	297.97	293.93	290.56	284.58	280.61	277.84	275.66	272.76	270.83
2, Good	302.69	293.10	286.30	281.20	277.25	274.11	268.53	264.83	262.16	260.08	257.38	255.51
2 & 3	279.46	270.66	264.37	259.68	256.09	253.03	247.93	244.54	241.97	240.21	237.56	235.93
3, Average	263.59	255.34	249.47	244.98	241.64	238.84	233.81	230.65	228.29	226.55	224.17	222.59
3 & 4	241.55	233.88	228.54	224.37	221.26	218.78	214.34	211.27	209.14	207.54	205.30	203.93
4, Low	221.06	214.05	209.06	205.38	202.48	200.18	196.07	193.28	191.46	189.89	187.94	186.56

Convalescent Hospitals – Wood Frame

Quality Classification

	Class 1 Best Quality	Class 2 Good Quality	Class 3 Average Quality	Class 4 Low Quality
Foundation (18% of total cost)	Reinforced concrete.	Reinforced concrete.	Reinforced concrete.	Reinforced concrete.
Floor Structure (10% of total cost)	Reinforced concrete.	Reinforced concrete.	Reinforced concrete.	Reinforced concrete.
Roof & Cover (7% of total cost)	Standard wood frame and composition roofing with large colored rock or tile. 4' to 6' sealed overhang.	Standard wood frame and composition roofing with large colored rock or asbestos shingles. 3' to 4' sealed overhang.	Standard wood frame and composition roofing small colored rock. 4' exposed or 3' sealed overhang.	Standard wood frame and composition with pea gravel or composition shingles. 2' exposed overhang.
Exterior Finishes				
Walls (12% of total cost)	Wood siding or stucco with extensive trim.	Stucco or wood siding.	Stucco or good plywood siding.	Stucco.
Front (5% of total cost)	Brick or stone veneer and float glass in aluminum frame.	Stone or brick veneer. Small amount of float glass in aluminum frame.	Rustic siding or some brick or stone veneer.	Stucco and small amount siding.
Windows (5% of total cost)	Good aluminum sliding glass doors or anodized aluminum casement windows.	Good aluminum sliding glass doors with aluminum jalousie window for vent.	Average aluminum sliding type. 6' sliding glass door in bedrooms.	Low cost aluminum sliding type.
Doors (5% of total cost)	Anodized aluminum entry doors.	Aluminum and glass entry doors.	Good hardwood slab door with side lights or aluminum and glass entry door.	Wood slab door.
Interior Finishes				
Floors (5% of total cost)	Resilient tile. Good carpet in entry and day room. Quarry tile or epoxy type floor in kitchen.	Resilient tile. Good carpet in day room. May have coved sheet vinyl in kitchen.	Composition tile. Day room may have average grade carpet.	Low cost tile. Composition tile in kitchen. Concrete in storage areas.
Walls (10% of total cost)	Gypsum wallboard and enamel. Large amount of plastic wall cover. Some wood paneling or glass walls in entry and day rooms. 4' ceramic tile wainscot in kitchen. Gypsum wallboard and paint in storage areas.	Gypsum wallboard and enamel. Gypsum wallboard and some plastic wall cover or wood paneling in entry, bedrooms, dining room, day room, and kitchen. Gypsum wallboard and paint in storage areas.	Gypsum wallboard and enamel. Some plastic wall cover or wood paneling in entry and day room. Gypsum wallboard and paint in storage areas.	Gypsum wallboard and texture. Gypsum wallboard and enamel in kitchen. Unfinished gypsum wallboard in storage areas.
Ceilings (8% of total cost)	Acoustical tile. Washable acoustical tile in kitchen. Gypsum wallboard and paint in storage areas.	Gypsum wallboard and enamel. Acoustical tile in corridor. Open beam ceilings with wood decking in entry and day room. Gypsum wallboard and paint in storage areas.	Gypsum wallboard & enamel or acoustical texture. Open beams with wood decking. Gypsum wallboard and paint in storage areas.	Gypsum wallboard and texture. Gypsum wallboard and acoustical texture in entry. Gypsum wallboard and enamel in kitchen. Unfinished gypsum in storage areas.
Bath Finish Detail				
Main Bath & Showers (10% of total cost)	Ceramic tile floors. Ceramic tile walls. Gypsum wallboard and enamel ceilings.	Ceramic tile floors. 6' ceramic tile wainscot. Gypsum wallboard and enamel walls and ceilings.	Ceramic tile floors. 4' ceramic tile wainscot. Gypsum wallboard and enamel walls and ceilings.	Resilient tile floors. Gypsum wallboard and enamel walls and ceilings.
Toilet Rooms (5% of total cost)	Vinyl or ceramic tile floors. 4' ceramic tile wainscot. Gypsum wallboard and enamel walls and ceilings.	Vinyl floors. 4' plastic or ceramic tile wainscot.	Resilient tile or linoleum floors. 4' plastic wainscot.	Composition tile floors. Gypsum wallboard and enamel walls and ceilings.

Note: Use the percent of total cost to help identify the correct quality classification.

Square foot costs include the following components: Foundations as required for normal soil conditions. Floor, wall and roof structures. Interior floor, wall and ceiling finishes. Exterior wall finish and roofing. Interior partitions. Cabinets, doors and windows. Basic electrical systems and all plumbing. Permits and fees. Contractor's mark-up. Cost of nurses' built-in station desks and cabinets. Nurses' call system. Emergency lighting system. Cubicle curtain tracks. Kitchen range hood.

The in-place cost of these extra components should be added to the basic building cost to arrive at the total structure cost. See the section "Additional Costs for Commercial Buildings" on page 236 to 248. Heating and air conditioning systems. Elevators. Fire sprinklers. Dumbwaiters. Fireplaces. *Square foot costs do not include the following items:* Drapes. Cubicle curtains. Incinerators. Laundry or kitchen equipment, other than the range hood.

Convalescent Hospitals – Wood Frame

Estimating Procedure

1. Use these figures to estimate institutions such as nursing or rest homes which have facilities limited to bed care for people unable to care for themselves.
2. Establish the building quality class by applying the information on page 169.
3. Compute the floor area. This should include everything within the exterior walls and all insets outside the main walls but under the main roof. Areas such as storage rooms which have a substantially inferior finish or any area with a slightly interior finish and which does not perform a function related to the hospital operation should be estimated at 1/2 to 3/4 of the square foot costs listed.
4. Estimate second stories and basements at 100% of the first floor cost if they are of the same general quality. A storage area of the same or slightly inferior quality located in a basement, second floor or separate building should be estimated at 100% of the building square foot cost.
5. Multiply the appropriate square foot cost below by the floor area.
6. Multiply the total cost by the location factor on page 7 or 8.
7. Add the cost of heating and air conditioning systems, fire sprinklers, drapes, therapy facilities, cubicle curtains, incinerators, elevators, dumbwaiters, fireplaces, laundry equipment and kitchen equipment other than a range hood. See the section beginning on page 236.

Convalescent Hospital, Class 4

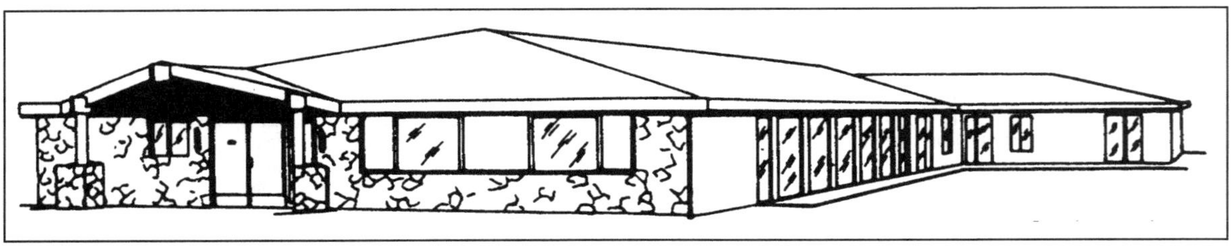

Convalescent Hospital, Class 3

Square Foot Area

Quality Class	5,000	6,000	7,000	8,000	9,000	10,000	12,500	15,000	17,500	20,000	25,000	30,000
Exceptional	235.13	227.69	222.36	218.26	215.06	212.52	207.85	204.73	202.37	200.65	198.11	196.45
1, Best	222.36	215.36	210.32	206.49	203.47	201.03	196.62	193.62	191.48	189.78	187.41	185.84
1 & 2	208.99	202.43	197.60	194.06	191.19	188.97	184.79	181.92	179.94	178.41	176.17	174.47
2, Good	197.68	191.46	186.95	183.56	180.83	178.75	174.80	172.10	170.16	168.69	166.62	165.17
2 & 3	184.64	178.79	174.63	171.41	168.88	166.93	163.20	160.78	158.97	157.57	155.60	154.26
3, Average	172.23	166.86	162.95	159.92	157.63	155.76	152.34	150.00	148.25	147.08	145.23	143.99
3 & 4	154.97	150.15	146.57	143.88	141.83	140.15	137.04	134.97	133.43	132.29	130.60	129.51
4, Low	145.14	140.59	137.18	134.72	132.79	131.24	128.35	126.42	124.90	123.89	122.32	121.28

Funeral Homes

Quality Classification

	Class 1 Best Quality	Class 2 Good Quality	Class 3 High Average Quality	Class 4 Low Average Quality	Class 5 Low Quality
Foundation (15% of total cost)	Reinforced concrete.	Reinforced concrete.	Reinforced concrete.	Reinforced concrete.	Reinforced concrete.
Floor Structure (7% of total cost)	Standard wood frame or reinforced concrete.	Standard wood frame or reinforced concrete.	Standard wood frame or reinforced concrete.	Standard wood frame or reinforced concrete.	Standard wood frame or reinforced concrete.
Wall Structure (12% of total cost)	Standard wood frame.	Standard wood frame.	Standard wood frame.	Standard wood frame.	Standard wood frame.
Roof & Cover (8% of total cost)	Standard wood frame with slate and 3' sealed overhang.	Standard wood frame with bar tile and 3' sealed overhang.	Standard wood frame with heavy shake and 3' sealed overhang.	Standard wood frame with composition roofing and large colored rock, asbestos shingles or medium shake, 3' exposed overhang.	Standard wood frame with composition roofing and small colored rock or good composition shingles, 2' exposed overhang.
Exterior Finishes					
Walls (5% of total cost)	Brick or stone veneer.	Good wood siding, brick or stone veneer.	Wood siding or stucco with extensive wood trim.	Stucco or wood siding.	Stucco.
Front (5% of total cost)	Brick or stone veneer. Some tinted float glass in bronze frames.	Brick or stone veneer. Some tinted float glass in anodized aluminum frames.	Brick or stone veneer. Some tinted float glass in anodized aluminum frames.	Stone or brick veneer. Some tinted crystal or float glass in wood frame.	Wood siding or stucco with stone brick veneer. Some colored veneer windows with wood frame (bottle glass type).
Windows (5% of total cost)	Good anodized aluminum windows. Extensive stained glass in chapel.	Good aluminum sliding type. Stained glass in chapel.	Good aluminum sliding type. Leaded glass in chapel.	Good aluminum sliding type. Mock leaded glass in chapel.	Average aluminum sliding type. Colored plastic in chapel.
Doors (5% of total cost)	Elaborate hardwood entry doors.	Custom-built hardwood entry doors.	Custom hardwood or anodized aluminum and glass.	Good hardwood or anodized aluminum and glass doors.	Good hardwood slab or aluminum and glass entry doors.
Interior Finishes					
Floors (6% of total cost)	Excellent carpet. Good tile in kitchen.	Excellent carpet. Good carpet in living quarters. Good tile in kitchen.	Good carpet. Good tile in kitchen.	Good carpet. Average carpet in living quarters. Tile in kitchen.	Average carpet. Composition tile in hallways and living quarters.
Walls (12% of total cost)	Hardwood paneling, extensive brick or stone work in lobby and chapel. Gypsum wallboard and texture in living quarters.	Good hardwood veneer with brick or stone work in lobby and chapel. Gypsum wallboard and texture in living quarters.	Gypsum wallboard and embossed wallpaper or fabric textured wallcover. Hardwood in chapel. Gypsum wallboard and texture in living quarters.	Gypsum wallboard and wallpaper. Hardwood veneer in chapel.	Gypsum wallboard and texture. Some hardwood veneer in chapel.
Ceilings (8% of total cost)	Acoustical plaster. Exposed ornamental beams with ornamental hardwood decking or ornamental acoustical plaster in chapel.	Acoustical plaster. Exposed ornamental beams with ornamental wood decking or acoustical plaster in chapel.	Gypsum wallboard and heavy acoustical texture. Exposed architectural glu-lams w/ deck in chapel. Gypsum wallboard and texture in living quarters.	Gypsum wallboard and acoustical texture. Exposed wood beam with wood deck in chapel.	Gypsum wallboard and acoustical texture. Exposed beam with wood deck in chapel.
Special Lighting					
Fixtures (3% of total cost)	Two expensive chandeliers in lobby (4' to 6' diameter). Recessed and spotlighting in chapel.	Two expensive chandeliers in lobby (4' to 6' diameter). Recessed and spotlighting in chapel.	Two good chandeliers in lobby (4' diameter). Recessed and spotlighting in chapel.	Two average chandeliers in lobby (3' diameter).	One average chandelier in lobby.
Bath Detail (9% of total cost)	Ceramic tile floors and walls. Gypsum wallboard and enamel ceilings. Excellent fixtures.	Ceramic tile floors. 4' ceramic tile wainscot. Gypsum wallboard and enamel ceilings. Excellent fixtures.	Ceramic tile floors. Gypsum wallboard and enamel walls and ceilings. Good fixtures.	Good tile floors. Gypsum wallboard and enamel walls and ceilings. Average fixtures.	Tile floors. Gypsum wallboard and enamel walls and ceilings. Average fixtures.

Note: Use the percent of total cost to help identify the correct quality classification.

Square foot costs include the following components: Foundations as required for normal soil conditions. Floor, wall and roof structures. Interior floor, wall and ceiling finishes. Exterior wall finish and roof cover. Interior partitions. Cabinets, doors and windows. Basic electrical system and plumbing system. Special lighting fixtures. Bath detail. Permits and fees. Contractor's mark-up.

Funeral Homes – Wood Frame

Estimating Procedure

1. Use these figures to estimate mortuaries and buildings designed for care of the dead and funeral services.
2. Establish the building quality class by applying the information on page 171.
3. Compute the first floor area under the main roof and inside the exterior walls. Include lobby, chapel and living sections, but exclude garages, whether internal, external or attached.
4. Add to the first floor area between 1/3 and 1/4 of the actual garage area. Use the higher value for better quality garages.
5. Add to the first floor area all finished basement and second floor area if it is similar in quality to the first floor area. Inferior quality second floor area should be reclassified as to quality and calculated separately. Unfinished funeral home basements should be estimated from the figures on page 237.
6. Finished space within the area formed by the roof but with an exterior wall height less than 7 feet (half story area) should be included at 1/3 to 1/2 of the actual floor area. Exclude half story area when selecting a square foot cost from the table.
7. Multiply the sum from steps 3 to 6 above by the appropriate square foot cost below.
8. Add 8% if the exterior walls are brick or block and 12% if the exterior walls are concrete. Add 2% if the building has more than 16 corners or if the length is more than twice with width.
9. Multiply the building cost by the location factor listed on page 7 or 8.
10. Add the cost of heating and cooling systems, elevators, fireplaces, pews, drapes, signs and yard improvement. See the section beginning on page 236.

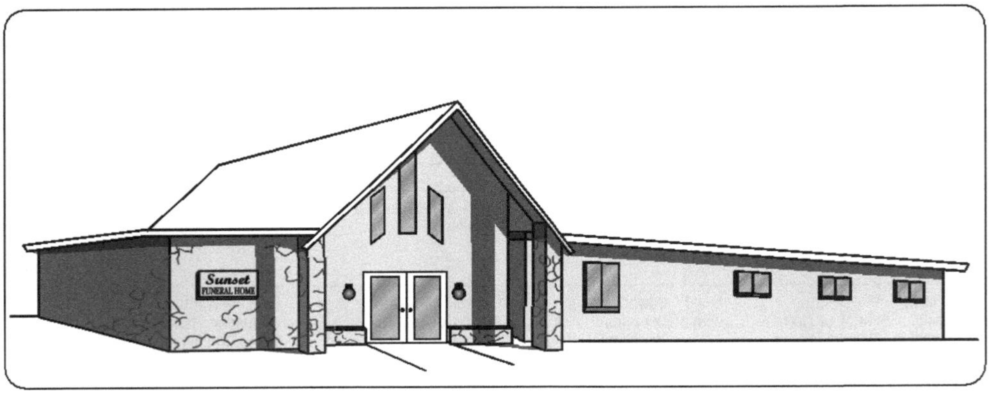

Funeral Home, Class 5

Square Foot Area

Quality	2,000	2,500	3,000	4,000	5,000	7,500	10,000	12,500	15,000	20,000	25,000	30,000
Exceptional	—	—	—	—	217.10	209.52	206.10	203.63	202.05	199.96	198.44	197.34
1, Best	—	—	—	—	207.51	199.96	195.98	193.42	191.65	189.34	187.79	186.77
1 & 2	—	—	—	—	193.24	186.24	182.53	180.09	178.50	176.21	174.84	173.84
2, Good	—	—	—	—	183.18	176.52	172.98	170.71	169.21	167.10	165.78	164.82
2 & 3	—	—	191.46	179.34	172.78	165.94	163.89	161.67	160.64	159.65	158.17	156.82
3, Average	—	—	180.92	169.52	163.33	154.25	153.59	152.38	151.81	151.51	151.70	151.99
3 & 4	186.35	176.38	169.66	161.24	156.16	149.27	145.76	143.61	142.22	140.38	139.31	138.48
4, Low	176.17	166.74	160.41	152.45	147.65	141.16	137.88	135.80	134.44	132.75	131.72	130.96

Ecclesiastic Buildings

Quality Classification

	Class 1 Best Quality	Class 2 Good Quality	Class 3 High Average Quality	Class 4 Low Average Quality	Class 5 Low Quality
Foundation (12% of total cost)	Reinforced concrete.	Reinforced concrete.	Reinforced concrete.	Reinforced concrete.	Block masonry or reinforced concrete.
Floor Structure (6% of total cost)	Engineered wood frame or reinforced concrete.	Engineered wood frame or reinforced concrete.	Engineered wood frame or reinforced concrete.	Standard wood frame or reinforced concrete.	Standard wood frame or reinforced concrete.
Wall Structure (13% of total cost)	Engineered wood frame. First floor wall is 16' high.	Engineered wood frame. First floor wall is 14' high.	Engineered wood frame. First floor wall is 12' high.	Standard wood frame. First floor wall is 10' high.	Standard wood frame. First floor wall is 8' high.
Roof & Cover (8% of total cost)	Steep pitch glu-lam or steel frame with slate, concrete tile or standing seam sheet metal surface, Closed decorative soffit over windows and doors. Includes an attached clarion tower or steeple.	Steep pitch glu-lam wood frame over sanctuary. High pitch over other rooms. Concrete tile or heavy shake surface. Closed soffit over windows and doors. With a tower or steeple.	High pitch wood frame over sanctuary. Heavy shake or architectural composition tile surface, Closed soffit over entrances. Includes a roof steeple.	Standard wood frame with wood or composition shingle surface. Open soffit over entrances.	Standard wood frame with inexpensive composition single surface.
Exterior Finish (5% of total cost)	Brick or stone veneer.	Good wood siding, brick or stone veneer.	Wood siding or stucco with extensive wood trim.	Stucco or wood siding.	Stucco.
Front (5% of total cost)	Brick or stone veneer surrounding tinted or faceted glass set in metal frame. Highly decorative or starkly original architectural design elements.	Brick or stone veneer. Some tinted or faceted glass set in metal frame. Decorative design elements at the main entrance.	Brick or stone veneer. Some tinted float glass set in a metal frame.	Some masonry veneer or applied wood trim accents. Tinted glass set in a wood frame. Wall-mounted or roof-mounted signage.	Wood siding or stucco with applied wood trim. Wall-mounted signage.
Doors and Windows (8% of total cost)	Elaborate hardwood entry doors at all exits. Large insulated float glass windows.	Custom-built hardwood doors at sanctuary exits. Insulated float glass windows	Custom hardwood or insulated metal doors with glass insets. Commercial grade windows.	Good hardwood or insulated metal doors. Good quality windows.	Average hardwood or metal entry doors. Residential quality windows.
Floors (7% of total cost)	Terrazzo or excellent carpet. Good sheet vinyl in food service rooms.	Excellent carpet in the sanctuary. Good carpet in meeting rooms. Sheet vinyl or tile in other rooms.	Good carpet in the sanctuary. Good sheet vinyl or tile in meeting and food service rooms.	Good carpet in the sanctuary. Average sheet vinyl, tile or coated concrete in meeting and food service rooms.	Average carpet in the sanctuary. Composition tile or coated concrete in other rooms.
Walls (13% of total cost)	Decorative paneling with extensive brick or stone veneer in the narthex and sanctuary. Embossed or textured gypsum wallboard in hallways and meeting rooms.	Good paneling with brick or stone accents in the narthex and sanctuary. Textured or vinyl-covered gypsum wallboard in other rooms.	Textured or embossed gypsum wallboard or good paneling in the sanctuary. Textured gypsum wallboard in other rooms.	Textured or embossed gypsum wallboard or inexpensive paneling in the sanctuary Gypsum wallboard in other rooms.	Gypsum wallboard throughout.
Ceilings (7% of total cost)	Decorative exposed beams with ornamental accents, Ornamental acoustic plaster in the sanctuary.	Exposed beams with acoustical plaster or tile in the sanctuary. Acoustical tile in other rooms.	Exposed beams or trusses in the sanctuary, Ceiling finished with acoustic tile. Textured gypsum wallboard in other rooms.	Textured gypsum wallboard or acoustic tile in the sanctuary. Gypsum wallboard in other rooms.	Gypsum wallboard.
Lighting Fixtures (4% of total cost)	Highly decorative chandeliers in the narthex. Recessed and indirect lighting with spotlights in the sanctuary. Decorative LED ceiling fixtures in other rooms.	Decorative chandeliers in the narthex. Recessed lighting and spotlights in the sanctuary. Good quality LED fixtures in other rooms.	Decorative chandeliers or suspended fluorescent fixtures in the narthex. Recessed lighting in the sanctuary. Standard LED fixtures in other rooms.	Decorative LED fixtures throughout.	Standard LED fixtures.
Chancel (5% of total cost)	Terrazzo, marble or good hardwood on a platform raised three steps above the nave. Two highly decorative altars or lecterns. Ample choir seating. Hardwood or marble railings	Excellent carpet on a raised platform. Decorative altar or lectern. Limited choir seating. Hardwood railings.	Good carpet on a raised platform. Decorative altar or lectern. Limited choir seating. Hardwood railings.	Good carpeted on a raised platform. Simple altar or lectern. Limited or no choir seating. Simple railings.	Wood or simulated wood pulpit on a raised platform.
Bath Detail (7% of total cost)	Simulated marble or terrazzo floors and walls. Gypsum wallboard and enamel ceilings. Excellent fixtures.	Terrazzo or ceramic tile floors. Ceramic tile wainscot. Gypsum wallboard and enamel ceilings. Excellent fixtures.	Ceramic tile floors. Plastic-coated wainscot. Gypsum wallboard and enamel walls and ceilings. Good fixtures.	Low cost ceramic tile or good vinyl floors. Gypsum wallboard and enamel walls and ceilings. Average fixtures.	Resilient tile floors. Gypsum wallboard and enamel walls and ceilings. Minimum fixtures.

Square foot costs include the following components: Foundations as required for normal soil conditions. Floor, wall and roof structures. Interior floor, wall and ceiling finishes. Exterior wall finish and roof cover. Interior partitions. Cabinets, doors and windows. Electrical, sound and plumbing systems. Special lighting fixtures. Building permits. Contractor's markup.

Ecclesiastic Buildings

Estimating Procedure

1. Use these figures to estimate religious buildings designed for worship or devotional purposes.
2. Establish the building quality class by applying the information on page 173.
3. Compute the first floor area under the main roof and inside the exterior walls. Include narthex (lobby), chancel and nave (sanctuary), meeting rooms, food preparation rooms and bathrooms but exclude garages, whether internal, external or attached.
4. Add to the first floor area between 1/4 and 1/3 of the actual garage area. Add to the first floor area 1/3 to 2/3 of any horizontal balcony area. Use higher values for better interior finish in the garage or balcony.
5. Add to the first floor area all finished basement (undercroft) and second floor area if similar in quality to the first floor area. Inferior quality second floor area should be reclassified as to quality and calculated separately. Unfinished ecclesiastic building basements should be estimated with the figures on page 237.
6. Finished space within the area formed by the roof but with an exterior wall height less than 7 feet (half story area) should be included at 1/3 to 1/2 of the actual floor area.
7. Multiply the sum from steps 3 to 6 above by the appropriate square foot cost below.
8. Add 4% if the building has more than 10 corners. Add 2% if the length is more than twice with the width. Add 8% if the exterior walls are masonry. Add 12% if the exterior walls are concrete.
9. Multiply the building cost by the location factor listed on page 7 or 8.
10. Add the cost of heating and cooling systems, elevators, fireplaces, pews, drapes, signs and yard improvement. See the section beginning on page 236.

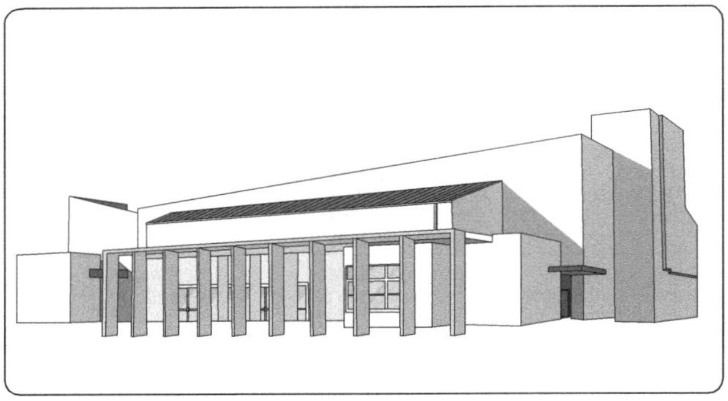

Ecclesiastic Building, Class 1 & 2

Ecclesiastic Building, Class 5

Square Foot Area

Quality	2,000	2,500	3,000	4,000	5,000	7,500	10,000	12,500	15,000	20,000	25,000	30,000
Exceptional	349.27	336.30	323.24	307.10	297.83	284.94	277.79	273.72	271.36	267.31	265.68	263.92
1, Best	319.42	302.32	290.64	276.16	267.74	256.16	249.80	246.10	243.92	240.38	238.87	237.29
1 & 2	290.42	274.91	264.31	251.14	243.46	232.86	227.13	223.77	221.79	218.59	217.15	215.76
2, Good	263.67	249.58	239.96	228.02	221.01	211.36	206.22	203.17	201.32	198.51	197.16	195.93
2 & 3	239.01	226.22	217.52	206.74	200.29	191.53	186.97	184.17	182.45	179.94	178.68	177.59
3, Average	216.26	204.68	196.88	187.09	181.25	173.29	169.16	166.63	165.06	162.83	161.66	160.69
3 & 4	195.29	184.89	177.78	168.99	163.69	156.49	152.80	150.51	149.07	147.11	146.01	145.14
4, Low	176.00	166.57	160.26	152.33	147.53	141.01	137.73	135.68	134.32	132.61	131.57	130.84

Self Service Restaurants

Quality Classification

	Class 1 Best Quality	Class 2 Good Quality	Class 3 Average Quality	Class 4 Low Quality
Foundation (15% of total cost)	Reinforced concrete.	Reinforced concrete.	Reinforced concrete.	Reinforced concrete.
Floor Structure (7% of total cost)	Reinforced concrete or standard wood frame.	Reinforced concrete or standard wood frame.	Reinforced concrete or standard wood frame.	Reinforced concrete or standard wood frame.
Wall Structure (12% of total cost)	Standard wood frame.	Standard wood frame.	Standard wood frame.	Standard wood frame.
Roof & Cover (9% of total cost)	Complex angular structure, with concrete tile, mission tile, or heavy shake.	Standard wood frame structure, low or medium pitch, with composition tar and gravel or medium shake.	Standard wood frame structure, low pitch shed or gable type with composition tar and gravel.	Standard wood frame structure, low pitch shed or gable type with composition or composition shingle.
Floor Finish (7% of total cost)	Terrazzo in dining areas. Quarry tile in kitchen.	Vinyl tile in dining area. Terrazzo or quarry tile in kitchen.	Resilient tile in dining area. Grease proof tile in kitchen.	Composition tile in dining area. Grease proof tile in kitchen.
Walls, *Exterior Finish* (12% of total cost)	Very good wood siding; ornamental brick or natural stone veneer. Minimum 60% of front and side walls have float glass in heavy metal frames or rabbeted timbers.	Good wood siding or plywood; ornamental brick or concrete block; good stucco with color finish and with 2' to 4' brick, or natural stone veneer bulkheads. 50% to 75% of front and side walls have float glass set in good wood or metal frames.	Stucco or average wood siding with 2' to 4' brick or stone veneer bulkheads. Float glass front from waist high to roof line running along service front. Glassed area from 40% to 70% of front and side walls.	Stucco, low cost wood siding or plywood. 4' to 6' fixed glass waist high with sliding aluminum window opening running along service front. Glassed area from 30% to 40% of front and side walls.
Interior Finish (11% of total cost)	Plaster with putty coat finish; plastic coated paneling.	Plaster with putty coat finish; plastic coated paneling.	Plaster putty coat finish; textured gypsum wallboard with portions of wall finished with natural wood or plastic coated veneers.	Painted plaster or gypsum wallboard.
Ceilings (10% of total cost)	Complex angular structure with acoustical plaster.	Multi-level structure with good plaster or acoustical tile.	Flat or low slope with textured gypsum wallboard or acoustical tile.	Flat or low slope with textured gypsum wallboard or acoustical tile.
Plumbing (9% of total cost)	Good fixtures with metal or marble toilet partitions. Full wall height ceramic tile wainscot. Numerous floor drains.	Commercial fixtures with metal toilet partitions. Ceramic tile wainscot. Numerous floor drains in kitchen and dishwasher area.	4 to 6 standard commercial fixtures with metal toilet partitions. Floor drains in kitchen and dishwashing area.	4 to 6 standard commercial fixtures with wood toilet partitions. Minimum floor drainage in kitchen area.
Lighting (8% of total cost)	Triple encased LED fixtures or decorative and ornate LED fixtures.	Triple encased LED fixtures or decorative LED fixtures.	Triple open or double encased louvered LED fixtures with some decorative LED fixtures.	Double open strip or low cost LED fixtures.

Note: Use the percent of total cost to help identify the correct quality classification.

Square foot costs include the following components: Foundations as required for normal soil conditions. Floor, wall and roof structures. Interior floor, wall and ceiling finishes. Exterior wall finish and roof cover. All glass and glazing. Interior partitions. Basic electrical systems and lighting fixtures. Rough and finish plumbing. Permits and fees. Contractor's mark-up.

The in-place cost of these extra components should be added to the basic building cost to arrive at the total structure cost. See the section beginning on page 236. Heating and air conditioning systems. Booths and counters. Kitchen equipment. Fire sprinklers. Exterior signs. Paving and curbing. Yard improvements. Canopies and overhang.

Self Service Restaurants – Masonry

Estimating Procedure

1. Use these figures to estimate the cost of self-service or drive-in restaurants. These restaurants specialize in rapid service, may or may not have an interior eating area, have a large amount of glass on the front and side walls and usually are well lighted.
2. Establish the structure quality class by applying the information on page 175.
3. Compute the building floor area. This should include everything within the exterior walls and all insets outside the main walls but under the main roof.
4. Multiply the floor area by the appropriate cost below.
5. Multiply the total cost by the factor listed on page 7 or 8.
6. Add the cost of heating and air conditioning systems, fire sprinklers, exterior signs, paving and curbing, yard improvements and canopies. See the section beginning on page 236.

Self Service Restaurant, Class 2

Self Service Restaurant, Class 3

Square Foot Area

Quality Class	400	500	600	700	800	900	1,000	1,100	1,200	1,300	1,400
1, Best	784.07	723.75	678.95	643.71	615.19	591.35	571.17	553.69	538.47	525.01	512.91
1 & 2	678.29	626.29	587.35	556.86	532.11	511.65	494.20	479.02	465.84	454.13	443.77
2, Good	601.30	555.14	520.66	493.70	471.72	453.47	438.01	424.72	412.85	402.60	393.38
2 & 3	520.09	480.17	450.35	427.07	408.09	392.32	378.89	367.26	357.24	348.11	340.22
3, Average	458.77	423.58	397.32	376.65	359.93	345.99	334.25	323.97	315.07	307.12	300.05
3 & 4	396.81	366.31	343.60	325.82	311.36	299.27	289.08	280.26	272.49	265.68	259.59
4, Low	344.93	318.38	298.62	283.15	270.61	260.08	251.25	243.58	236.83	230.89	225.62

Square Foot Area

Quality Class	1,500	1,600	1,700	1,800	2,000	2,200	2,400	2,600	2,800	3,000	3,400
1, Best	503.30	497.06	491.05	485.46	473.27	465.72	457.24	449.45	442.43	435.95	424.25
1 & 2	435.63	430.14	425.00	420.10	409.53	403.07	395.66	388.97	382.88	377.25	367.18
2, Good	385.86	381.11	376.51	372.19	362.85	357.17	350.60	344.68	339.28	334.25	325.35
2 & 3	333.83	329.65	325.74	322.02	313.90	308.86	303.23	298.16	293.44	289.20	281.44
3, Average	294.37	290.75	287.31	283.96	276.93	272.48	267.43	262.91	258.82	254.99	248.25
3 & 4	254.72	251.52	248.46	245.64	239.44	235.67	231.49	227.58	224.00	220.64	214.70
4, Low	221.35	218.72	215.92	213.53	208.14	204.85	201.19	197.73	194.61	191.73	186.58

Self Service Restaurants – Wood Frame

Estimating Procedure

1. Use these figures to estimate the cost of self-service or drive-in restaurants. These restaurants specialize in rapid service, may or may not have an interior eating area, have a large amount of glass on the front and side walls and usually are well lighted.
2. Establish the structure quality class by applying the information on page 175.
3. Compute the building floor area. This should include everything within the exterior walls and all insets outside the main walls but under the main roof.
4. Multiply the floor area by the appropriate cost below.
5. Multiply the total cost by the factor listed on page 7 or 8.
6. Add the cost of heating and air conditioning systems, booths and counters, kitchen equipment, fire sprinklers, exterior signs, paving and curbing, yard improvements and canopies. See the section beginning on page 236.

Self Service Restaurant, Class 1 & 2

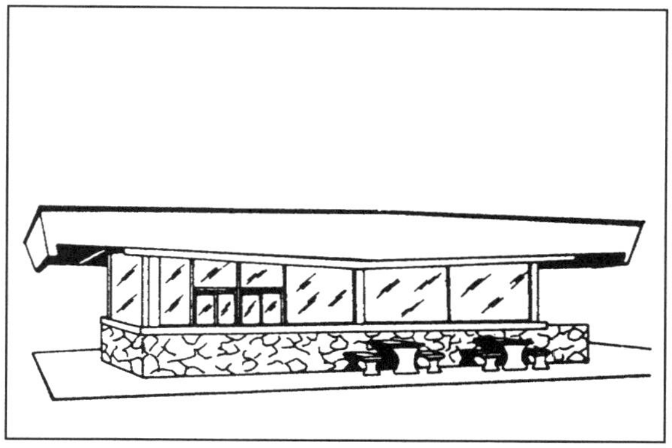

Self Service Restaurant, Class 1 & 2

Square Foot Area

Quality Class	400	500	600	700	800	900	1,000	1,100	1,200	1,300	1,400
1, Best	584.94	540.04	506.45	480.20	458.90	441.20	426.09	413.10	401.69	391.69	382.63
1 & 2	506.02	467.14	438.12	415.33	396.96	381.65	368.56	357.42	347.47	338.77	331.00
2, Good	446.20	411.92	386.33	366.26	350.05	336.57	325.03	315.16	306.40	298.78	291.94
2 & 3	385.77	356.11	333.99	316.65	302.64	290.92	280.97	272.40	264.92	258.32	252.39
3, Average	338.21	312.17	292.84	277.66	265.31	255.06	246.32	238.92	232.25	226.37	221.24
3 & 4	292.89	270.37	253.62	240.47	229.69	220.87	213.35	206.84	201.18	196.11	191.56
4, Low	254.43	234.83	220.27	208.86	199.61	191.89	185.37	179.68	174.74	170.36	166.44

Square Foot Area

Quality Class	1,500	1,600	1,700	1,800	2,000	2,200	2,400	2,600	2,800	3,000	3,400
1, Best	377.12	371.33	366.18	361.43	353.11	346.03	339.92	334.59	329.82	325.58	318.29
1 & 2	325.85	320.89	316.41	312.29	305.17	299.03	293.75	289.14	285.01	281.37	275.04
2, Good	287.67	283.18	279.27	275.67	269.30	263.96	259.25	255.21	251.57	248.31	242.80
2 & 3	248.74	244.94	241.51	238.36	232.94	228.27	224.22	220.63	217.53	214.72	209.94
3, Average	218.17	214.90	211.91	209.21	204.40	200.29	196.75	193.65	190.85	188.43	184.31
3 & 4	188.78	185.92	183.34	180.92	176.77	173.29	170.19	167.56	165.09	163.04	159.38
4, Low	164.01	161.50	159.29	157.21	153.59	150.52	147.85	145.56	143.49	141.65	138.43

Coffee Shop Restaurants

Quality Classification

	Class 1 Best Quality	Class 2 Good Quality	Class 3 Average Quality	Class 4 Low Quality
Foundation (15% of total cost)	Reinforced concrete.	Reinforced concrete.	Reinforced concrete.	Reinforced concrete.
Floor Structure (7% of total cost)	Reinforced concrete or standard wood frame.	Reinforced concrete or standard wood frame.	Reinforced concrete or standard wood frame.	Reinforced concrete or standard wood frame.
Wall Structure (12% of total cost)	Standard wood frame or concrete block.	Standard wood frame or concrete block.	Standard wood frame or concrete block.	Standard wood frame or concrete block.
Roof & Cover (9% of total cost)	Complex angular structure with concrete tile, mission tile, or heavy shake.	Standard wood frame structure, low or medium pitch, with composition tar and gravel roofing or medium shake.	Standard wood frame, low pitch shed or gable type with composition tar and gravel roofing.	Standard wood frame, low pitch shed or gable type with composition tar and gravel or roofing.
Floor Finish (7% of total cost)	Terrazzo or good carpeting in dining area. Quarry tile in kitchen. Terrazzo or natural flagstone in entry vestibule.	Good tile or average carpeting in dining area. Terrazzo or quarry tile in kitchen. Terrazzo entry vestibule.	Good tile in dining area. Grease proof tile in kitchen area. Terrazzo in entry vestibule.	Composition tile in dining area. Grease proof tile in kitchen area.
Exterior Wall Finish (12% of total cost)	Very good wood siding; ornamental brick or natural stone veneer. Minimum 60% of front and side walls in float glass in heavy metal frames or rabbeted timbers. Entry vestibule with decorative screens of ornamental concrete block or natural stone panels.	Good wood siding; ornamental brick or concrete blocks. Good stucco with color finish and with 2' to 4' brick, or natural stone veneer bulkheads. 50% to 75% of front and side walls have float glass set in good wood or metal frames. Double entry doors, recessed entries with decorative screen of concrete block.	Stucco or average wood siding with 2' to 4' brick or stone veneer bulkheads. 50% to 60% of front and side walls in float glass set in average quality wood or aluminum frames. Double entry doors of metal or wood.	Stucco, low cost wood siding or plywood. 40% to 60% of front and side walls in crystal or float glass set in low-cost wood or metal frames. Double entry doors.
Interior Finish (11% of total cost)	Select and matched wood paneling, decorative wallpaper or synthetic fabrics.	Plaster with putty coat finish. Wood or plastic coated veneer paneling. Decorative wallpapers or fabrics.	Plaster putty coat finish textured gypsum wallboard with portions of wall finished with natural wood or plastic coated veneers.	Painted plaster or gypsum wallboard.
Ceilings (10% of total cost)	Complex angular structure with acoustical plaster.	Multi-level structure with good plaster or acoustical tile.	Flat or low slope with textured gypsum wallboard or acoustical tile.	Flat or low slope with textured gypsum wallboard or acoustical tile.
Plumbing (9% of total cost)	Good fixtures with metal or marble toilet partitions. Full wall height ceramic tile wainscot. Numerous floor drains.	Commercial fixtures with metal toilet partitions; ceramic tile wainscot. Numerous floor drains in kitchen and dishwasher area.	4 to 6 standard commercial fixtures with metal toilet partitions. Floor drains in kitchen and dishwashing area.	4 to 6 standard commercial fixtures with wood toilet partitions. Minimum floor drainage in kitchen.
Lighting (8% of total cost)	Triple encased LED fixtures or decorative and ornate LED fixtures.	Triple encased LED fixtures or decorative LED fixtures.	Triple open or double encased louvered LED fixtures with some decorative LED fixtures.	Double open strip or low cost LED fixtures.

Note: Use the percent of total cost to help identify the correct quality classification.

Square foot costs include the following components: Foundations as required for normal soil conditions. Floor, wall and roof structures. Interior floor, wall and ceiling finishes. Exterior wall finish and roof cover. All glass and glazing. Interior partitions. Basic electrical systems and lighting fixtures. Rough and finish plumbing. Permits and fees. Contractor's mark-up.

The in-place cost of these extra components should be added to the basic building cost to arrive at the total structure cost. See the section beginning on page 236: Heating and air conditioning systems. Booths and counters. Kitchen equipment. Fire sprinklers. Exterior signs. Paving and curbing. Yard improvements. Canopies and overhang.

Coffee Shop Restaurants – Masonry

Estimating Procedure

1. Use these figures to estimate the cost of restaurants with counters, or with counter, booth and table seating. A lounge or bar may be part of the building. Higher quality coffee shops usually have cut-up or complex sloping ceiling and roof structures. Coffee shops have extensive exterior glass and good lighting.
2. Establish the structure quality class by applying the information on page 178.
3. Compute the building floor area. This should include everything within the exterior walls and all insets outside the main walls but under the main roof.
4. Multiply the floor area by the appropriate cost below.
5. Multiply the total cost by the factor listed on page 7 or 8.
6. Add the cost of heating and air conditioning systems, booths and counters, kitchen equipment, fire sprinklers, exterior signs, paving and curbing, yard improvements and canopies. See the section beginning on page 236.

Coffee Shop Restaurant, Class 1

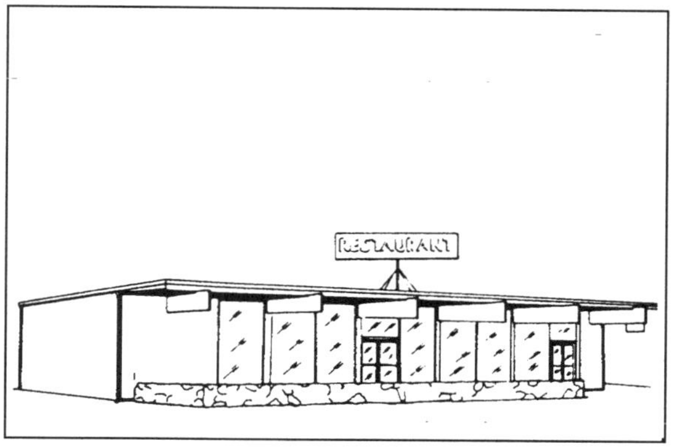

Coffee Shop Restaurant, Class 3

Square Foot Area

Quality Class	1,000	1,200	1,400	1,600	1,800	2,000	2,200	2,400	2,600	2,800	3,000
1, Best	542.87	502.50	473.48	451.28	433.96	419.80	408.25	398.38	390.03	382.85	376.44
1 & 2	470.40	435.54	410.28	391.09	375.98	363.82	353.66	345.19	338.07	331.73	326.27
2, Good	414.17	383.62	361.26	344.42	331.20	320.40	311.50	304.02	297.67	292.18	287.32
2 & 3	360.63	333.85	314.57	299.83	288.24	278.89	271.15	264.61	259.12	254.34	250.10
3, Average	317.62	294.13	277.07	264.14	253.89	245.64	238.87	233.09	228.17	224.02	220.31
3 & 4	274.75	254.48	239.70	228.47	219.71	212.55	206.71	201.72	197.41	193.83	190.55
4, Low	238.75	221.12	208.34	198.43	190.92	184.67	179.56	175.23	171.54	168.38	165.60

Square Foot Area

Quality Class	3,200	3,400	3,600	3,800	4,000	4,400	4,800	5,200	5,600	6,000	8,000
1, Best	366.60	362.42	358.84	355.43	352.33	346.75	341.92	337.63	333.79	330.40	317.43
1 & 2	317.11	313.60	310.42	307.46	304.75	299.98	295.75	292.00	288.74	285.78	274.62
2, Good	279.98	276.93	274.05	271.48	269.10	264.87	261.15	257.85	254.91	252.42	242.50
2 & 3	243.43	240.68	238.31	235.98	233.88	230.19	226.99	224.20	221.70	219.40	210.84
3, Average	214.70	212.36	210.12	208.14	206.35	203.03	200.23	197.73	195.49	193.56	185.93
3 & 4	185.71	183.63	181.81	180.15	178.53	175.70	173.23	171.05	169.09	167.40	160.84
4, Low	161.43	159.60	158.00	156.54	155.13	152.74	150.53	148.59	146.93	145.45	139.81

Coffee Shop Restaurants – Wood Frame

Estimating Procedure

1. Use these figures to estimate the cost of restaurants with counters, or with counter, booth and table seating. A lounge or bar may be part of the building. Higher quality coffee shops usually have cut-up or complex sloping ceiling and roof structures. Coffee shops have extensive exterior glass and good lighting.
2. Establish the structure quality class by applying the information on page 178.
3. Compute the building floor area. This should include everything within the exterior walls and all insets outside the main walls but under the main roof.
4. Multiply the floor area by the appropriate cost below.
5. Multiply the total cost by the factor listed on page 7 or 8.
6. Add the cost of heating and air conditioning systems, booths and counters, kitchen equipment, fire sprinklers, exterior signs, paving and curbing, yard improvements and canopies. See the section beginning on page 236.

Coffee Shop Restaurant, Class 3

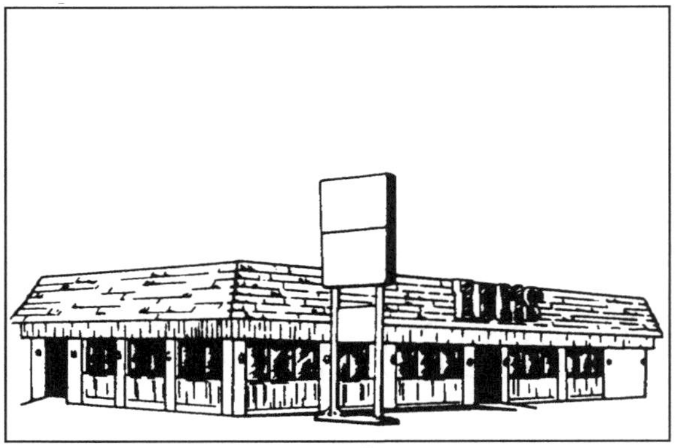

Coffee Shop Restaurant, Class 2

Square Foot Area

Quality Class	1,000	1,200	1,400	1,600	1,800	2,000	2,200	2,400	2,600	2,800	3,000
1, Best	412.09	381.25	359.06	342.20	328.98	318.36	309.51	302.16	295.81	290.38	285.58
1 & 2	356.73	330.06	310.80	296.31	284.86	275.65	267.99	261.63	256.13	251.39	247.36
2, Good	314.64	291.14	274.10	261.18	251.15	242.99	236.28	230.67	225.88	221.68	218.10
2 & 3	271.98	251.71	237.01	225.88	217.18	210.10	204.30	199.42	195.20	191.70	188.51
3, Average	238.67	220.86	208.03	198.23	190.53	184.35	179.29	174.99	171.34	168.20	165.46
3 & 4	206.42	191.00	179.84	171.41	164.79	159.38	155.08	151.34	148.20	145.41	143.03
4, Low	179.29	165.90	156.22	148.96	143.14	138.51	134.67	131.42	128.70	126.37	124.25

Square Foot Area

Quality Class	3,200	3,400	3,600	3,800	4,000	4,400	4,800	5,200	5,600	6,000	8,000
1, Best	278.46	275.34	272.63	270.05	267.67	263.48	259.79	256.52	253.67	251.06	241.18
1 & 2	240.85	238.23	235.82	233.59	231.63	227.94	224.75	221.93	219.44	217.21	208.70
2, Good	212.32	210.03	208.00	205.98	204.18	201.00	198.16	195.73	193.51	191.53	183.98
2 & 3	183.70	181.70	179.84	178.14	176.52	173.84	171.41	169.15	167.32	165.61	159.12
3, Average	161.12	159.33	157.80	156.31	154.91	152.50	150.36	148.44	146.82	145.29	139.63
3 & 4	139.36	137.88	136.44	135.22	134.05	131.89	130.05	128.38	126.99	125.69	120.68
4, Low	121.03	119.69	118.50	117.34	116.38	114.55	112.93	111.50	110.24	109.14	104.87

Conventional Restaurants

Quality Classification

	Class 1 Best Quality	Class 2 Good Quality	Class 3 Average Quality	Class 4 Low Quality
Foundation (15% of total cost)	Reinforced concrete.	Reinforced concrete.	Reinforced concrete.	Reinforced concrete.
Floor Structure (7% of total cost)	Reinforced concrete or standard wood frame.	Reinforced concrete or standard wood frame.	Reinforced concrete or standard wood frame.	Reinforced concrete or standard wood frame.
Wall Structure (12% of total cost)	Standard wood frame or concrete block.	Standard wood frame or concrete block.	Standard wood frame or concrete block.	Standard wood frame or concrete block.
Roof & Cover (9% of total cost)	Complex angular structure, with concrete tile, mission tile, or heavy shake.	Standard wood frame structure, low or medium pitch, with composition tar and gravel roofing or medium shake.	Standard wood frame, low pitch shed or gable type. composition tar and gravel roofing.	Standard wood frame, low pitch shed or gable type, with composition tar and gravel roofing.
Floor Finish (7% of total cost)	Terrazzo or good carpeting in dining area. Quarry tile in kitchen. Terrazzo or natural flagstone in entry vestibule.	Vinyl tile or average carpeting in dining area. Terrazzo or quarry tile in kitchen. Terrazzo in entry vestibule.	Good tile in dining area. Grease proof tile in kitchen area. Terrazzo in entry vestibule.	Composition tile in dining area. Grease proof tile in kitchen area.
Exterior Wall Finish (12% of total cost)	Very good wood siding, ornamental brick or natural stone veneer. 10% of front and side walls in float glass in heavy metal frames or rabbeted timbers. Entry vestibule with decorative screens of ornamental concrete block or natural stone panels.	Good wood siding or plywood; ornamental brick or concrete block. Good stucco with color finish and with 2' to 4' brick, or natural stone veneer bulkheads. 10% to 20% of front and side walls in float glass set in good wood or metal frames. Double entry doors, recessed entries with decorative screen of concrete block.	Stucco or average wood siding with 2' to 4' brick or stone veneer bulkheads. 10% to 20% of front and side walls in float glass set in wood or aluminum extrusions. Double entry doors of metal or wood.	Stucco, low cost wood siding or plywood. 10% to 20% of front & side walls in crystal float glass set in low-cost wood or metal frames. Double entry doors.
Interior Finish (11% of total cost)	Select and matched wood paneling, decorative wallpaper or synthetic fabrics.	Plaster with putty coat finish; wood or plastic coated veneer paneling; decorative wallpapers or fabrics.	Plaster putty coat finish; textured gypsum wallboard with portions of wall finished with natural wood or plastic coated veneer.	Painted plaster or gypsum wallboard.
Ceilings (10% of total cost)	Multi-level structure with acoustical plaster.	Multi-level structure with good plaster or acoustical tile.	Flat or low slope with textured gypsum wallboard or acoustical tile.	Flat or low slope with textured gypsum wallboard or acoustical tile
Plumbing (9% of total cost)	Good fixtures with metal or marble toilet partitions. Full wall height ceramic tile wainscot. Numerous floor drains.	Commercial fixtures with metal toilet partitions. Ceramic tile wainscot. Numerous floor drains in kitchen and dishwasher area.	4 to 6 standard commercial fixtures with metal toilet partitions. Floor drains in kitchen and dishwashing area.	4 to 6 standard commercial fixtures with wood toilet partitions. Minimum floor drainage in kitchen area.
Lighting (8% of total cost)	Decorative and ornate LED fixtures.	Decorative LED fixtures.	Double encased louvered and decorative LED fixtures.	Double open strip or low cost LED fixtures.

Note: Use the percent of total cost to help identify the correct quality classification.

Square foot costs include the following components: Foundations as required for normal soil conditions. Floor, wall and roof structures. Interior floor, wall and ceiling finishes. Exterior wall finish and roof cover. All glass and glazing. Interior partitions. Basic electrical systems and lighting fixtures. Rough and finish plumbing. Permits and fees. Contractor's mark-up.

The in-place cost of these extra components should be added to the basic building cost to arrive at the total structure cost. See the section beginning on page 236: Heating and air conditioning systems. Booths and counters. Kitchen equipment. Fire sprinklers. Exterior signs. Paving and curbing. Yard improvements. Canopies and overhang.

Conventional Restaurants – Masonry or Wood Frame

Estimating Procedure

1. Use these figures to estimate the cost of restaurants with limited use of glass as exterior walls and subdued lighting. Roof and ceiling structures are simple and conventional. Seating may be any combination of counter, stools, booths or tables. A lounge or bar is often a part of a building of this type.
2. Establish the structure quality class by applying the information on page 181.
3. Compute the building floor area. This should include everything within the exterior walls and all insets outside the main walls but under the main roof.
4. Multiply the floor area by the appropriate cost below.
5. Multiply the total cost by the factor listed on page 7 or 8.
6. Add the cost of heating and air conditioning systems, booths and counters, kitchen equipment, fire sprinklers, exterior signs, paving and curbing, yard improvements and canopies. See the section beginning on page 236.

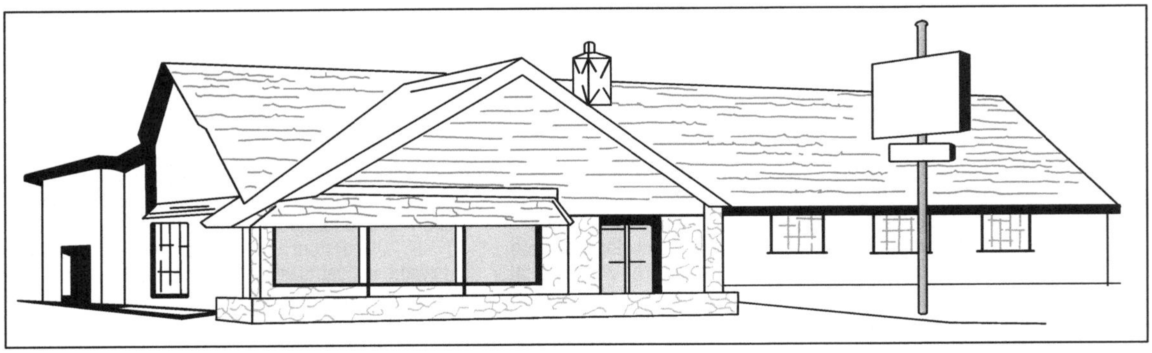

Conventional Restaurant, Class 2

Masonry – Square Foot Area

Quality Class	1,000	1,500	2,000	2,500	3,000	3,500	4,000	5,000	6,000	8,000	10,000
1, Best	562.78	503.27	470.13	448.38	432.86	421.09	411.76	397.92	387.99	374.32	365.24
1 & 2	486.78	435.35	406.62	387.83	374.34	364.23	356.19	344.13	335.61	323.73	315.94
2, Good	429.30	383.99	358.63	342.09	330.18	321.25	314.16	303.65	295.97	285.60	278.64
2 & 3	371.24	332.02	310.16	295.76	285.60	277.82	271.73	262.63	255.94	247.05	240.95
3, Average	325.80	291.38	272.08	259.62	250.61	243.70	238.38	230.37	224.67	216.66	211.43
3 & 4	281.73	251.91	235.27	224.53	216.66	210.80	206.15	199.18	194.15	187.38	182.83
4, Low	244.83	218.96	204.50	195.01	188.29	183.09	179.13	172.99	168.75	162.76	158.92

Wood Frame – Square Foot Area

Quality Class	1,000	1,500	2,000	2,500	3,000	3,500	4,000	5,000	6,000	8,000	10,000
1, Best	408.97	365.77	341.66	325.86	314.54	306.03	299.25	289.20	281.97	272.04	265.48
1 & 2	353.90	316.38	295.47	281.90	272.17	264.67	258.89	250.14	243.90	235.30	229.55
2, Good	311.99	279.00	260.60	248.62	239.94	233.40	228.26	220.62	215.08	207.49	202.46
2 & 3	269.87	241.33	225.36	214.99	207.58	202.00	197.44	190.81	186.07	179.53	175.12
3, Average	236.86	211.73	197.79	188.65	182.13	177.23	173.28	167.40	163.27	157.50	153.72
3 & 4	204.75	183.11	171.03	163.12	157.50	153.22	149.91	144.83	141.24	136.22	132.91
4, Low	177.88	159.08	148.60	141.75	136.92	133.13	130.15	125.83	122.68	118.32	115.49

"A-Frame" Restaurants

Quality Classification

	Class 1 Best Quality	Class 2 Good Quality	Class 3 Average Quality	Class 4 Low Quality
Foundation (19% of total cost)	Reinforced concrete.	Reinforced concrete.	Reinforced concrete.	Reinforced concrete.
Floor Structure (8% of total cost)	Reinforced concrete or standard wood frame.	Reinforced concrete or standard wood frame.	Reinforced concrete or standard wood frame.	Reinforced concrete or standard wood frame.
Roof & Cover (12% of total cost)	Wood frame or post and beam, high pitch with concrete tile or heavy wood shake.	Wood frame or post and beam, high pitch with aluminum shake or medium wood shake.	Standard wood frame or post and beam, high pitch with wood or composition shingles.	Standard wood frame or post and beam, high pitch with composition shingles.
Floor Finish (10% of total cost)	Terrazzo or good carpeting in dining area. Quarry tile in kitchen. Terrazzo or natural flagstone in entry vestibule.	Vinyl tile or average carpeting in dining area Terrazzo or quarry tile in kitchen, terrazzo in entry vestibule.	Good tile in dining area. Grease proof tile in kitchen area. Terrazzo in entry vestibule.	Composition tile in dining area. Grease proof tile in kitchen area.
Interior Finish (12% of total cost)	Select and matched wood paneling, decorative wallpapers or synthetic fabrics.	Plaster with putty coat finish, wood or plastic coated veneer paneling, decorative wallpapers or fabrics.	Plaster, putty coat finish; textured gypsum wallboard with portions of wall finished with natural wood or plastic coated veneer.	Painted plaster or gypsum wallboard.
End Walls (10% of total cost)	Very good wood siding; ornamental brick or natural stone veneer. 60% on front in float glass in heavy metal extrusions or rabbeted timbers. Entry vestibule with decorative screens of ornamental concrete block or natural stone panels.	Good wood siding; ornamental brick or concrete blocks. Good stucco with color finish and with 2' to 4' brick or natural stone veneer bulkheads. 50% of front in float glass set in good wood or metal frames. Double entry doors, recessed entries with decorative screen of concrete block.	Stucco or average wood siding with 2' to 4' brick or stone veneer bulkheads. 50% to 60% of front in float glass set in average quality wood or aluminum frames. Double entry doors of metal or wood.	Stucco, low cost wood siding or plywood. 40% to 60% of front in crystal glass set in low cost wood or metal frames.
Ceilings (9% of total cost)	Decorative wood.	Natural wood, good plaster or acoustical tile.	Painted wood, textured gypsum wallboard or tile.	Painted wood, textured gypsum wallboard or acoustical tile.
Plumbing (10% of total cost)	Good fixtures with metal or marble toilet partitions. Full wall height ceramic tile wainscot. Numerous floor drains.	Commercial fixtures with metal toilet partitions. Ceramic tile wainscot. Numerous floor drains in kitchen and dishwashing area.	4 to 6 standard commercial fixtures with metal toilet partitions. Floor drains in kitchen and dishwashing area.	4 to 6 standard commercial fixtures with wood toilet partitions. Minimum floor drainage in kitchen area.
Lighting (10% of total cost)	Triple encased or decorative and ornate LED fixtures.	Triple encased or decorative LED fixtures.	Triple open, double encased louvered or average LED fixtures.	Double open strip or low cost LED fixtures.

Note: Use the percent of total cost to help identify the correct quality classification.

Square foot costs include the following components: Foundations as required for normal soil conditions. Floor, wall and roof structures. Interior floor, wall and ceiling finishes. Exterior wall finish and roof cover. All glass and glazing. Interior partitions. Basic electrical systems and lighting fixtures. Rough and finish plumbing. Permits and fees. Contractor's mark-up.

The in-place cost of these extra components should be added to the basic building cost to arrive at the total structure cost. See the section beginning on page 236: Heating and air conditioning systems. Booths and counters. Kitchen equipment. Fire sprinklers. Exterior signs. Paving and curbing. Yard improvements. Canopies and overhang.

"A-Frame" Restaurants – Wood Frame

Estimating Procedure

1. Use these figures to estimate the cost of restaurants with a sloping roof that forms two or more exterior walls. Either conventional or self service may be used and a lounge or bar may be part of the building.
2. Establish the structure quality class by applying the information on page 183.
3. Compute the building floor area. This should include everything within the exterior walls and all insets outside the main walls but under the main roof.
4. Multiply the floor area by the appropriate cost below.
5. Multiply the total cost by the factor listed on page 7 or 8.
6. Add the cost of heating and air conditioning systems, booths and counters, kitchen equipment, fire sprinklers, exterior signs, paving and curbing, yard improvements and canopies. See the section beginning on page 236.

"A-Frame" Restaurant, Class 1

Square Foot Area

Quality Class	400	500	600	800	1,000	1,200	1,400	1,600	1,800	2,000	2,400
1, Best	525.57	467.26	427.88	377.87	347.33	326.60	311.55	300.10	291.00	283.71	272.64
1 & 2	454.52	404.18	370.15	326.90	300.42	282.46	269.43	259.55	251.69	245.43	235.78
2, Good	400.88	356.46	326.42	288.28	264.97	249.18	237.67	228.91	222.03	216.45	208.00
2 & 3	346.79	308.35	282.46	249.41	229.25	215.49	205.61	198.06	192.06	187.26	179.89
3, Average	304.41	270.66	247.93	218.95	201.19	189.18	180.48	173.87	168.61	164.35	157.93
3 & 4	263.00	233.92	214.14	189.14	173.90	163.44	155.92	150.20	145.68	142.01	136.44
4, Low	228.62	203.28	186.16	164.43	151.13	142.06	135.53	130.55	126.57	123.37	118.60

Square Foot Area

Quality Class	2,800	3,200	3,600	4,000	4,500	5,000	6,000	7,000	8,000	9,000	10,000
1, Best	264.57	258.58	253.73	249.72	245.61	240.97	236.72	232.73	229.56	226.96	224.87
1 & 2	228.87	223.62	219.44	215.97	212.34	209.42	204.76	201.30	198.58	196.33	194.50
2, Good	201.91	197.23	193.59	190.52	187.33	184.77	180.66	177.54	175.12	173.14	171.53
2 & 3	174.49	170.78	167.61	164.98	162.27	159.95	156.38	153.68	151.65	149.91	148.50
3, Average	153.33	149.75	147.01	144.66	142.26	140.34	137.17	134.75	132.96	131.47	130.23
3 & 4	132.39	129.42	126.96	124.91	122.86	121.13	118.54	116.50	114.87	113.61	112.53
4, Low	115.23	112.60	110.47	108.74	106.93	105.44	103.12	101.38	99.93	98.87	97.86

Theaters – Masonry or Concrete

Quality Classification

	Class 1 Best w/ Balcony	Class 2 Very Good w/ Balcony	Class 3 Good Quality	Class 4 Average Quality	Class 5 Low Quality
Foundation (12% of total cost)	Reinforced concrete.	Reinforced concrete.	Reinforced concrete.	Reinforced concrete.	Reinforced concrete.
Floor Structure (7% of total cost)	6" reinforced concrete on 6" rock fill or 2" x 12" joists, 16" o.c.	4" to 6" reinforced concrete on 6" rock fill or 2" x 10" joists, 16" o.c.	4" reinforced concrete on 6" rock fill or 2" x 10" joists, 16" o.c.	4" reinforced concrete on 6" rock fill or 2" x 8" joists, 16" o.c.	4" reinforced concrete on 6" rock fill or 2" x 6" joists, 16" o.c.
Wall Structure (15% of total cost)	8" reinforced concrete or 12" common brick.	8" reinforced concrete or 12" common brick.	8" reinforced concrete or 12" common brick.	8" reinforced concrete or 12" common brick.	8" reinforced concrete block or 6" reinforced concrete.
Roof Framing (7% of total cost)	2" x 10" joists 16" o.c. Trusses on heavy pilasters 20' o.c. 2" x 12" rafters or purlins 16" o.c.	2" x 10" joists 16" o.c. Trusses on heavy pilasters 20' o.c. 2" x 12" rafters or purlins 16" o.c.	2" x 10" joists 16" o.c. Trusses on heavy pilasters 20' o.c. 2" x 12" rafters or purlins 16" o.c.	2" x 10" joists 16" o.c. Trusses on heavy pilasters 20' o.c. 2" x 10" rafters or purlins 16" o.c.	2" x 10" joists 16" o.c. Trusses on pilasters 20' o.c. 2" x 10" rafters or purlins 16" o.c.
Roof Covering (5% of total cost)	5 ply composition roof on 1" sheathing with insulation.	5 ply composition roof on 1" sheathing with insulation.	5 ply composition roof on 1" sheathing with insulation.	5 ply composition roof on 1" sheathing with insulation.	4 ply composition roof on 1" sheathing.
Front (3% of total cost)	Highly ornamental stucco or plaster finishes, custom select stone veneers or highly ornamental terra cotta.	Ornamental stucco, custom select brick, natural stone veneers or marble.	Highly ornamental stucco or custom brick or natural stone or terra cotta veneers.	Ornamental stucco or select brick veneers or partial terra cotta veneers.	Plain or colored stucco or common brick or carrara glass veneers.
Floors, Entry (3% of total cost)	Custom terrazzo with intricate and ornamental designs, custom select natural stone, marble or very good carpet.	Custom terrazzo with highly ornamental designs, natural stone veneers, marble or very good carpet.	Custom designed terrazzo with ornamental designs with portions natural stone veneers, marble or good carpet.	Colored terrazzo with designs, or average carpet.	Colored concrete with some plain colored terrazzo.
Floors, Interior (5% of total cost)	Concrete with carpet throughout.	Concrete with carpet throughout.	Concrete with carpet throughout.	Concrete with carpet throughout.	Concrete with carpet runners at aisles.
Restrooms (8% of total cost)	Per code requirement capacity. Custom select ornamental ceramic tile, terrazzo or natural stone on floors and walls.	Per code requirement capacity. Custom ceramic tile or terrazzo on floor and walls.	Per code requirement capacity. Ceramic tile on floors and walls or terrazzo floors and walls.	Per code requirement capacity. Ceramic tile on walls and floors or terrazzo on floors.	Per code requirement capacity. Ceramic or vinyl tile on floors.
Walls, Interior (11% of total cost)	Painted or custom select canvas backed or custom molded cloth tapestry wallpapers or ornamental plaster finished with detailed sirocco type moldings and trim, walls with selected matched wood veneers.	Painted or custom canvas backed, custom molded cloth tapestry wallpapers or ornamental plaster finished with custom moldings and trimmings. Portions of wall select matched wood veneers.	Painted or finished with custom canvas backed wallpapers or molded tapestry finished wall-papers or select wood veneer matched full height at lobby.	Painted or finished with durable canvas or select quality wood veneers on gypsum board or plaster.	Painted and papered with durable canvas materials with portion wood veneers on gypsum board or plaster.
Ceiling (6% of total cost)	Suspended acoustical with highly ornate moldings and trim with acoustical baffles.	Suspended acoustical with highly ornate moldings and trim with acoustical baffles.	Suspended acoustical with ornate plaster cove moldings and trim with acoustical baffles.	Suspended acoustical with plaster moldings and sound baffles.	Suspended acoustical tile.
Lighting (10% of total cost)	LED or recessed lights throughout. Interior with ornate chandelier type fixtures throughout. All dimmer controlled.	LED or recessed lights throughout. Interior with ornate chandelier type fixtures throughout. All dimmer controlled.	LED fixtures with chandelier and recessed lights at theater area, dimmer controlled.	LED fixtures in lobby with chandelier type fixtures in theater area, recessed lighting, dimmer controlled.	LED fixtures with recessed fixtures, dimmer controlled.
Seating (8% of total cost)	Main floor and balcony seating.	Main floor and balcony seating.	May or may not have balcony.	May or may not have balcony.	May or may not have balcony.

Note: Use the percent of total cost to help identify the correct quality classification.

Square foot costs include the following components: Foundations as required for normal soil conditions. Floor, wall, and roof structures. Interior floor, wall and ceiling finishes. Exterior wall finish and roof cover. Display fronts. Interior partitions. All doors. Ticket booth. Basic lighting and electrical systems. Rough and finish plumbing. A mezzanine floor projection booth. A frame-work for mounting a picture screen. A balcony in auditorium type theaters. Permits and fees. Contractor's mark-up.

Theaters with Balcony – Masonry or Concrete

Estimating Procedure

1. Establish the structure quality class by applying the information on page 185.
2. Compute the building floor area. This should include everything within the main walls and all insets outside the main walls but under the main roof.
3. Add to or subtract from the square foot cost below the appropriate amount from the Wall Height Adjustment Table on page 190 if the wall height is more or less than 28 feet.
4. Multiply the adjusted square foot cost by the building floor area.
5. Deduct, if appropriate, for common walls, using the figures on page 190.
6. Multiply the total cost by the location factor listed on page 7 or 8.
7. Add the cost of heating and air conditioning systems, fire extinguishers, exterior signs, paving and curbing. See the section beginning on page 236.

Length less than twice width – Square Foot Area

Quality Class	5,000	6,000	7,000	8,000	9,000	10,000	12,000	15,000	20,000	24,000	30,000
1 Best	415.79	400.60	388.95	379.76	372.45	366.23	356.46	345.87	334.20	327.73	320.77
1 & 2	400.18	385.47	374.32	365.59	358.42	352.44	343.09	332.80	321.58	315.42	308.71
2, Very Good	396.69	382.05	371.01	362.36	355.28	349.32	340.06	329.90	318.82	312.66	306.02
2 & 3	380.52	366.63	356.11	347.71	340.94	335.23	326.31	316.61	305.94	300.03	293.63
3, Good	371.71	358.10	347.65	339.54	332.82	327.39	318.64	309.14	298.72	292.98	286.80
3 & 4	354.18	341.17	331.37	323.63	317.26	311.98	303.63	294.54	284.62	279.17	273.20
4, Average	344.17	331.53	321.93	314.43	308.28	303.14	295.02	286.27	276.63	271.29	265.50
4 & 5	325.46	313.50	304.43	297.26	291.48	286.60	278.96	270.72	261.63	256.59	251.07
5, Low	308.78	297.40	288.99	282.08	276.59	272.01	264.68	256.75	248.25	243.44	238.31

Length between 2 and 4 times width – Square Foot Area

Quality Class	5,000	6,000	7,000	8,000	9,000	10,000	12,000	15,000	20,000	24,000	30,000
1, Best	442.70	426.35	414.06	404.39	396.51	389.92	379.43	368.22	355.71	348.95	341.48
1 & 2	426.00	410.46	398.60	389.15	381.59	375.24	365.18	354.35	342.41	335.77	328.68
2, Very Good	422.44	406.92	395.15	385.91	378.34	372.01	362.11	351.34	339.51	333.02	325.92
2 & 3	405.19	390.26	379.06	370.14	362.89	356.89	347.30	337.01	325.64	319.34	312.56
3, Good	395.77	381.21	370.16	361.51	354.37	348.56	339.22	329.23	318.04	311.94	305.27
3 & 4	374.11	360.34	349.94	341.74	335.06	329.45	320.71	311.16	300.63	294.91	288.66
4, Average	366.69	353.26	343.02	334.97	328.41	322.91	314.35	304.98	294.73	289.08	282.89
4 & 5	346.62	333.83	324.26	316.59	310.42	305.25	297.14	288.28	278.54	273.17	267.39
5, Low	328.80	316.78	307.70	300.41	294.49	289.57	281.90	273.50	264.25	259.23	253.72

Length more than 4 times width – Square Foot Area

Quality Class	5,000	6,000	7,000	8,000	9,000	10,000	12,000	15,000	20,000	24,000	30,000
1, Best	471.47	454.13	441.09	430.85	422.40	415.38	404.28	392.14	378.89	371.45	363.46
1 & 2	453.61	437.13	424.58	414.64	406.51	399.77	388.96	377.49	364.63	357.51	349.82
2, Very Good	449.68	433.34	420.86	411.07	402.93	396.24	385.69	374.13	361.49	354.37	346.75
2 & 3	431.83	413.98	404.07	394.61	387.00	380.41	370.33	359.22	347.04	340.21	332.90
3, Good	421.30	405.92	394.14	384.99	377.52	371.13	361.20	350.45	338.51	331.88	324.85
3 & 4	401.58	386.91	375.80	366.99	359.79	353.84	344.37	334.09	322.72	316.43	309.63
4, Average	391.68	377.36	366.57	357.99	350.96	345.08	335.85	325.91	314.75	308.66	302.04
4 & 5	368.90	355.43	345.15	337.10	330.55	325.06	316.30	306.92	296.51	290.75	284.41
5, Low	349.77	337.27	327.58	319.90	313.60	308.47	300.15	291.29	281.33	275.93	269.95

Theaters without Balcony – Masonry or Concrete

Length Less Than Twice Width

Estimating Procedure

1. Establish the structure quality class by applying the information on page 185.
2. Compute the building floor area. This should include everything within the main walls and all insets outside the main walls but under the main roof.
3. Add to or subtract from the square foot cost below the appropriate amount from the Wall Height Adjustment Table on page 190 if the wall height is more or less than 20 feet.
4. Multiply the adjusted square foot cost by the building floor area.
5. Deduct, if appropriate, for common walls, using the figures on page 190.
6. Multiply the total cost by the location factor listed on page 7 or 8.
7. Add the cost of heating and air conditioning systems, fire extinguishers, exterior signs, paving and curbing. See the section beginning on page 236.

Theater Without Balcony, Class 4

Square Foot Area

Quality Class	3,000	3,500	4,000	5,000	6,000	7,000	8,000	10,000	12,000	15,000	20,000
3, Good	241.47	233.80	227.90	219.12	213.00	208.43	204.88	199.59	195.82	191.82	187.62
3 & 4	231.79	224.38	218.76	210.38	204.48	200.08	196.58	191.62	187.94	184.13	179.92
4, Average	228.09	220.83	215.18	206.92	201.19	196.82	193.48	188.49	184.98	181.09	177.04
4 & 5	220.22	213.23	207.77	199.79	194.19	189.97	186.76	181.96	178.56	174.89	170.98
5, Low	213.00	206.25	201.04	193.26	187.92	183.89	180.73	176.07	172.76	169.23	165.46

Theaters without Balcony – Masonry or Concrete

Length Between 2 and 4 Times Width

Estimating Procedure

1. Establish the structure quality class by applying the information on page 185.
2. Compute the building floor area. This should include everything within the main walls and all insets outside the main walls but under the main roof.
3. Add to or subtract from the square foot cost below the appropriate amount from the Wall Height Adjustment Table on page 190 if the wall height is more or less than 20 feet.
4. Multiply the adjusted square foot cost by the building floor area.
5. Deduct, if appropriate, for common walls, using the figures on page 190.
6. Multiply the total cost by the location factor listed on page 7 or 8.
7. Add the cost of heating and air conditioning systems, fire extinguishers, exterior signs, paving and curbing. See the section beginning on page 236.

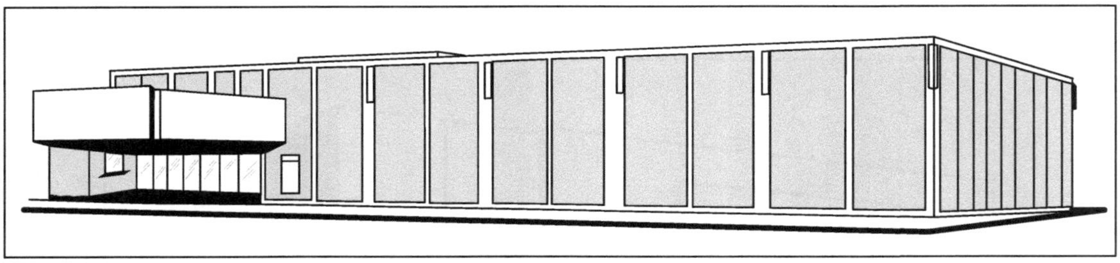

Theater Without Balcony, Class 3

Square Foot Area

Quality Class	3,000	3,500	4,000	5,000	6,000	7,000	8,000	10,000	12,000	15,000	20,000
3, Good	260.38	249.35	241.37	230.57	223.76	219.12	215.78	211.53	209.04	206.82	204.97
3 & 4	250.02	239.42	231.73	221.30	214.80	210.43	207.33	203.21	200.68	198.43	196.70
4, Average	246.29	235.82	228.24	218.08	211.60	207.23	204.10	200.10	197.72	195.58	193.83
4 & 5	237.69	227.69	220.34	210.49	204.36	200.06	197.12	193.14	190.75	188.78	187.04
5, Low	230.31	220.46	213.34	203.83	197.83	193.79	190.81	187.13	184.83	182.83	181.07

Theaters without Balcony – Masonry or Concrete

Length More Than 4 Times Width

Estimating Procedure

1. Establish the structure quality class by applying the information on page 185.
2. Compute the building floor area. This should include everything within the main walls and all insets outside the main walls but under the main roof.
3. Add to or subtract from the square foot cost below the appropriate amount from the Wall Height Adjustment Table on page 190 if the wall height is more or less than 20 feet.
4. Multiply the adjusted square foot cost by the building floor area.
5. Deduct, if appropriate, for common walls, using the figures on page 190.
6. Multiply the total cost by the location factor listed on page 7 or 8.
7. Add the cost of heating and air conditioning systems, fire extinguishers, exterior signs, paving and curbing. See the section beginning on page 236.

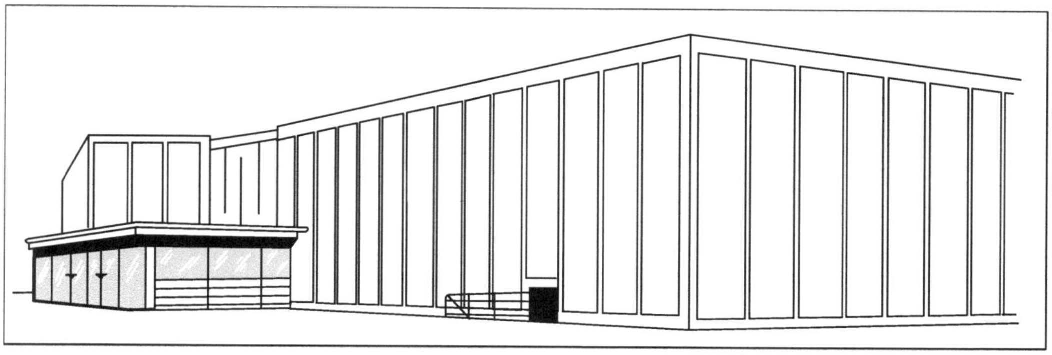

Theater Without Balcony, Class 3 & 4

Square Foot Area

Quality Class	3,000	3,500	4,000	5,000	6,000	7,000	8,000	10,000	12,000	15,000	20,000
3, Good	272.09	263.27	256.37	246.35	239.35	234.29	230.19	224.31	220.20	215.78	211.10
3 & 4	259.99	251.52	244.98	235.45	228.80	223.97	220.06	214.39	210.44	206.19	201.68
4, Average	257.01	248.71	242.09	232.74	226.15	221.21	217.47	211.88	207.94	203.87	199.44
4 & 5	248.07	240.01	233.77	224.59	218.32	213.51	210.01	204.53	200.75	196.79	192.46
5, Low	240.13	232.19	226.20	217.43	211.17	206.65	203.15	198.02	194.25	190.44	186.22

Theaters – Masonry or Concrete

Wall Height Adjustment

Add or subtract the appropriate amount listed in this table to or from the square foot of floor cost for each foot of wall height more or less than 28 feet, if adjusting for a theater with balcony, or 20 feet, if adjusting for a theater without a balcony.

Square Foot Area

Quality Class	3,000	3,500	4,000	5,000	6,000	7,000	8,000	10,000	12,000	15,000	20,000
1, Best	7.14	6.92	6.76	6.53	6.37	6.24	6.05	5.88	5.81	5.75	5.62
2, Very Good	6.76	6.63	6.51	6.24	5.94	5.81	5.79	5.64	5.54	5.37	5.30
3, Good	6.53	6.33	6.14	5.84	5.72	5.60	5.51	5.36	5.33	5.22	5.04
4, Average	6.08	5.80	5.75	5.52	5.37	5.33	5.08	5.04	4.96	4.83	4.73
5, Low	5.72	5.51	5.36	5.14	5.06	4.95	4.79	4.65	4.59	4.54	4.40

Perimeter (Common) Wall Adjustment

A common wall exists when two buildings share one wall. Adjust for common walls by deducting the linear foot costs below from the total structure cost. In some structures one or more walls are not owned at all. In this case, deduct the "No Ownership" cost per linear foot of wall not owned. For Common Wall, deduct $540 per linear foot. For no Wall Ownership, deduct $1,140 per linear foot.

Theaters – Wood Frame

Quality Classification

	Class 1 Best Quality	Class 2 Good Quality	Class 3 Average Quality	Class 4 Low Quality
Foundation (12% of total cost)	Reinforced concrete.	Reinforced concrete.	Reinforced concrete.	Reinforced concrete.
Floor Structure (7% of total cost)	4" reinforced concrete on 6" rock fill or 2" x 10" joists, 16" o.c.	4" reinforced concrete on 6" rock fill or 2" x 8" joists, 16" o.c.	4" reinforced concrete on 6" rock fill or 2" x 6" joists, 16" o.c.	4" reinforced concrete on 4" rock fill or 2" x 6" joists, 16" o.c.
Wall Structure (15% of total cost)	2" x 6", 16" o.c.	2" x 4" or 2" x 6", 16" o.c.	2" x 4", 16" o.c. up to 14' high, 2" x 6", 16" o.c. over 14' high.	2" x 4", 16" o.c. up to 14' high, 2" x 6", 16" o.c. over 14' high.
Roof Framing (7% of total cost)	2" x 10" joists, 16" o.c. Trusses on heavy pilasters, 20' o.c. 2" x 12" rafters or purlins, 16" o.c.	2" x 10" joists, 16" o.c. Trusses on heavy pilasters, 20' o.c. 2" x 10" rafters or purlins.	2" x 10" joists, 16" o.c. Steel trusses on pilasters, 20' o.c. 2" x 10" rafters or purlins, 16" o.c.	2" x 10" joists, 16" o.c. Wood trusses, 2" x 8" purlins, 16" o.c.
Roof Covering (5% of total cost)	5 ply composition roof on 1" x 6" sheathing with insulation.	5 ply composition roof on 1" x 6" sheathing with insulation.	4 ply composition roof on 1" x 6" sheathing.	4 ply composition roof on 1" x 6" sheathing.
Front (3% of total cost)	Highly ornamental stucco or custom brick or natural stone or terra cotta veneers.	Ornamental stucco or select brick veneers or partial terra cotta veneers.	Plain or colored stucco or common brick or ornamental wood.	Plain stucco.
Floors, Entry & Lobby (3% of total cost)	Custom designed terrazzo with ornamental designs with portions natural stone veneers, marble or good carpet.	Colored terrazzo with designs or average carpet.	Colored concrete and portions terrazzo, plain colored.	Plain or colored concrete.
Floors, Interior (5% of total cost)	Concrete with carpet throughout.	Concrete with carpet throughout.	Concrete with carpet runners at aisles.	Plain or colored concrete.
Restrooms (8% of total cost)	As per code requirement capacity. Ceramic tile on floors and walls or terrazzo floors and walls.	As per code requirement capacity. Ceramic tile on walls and floors or terrazzo on floors.	As per code requirement capacity. Ceramic or vinyl tile on floors.	As per code requirement capacity. Plain concrete floors and walls, painted.
Walls, Interior (11% of total cost)	Painted and finished with custom backed wallpapers or molded tapestry finished wallpapers or select wood veneer, matched full height at lobby.	Painted and finished with durable canvas or wood veneers, select quality on gypsum wallboard or plaster.	Painted and papered with durable canvas materials with portion wood veneers on gypsum wallboard or plaster.	Painted with or without stencil type painted molded designs on gypsum wallboard taped and textured.
Ceilings (6% of total cost)	Suspended acoustical, ornate cove moldings and trim with acoustical baffles.	Suspended acoustical with plaster moldings and sound baffles.	Suspended acoustical tile.	Gypsum wallboard taped, textured and painted.
Lighting (10% of total cost)	LED fixtures with chandelier fixtures, recessed at theater area, dimmer controlled.	LED fixtures in lobby with chandelier type fixtures in theater area, recessed lighting, dimmer controlled.	LED recessed fixtures, dimmer controlled.	Plain LED fixtures, with dimmers.
Seating (8% of total cost)	Main floor and balcony seating.	Main floor and balcony seating.	May or may not have balcony.	May or may not have balcony.

Note: Use the percent of total cost to help identify the correct quality classification.

Square foot costs include the following components: Foundations as required for normal soil conditions. Floor, wall, and roof structures. Interior floor, wall and ceiling finishes. Exterior wall finish and roof cover. Display fronts. Interior partitions. All doors. Ticket booth. Basic lighting and electrical systems. Rough and finish plumbing. A mezzanine floor projection booth. A frame-work for mounting a picture screen. A balcony in auditorium type theaters. Permits and fees. Contractor's mark-up.

Theaters – Wood Frame

Length Less Than Twice Width

Estimating Procedure

1. Establish the structure quality class by applying the information on page 191.
2. Compute the building floor area. This should include everything within the main walls and all insets outside the main walls but under the main roof.
3. Add to or subtract from the square foot cost below the appropriate amount from the Wall Height Adjustment Table on page 195 if the wall height is more or less than 20 feet.
4. Multiply the adjusted square foot cost by the building floor area.
5. Deduct, if appropriate, for common walls, using the figures on page 195.
6. Multiply the total cost by the location factor listed on page 7 or 8.
7. Add the cost of heating and air conditioning systems, fire extinguishers, exterior signs, paving and curbing. See the section beginning on page 236.

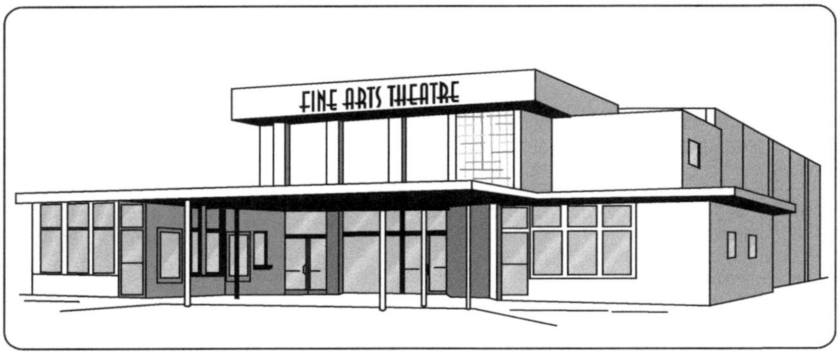

Theater, Class 4 Front, Class 3 Rear

Square Foot Area

Quality Class	3,000	3,500	4,000	5,000	6,000	7,000	8,000	10,000	12,000	15,000	20,000
1, Best	151.65	146.82	143.09	137.57	133.84	130.89	128.67	125.35	123.07	120.54	117.79
1 & 2	146.91	142.23	138.55	133.31	129.60	126.78	124.57	121.38	119.14	116.71	114.06
2, Good	144.92	140.34	136.76	131.54	127.83	125.12	123.00	119.83	117.51	115.18	112.57
2 & 3	140.34	135.82	132.35	127.41	123.75	121.10	119.05	115.95	113.74	111.46	108.93
3, Average	137.43	133.13	129.71	124.78	121.24	118.68	116.63	113.65	111.48	109.25	106.80
3 & 4	133.34	129.15	125.82	121.07	117.63	115.12	113.10	110.24	108.14	105.93	103.55
4, Low	129.51	125.40	122.22	117.46	114.22	111.78	109.86	107.01	105.03	102.90	100.59

Theaters – Wood Frame

Length Between 2 and 4 Times Width

Estimating Procedure

1. Establish the structure quality class by applying the information on page 191.
2. Compute the building floor area. This should include everything within the main walls and all insets outside the main walls but under the main roof.
3. Add to or subtract from the square foot cost below the appropriate amount from the Wall Height Adjustment Table on page 195 if the wall height is more or less than 20 feet.
4. Multiply the adjusted square foot cost by the building floor area.
5. Deduct, if appropriate, for common walls, using the figures on page 195.
6. Multiply the total cost by the location factor listed on page 7 or 8.
7. Add the cost of heating and air conditioning systems, fire extinguishers, exterior signs, paving and curbing. See the section beginning on page 236.

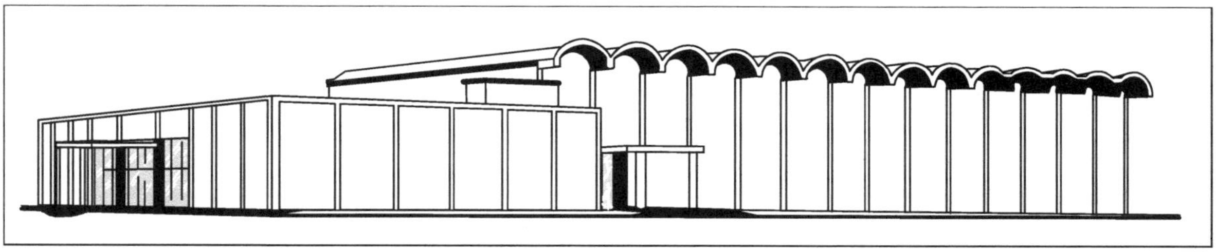

Theater, Class 3

Square Foot Area

Quality Class	3,000	3,500	4,000	5,000	6,000	7,000	8,000	10,000	12,000	15,000	20,000
1, Best	161.59	156.54	152.50	146.62	142.58	139.48	137.08	133.53	131.02	128.36	125.42
1 & 2	156.43	151.47	147.64	142.01	138.01	135.02	132.69	129.24	126.81	124.25	121.39
2, Good	154.31	149.39	145.56	140.03	136.10	133.14	130.86	127.46	125.08	122.50	119.78
2 & 3	149.29	144.53	140.82	135.46	131.65	128.84	126.63	123.35	121.03	118.52	115.83
3, Average	146.48	141.81	138.22	132.93	129.18	126.42	124.26	121.07	118.75	116.29	113.68
3 & 4	142.03	137.46	132.75	128.66	125.23	122.57	120.42	117.32	115.18	112.69	110.24
4, Low	137.64	133.31	129.92	124.90	121.38	118.80	116.75	113.72	111.53	109.31	106.85

Theaters – Wood Frame

Length More Than 4 Times Width

Estimating Procedure

1. Establish the structure quality class by applying the information on page 191.
2. Compute the building floor area. This should include everything within the main walls and all insets outside the main walls but under the main roof.
3. Add to or subtract from the square foot cost below the appropriate amount from the Wall Height Adjustment Table on page 195 if the wall height is more or less than 20 feet.
4. Multiply the adjusted square foot cost by the building floor area.
5. Deduct, if appropriate, for common walls, using the figures on page 195.
6. Multiply the total cost by the location factor listed on page 7 or 8.
7. Add the cost of heating and air conditioning systems, fire extinguishers, exterior signs, paving and curbing. See the section beginning on page 236.

Theater, Class 3

Square Foot Area

Quality Class	3,000	3,500	4,000	5,000	6,000	7,000	8,000	10,000	12,000	15,000	20,000
1, Best	171.91	166.40	162.19	156.00	151.61	148.32	145.72	142.06	139.36	136.55	133.37
1 & 2	166.43	161.19	157.07	151.03	146.83	143.73	141.22	137.57	135.01	132.25	129.24
2, Good	164.28	159.04	154.98	149.09	144.88	141.80	139.36	135.77	133.17	130.55	127.59
2 & 3	159.02	153.94	150.09	144.30	140.26	137.27	134.88	131.37	129.01	126.37	123.48
3, Average	156.07	151.03	147.21	141.59	137.67	134.67	132.34	129.01	126.55	123.94	121.15
3 & 4	151.20	146.40	142.66	137.23	133.37	130.55	128.28	124.90	122.61	120.06	117.40
4, Low	147.69	142.98	139.36	134.07	130.32	127.47	125.30	122.13	119.83	117.34	114.69

Theaters – Wood Frame

Wall Height Adjustment

Add or subtract the amount listed in this table to or from the square foot of floor cost for each foot of wall height more or less than 20 feet.

Square Foot Area

Quality Class	3,000	3,500	4,000	5,000	6,000	7,000	8,000	10,000	12,000	15,000	20,000
1, Best	4.02	3.91	3.85	3.67	3.64	3.52	3.45	3.40	3.29	3.21	3.12
2, Good	3.88	3.80	3.74	3.50	3.40	3.34	3.29	3.21	3.17	3.10	3.04
3, Average	3.70	3.61	3.51	3.34	3.29	3.21	3.17	3.10	3.04	3.00	2.88
4, Low	3.41	3.29	3.21	3.12	3.09	3.04	3.00	2.88	2.79	2.77	2.63

Perimeter (Common) Wall Adjustment

A common wall exists when two buildings share one wall. Adjust for common walls by deducting the linear foot costs below from the total structure cost. In some structures one or more walls are not owned at all. In this case, deduct the "No Ownership" cost per linear foot of wall not owned.

For common wall, deduct $209 per linear foot. For no wall ownership, deduct $417 per linear foot.

Mobile Home Parks

Quality Classification

	Class 1 Best Quality	Class 2 Good Quality	Class 3 Average Quality	Class 4 Low Quality
Engineering, Plans, Permits, Surveying (10% of total cost)	Good planning, necessary permits, good engineering; designed by architect.	Good planning, necessary permits, good engineering; designed by architect.	Average planning, necessary permits, engineered and designed.	Fair planning, necessary permits, minimum surveying.
Grading (10% of total cost)	Fully graded.	Fully graded.	Fully graded.	Minimum site leveling; grades not engineered; road grading.
Street Paving (10% of total cost)	2" thick asphalt surface on good base, concrete curbs, 30' width.	2" thick asphalt surface on good base, concrete curbs, 25' width.	20' roads, 2" asphalt on rock base; concrete or wood edging.	Narrow streets, 2" asphalt on ground; no curbs or edging.
Patios & Walks (5% of total cost)	Patios, 300 to 500 S.F. of good concrete. Walks to utility rooms, pools and recreation areas.	Patios, 200 to 300 S.F. of good concrete. Walks to utility rooms, pools and recreation areas.	Patios, approximately 150 S.F. average concrete or average grade asphalt. Walks to utility buildings.	Some patios, concrete or asphalt paving. No walks.
Trailer Pad & Parking (3% of total cost)	Concrete or good asphalt pad and driveway.	Asphalt under trailer and extended to one side for driveway.	Gravel under trailer and small asphalt driveway.	Gravel under trailer.
Sewer (10% of total cost)	8" lines, 10" mains. Meets all code requirements. Storm drain system.	8" lines, 10" mains. Meets all code requirements.	4" to 6" and 8" lines. Meets code requirements in most areas.	3" to 6" lines, inadequate. Below good code requirements.
Water (10% of total cost)	Engineered system for equalized pressure throughout park. Sprinkler system in common areas.	Adequate line size, designed and properly sized for equalized pressure.	Adequate line size; has required valves at each space.	Small lines; has required valves at each space.
Gas (10% of total cost)	Supplied to each space, sized to code requirements.	Supplied to each space, sized to code requirements.	None except in utility buildings and recreation buildings	None except in utility buildings.

Quality Classification continued

	Class 1 Best Quality	Class 2 Good Quality	Class 3 Average Quality	Class 4 Low Quality
Electric (10% of total cost)	Underground service, designed for larger modern trailers with adequate size to enlarge to take care of future needs. Approximately 100 amp or more. Speaker system, underground television system to each space.	Underground service, designed for larger modern trailers with facilities to enlarge capacity to 100 amp. Approximately 70 amp service or more. Speaker system, underground television system to each space.	Underground services, not designed for more capacity. Approximately 30 amp service or more. Speaker system.	Overhead system wired for 15 amp service at each space.
Outdoor Lighting (4% of total cost)	Lamp post each five spaces, ornate type.	Lamp post each five spaces, inexpensive type.	Overhead street lights at each corner.	Few overhead street lights.
Telephone (8% of total cost)	Underground to each space.	Underground to each space.	None.	None.
Sign (1% of total cost)	Large expensive sign.	Good sign.	Average sign.	Inexpensive sign.
Garbage (1% of total cost)	Built-in ground.	Built-in ground.	None.	None.
Mail Boxes (1% of total cost)	Good mail box and post each space.	Inexpensive mail box and post each space.	None.	None.
Fences and Gates (3% of total cost)	Good wood or cyclone. Ornamental fence or wall in front.	Good wood or cyclone. Block wall in front.	Inexpensive wood or wire.	None or inexpensive wire.
Pools (4% of total cost)	Good quality.	Good quality, adequate size for park.	Small with few extras.	None or inexpensive.
Utility Building (See page 185)	Wood frame and good stucco. Board batt redwood siding or concrete block exterior. Best composition shingle or tar and rock roofing. Good interior plaster or gypsum wallboard. Well finished concrete floors with vinyl tile. Good lighting. Good heating. Showers ceramic tile or fiberglass walls with ceramic tile floor. Glass shower doors. Good quality plumbing fixtures. Good workmanship throughout.	Wood frame and good stucco or concrete block exterior. Thick butt composition shingles or tar and gravel roof. Good exterior or plaster or gypsum wallboard. Well finished concrete floors. Good lighting. Good heating. Showers ceramic tile walls with ceramic tile base. Good quality plumbing fixtures. Good workmanship throughout.	Wood frame, average stucco exterior. Composition shingle or roll roofing. Gypsum wallboard taped and textured or plaster interior. Average concrete floors. Average lighting. Fair heating. Metal stall with showers or showers with enameled cement plaster walls and tile floor with tile base. Average plumbing fixtures. Average workmanship throughout.	Wood frame, fair stucco or fair siding exterior. Plastic interior. Composition roll roofing. Fair concrete finish. Fair lighting. Inexpensive heating. Showers enameled cement plaster walls and tile floors. Fair plumbing fixtures and fair workmanship throughout.
Recreation Building (See page 185)	Wood frame and stucco. Board and batt redwood siding or concrete block exterior. Best composition shingles or tar and rock roofing. Good interior plaster or gypsum wallboard taped, textured and painted. Well finished concrete floors with vinyl tile. Good heating. Good lighting. Rest room for each sex containing at least one each of the following fixtures: Shower, water closet & lavatory. Showers ceramic tile floors and walls or fiberglass walls and tile floor with glass shower doors. Good quality plumbing fixtures. Kitchen sink, range, refrigerator, cabinets and drainboard of formica or equal material. Large glass area in community room.	Wood frame and good stucco or concrete block exterior. Thick butt composition shingles or tar and gravel roof. Good exterior or plaster or gypsum wallboard. Well finished concrete floors. Good heating. Good lighting. Ceramic tile stall showers with ceramic tile base. Good quality plumbing fixtures. Kitchen with tile drain board and some hardwood cabinets. Small office area. Large glass windows in community room.	Wood frame, average stucco or siding exterior. Composition shingle or roll roofing. Gypsum wallboard taped and textured or plaster interior. Average concrete floors. Average lighting. Average heating. Showers with enameled cement plaster walls and ceramic tile base. Water closets and lavatories. One rest room for each sex with at least 1 each shower, water closet and lavatory. Average grade of plumbing fixtures. Ceiling of gypsum wallboard. Average workmanship throughout.	None.

Note: Use the percent of total cost to help identify the correct quality classification.

Mobile Home Parks

Estimating Procedure

1. Establish the park quality class by applying the information on pages 195 and 196.
2. Compute the square foot area per home space. This should include the mobile home space, streets, recreation and other community use areas but exclude excess land not improved or not in use as a part of the park operation. Divide this total area by the number of home spaces. The result is the average area per home space.
3. Multiply the appropriate cost below by the number of home spaces.
4. Determine the quality class and area of recreation and utility buildings. Compute the total cost of these buildings and add this amount to or subtract it from the total from step 3 to adjust for more or fewer buildings than included in the quality specification.
5. Multiply the total building costs by the location factor listed on page 7 or 8.
6. Add the cost of septic tank systems, wells, and covered areas built at the individual spaces.

Space costs with community facilities include the cost of the following components: Grading associated with a level site under normal soil conditions. Street paving and curbs. Patios and walks. Pads and parking paving. Sewer, electrical, gas and water systems including normal hook-up costs. Outdoor lighting. Signs. Mail boxes. Fences and gates. Contractor's mark-up.

Space costs with community facilities and buildings include the cost of all the above components plus these components in amounts proportionate to the size of the park: Recreation, administrative and utility buildings adequate for the size of the park. Recreation facilities such as pools, shuffle board courts, playground equipment, fire pits, etc. Telephones. Restrooms.

The cost of the following components are not included in the basic building cost: Septic tank systems. Wells. Structures or covered areas on individual spaces. The cost of grading beyond that associated with a level site.

Parks Without Community Facilities or Buildings – Square Foot Area

Quality Class	1,500	2,000	2,500	3,000	3,500	4,000	4,500	5,000	5,500	6,000	6,500
1, Best	—	—	—	—	—	17,924	18,255	18,961	19,865	20,079	20,079
1 & 2	—	—	—	16,227	17,049	17,404	17,727	18,152	18,471	19,013	19,426
2, Good	—	—	12,275	13,347	13,758	14,349	14,855	15,385	15,809	16,318	—
2 & 3	7,481	8,230	9,006	9,583	10,107	10,337	10,601	10,761	10,920	—	—

Parks With Community Facilities and Buildings – Square Foot Area

Quality Class	1,500	2,000	2,500	3,000	3,500	4,000	4,500	5,000	5,500	6,000	6,500
1, Best	—	—	—	—	—	26,229	27,492	28,325	29,128	29,934	30,415
2, Good	—	—	—	24,827	25,392	25,973	26,472	26,928	27,629	27,937	28,265
3, Average	—	—	19,979	19,977	20,962	21,300	21,545	21,821	22,054	22,258	—
4, Low	13,574	13,761	14,154	14,770	15,288	15,601	15,887	16,066	16,257	—	—

Square Foot Costs for Building Alone

	Recreational Buildings	Utility Buildings
1, Best	$87.76 to $127.80	$85.20 to $102.50
2, Good	80.00 to 93.90	63.00 to 85.69
3, Average	62.54 to 91.98	59.84 to 64.03
4, Low	—	55.26 to 62.13

Service Stations – Wood, Masonry or Painted Steel

Quality Classification

	Wood Frame	Masonry or Concrete	Painted Steel, Good	Painted Steel, Average	Painted Steel, Low
Foundation & Floor (25% of total cost)	Reinforced concrete.	Reinforced concrete.	Reinforced concrete.	Reinforced concrete.	Reinforced concrete.
Walls (15% of total cost)	Wood frame 2 x 4, 16" o.c.	8" concrete block.	Steel frame.	Steel frame.	Steel frame.
Roof Structure (6% of total cost)	Light wood frame, flat or shed type.	Light wood frame, flat or shed type.	Steel frame, flat or shed type.	Steel frame, flat or shed type.	Steel frame, flat or shed type.
Exterior Finish (10% of total cost)	Painted wood siding or stucco.	Painted concrete block.	Painted steel.	Painted steel.	Painted steel.
Roof Cover (5% of total cost)	Composition.	Composition.	Steel deck.	Steel deck.	Steel deck.
Glass Area (5% of total cost)	Small area, painted wood frames.	Small area, painted steel frames.	Large area, painted steel frames.	Average area, painted steel frames.	Small area, painted steel frames.
Lube Room Doors (5% of total cost)	Folding steel gate.	Folding steel gate.	Painted steel sectional roll up.	Painted steel sectional roll up.	Folding steel gate.
Floor Finish (5% of total cost)	Concrete.	Concrete.	Concrete, colored concrete in office.	Concrete.	Concrete.
Interior Wall Finish (5% of total cost)	Exposed studs, painted.	Concrete block, painted.	Exposed structure painted. Painted steel panels in office.	Exposed structure painted. Painted steel panels in office.	Exposed structure painted. Painted steel panels in office.
Ceiling Finish (3% of total cost)	Exposed structure painted.	Exposed structure painted.	Exposed structure painted. Painted steel panels in office.	Exposed structure painted. Painted steel panels in office.	Exposed structure painted.
Rest Room Finish (5% of total cost)	Wallboard and paint walls and ceilings.	Concrete block and paint walls, wallboard and paint ceilings.	Ceramic tile floors, 4' ceramic tile wainscot, painted steel ceilings.	Ceramic tile floors, 4' ceramic tile wainscot, painted steel ceilings.	Concrete floors, painted steel walls, painted steel ceilings.
Rest Room Fixtures (8% of total cost)	4 low cost fixtures.	4 low cost fixtures.	5 average cost fixtures.	5 average cost fixtures.	4 low cost fixtures.
Exterior Appointments (3% of total cost)	None.	None.	2' overhang on 3 sides, 3' raised walk on 3 sides, fluorescent soffit lights on 3 sides.	1' overhang on 2 sides, 3' raised walk on 2 sides.	None.

Note: Use the percent of total cost to help identify the correct quality classification.

Square foot costs include the cost of the following components: Foundations as required for normal soil conditions. Floor, wall and roof structure. Interior floor, wall and ceiling finishes as described above. Interior partitions. Exterior finish and roof cover. A built-in work bench, tire rack and shelving. Electrical services and fixtures contained within the building. Air and water lines within the building. That portion of rough plumbing serving the building and plumbing fixtures within the building. Roof overhangs and raised walks as described above. Lube room doors. Permits and fees. Contractor's mark-up.

The in-place cost of these extra components should be added to the basic building cost to arrive at the total structure cost. See the section "Additional Costs for Service Stations" beginning on Page 204. Canopies. Pumps, dispensers and turbines. Air and water services outside the building. Island lighters. Gasoline storage tanks. Hoists. Compressors. Yard lights. Signs. Paving. Curbs and fences. Miscellaneous equipment and accessories. Island office and storage buildings. Site improvements. Heating and cooling systems

Land improvement costs: Most service stations sites require an expenditure of $10,000 or more for items such as leveling, excavation, curbs, driveways, relocation of power poles, replacement of sidewalks with reinforced walks and street paving.

Service Stations – Wood, Masonry or Painted Steel

Estimating Procedure

1. Establish the structure quality class by applying the information on page 198.
2. Compute the building floor area.
3. Multiply the square foot cost by the building floor area.
4. Multiply the total cost by the location factor listed on page 7 or 8.
5. Add the cost of appropriate equipment and fixtures from the section "Additional Costs for Service Stations" beginning on page 204.

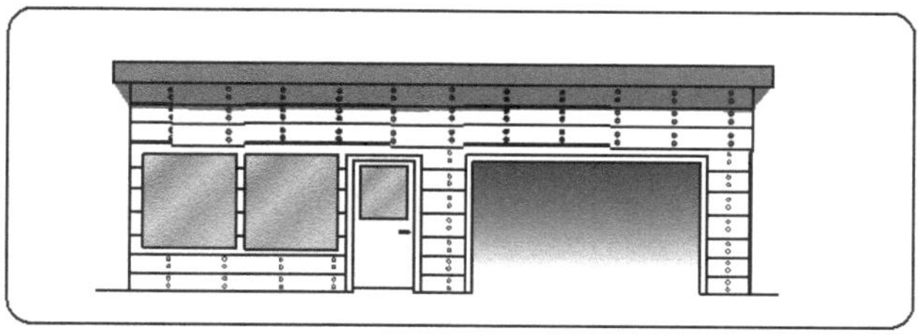

Wood Frame

Masonry

Painted Steel, Good

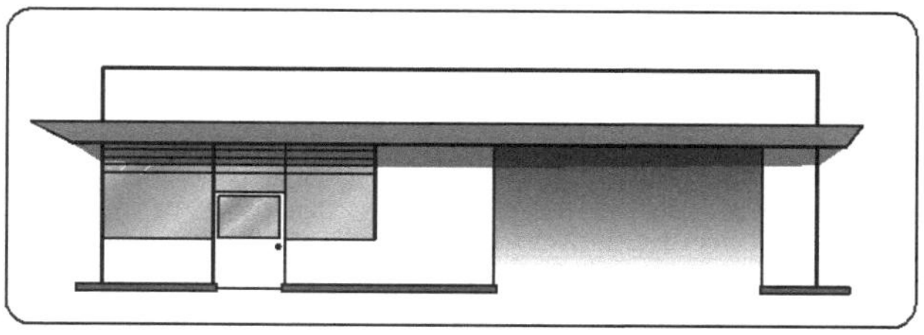

Painted Steel, Average

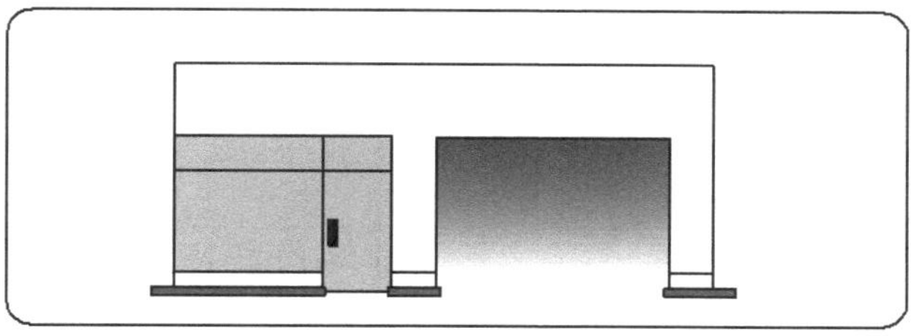

Painted Steel, Low

Square Foot Area

Quality Class	500	600	700	800	900	1,000	1,100	1,200	1,300	1,400	1,800
Wood Frame	151.35	139.02	130.01	123.47	118.32	114.18	110.79	108.03	105.60	103.54	97.65
Masonry or Concrete	237.27	217.35	203.47	193.09	184.98	178.64	173.38	168.96	165.25	161.98	152.76
Painted Steel, Good	258.14	236.89	221.84	210.50	201.73	194.73	188.89	184.12	180.06	176.67	166.56
Painted Steel, Avg.	231.27	213.82	201.17	189.99	182.01	174.70	168.54	161.80	157.74	153.78	150.30
Painted Steel, Low	202.59	185.96	174.12	165.27	158.36	152.81	148.31	144.55	141.36	137.82	130.77

Service Stations – Porcelain Finished Steel

Quality Classification

	Good Quality	Average Quality	Low Quality
Foundation & Floor (20% of total cost)	Reinforced concrete.	Reinforced concrete.	Reinforced concrete.
Walls (15% of total cost)	Steel frame.	Steel frame.	Steel frame.
Roof Structure (8% of total cost)	Steel frame, flat or shed type.	Steel frame, flat or shed type.	Steel frame, flat or shed type.
Exterior Finish (10% of total cost)	Porcelain and steel.	Porcelain and steel.	Porcelain and steel.
Roof Cover (6% of total cost)	Steel deck.	Steel deck.	Steel deck.
Glass Area (7% of total cost)	Large area, aluminum frames.	Large area, aluminum frames.	Average area, painted steel frames.
Lube Room Doors (3% of total cost)	Aluminum and glass sectional roll up.	Aluminum and glass sectional roll up.	Painted steel and glass sectional roll up.
Floor Finish (5% of total cost)	Concrete floors, ceramic tile in office.	Concrete floors, colored concrete in office.	Concrete floors, colored concrete in office.
Interior Wall Finish (5% of total cost)	Porcelain steel panels. Painted steel panels in office.	Exposed structure painted. Painted steel panels in office.	Exposed structure painted.
Ceiling Finish (3% of total cost)	Exposed structure painted. Porcelain steel panels in office.	Exposed structure painted. Painted steel panels in office.	Exposed structure painted. Painted steel panels in office.
Rest Room Finish (5% of total cost)	Ceramic tile floors, 8' ceramic tile or porcelain panel. Painted steel ceiling.	Ceramic tile floors, 5' ceramic tile wainscot. Painted steel ceiling.	Ceramic tile floors, 5' ceramic tile wainscot. Painted steel ceiling.
Rest Room Fixtures (10% of total cost)	5 good fixtures.	5 good fixtures.	5 good fixtures.
Exterior Appointments (3% of total cost)	3' to 4' overhang on 3 sides, 6' x 8' sign pylon, 3' raised walk on 3 sides, fluorescent soffit lights on 3 sides.	3' to 4' overhang on 3 sides, 6' x 8' sign pylon, 3' raised walk on 3 sides, fluorescent soffit lights on 3 sides.	3' raised walk on 3 sides, fluorescent soffit lights on 3 sides.

Note: Use the percent of total cost to help identify the correct quality classification.

Square foot costs include the cost of the following components: Foundations as required for normal soil conditions. Floor, wall and roof structure. Interior floor, wall and ceiling finishes as described above. Interior partitions. Exterior finish and roof cover. A built-in work bench, tire rack and shelving. Electrical services and fixtures contained within the building. Air and water lines within the building. That portion of rough plumbing serving the building and plumbing fixtures within the building. Roof overhangs and raised walks as described above. Lube room doors. Permits and fees. Contractor's mark-up.

The in-place cost of these extra components should be added to the basic building cost to arrive at the total structure cost. **See the section "Additional Costs for Service Stations" beginning on page 204.** Canopies. Pumps, dispensers and turbines. Air and water services outside the building. Island lighters. Gasoline storage tanks. Hoists. Compressors. Yard lights. Signs. Paving. Curbs and fences. Miscellaneous equipment and accessories. Island office and storage buildings. Site improvements. Heating and cooling systems

Land improvement costs: Most service station sites require an expenditure of $10,000 or more for items such as leveling, excavation, curbs, driveways, relocation of power poles, replacement of sidewalks with reinforced walks and street paving.

Service Stations – Porcelain Finished Steel

Estimating Procedure

1. Establish the structure quality class by applying the information on page 200.
2. Compute the building floor area.
3. Multiply the square foot cost by the building floor area.
4. Multiply the total cost by the location factor listed on page 7 or 8.
5. Add the cost of appropriate equipment and fixtures from the section "Additional Costs for Service Stations" beginning on page 204.

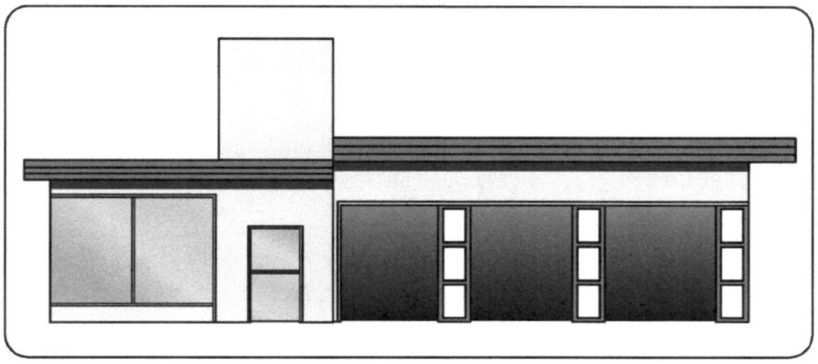

Good Quality

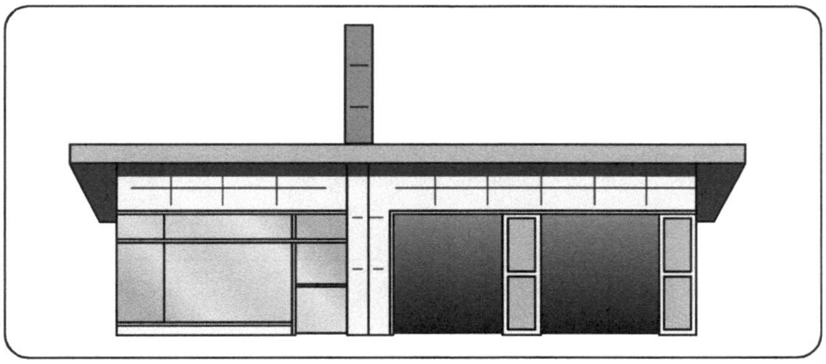

Average Quality

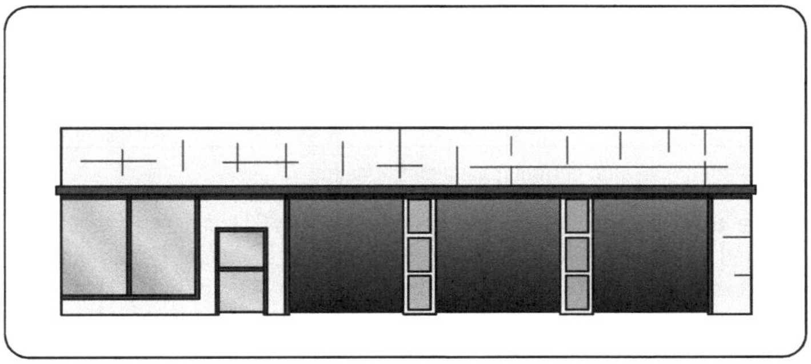

Low Quality

Square Foot Area

Quality Class	1,000	1,100	1,200	1,300	1,400	1,500	1,600	1,700	1,800	2,000	2,400
Good	242.08	233.70	227.06	221.61	217.26	213.79	210.72	208.22	206.12	202.88	198.71
Average	231.65	223.59	217.16	212.06	207.89	204.42	201.61	199.19	197.20	194.03	190.12
Low	210.41	203.07	197.23	192.65	188.89	185.68	183.12	181.01	179.10	176.28	172.70

Service Stations – Ranch or Rustic Type

Quality Classification

	Best Quality	Good Quality	Average Quality	Low Quality
Foundation & Floor (20% of total cost)	Reinforced concrete.	Reinforced concrete.	Reinforced concrete.	Reinforced concrete.
Walls (12% of total cost)	Steel frame.	Steel frame.	Steel frame, wood frame or masonry.	Steel frame, wood frame or masonry.
Roof Structure (8% of total cost)	Steel frame, hip or gable type.	Steel frame, hip or gable type.	Steel or wood frame, hip or gable type.	Steel or wood frame, hip or gable type.
Exterior Finish (10% of total cost)	Natural stone veneer.	Used brick veneer.	Painted steel and masonry veneer.	Painted steel or wood siding.
Roof Cover (6% of total cost)	Shingle tile or mission tile.	Heavy wood shakes or shingle tile.	Wood shakes or tar and rock.	Composition shingle or tar and gravel.
Glass Area (7% of total cost)	Large area float glass in heavy aluminum frame.	Large area float glass in heavy aluminum frame.	Large area, painted steel frame.	Average area, painted steel frame.
Lube Room Doors (5% of total cost)	Painted steel or aluminum and glass sectional roll up.	Painted steel or aluminum and glass sectional roll up.	Painted steel sectional roll up.	Painted steel sectional roll up.
Floor Finish (5% of total cost)	Concrete floors, ceramic tile in office.	Concrete floors, ceramic tile in office.	Concrete floors.	Concrete floors.
Interior Wall Finish (5% of total cost)	Painted steel panels or gypsum wallboard and paint.	Painted steel panels or gypsum wallboard and paint.	Painted steel panels or gypsum wallboard and paint.	Painted steel panels or gypsum wallboard and paint.
Ceiling Finish (3% of total cost)	Painted steel panels.	Painted steel panels.	Painted steel panels, gypsum wallboard, or "V" rustic and paint.	Painted steel panels, gypsum wallboard or "V" rustic and paint.
Restroom Finish (5% of total cost)	Ceramic tile floors, ceramic tile walls, painted steel ceiling.	Ceramic tile floors, ceramic tile walls, painted steel ceiling.	Ceramic tile floors, 5' ceramic tile wainscot, painted steel ceiling.	Ceramic tile floors, 5' ceramic tile wainscot, painted steel ceiling.
Restroom Fixtures (10% of total cost)	5 good fixtures.	5 good fixtures.	5 good fixtures.	5 good fixtures.
Exterior Appointments (4% of total cost)	3' to 6' overhang on all sides, 3' raised walk on 3 sides, fluorescent soffit lights on all sides.	3' to 6' overhang on all sides, 3' raised walk on 3 sides, fluorescent soffit lights on all sides.	3' to 6' overhang on 3 sides, 6' x 8' sign pylon, 3' raised walk on 3 sides, fluorescent soffit lights on 3 sides.	2' to 3' overhang on 3 sides, 3' raised walk on 3 sides, fluorescent soffit lights on 3 sides.

Note: Use the percent of total cost to help identify the correct quality classification.

Square foot costs include the cost of the following components: Foundations as required for normal soil conditions. Floor, wall and roof structure. Interior floor, wall and ceiling finishes as described above. Interior partitions. Exterior finish and roof cover. A built-in work bench, tire rack and shelving. Electrical services and fixtures contained within the building. Air and water lines within the building. That portion of rough plumbing serving the building and plumbing fixtures within the building. Roof overhangs and raised walks as described above. Lube room doors. Permits and fees. Contractor's mark-up.

The in-place cost of these extra components should be added to the basic building cost to arrive at the total structure cost. See the section "Additional Costs for Service Stations" beginning on page 204. Canopies. Pumps, dispensers and turbines. Air and water services outside the building. Island lighters. Gasoline storage tanks. Hoists. Compressors. Yard lights. Signs. Paving. Curbs and fences. Miscellaneous equipment and accessories. Island office and storage buildings. Site improvements. Heating and cooling systems

Land improvement costs: Most service stations sites require an expenditure of $10,000 or more for items such as leveling, excavation, curbs, driveways, relocation of power poles, replacement of sidewalks with reinforced walks and street paving.

Service Stations – Ranch or Rustic Type

Estimating Procedure

1. Establish the structure quality class by applying the information on page 202.
2. Compute the building floor area.
3. Multiply the square foot cost by the building floor area.
4. Multiply the total cost by the location factor listed on page 7 or 8.
5. Add the cost of appropriate equipment and fixtures from the section "Additional Costs for Service Stations" beginning on page 204.

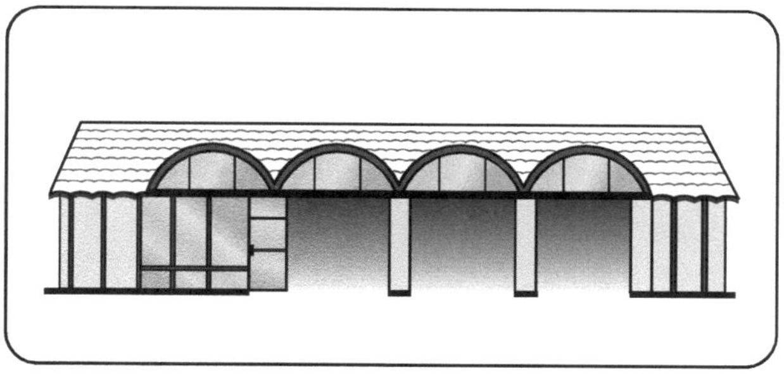

Best Quality

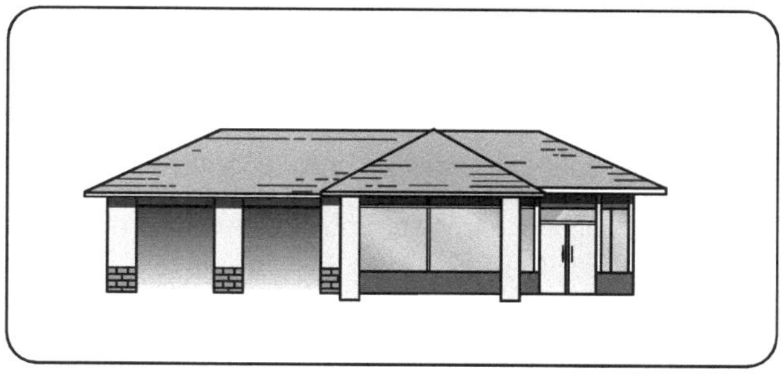

Good Quality

Average Quality

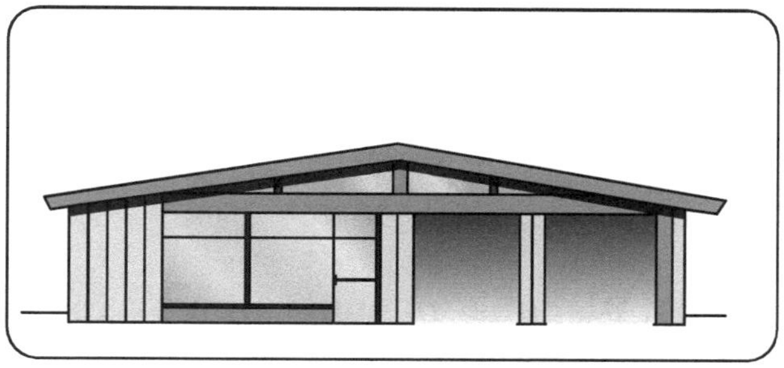

Low Quality

Square Foot Area

Quality Class	1,000	1,100	1,200	1,300	1,400	1,500	1,600	1,700	1,800	2,000	2,400
Best	268.84	259.34	251.79	245.77	240.97	236.88	233.63	230.87	228.56	225.00	220.57
Good	257.95	248.79	241.56	235.82	231.13	227.29	224.08	221.52	219.20	215.90	211.65
Average	247.56	238.73	231.87	226.28	221.79	218.09	215.11	212.51	210.41	207.07	203.07
Low	233.78	225.53	219.00	213.80	209.48	206.10	203.12	200.79	198.72	195.60	191.79

Additional Costs for Service Stations

A portion of the typical plumbing or electrical cost has been added to each item of equipment requiring these services. It will not be necessary except in rare instances to add extra cost for these items.

Canopies, cost per square foot

Type	Less than 500 S.F.	500 to 1,000 S.F.	Over 1,000 S.F.
Painted steel	$40.40 to $45.00	$37.00 to $45.30	$32.10 to $35.40
Porcelain and steel	46.30 to 47.80	40.30 to 44.80	37.40 to 40.20
Ranch style or gable roof type	52.10 to 56.70	41.40 to 48.90	37.90 to 41.50
Deluxe steel with illuminated plastic signs on sides or in gables. Also includes illuminated plastic island lighters.			
Round type, good steel	78.90 to 93.40		

Costs include cost of foundation, steel support column or columns, complete canopy, painting or porcelaining, light fixtures, and electrical service. Ranch or gable roof types include the cost of a rock or shake roof cover. Concrete pads under canopies or masonry trim on support columns are not included in these costs.

Island Office and Storage buildings, cost per square foot

				Area			
Type	Under 30	31 - 40	41 - 50	51 - 60	61 - 80	81 - 100	101 - 120
Steel and glass or concrete block	570.00	548.00	488.00	436.00	391.00	366.00	297.00
Wood frame with stucco and glass	458.00	430.00	358.00	335.00	257.00	237.00	225.00

These buildings are usually found at self-service stations. Add $2,160 per unit for any plumbing fixtures in these buildings. Steel island offices cost about $2,230

Pumps, Dispensers and Turbines, cost each

Type	Installed Cost	Type	Installed Cost
Single pump	$7,160	Blendomatic pump	$11,600
Twin pump	9,630	Blendomatic dispenser	9,530
Single dispenser	5,400	Turbine pump, 1/3 HP	2,010
Twin dispenser	9,030	Turbine pump, 3/4 HP	2,790

Installed cost includes the cost of the pump or dispenser, installation cost, electrical hookup cost, a portion of the piping cost and a portion of the island block cost. Concrete islands 4" to 6" thick cost from $13.80 to $16.60 per square foot.

All of the above pump and dispenser costs are for the computing type. Add for electronic remote control totalizer, per hose, $2,190. Add for vapor control system, per hose/dispenser, $2,280.

Dispenser cost does not include the cost of the pump. Turbine pump costs must be added. 1/3 HP turbines will serve a single product up to four dispensers. 3/4 HP turbines will serve a single product up to eight dispensers.

Additional Costs for Service Stations

Air and Water Services

Type	Air Only Equipment Cost	Air Only Installed Cost	Air and Water Equipment Cost	Air and Water Installed Cost
Underground disappearing hose type	$ 588	$ 868	$ 639	$1,310
Post type with auto inflator	$ 927	$1,369	$1,169	$1,865
Post type with auto inflator and disappearing hoses	$1,550	$2,106	$2,168	$2,891

Costs include cost of installation and a portion of the cost of air and water lines.

Island Lighters

Width	Length	4 Tubes	6 Tubes
42"	9'-5"	$2,600 ea.	$3,170 ea.
42"	11'-5"	3,180 ea.	4,050 ea.
42"	15'-6"	4,100 ea.	4,540 ea.
42"	19'-6"	5,050 ea.	5,380 ea.
36"	30'-0"	—	6,370 ea.

Cost includes foundation, davit poles or steel support columns and electrical service.

Cash Boxes complete, with pedestal $351 each

Gasoline Storage Tanks (Fiberglass)

Capacity in Gallons	Tank Cost	Installed Cost	Capacity in Gallons	Tank Cost	Installed Cost
110	$899	$1,585	**5,300**	$9,157	$11,911
150	1,115	1,895	**6,300**	9,349	12,271
280	1,340	2,180	**7,400**	11,888	13,288
550	1,775	2,769	**8,400**	12,239	13,397
1,000	3,236	5,746	**10,500**	12,304	15,943
2,000	6,224	7,629	**12,600**	15,604	19,877
4,000	8,344	12,533	—	—	—

Installed cost includes cost of tank, excavation (4' bury and soil disposal), placing backfill, fill box (concrete slab over tank), tank piping and vent piping.

Miscellaneous Lube Room Equipment

Air hose reel	$1,340	Pneumatic tube changer	$14,500
Water hose reel	1,390	Automatic lube equipment	10,700
Grease pit for trucks	$545 to $620 per L.F.	5 hose reel assembly	13,400

Yard Lights

LED floodlights. Costs include electrical connection and mounting on a building soffit. For pole mounted yard lights, add pole mounting costs from page 195. Cost per light fixture.

70 Watt	100 Watt	200 Watt	300 Watt	400 Watt
$548	$568	$597	$650	$728

Additional Costs for Service Stations

Vehicle Hoist

Type	Equipment Cost	Installed Cost
One post 8,000 lb. semi hydraulic hoist	$6,110	$13,100
One post 8,000 lb. fully hydraulic hoist	6,360	13,600
Two post 11,000 lb. semi hydraulic hoist	9,200	21,100
Two post 11,000 fully hydraulic hoist	9,290	20,900
Two post 11,000 lb. pneumatic hoist	11,900	19,150
Two post 24,000 lb. pneumatic hoist	16,700	28,400

Air Compressors

Horsepower	Equipment Cost	Installed Cost	Horsepower	Equipment Cost	Installed Cost
1/2	$2,890	$3,260	**2**	$3,110	$3,470
3/4	2,930	3,320	**3**	3,280	4,200
1	2,970	3,380	**5**	3,500	4,290
1-1/2	3,080	3,450	**7-1/2**	5,990	6,320

Costs include compressor and tank only.

Paving, cost per S.F.

Asphalt, 2" with 4" base	$3.40 to $4.36
Concrete 4", with base	4.16 to 5.80
Concrete 6", with base	5.28 to 6.50
Oil macadam	3.20
Pea gravel	1.63

Site Improvement

Vertical curb and gutter	$8.30 to $26.70 LF
Concrete apron	10.28 to 22.74 SF
6" reinforced concrete sidewalks	6.34 to 8.29 SF
Standard 4" sidewalk	5.11 to 5.65 SF

The above costs are normally included in land value.

Fencing and Curbing, cost per L.F.

Heavy 2 rail fence, 2" x 6"	$1.11 to $12.42
Rails on 4" x 4" posts 6' to 8' o.c.	11.21 to 13.40
Chain link 3' to 4' high	11.16 to 15.80
Solid board 3' to 4' high	11.32 to 12.94
Log barrier	9.90 to 17.12
Metal guard rail on wood posts	41.87 to 73.00
6" x 6" doweled wood bumper strip	10.73 to 13.40
6" x 6" concrete bumper strip	8.74 to 12.51
Cable railing on wood posts	11.04 to 13.35
6" x 12" concrete curb and gutter	19.90 to 22.98
6" concrete block walls, per S.F.	8.14 to 11.58

plus $9.80/LF for foundation

Service Station Signs, cost per square foot of sign area measured on one side

Painted sheet metal with floodlights	$68.39 to $90.75
Porcelain enamel with floodlights	72.65 to 97.72
Plastic with interior lights	82.92 to 126.20
Simple rectangular neon with painted sheet metal faces and a moderate amount of plain letters	89.89 to 160.36
Round or irregular neon with porcelain enamel faces and more elaborate lettering	126.34 to 192.04

All of the above sign costs are for single faced signs. Add 50% to these costs for double faced signs. Sign costs include costs of installation and normal electrical hookup. They do not include the post cost. See page 195. These costs are intended for use on **service station signs only** and are based on volume production. Costs of custom-built signs will be higher.

Rotators, cost per sign for rotating mount

Small signs	Less than 50 S.F.	$3,060 to $3,600
Medium signs	50 to 100 S.F.	3,140 to 6,390
Large signs	100 to 200 S.F.	6,310 to 11,600
Extra large signs	Over 200 S.F.	$66.30 per S.F. of sign area

Additional Costs for Service Stations

Post Mounting Costs

	Pole Diameter at Base					
Post Height	**4"**	**6"**	**8"**	**10"**	**12"**	**14"**
15	$1,481	$1,762	$2,594	$3,495	$5,440	$5,495
20	1,755	2,094	2,998	3,625	6,472	7,279
25	2,012	2,376	3,106	4,313	7,199	7,950
30	2,192	2,816	3,222	4,708	7,606	8,801
35	—	3,063	3,625	5,252	8,605	9,398
40	—	3,251	4,427	5,554	8,950	10,543
45	—	—	5,184	6,325	9,769	10,913
50	—	—	5,554	7,211	10,398	11,958
55	—	—	—	7,607	10,907	13,019
60	—	—	—	8,049	11,707	13,616
65	—	—	—	—	13,684	14,470

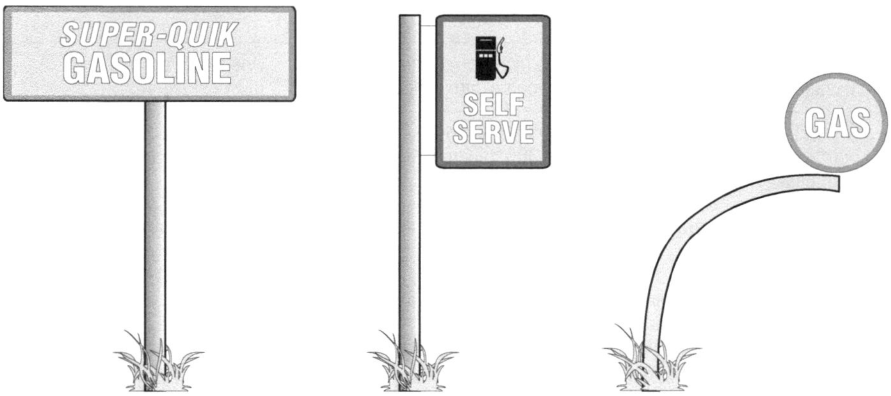

Horizontal Mount **Vertical Mount** **Cantilevered Mount**

If signs are mounted on separate posts, post mounting costs must be added. Post mounting costs include the installed cost of a galvanized steel post and foundation. On horizontally mounted signs, post height is the distance from the ground to the bottom of the sign. On vertically mounted signs, post height is the distance to the top of the post.

For cantilevered posts, use one and one-half to two times the conventional post cost.

All of the above post costs are for single posts. Use 90% of the single post costs for each additional post.

If signs are mounted on buildings or canopies and if, because of the extra weight of the sign, extra heavy support posts or foundations are required, 125% of the post mounting cost should be used.

For example, the cost of a 4' x 25' plastic sign mounted on a 15' by 6" post shared by an adjacent canopy might be estimated as follows:

Sign Cost (100 x $100)	$10,000
Post Cost ($1,762 x 1/2)	881
Total Cost	**$10,881**

If this sign were mounted on an 8" post 20' above the canopy with extra supports not needed, the cost might be estimated as follows:

Sign Cost (100 x $100)	$10,000
Post Cost ($2,855 x 1)	2,998
Total Cost	**$12,998**

Service Garage – Masonry or Concrete

Quality Classification

	Class 1 Best Quality	Class 2 Good Quality	Class 3 Average Quality	Class 4 Low Quality
Foundation (25% of total cost)	Reinforced concrete or masonry.	Reinforced concrete or masonry.	Reinforced concrete or masonry.	Unreinforced concrete or masonry.
Floor Structure (15% of total cost)	6" rock fill, 4" concrete with reinforcing mesh.	6" rock fill, 4" concrete with reinforcing mesh.	4" rock fill, 4" concrete with reinforcing mesh.	Unreinforced 4" concrete.
Walls (15% of total cost)	8" reinforced concrete block, 12" common brick.	8" reinforced concrete block, 6" reinforced concrete.	8" reinforced concrete block, 6" reinforced concrete or 8" common brick.	8" unreinforced concrete block or 8" clay tile.
Roof Structure (12% of total cost)	Glu-lams or steel trusses on heavy pilasters 20' o.c. 2" x 10" purlins 16" o.c.	Glu-lams or steel trusses on pilasters 20' o.c., 2" x 10" purlins 16" o.c.	Glu-lams or wood trusses with 2" x 8" purlins 16" o.c.	Glu-lams or light wood trusses, 2" x 8" rafters 24" o.c.
Roof Cover (8% of total cost)	5 ply built-up roof on wood sheathing, with small rock.	4 ply built-up roof on wood sheathing, with small rock.	4 ply built-up roof on wood sheathing.	4 ply built-up roof on wood sheathing.
Restrooms (10% of total cost)	Two rest rooms with three average fixtures each.	Two rest rooms with two average fixtures each.	One rest room with two low cost fixtures.	One rest room with two low cost fixtures.
Lighting (10% of total cost)	One LED fixture per 300 square feet of floor area.	One LED fixture per 300 square feet of floor area.	One LED fixture per 300 square feet of floor area.	One LED fixture per 300 square feet of floor area.
Windows (5% of total cost)	3% to 5% of wall area.	3% to 5% of wall area.	3% to 5% of wall area.	3% to 5% of wall area.

Note: Use the percent of total cost to help identify the correct quality classification.

Square foot costs include the cost of the following components: Foundations as required for normal soil conditions. Floor, wall and roof structures. Exterior wall finish and roof cover. Entry doors. Basic lighting and electrical systems. Rough and finish plumbing. Permits and fees. Contractor's mark-up.

The in-place cost of these extra components should be added to the basic building cost to arrive at the total structure cost. See page 236 to 248. Heating and air conditioning systems. Fire sprinklers. Interior finish costs. Interior partitions. Drive-through doors. Canopies and walks. Exterior signs. Paving and curbing. Miscellaneous yard improvements. Hoists, gas pump and compressor costs are listed in the section "Additional Costs for Service Stations" beginning on page 204.

Service Garage – Masonry or Concrete

Length Less Than Twice Width

Estimating Procedure

1. Use these figures to estimate buildings designed primarily for motor vehicle repair. Sales area should be figured separately. Use the costs for urban stores beginning on page 75.
2. Establish the building quality class by applying the information on page 208.
3. Compute the floor area.
4. If the wall height is more or less than 18 feet, add to or subtract from the square foot costs below the appropriate amount from the Wall Height Adjustment Table on page 212.
5. Multiply the adjusted square foot cost by the floor area.
6. Deduct for common walls or no wall ownership. Use the figures on page 212.
7. Multiply the total cost by the location factor on page 7 or 8.
8. Add the cost of heating and air conditioning systems, fire sprinklers, interior finish and partitions, drive-thru doors, canopies and walks, exterior signs, paving, curbing, and yard improvements. See page 236 to 248. Add the cost of hoists, pumps and compressors beginning on page 204.

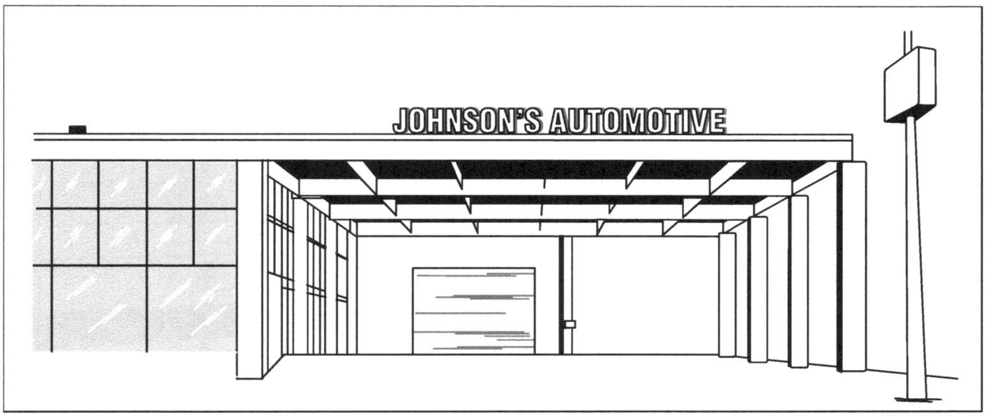

Service Garage (rear portion), Class 2

Square Foot Area

Quality Class	2,000	2,500	3,000	4,000	5,000	6,000	7,500	10,000	15,000	20,000	30,000
1, Best	113.93	103.45	96.01	86.10	79.55	74.93	69.89	64.41	58.25	54.61	50.55
1 & 2	109.35	99.38	92.14	82.64	76.35	71.86	67.10	61.95	55.89	52.46	48.48
2, Good	107.10	97.29	90.29	80.92	74.78	70.37	65.72	60.62	54.69	51.37	47.51
2 & 3	101.87	92.59	85.92	77.02	71.19	67.02	62.60	57.66	52.09	48.96	45.21
3, Average	98.98	89.88	83.38	74.76	69.06	65.03	60.73	55.93	50.54	47.48	43.91
3 & 4	93.62	84.97	78.84	70.70	65.37	61.58	57.45	52.94	47.77	44.81	41.49
4, Low	88.65	80.60	74.79	67.10	62.03	58.31	54.41	50.23	45.37	42.54	39.38

Service Garage – Masonry or Concrete

Length Between 2 and 4 Times Width

Estimating Procedure

1. Use these figures to estimate buildings designed primarily for motor vehicle repair. Sales area should be figured separately. Use the costs for urban stores beginning on page 75.
2. Establish the building quality class by applying the information on page 208.
3. Compute the floor area.
4. If the wall height is more or less than 18 feet, add to or subtract from the square foot costs below the appropriate amount from the Wall Height Adjustment Table on page 212.
5. Multiply the adjusted square foot cost by the floor area.
6. Deduct for common walls or no wall ownership. Use the figures on page 212.
7. Multiply the total cost by the location factor on page 7 or 8.
8. Add the cost of heating and air conditioning systems, fire sprinklers, interior finish and partitions, drive-thru doors, canopies and walks, exterior signs, paving, curbing, and yard improvements. See page 236 to 248. Add the cost of hoists, pumps and compressors beginning on page 204.

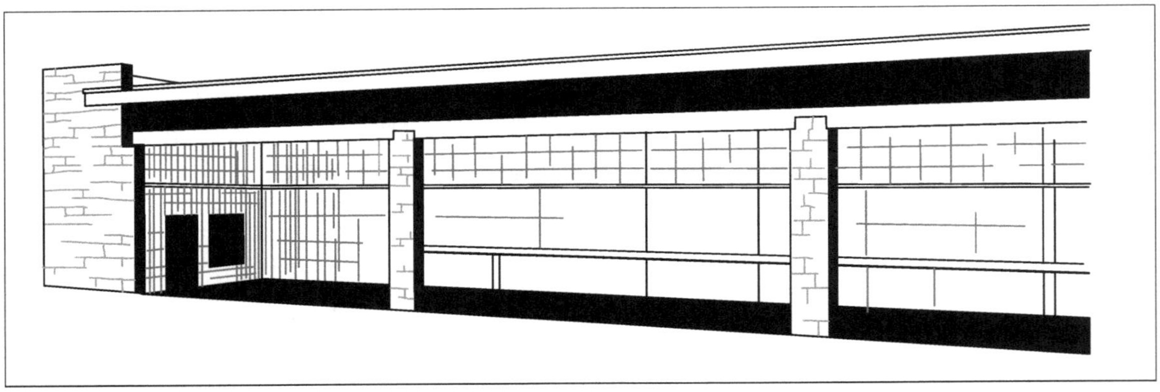

Service Garage, Class 3

Square Foot Area

Quality Class	2,000	2,500	3,000	4,000	5,000	6,000	7,500	10,000	15,000	20,000	30,000
1, Best	121.37	110.12	102.31	91.74	84.73	79.81	74.47	68.64	62.05	58.19	53.82
1 & 2	116.23	105.55	97.94	87.74	81.12	76.39	71.33	65.76	59.34	55.69	51.43
2, Good	113.76	103.33	95.98	85.96	79.39	74.80	69.80	64.30	58.12	54.54	50.40
2 & 3	108.24	98.26	91.22	81.74	75.60	71.10	66.33	61.09	55.24	51.86	47.90
3, Average	101.40	94.92	88.14	78.97	73.02	68.87	64.19	59.15	53.37	50.13	46.36
3 & 4	99.28	90.10	83.61	74.93	69.29	65.25	60.90	56.08	50.65	46.45	43.97
4, Low	94.02	85.43	79.20	71.07	65.76	61.83	57.67	53.21	48.04	45.02	41.64

Service Garage – Masonry or Concrete

Length More Than 4 Times Width

Estimating Procedure

1. Use these figures to estimate buildings designed primarily for motor vehicle repair. Sales area should be figured separately. Use the costs for urban stores beginning on page 75.
2. Establish the building quality class by applying the information on page 208.
3. Compute the floor area.
4. If the wall height is more or less than 18 feet, add to or subtract from the square foot costs below the appropriate amount from the Wall Height Adjustment Table on page 212.
5. Multiply the adjusted square foot cost by the floor area.
6. Deduct for common walls or no wall ownership. Use the figures on page 212.
7. Multiply the total cost by the location factor on page 7 or 8.
8. Add the cost of heating and air conditioning systems, fire sprinklers, interior finish and partitions, drive-thru doors, canopies and walks, exterior signs, paving curbing, and yard improvements. See page 236 to 248. Add the cost of hoists, pumps and compressors beginning on page 204.

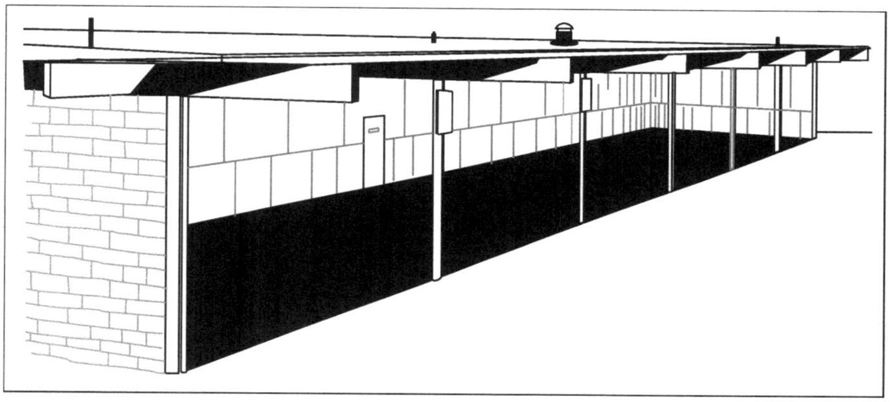

Service Garage, Class 3 & 4

Square Foot Area

Quality Class	2,000	2,500	3,000	4,000	5,000	6,000	7,500	10,000	15,000	20,000	30,000
1, Best	129.30	117.27	108.98	97.66	90.17	84.95	79.21	73.04	65.97	62.05	57.27
1 & 2	124.41	112.93	104.79	93.93	86.81	81.73	76.29	70.20	63.48	59.61	55.15
2, Good	121.29	110.05	102.10	91.46	84.60	79.55	74.37	68.47	61.93	58.12	53.76
2 & 3	115.52	104.82	97.36	87.11	80.62	75.91	70.81	65.25	59.03	55.39	51.27
3, Average	111.98	101.60	94.30	84.57	77.97	73.51	68.68	63.25	57.12	53.72	49.64
3 & 4	105.77	95.97	89.04	79.81	73.64	69.43	64.79	59.77	53.94	50.65	46.88
4, Low	105.53	90.83	84.31	75.58	69.80	65.78	61.31	56.47	51.06	47.90	44.37

Service Garage – Masonry or Concrete

Wall Height Adjustment

Add or subtract the amount listed in this table to or from the square foot of floor cost for each foot of wall height more or less than 18 feet.

Area	**2,000**	**2,500**	**3,000**	**4,000**	**5,000**	**6,000**	**7,500**	**10,000**	**15,000**	**20,000**	**30,000**
Cost	1.24	1.14	1.08	.90	.85	.81	.78	.38	.31	.28	.10

Perimeter (Common) Wall Adjustment

A common wall exists when two buildings share one wall. Adjust for common walls by deducting the linear foot costs below from the total structure cost. In some structures, one or more walls are not owned at all. In this case, deduct the "No Ownership" cost per linear foot of wall not owned.

For common wall, deduct $334 per linear foot.

For no wall ownership, deduct $663 per linear foot.

Service Garage – Wood Frame

Quality Classification

	Class 1 Best Quality	Class 2 Good Quality	Class 3 Average Quality	Class 4 Low Quality
Foundation (25% of total cost)	Concrete, heavily reinforced.	Reinforced concrete.	Masonry or reinforced concrete.	Masonry or concrete.
Floor Structure (12% of total cost)	4" reinforced concrete on 6" rock fill.	4" reinforced concrete on 6" rock fill.	4" concrete on 6" rock fill.	4" concrete on 4" rock fill.
Walls (12% of total cost)	2" x 4" studs 16" o.c. in walls 14' high; 2" x 6" studs 16" o.c. in walls over 14' high; 3" sill, double plate, adequate blocking and bracing.	2" x 4" studs 16" o.c. in walls 14' high; 2" x 6" studs 16" o.c. in walls over 14' high; 2" sill, double plate, adequate blocking and bracing.	2" x 4" studs 16" o.c. in walls to 14' high; 2" x 6" studs 16" o.c. in walls over 14' high; 2" sill, double plate, minimum blocking and bracing.	2" x 4" studs 24" o.c.; 2" x 4" sill, double 2" x 4" plate, minimum diagonal bracing.
Exterior (9% of total cost)	Good corrugated iron or board and batt.	Good corrugated iron or board and batt.	Average corrugated iron or board and batt.	Light corrugated iron or board and batt.
Roof Structures (12% of total cost)	Glu-lams, trusses or tapered steel girders on steel intermediate columns; 2" x 10" rafters 16" o.c.	Glu-lams, average wood trusses, tapered steel girders on steel intermediate columns; 2" x 8" purlins or rafters 16" o.c.	Glu-lams or light wood trusses, on wood posts 18' o.c.; 2" x 8" rafters on purlins 24" o.c.	Light trussed rafters, clear span in small buildings, post and beam support in large buildings.
Roof Cover (5% of total cost)	Good quality 4 ply composition roofing on wood sheathing.	Average quality 4 ply composition roofing on wood sheathing.	Average quality 3 ply composition roofing on wood sheathing, or good corrugated aluminum.	Light weight 3 ply composition roofing on wood sheathing, or heavy corrugated iron.
Rest Rooms (10% of total cost)	Two restrooms with three average fixtures each.	Two restrooms with three average fixtures each.	One restroom with two low cost fixtures.	One restroom with two low cost fixtures.
Lighting (10% of total cost)	One LED fixture per 300 square feet of floor area.	One LED fixture per 300 square feet of floor area.	One LED fixture per 300 square feet of floor area.	One LED fixture per 300 square feet of floor area.
Windows (5% of total cost)	3% to 5% of wall area.	3% to 5% of wall area.	3% to 5% of wall area.	3% to 5% of wall area.

Note: Use the percent of total cost to help identify the correct quality classification.

Square foot costs include the cost of the following components: Foundations as required for normal soil conditions. Floor, wall and roof structure. Exterior wall finish and roof cover. Entry doors. Basic lighting and electrical systems. Rough and finish plumbing. Permits and fees. Contractor's mark-up.

The in-place cost of these extra components should be added to the basic building cost to arrive at the total structure cost. See page 236 to 248. Heating and air conditioning systems. Fire sprinklers. Interior finish costs. Interior partitions. Drive-through doors. Canopies and walks. Exterior signs. Paving and curbing. Miscellaneous yard improvements. Hoists, gas pump and compressor costs are listed in the section "Additional Costs for Service Stations" beginning on page 204.

Service Garage – Wood Frame

Length Less Than Twice Width

Estimating Procedure

1. Use these figures to estimate buildings designed primarily for motor vehicle repair. Sales area should be figured separately. Use the costs for urban stores beginning on page 75.
2. Establish the building quality class by applying the information on page 213.
3. Compute the floor area.
4. If the wall height is more or less than 16 feet, add to or subtract from the square foot costs below the appropriate amount from the Wall Height Adjustment Table on page 217.
5. Multiply the adjusted square foot cost by the floor area.
6. Deduct for common walls or no wall ownership. Use the figures on page 217.
7. Multiply the total cost by the location factor on page 7 or 8.
8. Add the cost of heating and air conditioning systems, fire sprinklers, interior finish and partitions, drive-thru doors, canopies and walks, exterior signs, paving, curbing, and yard improvements. See page 236 to 248. Add the cost of hoists, pumps and compressors beginning on page 204.

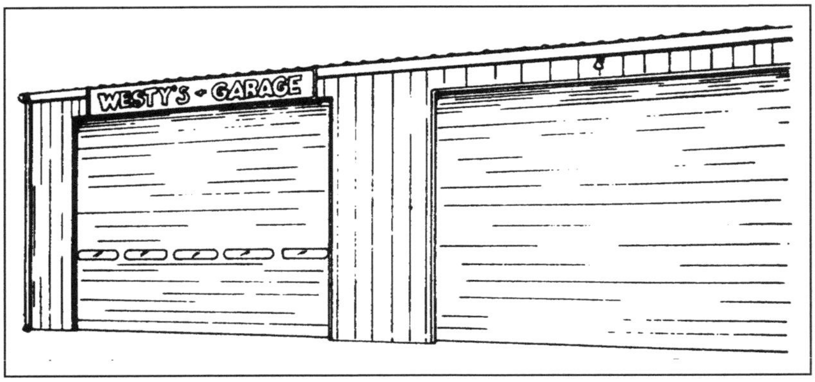

Service Garage, Class 3

Square Foot Area

Quality Class	2,000	2,500	3,000	4,000	5,000	6,000	7,500	10,000	15,000	20,000	30,000
1, Best	56.07	50.90	47.27	42.37	39.15	36.84	34.39	31.72	28.75	26.98	24.91
1 & 2	53.24	48.31	44.84	40.16	37.11	34.97	32.67	30.08	27.20	25.54	23.60
2, Good	51.38	46.63	43.29	38.79	35.88	33.76	31.52	29.07	26.25	24.67	22.86
2 & 3	48.35	43.83	40.70	36.47	33.75	31.69	29.64	27.29	24.76	23.19	21.46
3, Average	45.90	41.66	38.73	34.68	32.05	30.16	28.12	25.99	23.44	22.05	20.44
3 & 4	43.10	39.14	36.35	32.58	30.13	28.27	26.46	24.36	22.04	20.69	19.18
4, Low	40.54	36.76	34.21	30.58	28.27	26.67	24.91	22.88	20.66	19.48	18.04

Service Garage – Wood Frame

Length Between 2 and 4 Times Width

Estimating Procedure

1. Use these figures to estimate buildings designed primarily for motor vehicle repair. Sales area should be figured separately. Use the costs for urban stores beginning on page 75.
2. Establish the building quality class by applying the information on page 213.
3. Compute the floor area.
4. If the wall height is more or less than 16 feet, add to or subtract from the square foot costs below the appropriate amount from the Wall Height Adjustment Table on page 217.
5. Multiply the adjusted square foot cost by the floor area.
6. Deduct for common walls or no wall ownership. Use the figures on page 217.
7. Multiply the total cost by the location factor on page 7 or 8.
8. Add the cost of heating and air conditioning systems, fire sprinklers, interior finish and partitions, drive-thru doors, canopies and walks, exterior signs, paving, curbing, and yard improvements. See page 236 to 248. Add the cost of hoists, pumps and compressors beginning on page 204.

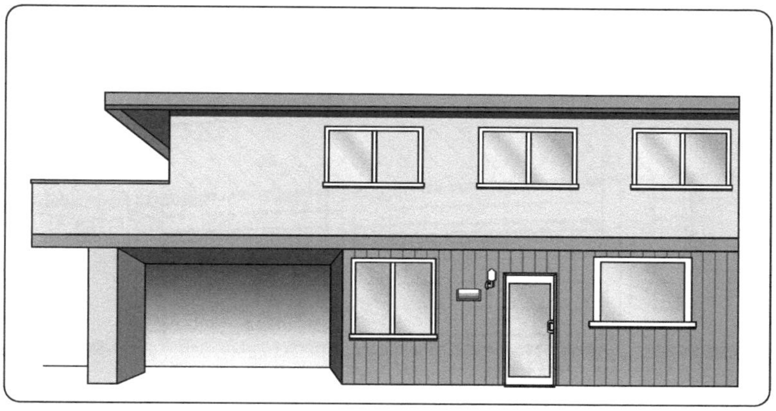

Service Garage (rear portion) Class 2 & 3

Square Foot Area

Quality Class	2,000	2,500	3,000	4,000	5,000	6,000	7,500	10,000	15,000	20,000	30,000
1, Best	59.63	54.16	50.29	45.09	41.66	39.26	36.63	33.76	30.50	28.63	26.53
1 & 2	56.66	51.46	47.72	42.84	39.59	37.28	34.83	32.05	28.99	27.20	25.16
2, Good	54.67	49.66	46.12	41.24	38.14	35.99	33.57	30.91	27.97	26.25	24.29
2 & 3	51.41	46.69	43.38	38.83	35.97	33.87	31.58	29.13	26.34	24.78	22.88
3, Average	49.09	44.63	41.38	37.11	34.33	32.31	30.15	27.80	25.10	23.59	21.82
3 & 4	45.78	41.53	38.59	34.62	31.98	30.13	28.09	25.88	23.37	21.97	20.31
4, Low	42.47	38.58	35.79	32.06	29.66	27.97	26.04	23.98	21.72	20.34	18.92

Service Garage – Wood Frame

Length More Than 4 Times Width

Estimating Procedure

1. Use these figures to estimate buildings designed primarily for motor vehicle repair. Sales area should be figured separately. Use the costs for urban stores beginning on page 75.
2. Establish the building quality class by applying the information on page 213.
3. Compute the floor area.
4. If the wall height is more or less than 16 feet, add to or subtract from the square foot costs below the appropriate amount from the Wall Height Adjustment Table on page 217.
5. Multiply the adjusted square foot cost by the floor area.
6. Deduct for common walls or no wall ownership. Use the figures on page 217.
7. Multiply the total cost by the location factor on page 7 or 8.
8. Add the cost of heating and air conditioning systems, fire sprinklers, interior finish and partitions, drive-thru doors, canopies and walks, exterior signs, paving, curbing, and yard improvements. See page 236 to 248. Add the cost of hoists, pumps and compressors beginning on page 204.

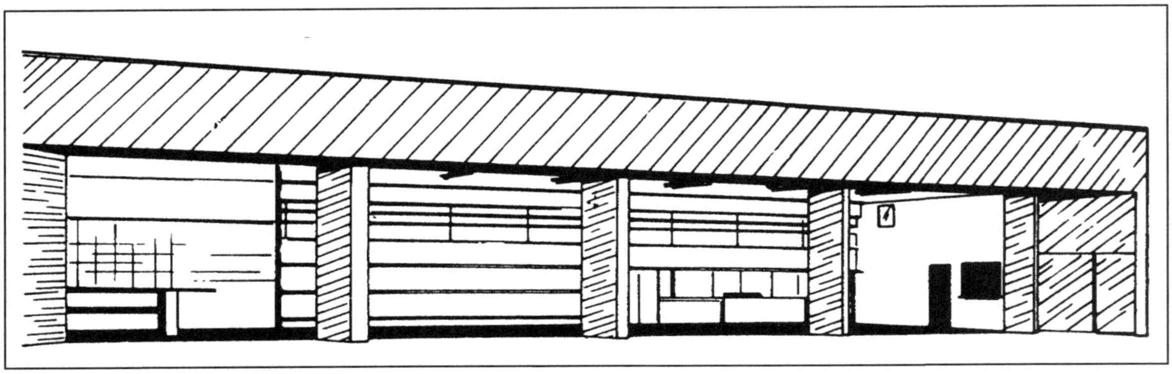

Service Garage, Class 3

Square Foot Area

Quality Class	2,000	2,500	3,000	4,000	5,000	6,000	7,500	10,000	15,000	20,000	30,000
1, Best	63.65	57.81	53.63	48.08	44.46	41.85	39.03	36.00	32.53	30.50	28.20
1 & 2	60.44	54.89	50.94	45.67	42.24	39.73	37.10	34.21	30.91	29.05	26.81
2, Good	58.44	53.01	49.23	44.11	40.76	38.44	35.88	33.05	29.86	28.02	25.88
2 & 3	54.94	49.92	46.39	41.58	38.45	36.17	33.76	31.14	28.11	26.43	24.36
3, Average	52.25	47.55	44.07	39.59	36.59	34.40	32.06	29.61	26.75	25.10	23.21
3 & 4	48.78	44.29	41.15	36.84	34.14	32.06	29.93	27.55	25.00	23.42	21.66
4, Low	45.56	41.35	38.40	34.42	31.83	29.91	27.91	25.82	23.30	21.84	20.18

Service Garage – Wood Frame

Wall Height Adjustment

Add or subtract the amount listed in this table to or from the square foot cost for each foot of wall height more or less than 16 feet.

Area	2,000	2,500	3,000	4,000	5,000	6,000	7,500	10,000	15,000	20,000	30,000
Cost	.73	.64	.60	.47	.40	.27	.22	.20	.14	.13	.09

Perimeter (Common) Wall Adjustment

A common wall exists when two buildings share one wall. Adjust for common walls by deducting the linear foot costs below from the total structure cost. In some structures one or more walls are not owned at all. In this case, deduct the "No Ownership" cost per linear foot of wall not owned.

For common wall, deduct $87 per linear foot.

For no wall ownership, deduct $170 per linear foot.

Auto Service Centers – Masonry or Concrete

Quality Classification

	Class 1 Best Quality	Class 2 Good Quality	Class 3 Average Quality	Class 4 Low Quality
Foundation (20% of total cost)	Reinforced concrete.	Reinforced concrete.	Reinforced concrete.	Reinforced concrete.
Floor Structure (10% of total cost)	4" reinforced concrete on 6" rock fill.	4" reinforced concrete on 6" rock fill.	4" reinforced concrete on 6" rock fill.	4" reinforced concrete on 6" rock fill.
Walls (15% of total cost)	8" reinforced decorative colored concrete block.	8" reinforced detailed concrete block.	8" reinforced concrete block.	8" reinforced concrete block.
Roof Structures (5% of total cost)	Steel open web joists, steel deck.	Glu-lams, 3" x 12" purlins 3' o.c., or wood open web joists (truss joists) 4' o.c.	Glu-lams, 3" x 12" purlins 3' o.c., or wood open web joists (truss joists) 4' o.c., 1/2" plywood sheathing.	Glu-lams, 3" x 12" purlins 3' o.c., or wood open web joists (truss joists) 4' o.c., 1/2" plywood sheathing.
Floor Finish (5% of total cost)	Concrete in work area, resilient tile in sales area.	Concrete in work area, resilient tile in sales area.	Concrete in work area, minimum grade tile in sales area.	Concrete.
Interior Wall Finish (5% of total cost)	Painted concrete block in work area; gypsum wallboard, texture and paint in sales area.	Painted concrete block in work area and sales area.	Unfinished in work area, painted concrete block in sales area	Unfinished.
Ceiling Finish (3% of total cost)	Open in work area, acoustical tile suspended in exposed grid in sales area.	Open in work area, acoustical tile suspended in exposed grid in sales area.	Open in work area. Celotex tile in sales area.	Open.
Exterior Finish (5% of total cost)	Decorative colored concrete block.	Colored or painted detailed block.	Painted concrete block.	Unpainted concrete block.
Display Front *(Covers about 25% of the exterior wall)* (7% of total cost)	1/4" float glass in good aluminum frame. Good aluminum and glass doors.	1/4" float glass in average aluminum frame. Aluminum and glass doors.	1/4" float glass in light aluminum frame. Wood and glass door.	Crystal glass in wood frame, wood door.
Roof and Cover (7% of total cost)	5 ply built-up roofing with insulation.	4 ply built-up roofing.	4 ply built-up roofing.	4 ply built-up roofing.
Plumbing (10% of total cost)	Two rest rooms with three good fixtures. Metal toilet partitions.	Two rest rooms with two average fixtures. Wood toilet partitions.	One rest room with two fixtures.	One rest room with two fixtures.
Electrical & Wiring (8% of total cost)	Conduit wiring with LED strips, 8' o.c.	Conduit wiring with LED strips, 8' o.c.	Conduit wiring, LED strips, 8' o.c.	Conduit wiring, LED fixtures, 10' o.c. or single tube fluorescent strips, 8' o.c.

Note: Use the percentage of total cost to help identify the correct quality classification.

Square foot costs include the cost of the following components: Foundations as required for normal soil conditions. Floor, wall and roof structures. Interior floor, wall and ceiling finishes (in sales area). Exterior wall finish and roof cover. Display windows. Interior partitions. Entry doors. Basic lighting and electrical systems. Rough and finish plumbing. Permits and fees. Contractor's mark-up.

Wall Height Adjustment: Add or subtract the amount listed in this table to or from the square foot of floor cost for each foot of wall height more or less than 16 feet.

Area	1,500	2,000	2,500	3,000	3,500	4,000	5,000	6,000	7,500	10,000	15,000
Cost	3.74	3.23	2.82	2.46	2.34	2.19	2.01	1.92	1.54	1.39	1.29

Perimeter Wall Adjustment: For common wall, deduct $235 per linear foot. For no wall ownership, deduct $503 per linear foot.

Auto Service Centers – Masonry or Concrete

Length Less Than Twice Width

Estimating Procedure

1. Use these figures to estimate buildings designed for selling and installing automobile accessories. The square foot costs below allow for a sales area occupying 25% of the building space. The sales area has finished floors, walls and ceiling as described in the quality classification. The remaining 75% of the building is service area and has no interior finish.
2. Establish the building quality class by applying the information on page 218.
3. Compute the floor area. This should include everything within the exterior walls.
4. If the wall height is more or less than 16 feet, add to or subtract from the square foot costs below the appropriate amount from the Wall Height Adjustment Table on page 218.
5. Deduct for common walls or no wall ownership. See page 218.
6. Multiply the total cost by the location factor on page 7 or 8.
7. Add the cost of heating and air conditioning systems, fire sprinklers, canopies, walks, exterior signs, paving, curbing, loading docks, ramps, and yard improvements. See page 236 to 248. Add the cost of service station equipment beginning on 204.

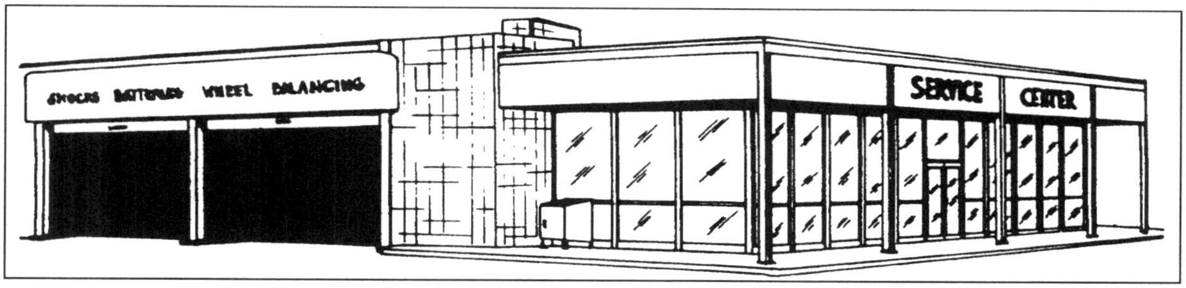

Auto Service Center, Class 2

Square Foot Area

Quality Class	1,500	2,000	2,500	3,000	3,500	4,000	5,000	6,000	7,500	10,000	15,000
Exceptional	222.97	203.97	191.55	182.56	175.68	170.43	162.23	156.32	149.94	142.90	134.67
1, Best	213.34	195.24	183.26	174.68	168.19	163.01	155.18	149.60	143.45	136.72	128.85
1 & 2	203.97	186.56	175.17	166.99	160.75	155.77	148.33	142.92	137.12	130.63	123.23
2, Good	197.93	181.04	169.98	162.08	156.03	151.19	143.98	138.80	133.14	126.81	119.47
2 & 3	188.72	172.64	162.11	154.50	148.75	144.12	137.32	132.25	126.90	120.91	114.00
3, Average	183.35	167.70	157.40	150.04	144.43	140.00	133.35	128.45	123.33	117.45	110.75
3 & 4	174.36	159.59	149.79	142.81	137.44	133.24	126.90	122.28	117.25	111.75	105.32
4, Low	167.10	152.91	143.56	136.93	131.72	127.70	121.67	117.21	112.46	107.10	100.92

Auto Service Centers – Masonry or Concrete

Length Between 2 and 4 Times Width

Estimating Procedure

1. Use these figures to estimate buildings designed for selling and installing automobile accessories. The square foot costs below allow for a sales area occupying 25% of the building space. The sales area has finished floors, walls and ceiling as described in the quality classification. The remaining 75% of the building is service area and has no interior finish.
2. Establish the building quality class by applying the information on page 218.
3. Compute the floor area. This should include everything within the exterior walls.
4. If the wall height is more or less than 16 feet, add to or subtract from the square foot costs below the appropriate amount from the Wall Height Adjustment Table on page 218.
5. Deduct for common walls or no wall ownership. See page 218.
6. Multiply the total cost by the location factor on page 7 or 8.
7. Add the cost of heating and air conditioning systems, fire sprinklers, canopies, walks, exterior signs, paving, curbing, loading docks, ramps, and yard improvements. See page 236 to 248. Add the cost of service station equipment beginning on 204.

Auto Service Center, Class 2

Square Foot Area

Quality Class	1,500	2,000	2,500	3,000	3,500	4,000	5,000	6,000	7,500	10,000	15,000
Exceptional	236.97	215.70	201.72	191.78	184.12	178.07	169.07	162.63	155.58	147.75	138.70
1, Best	226.77	206.39	193.11	183.44	176.25	170.48	161.83	155.62	148.85	141.32	132.86
1 & 2	216.48	197.05	184.25	175.17	168.23	162.70	154.50	148.51	142.14	134.97	126.69
2, Good	209.87	190.95	178.57	169.72	163.03	157.72	149.75	144.03	137.79	130.86	122.86
2 & 3	200.44	182.35	170.65	162.17	155.69	150.75	143.05	137.52	131.62	124.97	117.34
3, Average	194.09	176.62	165.16	157.01	150.77	145.79	138.45	133.16	127.35	121.00	113.59
3 & 4	185.29	168.73	157.81	149.91	144.03	139.33	132.24	127.20	121.70	115.55	108.49
4, Low	176.72	160.84	150.42	143.05	137.39	132.88	126.17	121.24	116.05	110.12	103.45

Auto Service Centers – Masonry or Concrete

Length More Than 4 Times Width

Estimating Procedure

1. Use these figures to estimate buildings designed for selling and installing automobile accessories. The square foot costs below allow for a sales area occupying 25% of the building space. The sales area has finished floors, walls and ceiling as described in the quality classification. The remaining 75% of the building is service area and has no interior finish.
2. Establish the building quality class by applying the information on page 218.
3. Compute the floor area. This should include everything within the exterior walls.
4. If the wall height is more or less than 16 feet, add to or subtract from the square foot costs below the appropriate amount from the Wall Height Adjustment Table on page 218.
5. Deduct for common walls or no wall ownership. See page 218.
6. Multiply the total cost by the location factor on page 7 or 8.
7. Add the cost of heating and air conditioning systems, fire sprinklers, canopies, walks, exterior signs, paving, curbing, loading docks, ramps, and yard improvements. See page 236 to 248. Add the cost of service station equipment beginning on 204.

Auto Service Center, Class 3

Square Foot Area

Quality Class	1,500	2,000	2,500	3,000	3,500	4,000	5,000	6,000	7,500	10,000	15,000
Exceptional	255.51	232.31	216.90	205.74	197.14	190.33	179.89	172.39	164.05	154.86	144.00
1, Best	244.37	222.27	207.45	196.78	188.53	181.96	172.13	164.84	156.92	148.16	137.74
1 & 2	233.13	212.05	198.10	187.78	179.92	173.63	164.22	157.34	149.75	141.31	131.38
2, Good	226.30	205.87	192.20	182.24	174.67	168.57	159.34	152.72	145.36	137.12	127.56
2 & 3	215.84	196.30	183.30	173.71	166.57	160.72	151.94	145.49	138.64	130.81	121.70
3, Average	209.55	190.47	177.93	168.73	161.63	156.04	147.59	141.35	134.58	126.92	118.05
3 & 4	199.89	181.81	169.72	161.07	154.18	148.79	140.80	134.82	128.36	121.12	112.69
4, Low	190.59	173.32	161.82	153.53	147.05	141.97	134.21	128.56	122.40	115.50	107.45

Industrial Structures Section

Section Contents

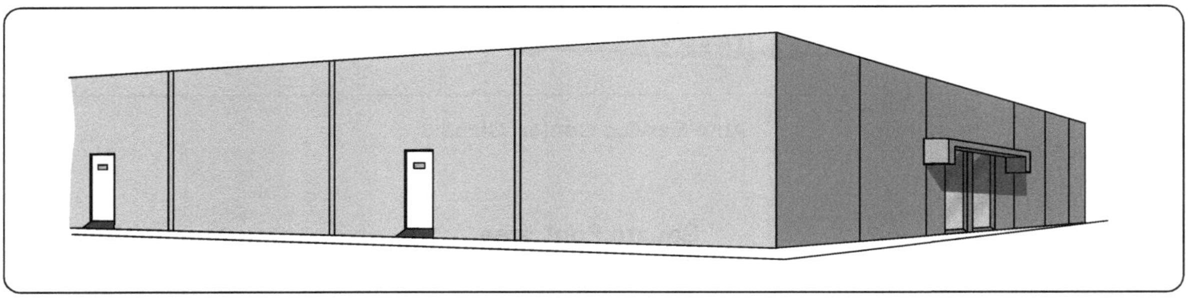

Warehouse, Class 3

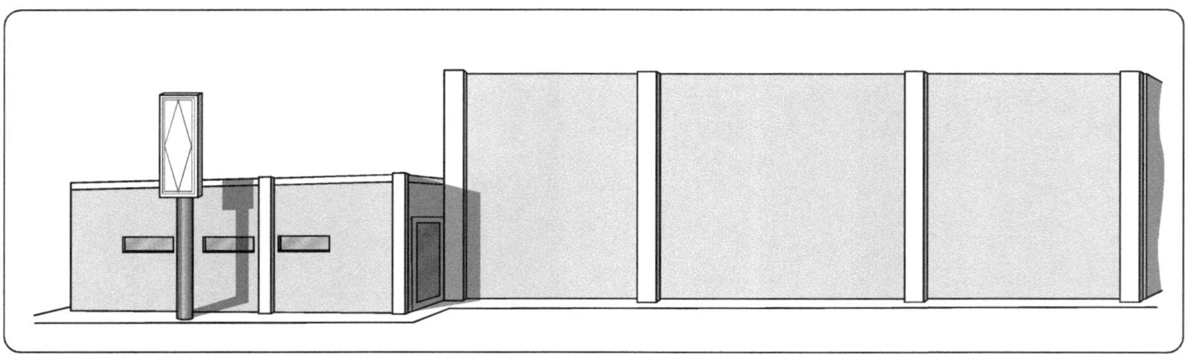

Light Industrial, Class 2

Industrial Buildings

Quality Classification

	Class 1 Best Quality	Class 2 Good Quality	Class 3 Average Quality	Class 4 Low Quality
Foundations (22% of total cost)	Continuous reinforced concrete.	Continuous reinforced concrete.	Continuous reinforced concrete.	Reinforced concrete pads under pilasters.
Floor Structure (15% of total cost)	6" rock base, 6" concrete with reinforcing mesh or bars.	6" rock base, 6" concrete with reinforcing mesh or bars.	6" rock base, 5" concrete with reinforcing mesh or bars.	6" rock base, 4" concrete with reinforcing mesh.
Wall Structure (25% of total cost)	8" reinforced concrete block or brick with pilasters 20' o.c., painted sides and rear exterior, front wall brick veneer.	8" reinforced concrete block or brick with pilasters 20' o.c., painted sides and rear exterior, stucco and some brick veneer on front.	8" reinforced concrete block or brick, unpainted.	8" reinforced concrete block or brick, unpainted.
Roof Structure (12% of total cost)	Glu-lams, wood or steel trusses on steel intermediate columns, span exceeds 70'.	Glu-lams, wood or steel trusses on steel intermediate columns, span exceeds 70'.	Glu-lams or steel beams on steel intermediate columns, short span.	Glu-lams on steel intermediate columns, short span.
Roof Cover (7% of total cost)	Panelized roof system, 1/2" plywood sheathing, 5 ply built-up roof.	Panelized roof system, 1/2" plywood sheathing, 4 ply built-up roof.	Panelized roof system, 1/2" plywood sheathing, 4 ply built-up roof.	Panelized roof system, 1/2" plywood sheathing, 4 ply built-up roof.
Skylights (1% of total cost)	48 S.F. of skylight per 2500 S.F. of floor area (1-6' x 8' skylight 40' to 50' o.c.).	32 S.F. of skylight per 2500 S.F. of floor area (1-4' x 8' skylight 40' to 50' o.c.).	24 S.F. of skylight per 2500 S.F. of floor area (1-4' x 6' skylight 40' to 50' o.c.).	10 S.F. of skylight per 2500 S.F. of floor area (1-2' x 4' skylight 40' to 50' o.c.).
Ventilators (2% of total cost)	1 large rotary vent per 2500 S.F. of floor area.	1 medium rotary vent per 2500 S.F. of floor area.	1 medium rotary vent per 2500 S.F. of floor area.	1 small rotary vent per 2500 S.F. of floor area.
Rest Rooms, Finish (3% of total cost)	Good vinyl asbestos tile floors enameled gypsum wallboard partitions.	Vinyl asbestos tile floors, enameled gypsum wallboard partitions.	Concrete floors, painted gypsum wallboard partitions.	Concrete floors, unfinished wallboard partitions.
Fixtures (5% of total cost)	2 rest rooms, 3 good fixtures in each.	2 rest rooms, 3 average fixtures in each.	2 rest rooms, 2 average fixtures in each.	1 rest room, 2 low cost fixtures in each.
Lighting (8% of total cost)	4' LED fixtures 10' x 12' spacing.	Low cost LED fixtures 12' x 20' spacing.	Low cost LED fixtures 20' x 20' spacing.	Low cost fixtures 20' x 30' spacing.

Note: Use the percent of total cost to help identify the correct quality classification.

Square foot costs include the following components: Foundations as required for normal soil conditions. Floor, wall and roof structures. Exterior wall finish and roof cover. Basic lighting and electrical systems. Rough and finish plumbing. A usual or normal parapet wall. Walk-through doors. Contractors' mark-up.

Warehouses

Estimating Procedure

1. Establish the structure quality class by applying the information on page 223.
2. Compute the building floor area.
3. Add to or subtract from the square foot cost below the appropriate amount from the Wall Height Adjustment Table (at the bottom of page 226) if the wall height is more or less than 20 feet.
4. Multiply the adjusted square foot cost by the building floor area.
5. Deduct, if appropriate, for common walls, using the figures at the bottom of page 226.
6. Multiply the total cost by the location factor on page 7 or 8.
7. Add the cost of heating and air conditioning equipment, fire sprinklers, interior offices, drive-through or delivery doors, canopies, interior partitions, docks and ramps, paving and curbing, and miscellaneous yard improvements. See the section beginning on page 236.

Length less than twice width – Square Foot Area

Quality Class	3,000	4,000	5,000	7,500	10,000	15,000	20,000	30,000	50,000	100,000	200,000
1, Best	184.38	168.74	158.24	142.18	132.79	121.78	115.34	107.69	100.20	92.73	87.48
1 & 2	171.66	157.13	147.39	132.43	123.67	113.47	107.48	100.38	93.29	86.35	81.50
2, Good	162.21	148.38	139.22	125.07	116.82	107.12	101.44	94.78	88.09	81.52	76.94
2 & 3	151.69	138.69	130.15	117.03	109.22	100.22	94.91	88.62	82.46	76.21	71.96
3, Average	142.76	130.62	122.41	110.12	102.80	94.36	89.36	83.35	77.52	71.84	67.74
3 & 4	133.52	122.24	114.56	103.02	96.17	88.26	83.50	78.07	72.53	67.09	63.26
4, Low	124.40	113.84	106.77	95.95	89.63	82.20	77.86	72.65	67.65	62.56	59.06

Length between 2 and 4 times width – Square Foot Area

Quality Class	3,000	4,000	5,000	7,500	10,000	15,000	20,000	30,000	50,000	100,000	200,000
1, Best	196.64	179.06	167.34	149.47	139.13	126.92	119.78	111.37	103.22	95.09	89.40
1 & 2	183.68	167.27	156.28	139.57	129.74	118.49	111.86	104.04	96.38	88.74	83.41
2, Good	173.38	157.89	147.48	131.82	122.58	111.90	105.53	98.27	91.04	83.79	78.71
2 & 3	162.21	147.74	138.03	123.19	114.63	104.73	98.75	91.94	85.07	78.40	73.64
3, Average	152.60	138.96	129.74	116.03	107.87	98.43	92.97	86.48	80.08	73.76	69.25
3 & 4	142.80	130.00	121.50	108.57	101.00	92.13	86.93	80.88	74.94	68.97	64.91
4, Low	132.69	120.80	112.89	100.86	93.81	85.75	80.84	75.18	69.57	64.04	60.25

Length more than 4 times width – Square Foot Area

Quality Class	3,000	4,000	5,000	7,500	10,000	15,000	20,000	30,000	50,000	100,000	200,000
1, Best	213.06	193.64	180.54	160.38	148.50	134.60	126.27	116.71	107.10	97.53	90.75
1 & 2	204.39	180.82	168.50	149.72	138.60	125.66	117.96	109.03	100.01	91.11	84.72
2, Good	193.34	170.81	159.20	141.43	131.01	118.76	111.38	102.95	94.50	86.07	80.09
2 & 3	175.91	159.83	148.95	132.34	122.57	111.08	104.25	96.28	88.43	80.50	74.93
3, Average	165.50	150.42	140.20	124.53	115.29	104.52	98.21	90.58	83.23	75.69	70.48
3 & 4	154.65	140.46	131.01	116.34	107.83	97.68	91.67	84.67	77.72	70.73	65.92
4, Low	143.99	130.86	122.01	108.41	100.33	91.01	85.40	78.89	72.38	65.92	61.37

Light Industrial Buildings

Estimating Procedure

1. Establish the structure quality class by applying the information on page 223.
2. Compute the building floor area.
3. Add to or subtract from the square foot cost below the appropriate amount from the Wall Height Adjustment Table (at the bottom of page 226) if the wall height is more or less than 20 feet.
4. Multiply the adjusted square foot cost by the building floor area.
5. Deduct, if appropriate, for common walls, using the figures at the bottom of page 226.
6. Multiply the total cost by the location factor on page 7 or 8.
7. Add the cost of heating and air conditioning equipment, fire sprinklers, interior offices, drive-through or delivery doors, canopies, interior partitions, docks and ramps, paving and curbing, and miscellaneous yard improvements. See the section beginning on page 236.

Length less than twice width – Square Foot Area

Quality Class	3,000	4,000	5,000	7,500	10,000	15,000	20,000	30,000	50,000	100,000	200,000
1, Best	182.66	168.59	159.17	144.70	136.21	126.24	120.43	113.51	106.69	99.88	95.11
1 & 2	170.39	157.30	148.44	135.00	127.08	117.83	112.34	105.93	99.50	93.16	88.64
2, Good	159.39	147.19	138.91	126.27	118.98	110.28	105.14	99.11	93.16	87.16	82.98
2 & 3	150.63	139.13	131.29	119.42	112.46	104.17	99.32	93.68	88.00	82.39	78.44
3, Average	140.28	129.59	122.30	111.14	104.76	97.00	92.52	87.27	81.98	76.78	73.07
3 & 4	131.46	121.37	114.61	104.14	98.07	90.97	86.69	81.72	76.87	71.92	68.48
4, Low	122.32	112.92	106.58	96.96	91.29	84.65	80.75	76.12	71.48	66.91	63.72

Length between 2 and 4 times width – Square Foot Area

Quality Class	3,000	4,000	5,000	7,500	10,000	15,000	20,000	30,000	50,000	100,000	200,000
1, Best	194.23	178.34	167.72	151.63	142.10	131.11	124.59	116.97	109.50	102.08	96.77
1 & 2	181.62	166.83	156.84	141.76	132.96	122.60	116.49	109.45	102.45	95.43	90.49
2, Good	170.53	156.59	147.30	133.09	124.81	115.09	109.40	102.67	96.18	89.63	85.04
2 & 3	159.31	146.30	137.57	124.32	116.51	107.55	102.19	95.95	89.68	83.65	79.42
3, Average	149.31	137.09	128.97	116.51	109.23	100.83	95.85	90.04	84.20	78.45	74.39
3 & 4	139.88	128.53	120.80	109.15	102.33	94.43	89.67	84.27	78.97	73.48	69.73
4, Low	130.07	119.48	112.34	101.59	95.19	87.88	83.46	78.45	73.37	68.41	64.91

Length more than 4 times width – Square Foot Area

Quality Class	3,000	4,000	5,000	7,500	10,000	15,000	20,000	30,000	50,000	100,000	200,000
1, Best	209.22	191.32	179.40	161.11	150.42	137.78	130.34	121.61	112.94	104.41	98.29
1 & 2	195.88	179.29	168.01	150.87	140.89	129.10	122.12	113.92	105.86	97.79	92.07
2, Good	183.49	167.87	157.37	141.38	131.90	120.82	114.36	106.69	99.11	91.51	86.18
2 & 3	171.89	157.26	147.41	132.40	123.49	113.24	107.08	99.92	92.90	85.78	80.83
3, Average	161.23	147.54	138.29	124.22	115.97	106.18	100.45	93.79	87.10	80.41	75.81
3 & 4	150.57	137.75	129.13	116.03	108.22	99.17	93.82	87.61	81.32	75.14	70.73
4, Low	140.19	128.25	120.33	108.00	100.72	92.28	87.40	81.58	75.70	69.98	65.88

Factory Buildings

Estimating Procedure

1. Establish the structure quality class by applying the information on page 223.
2. Compute the building floor area.
3. Add to or subtract from the square foot cost below the appropriate amount from the Wall Height Adjustment Table (at the bottom of this page) if the wall height is more or less than 20 feet.
4. Multiply the adjusted square foot cost by the building floor area.
5. Deduct, if appropriate, for common walls, using the figures at the bottom of this page.
6. Multiply the total cost by the location factor on page 7 or 8.
7. Add the cost of heating and air conditioning equipment, fire sprinklers, interior offices, drive-through or delivery doors, canopies, interior partitions, docks and ramps, paving and curbing, and miscellaneous yard improvements. See the section beginning on page 236.

Length less than twice width – Square Foot Area

Quality Class	3,000	4,000	5,000	7,500	10,000	15,000	20,000	30,000	50,000	100,000	200,000
1, Best	180.82	167.99	159.31	146.10	138.43	129.38	124.09	117.78	111.61	105.40	101.10
1 & 2	169.79	157.74	149.57	137.24	129.98	121.49	116.40	110.60	104.79	98.99	94.99
2, Good	159.94	148.62	140.94	129.31	122.39	114.43	109.75	104.23	98.70	93.28	89.45
2 & 3	149.83	139.19	132.07	121.10	114.65	107.18	102.69	97.59	92.46	87.39	83.86
3, Average	140.90	130.80	124.09	113.78	107.81	100.69	96.62	91.67	86.86	82.11	78.80
3 & 4	131.53	122.21	116.01	106.28	100.63	94.05	90.19	85.77	81.15	76.70	73.51
4, Low	122.26	113.55	107.68	98.75	93.48	87.48	83.89	79.59	75.45	71.30	68.41

Length between 2 and 4 times width – Square Foot Area

Quality Class	3,000	4,000	5,000	7,500	10,000	15,000	20,000	30,000	50,000	100,000	200,000
1, Best	191.01	176.63	166.96	152.40	143.77	133.76	127.82	121.01	114.23	107.41	102.55
1 & 2	179.14	165.67	156.73	142.88	134.94	125.46	119.97	113.48	107.12	100.69	96.27
2, Good	169.20	156.45	147.92	134.99	127.38	118.49	113.31	107.17	101.10	95.13	90.97
2 & 3	158.39	146.46	138.51	126.32	119.18	110.98	106.05	100.33	94.67	89.06	85.11
3, Average	148.80	137.58	130.05	118.70	111.97	104.12	99.60	94.21	88.92	83.62	79.98
3 & 4	139.06	128.61	121.56	110.98	104.73	97.42	93.10	88.02	83.05	78.17	74.67
4, Low	129.12	119.44	112.88	102.95	97.16	90.41	86.46	81.80	77.21	72.51	69.37

Length more than 4 times width – Square Foot Area

Quality Class	3,000	4,000	5,000	7,500	10,000	15,000	20,000	30,000	50,000	100,000	200,000
1, Best	204.46	188.44	177.66	161.11	151.37	139.99	133.20	125.26	117.32	109.40	103.96
1 & 2	192.16	177.17	166.92	151.38	142.19	131.53	125.09	117.68	110.28	102.82	97.67
2, Good	181.20	167.03	157.41	142.79	134.16	123.98	117.97	110.98	103.98	96.97	92.04
2 & 3	169.79	156.46	147.46	133.72	125.66	116.16	110.60	103.98	97.45	90.97	86.25
3, Average	159.44	147.13	138.63	125.77	118.16	109.22	103.89	97.68	91.51	85.34	80.96
3 & 4	148.98	137.34	129.44	117.30	110.26	101.92	97.00	91.24	85.49	79.74	75.73
4, Low	138.28	127.46	120.05	108.98	102.33	94.64	90.08	84.67	79.30	74.04	70.26

Wall Height Adjustment: Industrial building costs are based on a 20' wall height, measured from the bottom of the floor slab to the top of the roof cover. Add or subtract the amount listed to the square foot cost for each foot more or less than 20 feet.

Area	3,000	4,000	5,000	7,500	10,000	15,000	20,000	30,000	50,000	100,000	200,000
Cost	1.42	1.22	1.12	.98	.83	.75	.69	.56	.26	.08	.04

Perimeter Wall Adjustment: A common wall exists when two buildings share one wall. Adjust for common walls by deducting $182 per linear foot from the total structure cost. In some structures one or more walls are not owned at all. In this case, deduct $365 per liner foot of wall not owned.

Internal Offices

Internal offices are office areas built into the interior area of an industrial building. Add the square foot costs in this section to the basic building cost. The costs include floor finish, partition framing, wall finish, trim and doors, counter, ceiling structure, ceiling finish, and the difference between the cost of lighting and windows in an industrial building and the cost of lighting and windows in an office building.

Plumbing costs are not included in these internal office costs. If a building has fixtures in excess of those listed in the quality classification for the main building, add $1,530 to $1,900 for each extra fixture. For two-story internal offices, apply a square foot cost based on the total area of both floor levels.

	Best	Good	Average	Low Cost
Floor Finish (15% of total cost)	Carpet, some vinyl tile or terrazzo.	Resilient tile, some carpet.	Composition tile.	Colored concrete.
Wall Finish (25% of total cost)	Hardwood veneer, some textured cloth wall cover.	Gypsum board, texture and paint, some hardwood veneer.	Gypsum board, texture and paint.	Plywood and paint.
Ceiling Finish (15% of total cost)	Illuminated plastic ceilings.	Suspended "T" bar and acoustical tile.	Gypsum board, texture and paint.	Exposed or plywood and paint.
Doors - Interior (10% of total cost)	Good grade hardwood.	Good grade hardwood.	Low cost hardwood.	Standard paint grade.
Doors - Exterior (5% of total cost)	Pair aluminum entry doors with extensive sidelights.	Aluminum entry door with sidelights.	Wood store door.	Standard paint grade.
Windows (5% of total cost)	Average amount in good aluminum frame, fixed float glass in good frame in front.	Average amount in good aluminum frame, some fixed float glass.	Average amount of average cost aluminum sliding type.	Average amount of low cost aluminum sliding type.
Counters (15% of total cost)	Good grade hardwood counter, good shelving below.	Good grade plastic counter, average amount of shelving below.	Low cost plastic counter, average amount of shelving.	Paint grade counter, small amount of shelving.
Lighting (10% of total cost)	Illuminated ceilings.	Recessed LED fixtures.	Average amount of LED fixtures.	One LED fixture per each 150 S.F.
S.F. Costs	$68.40 to $95.80	$53.90 to $69.60	$43.60 to $57.10	$21.80 to $31.30

Note: Use the percent of total cost to help identify the correct quality classification.

The range of costs is primarily due to the density of partitions in the office area. Use the lower costs for offices with larger rooms and fewer partitions.

Wall Height Adjustment*	$.97 to $1.40	$.91 to $1.32	$.87 to $1.25	$.79 to $1.20

***Wall Height Adjustment:** Add or subtract the amount listed in this column to or from the square foot cost for each foot when the office wall height is more or less than 8 feet.

External Offices

External offices outside the main building walls have one side that is common to the main building. Square foot costs for external office areas should be estimated with the figures for general office buildings with interior suite entrances. See pages 139 to 142 or 147 to 150.

In selecting a square foot cost for the external office area as well as the main building area, the area of each portion should be used separately for area classification. The area and the perimeter formed by these individual areas is used for shape classification of the main area and for calculating the area and perimeter relationship of the office area. A "no wall ownership" deduction based on the office wall cost should be made for the length of the common side. Wall height adjustments should be made for each area individually.

Steel Buildings

Engineered steel buildings are constructed to serve as warehouses, factories, airplane hangars, garages, and stores, among other uses. They are generally either high-profile (4 in 12 rise roof) or low-profile (1 in 12 rise roof).

The square foot costs for the basic building include: Foundations as required for normal soil conditions. A 4 inch concrete floor with reinforcing mesh and a 2 inch sand fill. A steel building made up of steel frames or bents set 20 or 24 feet on centers, steel roof purlins 4-1/2 to 5-1/2 feet on centers, steel wall girts 3-1/2 to 4-1/2 feet on centers, post and beam type end wall frames, 26-gauge galvanized steel on ends, sides, and roof, window area equal to 2% of the floor area. Basic wiring and minimum lighting fixtures. One small or medium gravity vent per 2,500 square feet of floor area.

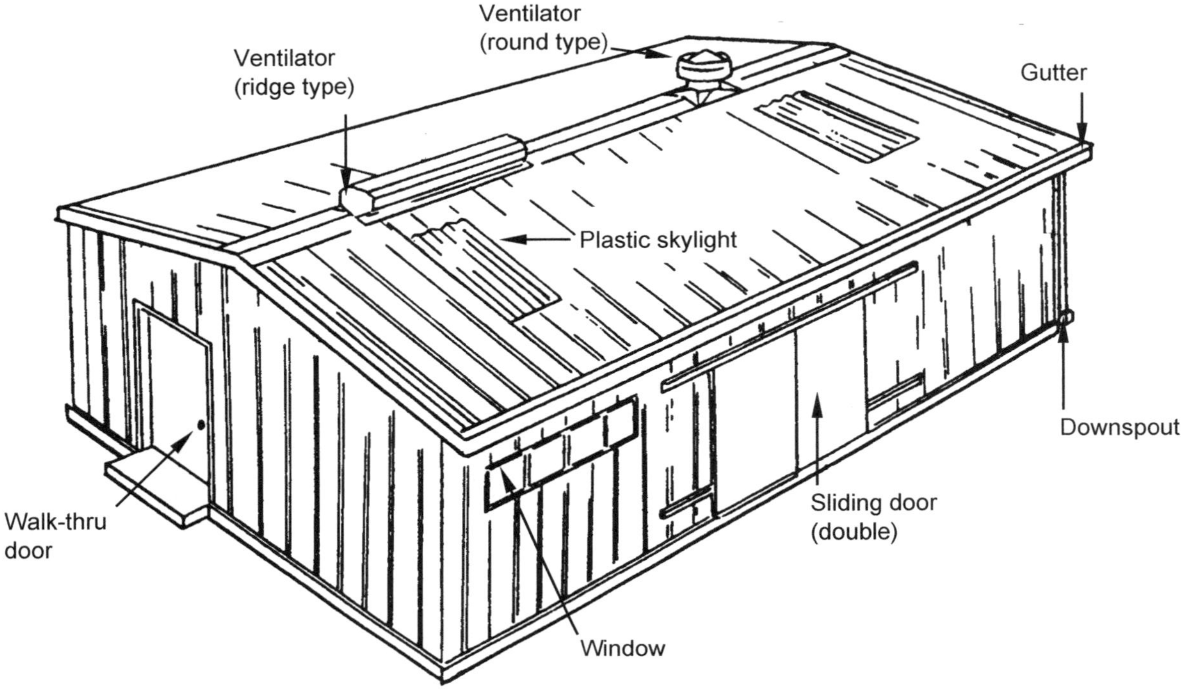

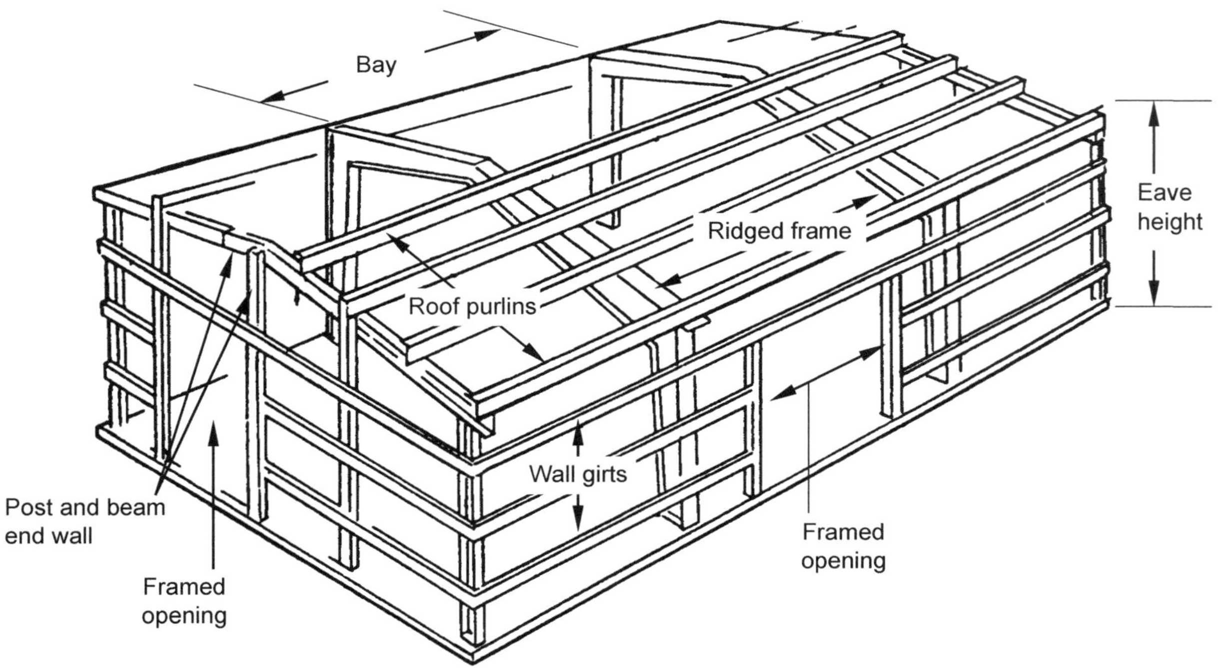

Steel Buildings

Estimating Procedure

1. Compute the building area.
2. Add to or subtract from the square foot cost below $.34 for each foot of wall height at the eave more or less than 14 feet. Subtract $.54 from the square foot cost below for low profile buildings.
3. Multiply the adjusted square foot cost by the building area.
4. Multiply the total by the appropriate live load adjustment. The basic building costs are for a 12 pound live load usually built where snow load is not a design factor. A 20 pound live load building is usually found in light snow areas, and a 30 pound live load building is usually found in heavy snow areas.
5. Add or subtract alternate costs from pages 230 to 232. Alternate costs reflect the difference between unit costs of components that are included in the basic building square foot costs and the unit costs of alternate components. Total alternate cost is found by multiplying the alternate cost by the area or number of alternate components.
6. Add the appropriate costs from pages 233 to 234, "Alternate Costs for Steel Buildings." These costs reflect the in-place cost of components that are not included in the basic square foot cost but are often found as part of steel buildings. The cost of items that alter or replace a portion of the building reflect the net added cost of the component in-place. The cost of the item that is replaced has been deducted from the total cost of the additive components. No further deduction is necessary.
7. Add the cost of heating and air conditioning systems, fire sprinklers, plumbing, additional electrical work, alternate interior partitions and offices, dock height additive costs, loading docks or ramps, paving and curbing, and miscellaneous yard improvements. See the section beginning on page 236.

High Profile Rigid Steel Buildings

Square Foot Cost

	Length			
Width	**60'**	**80'**	**120'**	**Over 140'**
20'	$43.20	$40.29	$39.32	$37.98
30'	40.15	37.90	34.99	32.49
40'	37.04	33.91	32.07	30.85
50' to 70'	31.12	30.52	27.55	26.98
80' to 140'	—	27.39	27.29	24.54

Deduct $.62 per square foot for low profile buildings.

Live Load Adjustment Factors (multiply base cost by factor shown)

	High Profile				Low Profile			
Building	**20 lb. Live Load**		**30 lb. Live Load**		**20 lb. Live Load**		**30 lb. Live Load**	
Width	**20' Bays**	**24' Bays**	**20' Bays**	**24' Bays**	**20' Bays**	**24' Bays**	**20' Bays**	**24' Bays**
20'	1.00	1.01	1.05	1.07	1.00	1.00	1.03	1.04
24'	1.01	1.02	1.06	1.08	1.00	1.00	1.07	1.09
28'	1.01	1.02	1.06	1.08	—	—	—	—
30'	1.01	1.02	1.06	1.08	1.01	1.02	1.09	1.10
32'	1.01	1.02	1.07	1.09	1.01	1.02	1.10	1.11
36'	1.01	1.02	1.08	1.10	1.01	1.02	1.10	1.11
40'	1.02	1.03	1.08	1.11	1.01	1.02	1.10	1.11
50'	1.04	1.06	1.10	1.15	1.02	1.03	1.10	1.12
60'	1.05	1.07	1.12	1.18	1.03	1.04	1.15	1.18
70'	1.07	1.09	1.14	1.20	1.03	1.04	1.15	1.19
80'	1.08	1.10	1.15	1.21	1.04	1.05	1.16	1.20
90'	1.09	1.10	1.16	1.22	1.04	1.05	1.17	1.21
100'	1.09	1.10	1.17	1.23	1.04	1.05	1.17	1.22
110'	1.10	1.11	1.18	1.24	1.05	1.06	1.18	1.24
120'	1.11	1.12	1.19	1.25	1.06	1.07	1.19	1.27
130'	—	—	—	—	1.07	1.08	1.20	1.30
140'	1.13	1.14	1.20	1.27	1.08	1.09	1.22	1.33
150'	—	—	—	—	1.09	1.10	1.24	1.35

Alternate Costs for Steel Buildings

Floors, cost difference per square foot

4" concrete with reinforced mesh and 2" sand fill	0
5" concrete with reinforced mesh and 2" sand fill	+.57
6" concrete with reinforced mesh and 2" sand fill	+1.11
6" concrete with 1/2" rebars, 12" x 12" o.c. and 2" sand fill	+2.20
3" plant mix asphalt on 4" untreated rock base	-.71
2" plant mix asphalt on 4" untreated rock base	-.91
Gravel (compacted)	-1.53
Dirt (compacted)	-2.20

Roof Covering,
cost difference per square foot

26 gauge galvanized steel	0
24 gauge galvanized steel	+.26
22 gauge galvanized steel	+.87
26 gauge colored steel	+.66
24 gauge colored steel	+1.09
.024 aluminum	+.64
.032 aluminum	+1.31
.032 colored aluminum	+1.98
No roof cover (just purlins)	-2.19

Wall Covering,
cost difference per square foot

26 gauge colored steel	+.63
24 gauge colored steel	+1.01
26 gauge galvanized steel	0
24 gauge galvanized steel	+.29
22 gauge galvanized steel	+.85
26 gauge aluminum and steel	+.55
24 gauge aluminum and steel	+.93
.024 aluminum	+.69
.032 aluminum	+1.32
.032 colored aluminum	+2.10
26 gauge galvanized steel insulated panel	+.82
26 gauge colored steel insulated panel	+1.60
26 gauge aluminum steel insulated panel	+1.32
26 gauge galvanized steel mono panel	+1.59
26 gauge colored steel mono panel	+2.47
No wall cover (just girts)	-2.58

These costs are to be added to or subtracted from the building square foot costs on page 217.

Open Endwall (includes ridged frame in lieu of post and beam), low profile building

	Cost Deduction Per Each Endwall Eave Heights						
Width	**10'**	**12'**	**14'**	**16'**	**18'**	**20'**	**24'**
20'	$1,096	$1,321	$1,595	$1,834	—	—	—
24'	1,595	1,667	2,111	2,192	—	—	—
30'	1,900	2,219	2,589	3,111	—	—	—
32'	2,033	2,378	2,827	3,280	3,511	3,954	5,989
36'	2,191	2,511	—	—	—	—	—
40'	2,363	2,778	3,348	3,812	4,157	4,902	6,046
50'	2,589	3,111	3,784	4,373	5,088	5,843	7,053
60'	3,078	3,398	4,102	4,861	6,057	7,067	8,901
70'	—	3,770	4,213	5,524	6,679	7,759	9,855
80'	—	3,959	4,861	6,072	7,309	8,288	11,476
90'	—	—	5,332	5,928	7,732	8,512	12,159
100'	—	—	4,325	5,788	7,268	8,552	12,514
110'	—	—	4,050	5,448	6,834	8,447	12,655
120'	—	—	3,627	4,730	5,967	8,447	12,655
130'	—	—	3,069	4,585	6,284	8,210	12,655
140'	—	—	2,258	3,812	6,099	7,692	12,275
150'	—	—	1,834	3,427	5,207	6,998	11,984

These costs are to be subtracted from the basic building cost.

Open Sidewalls, per linear foot of wall

Eave Height	**10'**	**12'**	**14'**	**16'**	**18'**	**20'**	**24'**
	$5.85	$7.10	$8.54	$9.56	$10.56	$11.52	$14.31

These costs are to be subtracted from the basic building cost.

Alternate Costs for Steel Buildings

Open Endwall (includes ridged frame in lieu of post and beam), high profile building

	Cost Deduction Per Each Endwall – Eave Heights						
Width	10'	12'	14'	16'	18'	20'	24'
20'	1,300	1,378	1,982	—	—	—	—
24'	1,502	2,107	2,443	—	—	—	—
28'	2,219	2,497	2,869	—	—	—	—
30'	2,402	2,656	3,160	—	—	—	—
32'	2,539	2,803	3,320	3,759	4,411	4,942	7,075
36'	2,883	3,124	3,760	—	—	—	—
40'	3,441	3,679	4,250	4,959	5,661	6,208	7,115
50'	4,357	4,800	5,566	6,179	6,692	7,144	9,314
60'	5,679	6,179	7,115	7,694	8,633	9,749	11,754
70'	6,642	8,185	8,380	9,856	10,815	11,895	14,530
80'	7,839	9,274	9,736	10,815	11,895	13,124	15,863
90'	—	11,331	12,833	13,598	13,631	16,178	19,684
100'	—	11,727	13,766	14,900	16,336	17,991	21,596
110'	—	—	17,799	19,826	21,647	23,055	27,113

Ridged Frame Endwall in lieu of post and beam, low profile building

	Cost Deduction Per Each Endwall – Eave Heights						
Width	10'	12'	14'	16'	18'	20'	24'
20'	1,239	1,296	1,425	—	—	—	—
24'	1,307	1,404	1,448	—	—	—	—
28'	1,425	1,554	1,554	—	—	—	—
30'	1,502	1,582	1,519	—	—	—	—
32'	1,582	1,700	1,753	1,979	2,219	2,350	2,724
36'	1,872	1,964	2,044	—	—	—	—
40'	2,059	2,219	2,350	2,497	2,748	2,803	3,466
50'	2,803	2,978	3,189	3,361	3,406	3,734	4,344
60'	3,506	3,760	3,946	4,264	4,465	4,503	5,183
70'	4,860	4,997	5,737	5,514	5,839	5,693	6,282
80'	5,514	5,657	6,282	6,282	6,761	7,235	7,732
90'	—	7,081	7,864	8,318	8,318	7,384	10,529
100'	—	8,750	9,498	9,122	10,569	11,001	12,050
110'	—	—	11,013	12,050	12,062	12,660	14,004

Ridged Frame Endwall in lieu of post and beam, high profile building

	Cost Deduction Per Each Endwall – Eave Heights						
Width	10'	12'	14'	16'	18'	20'	24'
20'	1,443	1,502	1,582	1,582	—	—	—
24'	1,540	1,554	1,582	1,700	—	—	—
30'	1,582	1,620	2,045	—	—	—	—
32'	1,958	1,979	2,100	2,140	2,259	2,286	2,539
36'	2,140	2,219	—	—	—	—	—
40'	2,203	2,350	2,472	2,591	2,724	2,789	3,189
50'	3,056	3,240	3,296	3,453	3,565	3,798	4,397
60'	3,865	4,025	4,250	3,466	4,465	4,637	5,152
70'	—	5,195	5,646	5,809	5,680	6,123	6,758
80'	—	6,282	6,854	7,081	7,319	7,559	7,864
90'	—	6,282	7,535	7,081	7,319	9,598	9,974
100'	—	—	10,578	10,801	11,013	11,199	11,819
110'	—	—	12,991	12,991	13,294	13,476	14,570
120'	—	—	13,567	13,862	14,410	14,716	15,187
130'	—	—	15,195	15,525	16,139	16,482	17,010
140'	—	—	16,996	17,931	18,327	17,613	20,286

Alternate Costs for Steel Buildings

Canopies (sidewall location), cost per linear foot

Type and Size	5'	6'	8'	10'	12'	15'
High profile 20' bays	50.30	57.30	71.80	83.30	91.50	103.60
High profile 24' bays	49.02	55.15	69.04	79.00	86.44	98.92
Low profile 20' bays	39.51	45.42	58.32	69.16	81.30	93.66
Low profile 24' bays	37.97	43.33	55.80	66.20	76.48	88.96
			Cost Per Each Canopy			
Basic end cost	376	456	633	844	1,129	1,460

The above costs are for 26 gauge galvanized metal canopy.

Canopies (endwall location), add the following to sidewall location costs above

Eave Height	10'	12'	14'	16'	18'	20'	24'
Cost	436	500	565	635	707	742	1,015

Sliding Sidewall Doors, cost per opening, including framed opening

Single Doors	10'	12'	14'	16'	18'	20'	24'
8' x 9'	1,901	1,901	1,864	2,109	2,122	2,293	2,293
10' x 9'	1,986	2,122	2,073	2,293	2,367	2,391	2,428
10' x 11'	—	2,073	2,085	2,367	2,428	2,428	2,465
10' x 13'	—	—	2,011	2,391	2,465	2,465	2,502
12' x 9'	2,710	2,809	2,710	3,176	3,079	3,127	3,066
12' x 11'	—	—	2,772	3,104	3,176	3,213	3,213
12' x 13'	—	—	2,710	3,176	3,213	3,238	3,274
12' x 15'	—	—	—	3,165	3,274	3,422	3,507
14' x 9'	3,127	3,042	2,967	3,349	3,422	3,458	3,472
14' x 11'	—	3,311	2,772	3,311	3,458	3,384	3,630
14' x 13'	—	—	2,956	3,507	3,729	3,422	3,691
14' x 15'	—	—	—	3,349	3,703	3,814	3,839
16' x 9'	3,349	3,630	3,593	3,814	3,875	3,875	3,839
16' x 11'	—	3,458	3,630	3,875	3,924	3,961	3,961
16' x 13'	—	—	3,410	3,985	—	—	—

The above costs are for 26 gauge colored sliding. Doors in eave heights listed at the top of each column. If door is in an endwall, use the lowest cost of size desired and add $375

Hollow Metal Walk-Thru Doors, cost per opening including framed opening

	1-3/8" Thick				1-3/4" Thick			
	Flush		Half Glass		Flush		Half Glass	
Door Size	Galvanized	Colored	Galvanized	Colored	Galvanized	Colored	Galvanized	Colored
2'0" x 6'8" single	$957	$1,038	1,178	$1,259	—	—	—	—
2'6" x 6'8" single	987	1,071	1,224	1,300	—	—	—	—
3'0" x 6'8" single	957	1,120	1,221	1,401	1,406	1,518	1,600	1,698
3'0" x 7'0" single	978	1,151	1,257	1,458	1,448	1,560	1,632	1,780
3'4" x 7'0" single	—	—	—	—	1,457	1,641	1,682	1,836
4'0" x 7'0" single	—	—	—	—	1,600	1,836	1,905	2,082
6'0" x 6'8" pair	1,612	1,905	1,974	2,220	2,220	2,524	2,732	2,924
6'0" x 7'0" pair	1,724	2,044	2,056	2,358	2,372	2,647	2,791	3,048
8'0" x 7'0" pair	—	—	—	—	2,538	2,814	3,048	3,433

Half glass doors include glazing.

Additives: When walk-thru door is used as a pilot in a sliding door, **add** $105 to the costs above.

Heavy duty lockset	Add $160 each	Panic hardware, pair	Add $1,530 pair
Door closer	Add $221 each	Weatherstripping, bronze and neoprene	Add $19.00/LF
Panic hardware, single	Add $770 each		

Alternate Costs for Steel Buildings

These costs are to be added to the basic building cost.

Framed Openings, cost per opening

Location	Eave Height 10'	12'	14'	16'	18'	20'	24'
Sidewalls	$541	$557	$586	$610	$803	$902	$1,360
Endwalls	776	810	835	857	934	1,060	1,214

These costs are for the standard widths of 8', 10', 12', 14', 16', 18', and 20'. If the width is other than standard, add $108 per opening. If there is only one opening per building, add $158.

Gutters and Downspouts, cost per linear foot

Eave gutter (4")	
24 gauge galvanized steel	$14.70
24 gauge colored steel	15.43
.032 aluminum	13.67
Valley gutter (6")	
12 gauge galvanized steel	29.29
24 gauge galvanized steel	22.31
Downspouts	
4" x 4" x 26 gauge galvanized steel	10.44
4" x 4" x 26 gauge colored steel	11.08
6" x 6" x 24 gauge galvanized steel	11.58
6" x 6" x 24 gauge colored steel	13.25

Door Hoods, cost each

Width of Door	Galvanized	Colored
12'	$257	$289
16'	334	368
20'	382	413

Louvers, cost each, including screens

Size	Fixed Galvanized	Fixed Color	Adjustable Galvanized	Adjustable Color
3' x 2'	$445	$462	$578	$605
3' x 3'	462	605	627	682
5' x 5'	561	763	860	960

Insulation, cost per square foot of surface area

Size and Type	Roof Thickness 1"	Roof 1-1/2"	Roof 2"	Wall Thickness 1"	Wall 1-1/2"	Wall 2"
.6 pound density						
White vinyl faced	$1.89	$1.98	$2.08	$1.98	$1.98	$2.40
Colored vinyl faced	2.18	2.31	2.46	2.40	2.46	2.81
Aluminum faced	2.40	2.46	2.81	2.60	2.81	2.99
.75 pound density						
White vinyl faced	1.98	2.08	2.31	2.08	2.31	2.46
Colored vinyl faced	2.40	2.40	2.60	2.46	2.60	2.99
Aluminum faced	2.60	2.60	2.99	2.81	2.99	3.17

Overhangs (sidewall location), cost per linear foot of overhang, 26 gauge galvanized

Size and Type	20' Bays 3'	20' Bays 4'	20' Bays 5'	24' Bays 3'	24' Bays 4'	24' Bays 5'
No soffit	34.50	41.82	48.20	31.73	37.86	42.32
Soffit	74.39	87.50	97.00	72.11	82.46	97.95
Color, soffit	83.17	87.50	99.76	74.87	87.50	97.00
Color, no soffit	36.90	45.07	52.40	34.50	41.34	48.55
Color, galvanized soffit	77.88	93.63	102.41	74.87	87.50	99.76
Color, color soffit	80.05	92.79	104.82	115.02	87.50	102.51

Overhangs (endwall location), cost per linear foot of overhang, 26 gauge galvanized

Size and Type	3'	4'	5'	6'
No soffit	22.11	36.66	35.94	42.07
Galvanized soffit	53.60	66.35	78.48	88.09
Color, soffit	66.95	65.15	81.60	91.10
Color, no soffit	24.28	31.26	39.54	47.35
Color, galvanized soffit	55.04	71.39	81.60	93.63
Color, color soffit	58.17	71.39	84.02	91.10

Alternate Costs for Steel Buildings

These costs are to be added to the basic building cost

Skylights, Polycarbonate, with curb

	2' x 2'	4' x 4'	4' x 8'
Single dome	$175	$462	$583
Double dome	208	525	677
Triple dome	258	606	980
Double, ventilating	501	859	1,114

Partitions, Interior, 26 gauge steel, cost per square foot of partition with two sides finished

Painted drywall finish	$4.06
Painted plywood, fire retardant	5.93

Ventilators, round type, includes screen (gravity type), cost each

Diameter	Stationary Galvanized	Stationary Colored	Stationary Aluminum	Rotary Galvanized	Rotary Colored	Rotary Aluminum
12"	$253	$274	$452	$403	$452	$569
16"	353	369	528	504	552	705
20"	425	451	605	615	648	803
24"	459	485	672	686	728	917

Ventilators, ridge type, includes screen and damper

Throat Size	4"	9"	12"	14"
Cost per linear foot, galvanized	$68	$100	$128	$147
Cost per linear foot, colored	70	111	138	158

Ventilator-Dampers, cost each

Diameter	12"	16"	20"	24"
Damper only	$101	$127	$132	$153
Dampers with cords and pulleys	186	233	287	331

Continuous Ridge Ventilator, includes screen and damper, cost per 10 foot unit

Size & Type	First 10' Galvanized	First 10' Color	Each Additional 10' Galvanized	Each Additional 10' Color
9" throat	$968	$1,104	$918	$1,021
10" throat	1,104	1,215	983	1,104
12" throat	707	1,588	1,300	1,390

Steel Sliding Windows, includes glass and screens, cost per window

3' x 2'6"	$470
6' x 2'6"	599
6' x 3'8"	683

Smoke and Heat Vents, automatic control, cost per 10 foot unit

Size & Type	First 10' Galvanized	First 10' Color	Each Additional 10' Galvanized	Each Additional 10' Color
9" ridge mounted	$3,556	$3,759	$2,939	$3,018
9" slope mounted	3,669	3,938	3,018	3,164

Add for operators:

One or two 10 foot sections	Add $117.00
Two to seven 10 foot sections	Add $245.00

Aluminum Industrial Windows, includes glass and screens, cost per window

Size and Type	Project Out	Fixed
3' x 2'6"	$488	$357
2' x 2'8"	522	393
6' x 2'6"	727	478
6' x 3'8"	870	607

Aluminum Sliding Windows, includes glass and screens, cost per window

Width	Height 2'	Height 2'6"	Height 3'	Height 3'6"	Height 4'
2'	$373	$408	$416	$439	$464
3'	408	434	438	464	481
4'	—	455	481	481	535
5'	—	482	500	613	634
6'	—	533	613	660	709

If window is fixed, deduct $6.05 per window. For mullions add $12.10 each.

Commercial, Industrial, and Public Structures

Typical Physical Lives in Years by Quality Class

	Masonry or Concrete						Wood or Wood and Steel Frame				
Building Type	**1**	**2**	**3**	**4**	**5**	**6**	**1**	**2**	**3**	**4**	**5**
Public Buildings	70	70	70	60	60	60	70	60	60	60	–
Urban Stores	70	70	70	60	60	60	70	60	60	60	–
Suburban Stores	70	60	60	60	60	–	60	50	50	45	–
Supermarkets	70	60	60	60	–	–	60	50	50	45	–
Small Food Stores	60	60	60	60	–	–	50	50	45	45	–
Discount Houses	70	60	60	60	–	–	60	50	50	45	–
Banks and Savings Offices	70	70	70	60	60	–	60	60	60	50	50
Department Stores	70	60	60	60	–	–	60	50	50	45	–
General Office Buildings	60	60	60	60	–	–	60	50	50	45	–
Medical-Dental Buildings	60	60	60	50	–	–	60	50	50	45	–
Convalescent Hospitals	60	60	60	50	–	–	55	50	50	45	–
Funeral Homes	70	70	70	60	60	–	60	60	60	50	50
Restaurants	70	60	60	60	–	–	60	50	50	45	–
Theaters	50	50	50	50	–	–	50	45	45	40	–
Service Garages	60	60	50	50	–	–	45	45	40	40	–
Auto Service Centers	50	50	45	45	–	–	–	–	–	–	–
Warehouses	55	55	50	50	–	–	–	–	–	–	–
Light Industrial Buildings	55	55	50	50	–	–	–	–	–	–	–
Factory Buildings	40	40	35	35	–	–	–	–	–	–	–

Service Stations located on main highways or in high land value areas can be expected to become obsolete in 20 years. Other service stations can be expected to become obsolete in 25 years. Reinforced Concrete Department Stores have a typical physical life of 80 years for class one structures and 70 years for lower quality class structures. Steel Buildings have a typical physical life of 50 years

Normal Percent Good Table

Average Life in Years

	20 Years		25 Years		30 Years		35 Years	
	Rem.	%	Rem.	%	Rem.	%	Rem.	%
Age	**Life**	**Good**	**Life**	**Good**	**Life**	**Good**	**Life**	**Good**
0	20	100	25	100	30	100	35	100
1	19	95	24	97	29	98	34	99
2	18	90	23	93	28	96	33	97
3	17	85	22	90	27	93	32	95
4	16	79	21	86	26	90	31	93
5	15	73	20	82	25	88	30	91
6	14	67	19	78	24	85	29	89
7	13	61	18	74	23	82	28	87
8	12	56	17	70	22	79	27	85
9	11	51	16	65	21	75	26	83
10	10	49	15	60	20	72	25	80
11	9	48	14	56	19	68	24	78
12	9	46	13	52	18	65	23	75
13	8	44	12	50	17	61	22	72
14	7	43	11	48	16	58	21	69
15	6	43	10	47	15	54	20	66
16	6	41	9	46	14	50	19	63
17	5	39	8	45	13	49	18	60
18	5	38	8	44	12	48	17	57
19	5	37	7	43	12	47	16	54
20	4	35	7	42	11	47	15	51
21	4	34	6	41	11	46	14	50
22	4	33	6	40	10	45	13	49
23	3	32	5	39	10	44	13	48
24	3	30	5	38	9	43	12	47
25	3	29	5	37	9	43	12	47
26	3	28	4	36	8	42	11	46
27	2	27	4	35	8	41	11	45
28	2	25	4	34	7	40	10	44
29	2	24	4	33	7	39	10	43
30	2	22	3	32	6	38	9	43
31	2	21	3	31	6	37	9	42
32	1	20	3	30	5	36	8	42
33	–	–	3	29	5	35	8	41
34	–	–	3	28	5	35	7	40
35	–	–	2	27	5	34	7	39
36	–	–	2	26	4	33	6	38
38	–	–	2	24	4	32	6	37
40	–	–	2	22	3	30	5	36
42	–	–	1	20	3	28	5	34
45	–	–	–	–	2	26	4	32
48	–	–	–	–	2	23	3	30
52	–	–	–	–	1	20	3	27
56	–	–	–	–	–	–	2	24
62	–	–	–	–	–	–	1	20

	40 Years		45 Years		50 Years		55 Years	
	Rem.	%	Rem.	%	Rem.	%	Rem.	%
Age	**Life**	**Good**	**Life**	**Good**	**Life**	**Good**	**Life**	**Good**
0	40	100	45	100	50	100	55	100
2	38	98	43	99	48	99	53	99
4	36	96	41	97	46	98	51	98
6	34	93	39	95	44	97	49	97
8	32	90	37	93	42	95	47	96
10	30	86	35	90	40	93	45	95
12	28	82	33	87	38	91	43	94
14	26	78	31	84	36	88	41	92
16	24	73	29	81	34	85	39	90
18	22	68	27	77	32	82	37	88
20	20	63	25	73	30	80	35	86
22	18	58	23	69	28	77	33	83
24	17	53	21	65	26	73	31	80
26	15	50	20	60	24	69	29	77
28	14	48	18	55	23	65	27	74
30	13	47	17	50	21	61	26	71
32	11	45	15	49	20	57	24	67
34	10	44	14	48	18	53	22	63
36	9	43	13	47	17	50	21	59
38	8	42	12	46	16	48	19	55
40	8	40	11	44	14	47	18	52
42	7	39	10	43	13	46	17	50
44	6	38	9	42	12	45	16	49
46	6	36	8	41	11	44	15	48
49	5	35	7	40	10	43	14	47
50	5	34	7	38	10	42	13	45
52	4	32	6	37	9	41	12	44
54	4	31	6	36	8	40	11	43
56	3	30	5	35	8	39	10	42
58	3	29	5	34	7	38	9	41
60	3	27	4	32	7	37	9	40
62	2	26	4	31	6	36	8	39
64	2	25	4	30	6	35	8	38
66	2	24	3	29	5	34	7	37
68	2	22	3	28	5	33	7	36
70	2	21	3	27	4	32	6	36
72	1	20	3	25	4	31	6	35
74	–	–	2	24	4	30	5	34
76	–	–	2	23	3	28	5	32
82	–	–	1	20	3	26	4	30
84	–	–	–	–	2	24	4	29
88	–	–	–	–	2	22	3	27
92	–	–	–	–	1	20	2	25
96	–	–	–	–	–	–	2	23
102	–	–	–	–	–	–	1	20

	60 Years		70 Years	
	Rem.	%	Rem.	%
Age	**Life**	**Good**	**Life**	**Good**
0	60	100	70	100
2	58	99	68	99
4	56	99	66	99
6	54	98	64	99
8	52	97	62	98
10	50	96	60	98
12	48	95	58	97
14	46	94	56	96
16	44	93	54	96
18	42	92	52	95
20	40	89	50	94
22	38	87	48	93
24	36	85	46	92
26	34	83	45	91
28	32	81	42	89
30	30	78	40	87
32	29	75	39	85
34	27	72	37	83
36	25	69	35	81
38	24	66	33	79
40	22	63	31	76
42	21	60	30	73
44	20	56	29	70
46	18	52	27	67
48	17	49	26	64
50	16	48	25	61
52	15	47	23	58
54	14	46	22	56
56	13	46	21	54
58	12	45	20	52
60	11	44	19	50
64	10	42	17	48
68	9	40	15	46
72	8	38	13	44
76	7	36	12	43
80	6	35	11	41
86	5	32	9	39
92	4	29	8	36
100	3	25	6	33
108	2	22	4	29
112	1	20	3	27
122	–	–	2	24
130	–	–	1	20

Additional Costs for Commercial, Industrial, and Public Structures

Section Contents

Additional Structure Costs

Basements

Cost includes concrete floor and walls, open ceiling, minimum lighting, no plumbing, and no wall finish. Cost per square foot of floor at 12' wall height.

Area	500	1,000	1,500	2,000	3,000	4,000	5,000	7,500	10,000	15,000	20,000
Cost	70.57	63.21	55.07	49.90	48.60	42.26	40.88	39.49	34.69	33.17	31.19

Add or subtract the amount listed in the table below to or from the square foot of floor cost for each foot of wall height more or less than 12 feet.

Wall Height Adjustment
Square Foot Area

Area	500	1,000	1,500	2,000	3,000	4,000	5,000	7,500	10,000	15,000	20,000
Cost	4.74	3.47	3.08	2.50	2.05	1.88	1.75	1.37	1.12	.87	.81

Canopies, per S.F. of canopy area

Light frame, flat roof underside, plywood and paint or cheap stucco supported by wood or light steel posts, 4" to 6" wood fascia.	$24.80 to $26.40
Average frame, underside of good stucco, flat roof, cantilevered from building or supported by steel posts, 6" to 12" metal fascia.	$27.60 to $37.10
Same as above but with sloping shake or tile roof.	$28.80 to $39.50
Corrugated metal on steel frame.	$24.90 to $37.10

Canopy Lights, per S.F. based on one row of lights for 5' canopy

Recessed spots (1 each 6 linear feet)	$3.65
Single tube fluorescent	6.75
Double tube fluorescent	9.40

Public Address Systems, speakers attached to building. No conduit included.

Base cost, master control	$1054 to $2,027
Per indoor speaker	217
Per outdoor speaker	436

Sound Systems, cost per unit

Voice only, per unit	$111 to $188
Music (add to above), small units	111 to 144
Music (add to above), large units	143 to 433

Larger installations cost the least per unit.

Docks for unloading trucks. Cost per S.F. of dock at 4' height

L x W	10'	20'	30'	50'	100'	200'
5'	41.73	37.42	34.38	31.15	28.86	26.69
10'	37.42	32.87	28.73	25.45	23.80	23.00
15'	33.22	27.82	24.50	21.37	19.76	18.28
20'	29.90	23.80	20.68	19.18	17.81	16.78

Cost includes compacted fill, three concrete walls, concrete floor, and rock base.

Intercommunication Systems

Master control, base cost	$2,296 to $6,808
Cost per station	175 to 260
Nurses call system, per station	259 to 478

Security Systems

Control panel	$178 to $356
Each door or window secured	35 to 79
Heat detectors, each	12 to 59
Smoke detectors, each	23 to 116
Motion detectors, each	23 to 47

Loading Ramps, cost per S.F. of ramp

Size	
Under 300 S.F.	$11.10
Over 300 S.F.	10.60

Dock Levelers and Lifts, cost each

Dock leveler, manual	$8,990
Dock leveler, mechanical	4,410
Powered platform dock leveler	
6' x 6' recessed	3,890
6' x 8' recessed	4,420
Electro-hydraulic, pit recessed scissor lift	
5,000 lb. capacity, 6' x 8'	11,690
10,000 lb. capacity, 8' x 10'	20,700
20,000 lb. capacity, 8' x 12'	34,080

Additional Structure Costs

Doors, with hardware

Exterior, commercial, cost per door			
Glass in wood (3' x 7')	$943	to	$1,474
1/4" plate in aluminum (3' x 7')	1,525	to	2,473
Automatic, tempered glass (3' x 7')	7,253	to	11,023
Residential type (3' x 7')	359	to	660
Interior, commercial and industrial, cost per S.F.			
Hollow core wood	$15.25	to	$15.97
Solid wood	15.35	to	19.88
Hollow core metal	33.48	to	39.63
Fire, cost per S.F.			
Hollow metal, 1-3/4"	$52.82	to	$69.99
Metal clad, rolling	43.23	to	70.80
Metal clad, swinging	61.81	to	88.38

Elevators, Freight, Electric, car and equipment, per shaft, car speed in feet per minute, 2 stop

Capacity	50 to 75	100 to 150	200
2,500 lbs	$74,470	—	—
3,000	78,600	89,900	103,000
3,500	94,000	94,000	106,000
4,000	88,260	98,300	113,400
5,000	94,700	107,000	121,500
6,000	104,300	117,700	131,200
8,000	117,000	131,000	148,000
10,000	135,000	148,000	170,000

For manual doors, **add** $6,100 for each stop. For power operated doors, **add** $9,300 for each additional stop. **Add** $9,390 per car for self-leveling cars. **Add** for double center opening doors, per stop $390. **Add** for deluxe cab (raised panel, interior, drop ceiling) $4,700.

Elevators, Freight, Hydraulic, 100 F.P.M.

Shaft, car, machinery	
2,500 lb. capacity	$69,900
6,000 lb. capacity	118,000
Cost per stop	
Manual doors	11,150
Automatic doors	24,500

Roll-Up Metal Warehouse Door with chain operator, cost each

10' x 10'	$2,677
12' x 12'	3,553
14' x 14'	3,900
Fusible link (add to above)	580
Motor controlled (add to above)	293

Draperies, cost per square yard of opening

	54" high	68" high	96" high
Minimum	$22.90	$23.00	$25.80
Good quality	51.60	52.80	61.80
Better quality	62.90	68.80	84.50

Escalators, cost per flight up or down

Total Rise	32" W	40" W	48" W
10' to 13'	$158,200	$161,800	$170,000
14'	163,000	170,100	184,400
15'	169,000	177,300	191,500
16'	173,700	189,100	193,600
17'	180,800	191,500	195,100
18'	184,400	197,500	195,400
19'	190,300	198,700	196,200
20'	197,400	204,600	202,200
21'	203,400	207,000	207,000

Add for glass side enclosure: $17,000 - $20,000.

Dumbwaiters, includes door, traction type

	1st Two Stops			Add'l. Stops
Hand operated, 25 fpm (no doors)				
25 lb.	$2,420	to	$4,410	$1,840
75 lb.	3,260	to	5,460	1,840
Electrical, with machinery above, floor loading				
100 lb., 50 fpm	$9,100	to	$14,590	$3,470
300 lb., 50 fpm	9,460	to	14,560	3,470
500 lb., 50 fpm	9,950	to	15,770	3,470
500 lb., 100 fpm	14,560	to	23,500	(5 stop)

Elevators, Passenger, Electric, car and machinery cost, per shaft*

Capacity	200 F.P.M., 5 Stops	350 F.P.M., 5 Stops	500 F.P.M., 5 Stops
2,000 lbs.	$155,000	$130,300	$278,000
2,500 lbs.	149,000	138,500	288,000
3,000 lbs.	150,000	147,000	290,000
3,500 lbs.	152,000	155,700	292,000
4,000 lbs.	153,000	165,000	296,000
4,000 lbs (Hospital)	155,000	169,000	303,000

*Add for each additional stop: 200 or 350 F.P.M. units, $8,000; 500 F.P.M. units, $13,200. Deduct for multi-shaft applications, $3,740 to $8,000 per additional shaft. Add for rear-opening door: $11,400 to base cost, plus $8,300 per door.

Elevator Shaft Costs. Add to the car and machinery cost. Costs will be higher in seismic zones and where a high water table exists.

Excavate, form and pour the elevator pit	$15,800
12" CMU wall around the elevator pit	12,600
12" CMU wall around the elevator shaft, 5 floors	39,600
Elevator pit access and ladder	2,100

Additional Structure Costs

Fill, compacted under raised floor, includes perimeter retaining wall but not slab, per C.F.

Up to 10,000 S.F.	$1.16 to $1.87
Over 10,000 to 50,000 S.F	.99 to 1.45

Fire Extinguishers, cost each

Fire hose and cabinet	$410 to $830
Extinguisher cabinets	110 to 240
Extinguishers, chemical	88 to 218
Extinguishers, carbon dioxide	250 to 490

Fire Escapes

Type	Unit	Cost
Second story	Each	$4,222 to $5,767
Additional floors	Per story	2,485 to 3,727

Fire Sprinklers, cost per S.F. of area served

	Wet Pipe System	
Area	**Normal**	**Special***
to 2,000	$5.10	$6.22
2,001 to 4,000	3.53	5.08
4,001 to 10,000	3.13	4.32
Over 10,000	2.75	3.92
	Dry Pipe System	
Area	**Normal**	**Special***
to 2,000	$5.40	$6.54
2,001 to 4,000	3.45	4.67
4,001 to 10,000	3.32	4.62
Over 10,000	3.06	4.12

Costs include normal installation, service lines, permit and valves.

*Special hazard systems are custom engineered to meet code or insurance requirements and are usually so identified by a metal plate attached to the riser.

Overhead Suspended Heaters, per unit

25 MBTU	$1,120 to $1,358
50	1,310 to 1,450
75	1,450 to 1,605
100	1,669 to 1,890
150	2,028 to 2,180
200	2,342 to 2,450
250	2,590 to 2,645

Fireplace

	1 Story	2 Story
Freestanding wood burning heat circulating prefab fireplace, with interior flue, base and cap.	$1,830	—
Zero-clearance, insulated prefab metal fireplace, brick face.	2,510	$3,430
5' base, common brick, on interior face.	3,460	3,880
6' base, common brick, used brick, face brick or natural stone on interior face with average wood mantle.	5,400	5,700
8' base, common brick, used brick or natural stone on interior face, raised hearth.	7,700	8,400

Electric Heating Units

Baseboard, per linear foot	$15.75 to $31.50
Add for thermostat	42.00
Cable in ceiling, per S.F.	2.38 to 3.06
Wall heaters, per K.W.	52.50 to 105.00

Heating and Cooling Systems

Cost per S.F. of Floor Area**

Type and Use	Heating Only	Heating & Cooling
Elementary schools	$8.42 to $13.07	$15.24 to $23.84
Secondary schools	9.01 to 13.96	16.38 to 25.40
Government offices	14.58 to 22.64	26.56 to 41.11
Libraries	9.97 to 18.18	18.18 to 28.01
Fire stations	8.83 to 16.01	13.87 to 21.56
Urban stores	5.54 to 8.59	10.10 to 15.65
Suburban stores	4.44 to 6.87	8.06 to 12.53
Small food stores	4.71 to 7.33	8.59 to 13.30
Supermarkets	5.52 to 8.52	10.05 to 15.54
Discount houses	4.07 to 6.34	7.31 to 11.55
Bank and savings	7.43 to 11.53	13.47 to 20.88
Department stores	5.21 to 8.07	9.48 to 14.73
Reinforced concrete	6.56 to 10.19	11.98 to 18.54
General offices		
Forced air	6.49 to 10.08	11.83 to 18.35
Hot & chilled water	—	13.68 to 21.23
Medical-Dental		
Forced air	7.06 to 10.98	12.85 to 20.04
Hot & chilled water	—	14.48 to 22.55
Convalescent hospitals		
Forced air	6.56 to 10.19	11.98 to 18.52
Hot & chilled water	—	14.07 to 21.86
Funeral homes	9.46 to 14.62	17.18 to 26.57
Ecclesiastic buildings	7.30 to 11.34	13.28 to 20.44
Restaurants	9.73 to 15.14	17.84 to 27.54
Theaters	6.51 to 10.13	11.89 to 18.41
Industrial buildings	2.81 to 6.99	—
Interior offices	3.20 to 4.48	5.79 to 9.01

**Use the higher figures where more heating and cooling density is required.

Additional Structure Costs

Kitchen Equipment, cost per linear foot of stainless steel fixture

Work tables	$877 to $1,067
Serving fixtures	363 to 2,029

Mezzanines, cost per S.F. of floor

Unfinished (min. lighting and plumbing)	2.75 to 32.60
Store mezzanines	45.10 to 57.20
Office mezzanines (without partitions)	48.90 to 63.50
Office mezzanines (with partitions)	63.50 to 99.00

Costs include floor system, floor finish, stairways, lighting, and partitions where applicable.

Seating, cost per seat space

Theater, economy	$193
Theater, lodge	354
Pews, bench type	94
Pews, seat type	130

Partitions, cost per S.F. of surface

Gypsum on wood frame, (finished both sides) 2" x 4" wood studs, 24" on center with 1/2" gypsum board, taped, textures and painted.	$6.28
Plaster on wood frame (finished both sides) 2" x 4" wood studs, 24" on center with 2 coats plaster over gypsum lath, painted with primer and 1 coat enamel.	$12.20

Pneumatic Tube Systems

Twin tube, two station system	
2-1/4" round, 500 to 1,500 feet	$21,190 to $38,400
3" round, 500 to 1,500 feet	21,800 to 42,060
4" round, 500 to 1,500 feet	22,700 to 48,400
4" x 7" oval, 500 to 1,500 feet	35,800 to 59,700
Automatic System, twin tube, cost per station	
4" round, 500 to 1,500 feet	$26,900 to $35,800
4" x 7" oval, 500 to 1,500 feet	36,100 to 38,700

Skylights, Plastic Rectangular Domes, cost per unit

	Single Plastic Panel		Double Plastic Panel	
Size	Skylight Only	With 4" or 9" Insulated Curb	Skylight Only	With 4" or 9" Insulated Curb
16" x 16"	203	344	246	363
16" x 24"	231	389	278	381
16" x 48"	263	449	363	539
24" x 24"	263	417	389	411
24" x 32"	278	449	425	539
24" x 48"	324	508	449	611
28" x 92"	554	821	876	1,015
32" x 32"	278	449	389	539
32" x 48"	376	522	499	620
32" x 72"	476	740	813	938
39" x 39"	389	522	522	613
39" x 77"	581	821	1,015	1,089
40" x 61"	539	740	813	1,104
48" x 48"	417	611	611	753
48" x 64"	581	911	911	1,089
48" x 72"	674	1,048	1,048	1,231
48" x 92"	882	1,128	1,349	1,714
48" x 122"	1,242	1,489	1,704	1,987
58" x 58"	674	926	1,089	1,252
60" x 72"	820	1,095	1,274	1,457
60" x 92"	1,074	1,372	1,396	1,600
64" x 64"	858	1,089	1,182	1,880
77" x 77"	1,231	1,549	1,987	2,177
94" x 94"	2,188	2,554	3,583	3,973

Triple dome skylights cost about 30% more than double dome skylights.

Additional Structure Costs

Plastic Circular Dome Skylights, cost each

	Single Plastic Panel		Double Plastic Panel		Additives	
Size	4" Curb	9" Curb	4" Curb	9" Curb	Ceiling Dome	Wall Liner
30"	861	879	1,020	1,067	414	414
36"	879	919	1,067	1,148	527	442
48"	1,166	1,275	1,491	1,567	706	527
60"	1,477	1,540	1,982	2,007	1,030	543
72"	2,210	2,186	2,944	3,069	1,314	655
84"	3,031	3,233	4,244	4,471	1,882	774
96"	4,459	4,661	6,212	6,644	2,060	919

The above costs are for single skylights. For three or more, deduct 20%.

Plastic Pyramid Skylights, cost each

		Installed Cost	
Size	Height	2 or Less	3 or More
39" x 39"	34"	$1,370	$1,130
48" x 48"	42"	1,940	1,610
58" x 58"	49"	2,140	2,340

Plastic Continuous Vaulted Skylights, cost per L.F.

Width	Single Panel	Double Panel
16"	$130	$191
20"	138	199
24"	162	227
30"	198	273
36"	213	286
42"	227	298
48"	273	331
54"	286	357
60"	298	380
72"	331	408
84"	394	534

Ventilators, Roof, Power Type, cost each

Throat Dia.	2 or Less	3 or More	Add for Insulated Curb
6"	$616	$610	$110
8"	996	900	155
10"	1,200	1,120	189
12"	1,330	1,300	194
18"	1,450	1,370	208
24"	1,660	1,520	256
30"	2,830	2,580	271
36"	3,050	2,620	278
48"	6,690	5,830	342

Above costs are for a single-speed motor installation. Dampers and bird screens are included. Add: Explosive-proof units, add $420 each. Two-speed motors, add $760 to $1,130 each. Plastic coating, depending on size of unit, add $156 to $237 each.

Plastic Ridge Type Skylights, cost per linear foot

Width*	Single Panel	Double Panel
18"	$295	$435
24"	326	528
30"	405	602
36"	514	698
42"	540	840
48"	557	1,080

*Width is from ridge to curb following slope of roof.

Wire Glass Skylights, Exterior Aluminum Frame, cost each

24" x 48"	$495
24" x 72"	600
24" x 96"	790
48" x 48"	798
48" x 72"	980
48" x 96"	1,200

Ventilators, Roof, Gravity Type, cost each

Throat Dia.	2 or Less	3 or More	Add for Insulated Curb
8"	$269	$253	$131
12"	294	270	146
18"	450	440	206
24"	535	513	221

Heat and Smoke Vents, cost each

Size	Plastic Dome Lid	Aluminum Covered Lid
32" x 32"	$1,120	$1,270
32" x 48"	1,340	1,390
50" x 50"	1,520	1,580
50" x 62"	1,730	1,990
50" x 74"	1,580	2,060
50" x 92"	2,080	2,295
50" x 98"	2,420	2,450
62" x 104"	3,100	3,170
74" x 104"	3,450	3,800

Additional Structure Costs

Walk-In Boxes, cost per S.F. of floor area

Temperature Range	50	100	200	300	400	500	600
Over 45°	212	144	111	95	91	83	83
25.50° to 45°	237	168	132	112	100	95	91
0° to 25°	244	208	152	132	112	105	99
-25.50° to 0°	306	250	200	167	144	134	123

Cost Includes:
Painted wood exterior facing, insulation as required for temperature, interior plaster, one 4 x 7 door per 300 S.F. of floor area. Costs are based upon 8' exterior wall height. Costs do not include machinery and wiring. Figure refrigeration machinery at $2,300.00 per ton capacity.

Material Handling Systems

Belt type conveyors, 24" wide.	
Horizontal sections, per linear foot	$255
Elevating, descending sections, per flight	417
Mail conveyors, automatic, electronic.	
Horizontal, per linear foot	2,000
Vertical, per 12' floor	24,100
Mail chutes, cost per floor, 5" x 14", aluminum	1,270
Linen chutes, 18 gauge steel, 30" diameter, per 10' floor	2,110
Disinfecting and sanitizing unit, each	850

Display Fronts

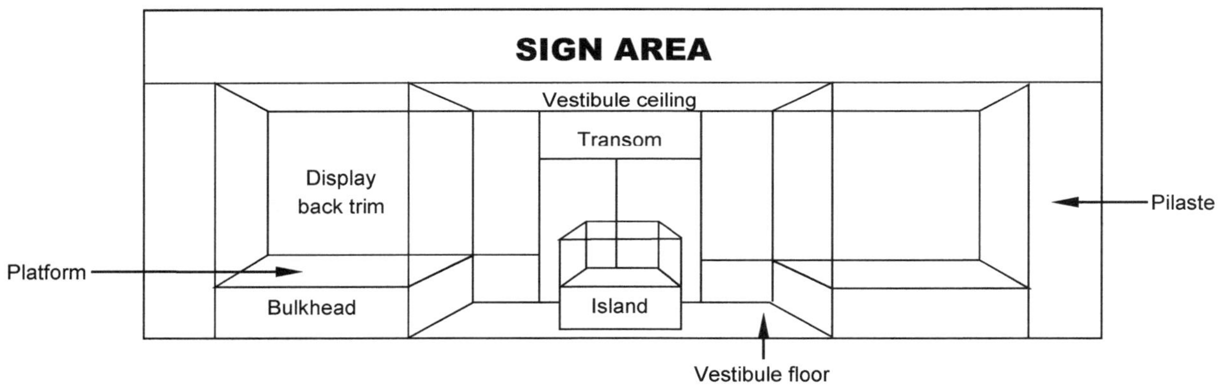

Display front costs may be estimated by calculating the in-place cost of each component or by estimating a cost per linear foot of bulkhead and multiplying by the bulkhead length. This section contains data for both methods. For most fronts, the cost per linear foot method is best suited for rapid preliminary estimates.

Bulkhead length is the distance from the inside of the pilaster and following along the bulkhead or glass to the inside of the opposite pilaster. This measurement includes the distance across entryways.

The cost per linear foot of bulkhead is estimated using the storefront specifications and costs in this section. This manual suggests linear foot costs for four quality types: low cost, average, good, and very good. Costs are related for each quality type in terms of flat or recessed type fronts.

Recessed type fronts include all components described in the specifications. Flat front costs do not include the following components: vestibule floor, vestibule and display area ceiling framing, back trim, display platform, lighting. The cost of automatic door openers is not included in front costs.

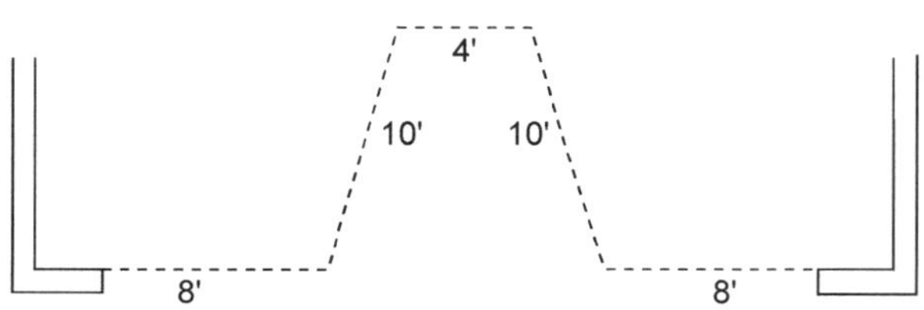

Bulkhead or storefront length 8'+10'+4'+10'+8' = 40'

Display Fronts

Display Front, Best

Display Front, Best

Display Front, Good

Display Front, Good

Display Front, Average

Display Front, Average

Display Front, Low Cost

Display Front, Low Cost

Display Fronts

All front costs are based upon a height of 10 feet from the floor level to the top of the display window. Variations from this standard should be adjusted by using the display window height adjustment costs shown with the front foot costs. These amounts are added or deducted for each foot of variation from the standard.

Bulkhead height variations will not require adjustment. Cost differentials, due to variations in bulkhead height, will be compensated for by equal variations in display window height if overall heights are equal.

Sign areas are based upon 4 foot heights. Cost adjustments are given for flat type and for recessed type fronts. A cost range is given for recessed type fronts because deeply recessed fronts will have lower linear foot costs for sign area components than will moderately recessed fronts because this cost is spread over a longer distance when recesses are deep.

Display island costs are estimated by applying 60 to 80 percent of the applicable linear foot cost to the island bulkhead length. Window height adjustments should be made, but sign height adjustments will not be necessary.

Components	Best	Good	Average	Low Cost
Bulkhead *(0 to 4' high)* (10% of total cost)	Vitrolite domestic marble or stainless steel.	Black Carrera flagstone, terrazzo or good ceramic tile.	Average ceramic tile, Roman brick or imitation flagstone.	Stucco, wood or common brick.
Window Area (30% of total cost)	Bronze or stainless steel. 1/4" float glass with mitered joints.	Heavy aluminum. 1/4" float glass, some mitered joints.	Aluminum. 1/4" float glass.	Light aluminum with wood stops. Crystal or 1/4" float glass.
Sign Area *(4' high)* (10% of total cost)	Vitrolite, domestic marble or stainless steel.	Black Carrera flagstone, terrazzo or good ceramic tile.	Average ceramic tile, Roman brick or imitation flagstone.	Stucco.
Pilasters (5% of total cost)	Vitrolite, domestic marble.	Black Carrera flagstone, terrazzo or good ceramic tile.	Average ceramic tile, Roman brick or imitation flagstone.	Stucco.
Vestibule Floor* (5% of total cost)	Decorative terrazzo.	Decorative terrazzo.	Plain terrazzo.	Concrete.
Vestibule and Display Area Ceilings* (10% of total cost)	Stucco or gypsum wallboard and texture.	Stucco or gypsum wallboard and texture.	Stucco or gypsum wallboard and texture.	Gypsum wallboard and texture.
Back Trim* (5% of total cost)	Hardwood veneer on average frame.	Gypsum wallboard and texture or light frame.	None.	None.
Display Platform Cover* (10% of total cost)	Excellent carpet.	Good carpet.	Average carpet.	Plywood with tile.
Lighting* (10% of total cost)	1 recessed spot per linear foot of bulkhead.	1 recessed spot per linear foot of bulkhead.	1 exposed spot per 2 linear feet of bulkhead.	1 exposed spot per 4 linear feet of bulkhead.
Doors (5% of total cost)	3/4" glass double doors.	Good aluminum and glass double door or single 3/4" glass door.	Average aluminum and glass double door.	Wood and glass.
Note: Use the percent of total cost to help identify the correct quality classification.				
Costs, Flat Fronts	$1,242/linear foot.	$830/linear foot.	$541/linear foot.	$473/linear foot.
Costs, Recessed Fronts	$1,333/linear foot.	$1,134/linear foot.	$645/linear foot.	$523/linear foot.
Display Window *Adjustment per Foot of Height*	$52.30/linear foot.	$48.60/linear foot.	$46.20/linear foot.	$45.20/linear foot.
Flat Front, Sign Area *Adjustment per Foot of Height*	$51.90/linear foot.	$32.70/linear foot.	$12.99/linear foot.	$4.32/linear foot.
Recessed Front, Sign Area *Adjustment per Foot of Height*	$18.30 to $24.35/ linear foot.	$11.20 to $14.28/ linear foot.	$4.88 to 6.98/ linear foot.	$2.59 to $2.87/ linear foot.

*Not included in flat front costs.

Display Fronts

Lighting, cost per fixture

LED strip, 4' single	49.72	to	96.93
Fluorescent exposed, 4' single	38.29	to	48.49
Fluorescent recessed, 4' single	90.59	to	116.80

Bulkhead Walls, cost per S.F. of wall

Up to 5' high, nominal 6" thick			
Concrete	$18.10	to	$26.60
Concrete block	12.61	to	18.80
Wood frame	7.80	to	13.70

Ceiling, cost per S.F. of floor

Dropped ceiling framing	$2.62	to	$3.68
Acoustical tile on wood strips	4.06	to	5.01
Acoustical plaster including lath	3.77	to	5.52
Gypsum board, texture and paint	2.93	to	4.33
Plaster and paint including lath	4.10	to	5.77

Entrances, cost per entrance

Aluminum and 1/4" float glass			
Single door, 3' x 7'	$1,960	to	$3,240
Double door, 6' x 7'	3,470	to	4,890
Stainless steel and 1/4" float glass			
Single door, 3' x 7'	3,870	to	6,080
Double door, 6' x 7'	5,340	to	8,700
3/4" tempered glass			
Single door, 3' x 7'	4,410	to	5,820
Double door, 6' x 7'	6,790	to	8,600

Includes door, glass, lock handles, hinges, sill and frame.

Exterior Wall Finish, cost per S.F. of wall

Aluminum sheet baked enamel finish	$5.35	to	$8.60
Brick veneer			
Common brick	$10.50	to	$13.70
Roman	13.60	to	20.00
Norman	13.60	to	20.10
Glazed	16.60	to	24.90
Carrera glass			
Black	26.12	to	40.10
Red	28.20	to	43.00
Flagstone veneer			
Imitation	17.93	to	23.00
Natural	27.68	to	44.05
Marble			
Plain colors	40.67	to	53.50
With color variations	74.26	to	87.20
Stucco	4.11	to	5.30
Terrazzo	26.96	to	36.80
Tile, ceramic	17.05	to	24.90

Display Platforms,
cost per S.F. of platform area

Framing up to 5' high	$8.10	to	$12.60
Hardwood cover	4.92	to	8.00
Plywood cover	3.30	to	4.52

Glass and Window Frames,
per S.F. of glass

Glass only installed			
1/4" float	$10.02	to	$13.86
1/4" float, tempered	13.20	to	17.72
1/4" float, colored	10.24	to	17.72
Store front, 1/4" glass in aluminum frame			
Anodized, 8' high	33.24	to	45.47
Anodized, 6' high	39.54	to	53.41
Anodized, 3' high	49.50	to	61.90
Satin bronze, 6' high	48.52	to	60.00
Satin bronze, 3' high	56.76	to	70.10

Satellite Receiver Systems

Satellite receiver systems are common in mountain and rural areas, where TV reception is limited. They are also often installed in residential or commercial areas, for homes, motels or hotels, restaurants and businesses. Installed cost for an all-automatic, motorized system, including wiring to one interior outlet.

Exterior disk plus electronics	$230	to	$610

Signs

Lighted Display Signs, Cost per S.F. of sign area

Painted sheet metal with floodlights	$102.00 to $131.00
Porcelain enamel with floodlights	109.00 to 147.00
Plastic with interior lights	123.30 to 192.00
Simple rectangular neon with painted sheet metal faces and a moderate amount of plain letters	130.00 to 235.00
Round or irregular neon with porcelain enamel faces and more elaborate lettering	192.00 to 271.00
Channel letters - individual neon illuminated metal letters with translucent plastic faces, (per upright inch, per letter)	16.40 to 24.60

All of the above sign costs are for double-faced signs. Use 2/3 of those costs for single-faced signs. Sign costs include costs of installation and normal electrical hookup. They do not include the cost of a post. If signs are mounted on separate posts, post mounting costs must be added. These costs are for custom-built signs (one-at-a-time orders). Mass-produced signs will have lower costs.

Post Mounting Costs for Signs

Horizontally Mounted — Vertically Mounted — Cantilever

Post Mounting Costs

Post Height	Pole Diameter at Base 4"	6"	8"	10"	12"	14"
15	$1,481	$1,762	$2,594	$3,495	$5,440	$5,495
20	1,755	2,094	2,998	3,625	6,472	7,279
25	2,012	2,376	3,106	4,313	7,199	7,950
30	2,192	2,816	3,222	4,708	7,606	8,801
35	—	3,063	3,625	5,252	8,605	9,398
40	—	3,251	4,427	5,554	8,950	10,543
45	—	—	5,184	6,325	9,769	10,913
50	—	—	5,554	7,211	10,398	11,958
55	—	—	—	7,607	10,907	13,019
60	—	—	—	8,049	11,707	13,616
65	—	—	—	—	13,684	14,470

If signs are mounted on separate posts, post mounting costs must be added. Post mounting costs include the installed cost of the post and foundation. On horizontally mounted signs, post height is the distance from the ground to the bottom of the sign. On vertically mounted signs, post height is the distance to the top of the post. For cantilevered posts, use one and one-half to two times the conventional post cost.

All of the above post costs are for single posts. Use 90% of the single post costs for each additional post.

Signs

If signs are mounted on buildings or canopies and if, because of the extra weight of the sign, extra heavy support posts or foundations are required, 125% of the post mounting cost should be used.

Sample calculations:

For example, the cost of a 4' x 25' plastic sign mounted on a 15' by 6" post shared by an adjacent canopy might be estimated as follows:

Sign cost 100 S.F. at $100.00	$10,000
Post cost $1,678 x 1/2	881
Total cost	**$10,881**

If this sign were mounted on a 8" post 20' high above the canopy with extra supports not needed, the cost might be estimated as follows:

Sign cost	$10,000
Post cost $2,855 x 1	2,998
Total cost	**$12,998**

Rotators, cost per sign for rotating mount

Small signs	Less than 50 S.F.	$3,190 to $3,840
Medium signs	50 to 100 S.F.	3,350 to 6,830
Large signs	100 to 200 S.F.	3,440 to 12,400
Extra large signs	Over 200 S.F.	$70.07 per S.F. of sign area

Yard Improvements

Bumpers, pre-cast concrete or good quality painted wood.

Typical 4-foot lengths $13.80 to $20.50 per linear foot.

Asphaltic Concrete Paving, cost per square foot

	Under 1,000	**1,000 to 2,000**	**2,000 to 5,000**	**5,000 to 12,000**	**12,000 or more**
2", no base	$4.70	$2.32	$2.00	$1.74	$1.57
3", no base	5.18	2.95	2.44	2.22	2.07
4", no base	5.68	3.61	3.09	2.95	2.41
2", 4" base	8.00	4.52	3.72	3.45	3.06
3", 6" base	8.76	5.28	4.52	4.20	3.82

Concrete Paving, small quantities, per S.F.

4" concrete (without wire mesh) no base	$4.70
4" concrete (without wire mesh) 4" base	5.32
5" concrete (with wire mesh) no base	5.98
6" concrete (with wire mesh) no base	5.45
Add for color	.66

Gates, coin or card operated

Single gate, 10' arm	$4,100 to $6,830
Lane spikes, 6' wide	1,646 to 2,060

Striping

Parking spaces		
Single line between spaces	$ 8.70	per space
Double line between spaces	13.50	per space
Pavement line marking, 4"	.49	per LF

Curbs, per linear foot

Asphalt 6" high berm	$8.30 to $9.95
Concrete 6" wide x 12" high	13.40 to 16.75
Concrete 6" wide x 18" high	17.20 to 21.06
Wood bumper rail 6" x 6"	14.40 to 16.50

Yard Improvements

Lighting

Fluorescent arm-type fixtures

	One arm unit, pole height			Two arm unit, pole height			
	12 ft.	**16 ft.**	**30 ft.**	**12 ft.**	**16 ft.**	**30 ft.**	**Add for 3-fixture unit**
2 tube, 60"	$2,067	$2,247	—	$2,622	$2,647	—	$1,046
4 tube, 48"	2,103	2,485	—	2,715	3,155	—	1,223
4 tube, 72"	2,390	2,647	3,043	3,260	3,185	3,653	1,406
6 tube, 72"	2,647	2,866	3,276	3,260	3,453	4,201	1,651
4 tube, 96"	2,756	2,968	3,440	3,390	3,653	4,192	1,650
6 tube, 96"	2,853	3,153	3,680	3,455	4,028	4,626	1,764

Mercury vapor

	1-fixture unit, pole height			2-fixture unit, pole height			4-fixture unit, pole height		
	12 ft.	**16 ft.**	**30 ft.**	**12 ft.**	**16 ft.**	**30 ft.**	**12 ft.**	**16 ft.**	**30 ft.**
400 watt	$3,564	$3,706	$4,996	$5,015	$5,375	$6,626	$9,007	$9,924	$11,268
1,000 watt	—	4,660	5,206	—	6,287	7,921	—	—	13,020

Mall and garden lighting

	1-fixture unit, pole height		
	3 ft.*	**5 - 7 ft.**	**10 ft.**
LED 200 watt	$993	$1,454	$1,540
Mercury vapor 100-250 watt	1,040	1,710	3,590

*Area lights usually found in service stations.

Chain Link Fences, 9 gauge 2" mesh with top rail line posts 10' on center, per linear foot.

Height	Under 150 LF	150 to 1,000 LF	1,000 to 2,000 LF	Over 2,000 LF	3 strands barbed wire	Add for line posts 8' or less o.c.	Deduct For: No top rail	Deduct For: #11 gauge wire
4'	$23.57	$19.60	$18.29	$17.70	$1.89	$2.64	$3.99	$2.17
5'	24.86	20.65	19.29	18.67	1.89	3.00	3.99	2.37
6'	26.04	21.49	20.53	19.55	1.89	3.13	3.99	2.40
8'	28.31	23.60	22.35	21.14	1.89	3.38	3.99	2.64
10'	32.69	27.03	25.18	24.59	1.89	3.61	3.99	3.03
12'	37.31	31.39	29.74	28.24	1.89	3.99	3.99	3.47

Add for gates, per square foot of frame area $8.70 to $11.10

Add for gate hardware per walkway gate $39.20 to $48.80
per driveway gate $76.30 to $93.90

Drainage, per linear foot, runs to 50'

6" non-reinforced concrete pipe	$9.00 to $10.40
8" non-reinforced concrete pipe	12.18 to 14.40
10" non-reinforced concrete pipe	12.21 to 25.10
12" non-reinforced concrete pipe	16.60 to 31.00

The cost varies with the depth of the pipe in the ground.

Drainage Items

Size and Type	Cost	
Catch basin 4' x 4' x 4' deep	$2,815	ea.
For each additional foot in depth, add	151	ea.
Drop inlets	670 to 1,150	ea.
Manholes 4' diameter x 6' deep	2,060	ea.
For each additional foot in depth, add	191	per ft.

Wood Fences, per linear foot

4' high Redwood board	$20.90
6' high Redwood board	26.80
4' basketweave	27.40
6' basketweave	27.80
4' board and batt	15.00
6' board and batt	15.00
3' split rail	7.98
4' split rail	9.45
3' two rail Redwood picket	12.18
5' three rail Redwood picket	14.80
Typical 3' wide gate, 4' to 6' high	141.80

Agricultural Structures Section

Section Contents

General Purpose Barns

Quality Classification

General Purpose Barn, Class 1

General Purpose Barn, Class 3

Component	Class 1 Good Quality	Class 2 Average Quality	Class 3 Low Quality
Foundation (20% of total cost)	Continuous concrete.	Concrete or masonry piers.	Redwood or cedar mudsills.
Floor (5% of total cost)	Concrete.	Dirt, leveled & compacted.	Dirt, leveled & compacted.
Wall Structure (25% of total cost)	Good wood frame, 10' eave height.	Average wood frame, 10' eave height.	Light wood frame, 10' eave height.
Exterior Wall Cover (25% of total cost)	Good wood siding, painted.	Standard gauge corrugated iron, aluminum or average wood siding.	Light aluminum or low cost boards.
Roof Construction (9% of total cost)	Medium to high pitch, good wood trusses.	Medium to high pitch, average wood trusses.	Medium to high pitch, 2" x 4" rafters 24" to 36" o.c. or light wood trusses.
Roof Cover (5% of total cost)	Wood shingles.	Standard gauge corrugated iron or aluminum.	Light aluminum.
Electrical (8% of total cost)	Four outlets per 1,000 S.F.	Two outlets per 1,000 S.F.	None.
Plumbing (3% of total cost)	Two cold water outlets.	One cold water outlet.	None.

Note: Use the percent of total cost to help identify the correct quality classification.

Square Foot Area

Quality Class	1,000	2,000	3,000	4,000	5,000	6,000	7,000	8,000	9,000	10,000	11,000
1, Good	58.92	54.28	50.26	48.26	46.29	44.23	42.81	41.84	41.01	40.21	39.51
2, Average	43.73	39.46	36.55	35.11	33.73	32.52	31.58	31.00	30.48	29.97	29.48
3, Low	28.51	24.62	22.87	21.89	21.23	20.80	20.37	20.13	19.85	19.69	19.47

Hay Storage Barns

Quality Classification

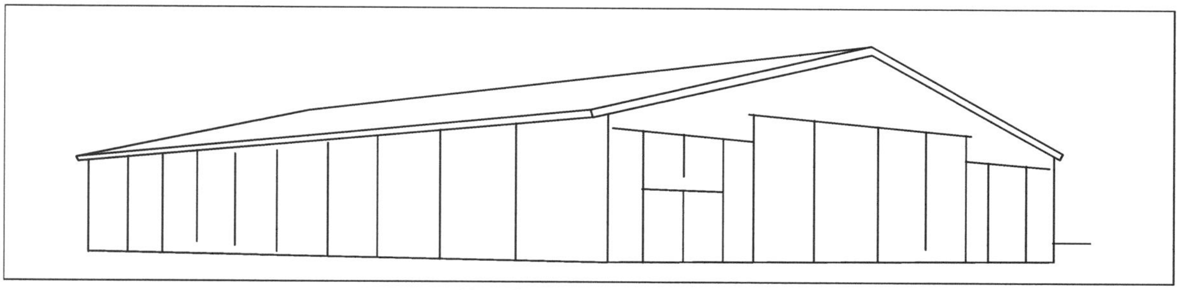

Hay Storage Barn, Class 2

Length between one and two times width

Component	Class 1 Good Quality	Class 2 Average Quality	Class 3 Low Quality
Foundation (25% of total cost)	Continuous concrete.	Concrete or masonry piers.	Redwood or cedar mudsills.
Floor (5% of total cost)	Concrete.	Dirt, leveled & compacted.	Dirt, leveled & compacted.
Wall Structure (25% of total cost)	Good wood frame, 20' eave height.	Average wood frame, 20' eave height.	Light wood frame, 20' eave height.
Exterior Wall Cover (30% of total cost)	Good wood siding, painted.	Standard gauge corrugated iron or aluminum.	Light aluminum or low cost boards.
Roof Construction (10% of total cost)	Low to medium pitch, good wood trusses.	Low to medium pitch, average wood trusses.	Low to medium pitch, 2" x 4" rafters 24" to 36" o.c. or light wood trusses.
Roof Cover (5% of total cost)	Wood shingles.	Standard gauge corrugated iron or aluminum.	Light aluminum.
Electrical	None.	None.	None.
Plumbing	None.	None.	None.

Note: Use the percent of total cost to help identify the correct quality classification.

Square Foot Area

Quality Class	1,000	2,000	3,000	4,000	5,000	6,000	7,000	8,000	9,000	10,000	11,000
1, Good	45.74	41.17	38.14	35.74	34.28	33.01	32.16	31.32	30.72	30.08	29.45
2, Average	28.10	25.15	23.30	22.05	21.22	20.48	19.75	19.09	18.68	18.31	18.10
3, Low	24.42	21.92	20.48	19.39	18.49	17.76	17.21	16.81	16.54	16.02	15.66

Feed Barns

Quality Classification

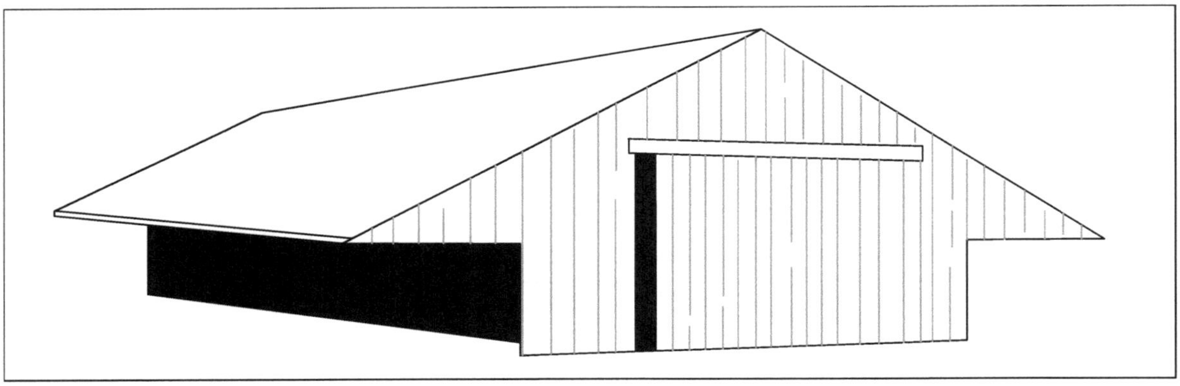

Feed Barn, Class 2

Component	Class 1 Good Quality	Class 2 Average Quality	Class 3 Low Quality
Foundation (20% of total cost)	Continuous concrete.	Concrete or masonry piers.	Redwood or cedar mudsills.
Floor (5% of total cost)	Concrete in center section.	Concrete in center section.	Dirt.
Wall Structure (25% of total cost)	Good wood frame, 8' eave height at drip line.	Average wood frame, 8' eave height at drip line.	Light wood frame, 8' eave eight at drip line.
Exterior Wall Cover (25% of total cost)	Open sides, good siding painted on ends.	Open sides, standard gauge corrugated iron, aluminum or average wood siding on ends.	Open sides and ends.
Roof Construction (9% of total cost)	Medium to low pitch, good wood trusses.	Medium to low pitch, average wood trusses.	Medium to low pitch, 2" x 4" rafters 24" to 36" o.c. or light wood trusses.
Roof Cover (5% of total cost)	Wood shingles.	Standard gauge corrugated iron or aluminum.	Light gauge corrugated iron or aluminum.
Electrical (7% of total cost)	Four outlets per 1,000 S.F.	Two outlets per 1,000 S.F.	None.
Plumbing (4% of total cost)	Two cold water outlets.	One cold water outlet.	None.

Note: Use the percent of total cost to help identify the correct quality classification.

Square Foot Area

Quality Class	1,000	2,000	3,000	4,000	5,000	6,000	7,000	8,000	9,000	10,000	11,000
1, Good	28.70	27.14	26.44	26.02	25.70	25.47	25.28	25.16	25.08	25.03	24.99
2, Average	24.76	23.51	22.77	22.26	21.92	21.70	21.63	21.53	21.40	21.35	21.31
3, Low	15.66	14.97	14.41	13.99	13.77	13.68	13.59	13.54	13.48	13.45	13.40

Shop Buildings

Quality Classification

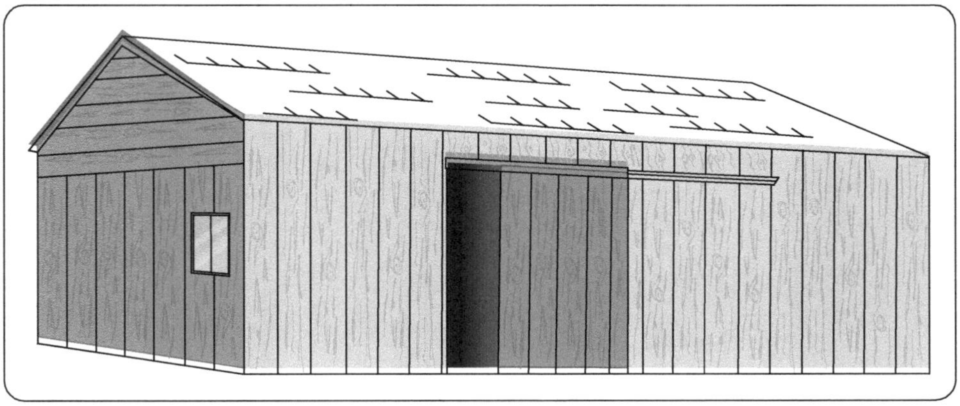

Shop, Class 2

Length between one and two times width

Component	Class 1 Good Quality	Class 2 Average Quality	Class 3 Low Quality
Foundation (20% of total cost)	Continuous concrete.	Light concrete.	Light concrete.
Floor (5% of total cost)	Concrete.	Concrete.	Concrete.
Wall Structure (20% of total cost)	Good wood frame, 15' eave height.	Average wood frame, 15' eave height.	Light wood frame, 15' eave height.
Exterior Wall Cover (25% of total cost)	Good wood siding, painted.	Standard gauge corrugated iron, aluminum or average wood siding.	Light aluminum or low cost boards.
Roof Construction (5% of total cost)	Low to medium pitch, good wood trusses.	Low to medium pitch, average wood trusses.	Low to medium pitch, 2" x 4" rafters 24" to 36" o.c. or light wood trusses.
Roof Cover (5% of total cost)	Wood shingles.	Standard gauge corrugated iron or aluminum.	Light gauge corrugated iron or aluminum.
Electrical (5% of total cost)	Four outlets per 1,000 S.F.	Two outlets per 1,000 S.F.	Two outlets per 1,000 S.F.
Plumbing (7% of total cost)	Two cold water outlets.	One cold water outlet.	None.
Doors (5% of total cost)	One drive-thru door per 1,000 S.F. plus one walk-thru door.	One average sliding or swinging door per 2,000 S.F.	One light sliding or swinging door per 2,000 S.F.
Windows (3% of total cost)	Five percent of floor area.	None or few low cost.	None.

Note: Use the percent of total cost to help identify the correct quality classification.

Square Foot Area

Quality Class	1,000	1,500	2,000	2,500	3,000	4,000	5,000	6,000	8,000	10,000
1, Good	48.40	48.40	45.84	43.86	41.95	40.65	39.34	38.12	36.78	35.38
2, Average	41.95	38.73	36.13	34.87	33.57	32.05	30.77	30.13	29.47	28.82
3, Low	33.55	30.76	28.82	27.28	26.06	25.46	24.74	23.48	23.45	22.84

Machinery and Equipment Sheds

Quality Classification

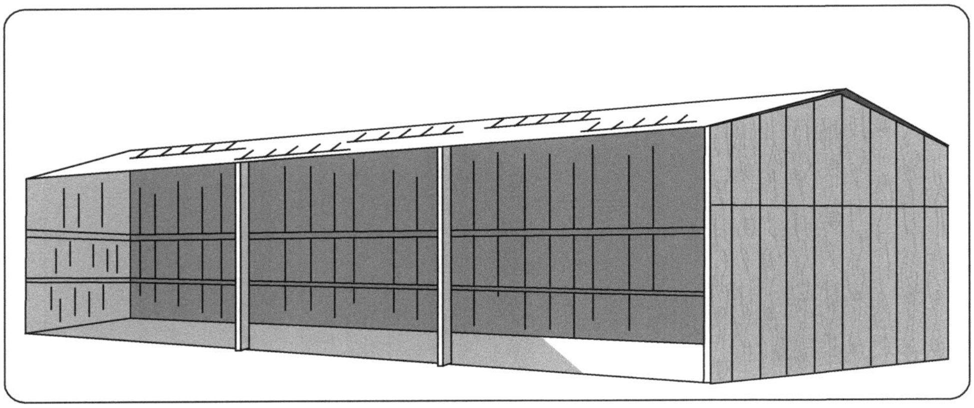

Equipment Shed, Class 3

Usually elongated, width between 15 and 30 feet, any length

Component	Class 1 Good Quality	Class 2 Average Quality	Class 3 Low Quality
Foundation (22% of total cost)	Continuous concrete.	Concrete or masonry piers.	Redwood or cedar mudsills.
Floor (5% of total cost)	Concrete.	Concrete.	Dirt, leveled & compacted
Wall Structure (25% of total cost)	Good wood frame, 10' eave height.	Average wood frame, 10' eave height.	Light wood frame, 10' eave height.
Exterior Wall Cover (30% of total cost)	Good wood siding, painted.	Standard gauge corrugated iron, aluminum or average wood siding.	Light aluminum or low cost boards.
Roof Construction (10% of total cost)	Low to medium pitch, gable or shed type, good wood framing.	Low to medium pitch, gable or shed type, average wood framing.	Low to medium pitch, shed type, light wood framing.
Roof Cover (5% of total cost)	Wood shingles.	Standard gauge corrugated iron or aluminum.	Light aluminum.
Electrical (3% of total cost)	Four outlets per 1,000 S.F.	Two outlets per 1,000 S.F.	None.

Note: Use the percent of total cost to help identify the correct quality classification.

All Sides Closed – Square Foot Area

Quality Class	500	1,000	1,500	2,000	2,500	3,000	3,500	4,000	4,500	5,000	6,000
1, Good	39.83	35.95	33.43	32.80	32.05	32.07	31.70	31.29	31.04	30.82	30.43
2, Average	30.74	26.82	25.51	25.01	24.40	24.40	23.81	23.61	23.51	23.38	23.15
3, Low	20.97	18.80	17.49	16.88	17.17	17.13	16.36	16.22	16.02	15.82	15.63

One Side Open – Square Foot Area

Quality Class	500	1,000	1,500	2,000	2,500	3,000	3,500	4,000	4,500	5,000	6,000
1 Good	32.54	30.65	29.58	28.33	27.56	27.20	26.96	26.68	26.58	26.44	26.33
2, Average	28.13	24.12	22.27	21.63	20.99	20.80	20.48	20.34	20.21	20.01	19.86
3, Low	18.40	15.45	14.53	14.00	13.77	13.61	13.48	13.34	13.20	13.09	13.00

Small Sheds

Quality Classification

Usually elongated, width between 6 and 12 feet, any length

Component	Class 1 Good Quality	Class 2 Average Quality	Class 3 Low Quality
Foundation (25% of total cost)	Continuous concrete.	Concrete or masonry piers.	Redwood or cedar mudsills.
Floor (5% of total cost)	Concrete.	Boards.	Dirt, leveled & compacted.
Wall Structure (25% of total cost)	Good wood frame, 8' eave height.	Average wood frame, 8' eave height.	Light wood frame, 8' eave height.
Exterior Wall Cover (30% of total cost)	Good wood siding, painted.	Standard gauge corrugated iron, aluminum or average wood siding.	Light aluminum or low cost boards.
Roof Construction (10% of total cost)	Low to medium pitch, gable or shed type, good wood framing.	Low to medium pitch, gable or shed type, average wood framing.	Low to medium pitch, shed type, light wood framing.
Roof Cover (5% of total cost)	Wood shingles.	Standard gauge corrugated iron or aluminum.	Light aluminum.
Electrical	None.	None.	None.

Note: Use the percent of total cost to help identify the correct quality classification.

All Sides Closed – Square Foot Area

Quality Class	50	60	80	100	120	150	200	250	300	400	500
1, Good	51.17	45.93	43.86	41.10	38.22	35.34	33.24	31.95	30.43	29.76	29.06
2, Average	41.74	37.50	33.95	31.27	29.74	28.33	27.01	25.56	24.06	23.42	22.71
3, Low	29.74	26.88	24.04	20.55	19.75	18.46	17.73	17.01	18.14	15.62	14.84

One Side Open – Square Foot Area

Quality Class	50	60	80	100	120	150	200	250	300	400	500
1, Good	39.16	35.16	33.95	31.88	29.74	28.33	27.14	25.33	24.10	22.72	22.05
2, Average	31.29	29.06	26.88	25.33	24.10	22.71	21.21	19.75	19.02	18.47	18.24
3, Low	21.17	19.75	18.47	17.01	15.51	14.84	13.77	12.95	12.34	11.52	11.28

Pole Barns

These prices are for pole barns with a low pitch corrugated iron or aluminum covered roof supported by light wood trusses and poles 15' to 20' o.c. The gable end is enclosed and the roof overhangs about 2' on two sides. Wall height is 18 feet. Where sides are enclosed, the wall consists of a light wood frame covered with corrugated metal.

All Sides Open – Side Length

End Width	34	51	68	85	102	119	136	153	170	187
20	12.52	12.07	11.81	11.72	11.63	11.50	11.48	11.48	11.47	11.46
25	11.75	11.30	11.09	10.98	10.94	10.83	10.80	10.78	10.77	10.76
30	11.22	10.78	10.64	10.49	10.44	10.37	10.36	10.32	10.32	10.32
35	10.80	10.44	10.26	10.12	10.06	9.99	9.98	9.97	9.93	9.93
40	10.67	10.26	10.06	9.93	9.89	9.84	9.82	9.73	9.73	9.70
45	10.45	10.06	9.89	9.73	9.67	9.64	9.61	9.59	9.56	9.56
50	10.21	9.83	9.61	9.55	9.45	9.44	9.39	9.38	9.38	9.36
60	10.17	9.76	9.56	9.45	9.40	9.38	9.36	9.34	9.29	9.29
70	10.06	9.64	9.50	9.40	9.36	9.29	9.28	9.25	9.23	9.23
80	9.99	9.61	9.44	9.36	9.28	9.25	9.18	9.18	9.15	9.10

Ends and One Side Closed, One Side Open – Side Length

End Width	34	51	68	85	102	119	136	153	170	187
20	22.58	19.88	18.54	17.83	17.38	17.06	16.78	16.54	16.40	16.33
25	20.65	18.22	17.06	16.33	15.83	15.55	15.39	15.18	15.02	14.88
30	19.39	17.09	15.95	15.36	14.85	14.61	14.41	14.24	14.09	13.98
35	18.39	16.27	15.18	14.55	14.15	13.86	13.70	13.59	13.41	13.30
40	17.72	15.66	14.63	14.01	13.70	13.35	13.18	13.04	12.90	12.84
45	17.26	15.23	14.18	13.68	13.21	13.02	12.81	12.68	12.53	12.47
50	16.78	14.73	13.80	13.21	12.85	12.60	12.47	12.29	12.16	12.12
60	16.33	14.31	13.43	12.85	12.53	12.29	12.12	11.93	11.86	11.76
70	15.95	14.01	13.07	12.57	12.29	12.05	11.80	11.71	11.61	11.48
80	15.58	13.62	12.81	12.32	11.92	11.72	11.54	11.46	11.33	11.22

All Sides Closed – Side Length

End Width	34	51	68	85	102	119	136	153	170	187
20	27.08	24.34	22.96	22.13	21.60	21.20	20.96	20.73	20.58	20.42
25	24.19	21.75	20.49	19.79	19.26	18.95	18.67	18.44	18.33	18.22
30	22.14	19.91	18.83	18.11	17.71	17.42	17.18	17.05	16.87	16.76
35	20.85	18.76	17.66	17.07	16.60	16.30	16.08	15.97	15.80	15.71
40	19.79	17.84	16.79	16.22	15.80	15.55	15.36	15.18	15.02	14.90
45	19.00	17.07	16.07	15.53	15.18	14.85	14.67	14.53	14.42	14.28
50	18.25	16.40	15.50	14.88	14.53	14.29	14.14	13.96	13.84	13.76
60	17.44	15.63	14.73	14.25	13.89	13.68	13.50	13.32	13.21	13.10
70	16.90	15.18	14.30	13.84	13.46	13.21	13.04	12.88	12.81	12.73
80	16.30	14.61	13.84	13.32	12.92	12.75	12.60	12.47	12.38	12.30

Side sheds tying into one side of a pole barn are priced as follows. The shed consists of one row of poles 14' to 16' high, spaced 15' to 20' o.c. A light wood truss covered with a low pitch sheet metal roof spans the distance between the poles and the barn side. If the sides are open, the cost will be between $10.70 and $14.10 per square foot of area covered. If all sides are enclosed with sheet metal and a light wood frame, the square foot cost will be $16.90 to $22.30.

Low Cost Dairy Barns

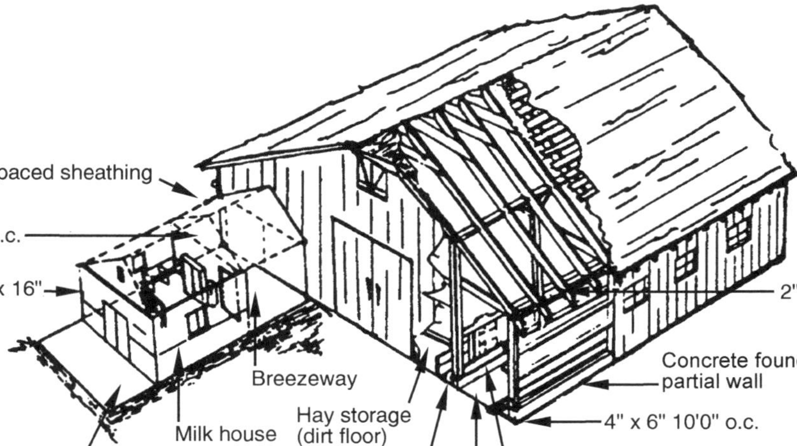

	Milk House	Dairy Barn
Foundation (20% of total cost)	Concrete.	Light concrete.
Floors (15% of total cost)	Concrete slab.	Concrete cow stands.
Walls (30% of total cost)	6" or 8" concrete block 36" high 2" x 4" @16" o.c. framing above.	Box frame, 4" x 6", 10' o.c.
Roof (15% of total cost)	Average wood frame, corrugated iron or aluminum cover.	Average wood frame, wood shingles, corrugated iron or aluminum cover.
Windows (5% of total cost)	Metal sash or metal louvers, on 5% of wall area.	Barn sash.
Interior (10% of total cost)	Smooth finish plaster.	Unfinished, wood stanchions.
Electrical (3% of total cost)	Minimum grade, fair fixtures.	None.
Plumbing (2% of total cost)	One wash basin.	None.
Square Foot Cost	$66.20 to $82.80 (including breezeway).	$35.80 to $39.50 (exclusive of milking equipment).

Stanchion Dairy Barns

Component parts of this dairy:
A Milking barn
B Feed room
C Milk, wash and equipment room

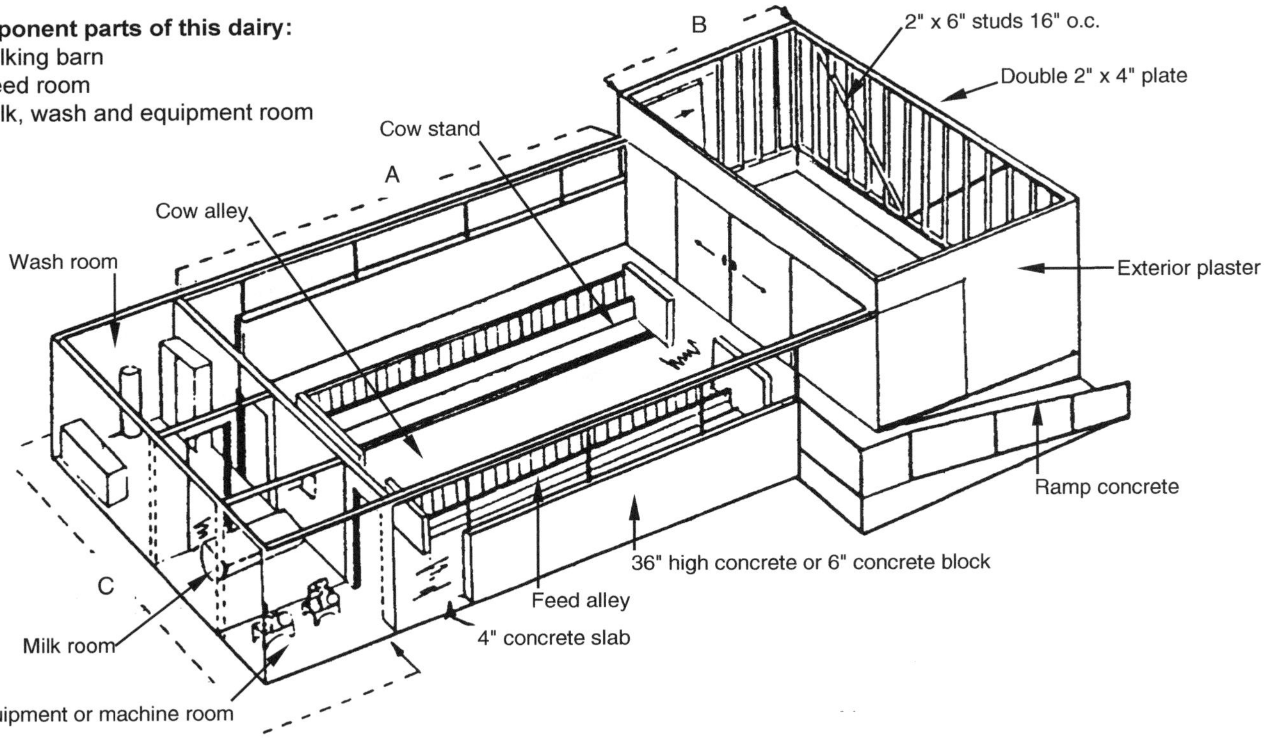

	Feed Room	Milk, Wash and Equipment Room	Milking Barn
Foundation (20% of total cost)	Reinforced concrete.	Reinforced concrete.	Reinforced concrete.
Floors (15% of total cost)	Concrete slab.	Concrete slab.	Concrete, well formed gutters and mangers.
Walls (30% of total cost)	2" x 4" or 2" x 6" @16" o.c. framing.	6" or 8" concrete block 36" high with 2" x 6" @16" o.c. framing above.	6" or 8" concrete block or reinforced concrete 36" high with 2" x 6" @ 16" o.c. framing above.
Roofs (15% of total cost)	Average wood frame, corrugated iron or aluminum cover.	Average wood frame, corrugated iron or aluminum cover.	Average wood frame, iron or aluminum cover.
Windows (5% of total cost)	None.	Metal sash or metal louvers on 10% of wall area.	Metal sash or metal louvers.
Interior (5% of total cost)	Unfinished.	Smooth finish plaster, cove base.	Smooth plaster 36" high, metal stanchions.
Electrical (4% of total cost)	Conduit, average fixtures.	Conduit, average fixtures.	Conduit, average fixtures.
Plumbing (6% of total cost)	None.	One wash basin with floor drains.	Floor drains and hose bibs.
Square Foot Costs	$28.60 to $47.00.	$69.60 to $83.30 (including breezeway).	$51.00 to $59.30 (exclusive of milking equipment).

Walk-Through Dairy Barns

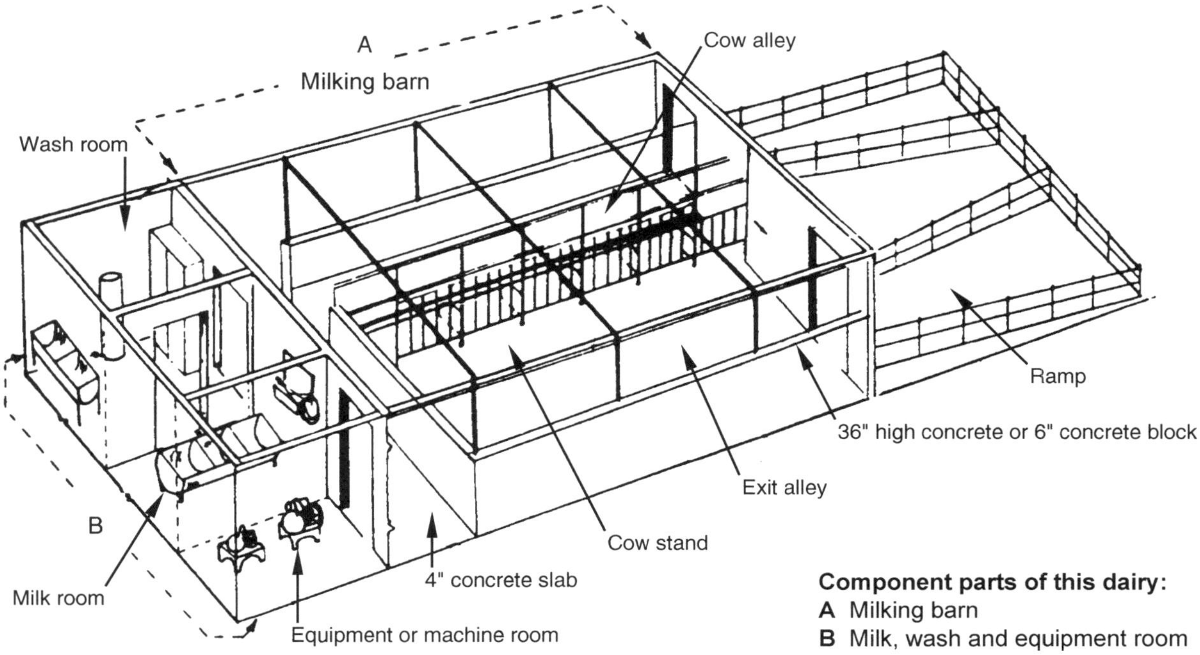

Component parts of this dairy:
A Milking barn
B Milk, wash and equipment room

	Milking Barn	Milk, Wash, and Equipment Room
Foundation (20% of total cost)	Reinforced concrete.	Reinforced concrete.
Floors (15% of total cost)	Concrete, well-formed gutters and mangers.	Concrete slab
Walls (30% of total cost)	6" or 8" concrete block or reinforced concrete 36" high with 2" x 6" @16" o.c. framing above or all concrete block.	6" or 8" concrete block 36" high with 2" x 6" @16" o.c. framing above, or all concrete block.
Roofs (15% of total cost)	Average wood frame, corrugated iron or aluminum cover.	Average wood frame, corru-gated iron or aluminum cover.
Windows (5% of total cost)	Metal sash or metal louvers.	Metal sash or metal louvers on 10% of wall area.
Interior (5% of total cost)	Smooth plaster 36" high, metal stanchions.	Smooth finish plaster, cove base.
Electrical (4% of total cost)	Conduit, average fixtures.	Conduit, average fixtures.
Plumbing (6% of total cost)	Floor drains and hose bibs.	One wash basin, floor drains.
Square Foot Costs	$56.80 to $61.30 (exclusive of milking equipment).	$58.90 to $63.40 (including breezeway).

Modern Herringbone Barns

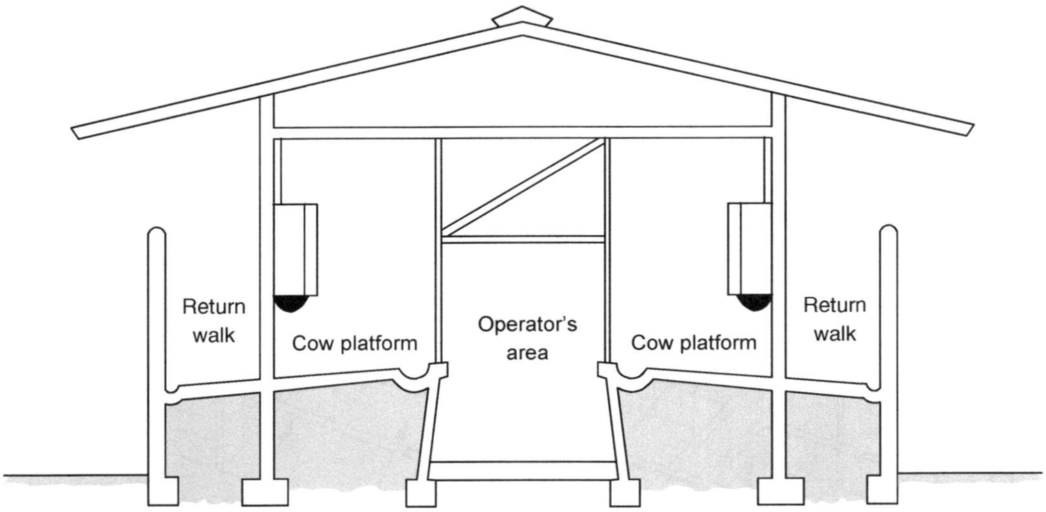

Cross Section

Milking Barn

Foundation (15% of total cost)	6" Reinforced concrete.
Floors (15% of total cost)	Concrete, well-formed gutters and mangers.
Walls (30% of total cost)	6" or 8" concrete block or reinforced concrete 60" high with 2" x 6" @16" o.c. framing above or all concrete block.
Roof (15% of total cost)	Average wood frame, corrugated iron or steel beams. steel roofing, skylights.
Windows (5% of total cost)	Metal sash or metal louvers.
Interior (5% of total cost)	Smooth plaster on entire block walls or tile and plaster.
Electrical (4% of total cost)	Conduit, average fixtures. Not major supply for milking equipment.
Plumbing (6% of total cost)	Floor drains and hose bibs.
Stanchions (5% of total cost)	Metal stanchions.
Cost per SF $58.40	Excluding stalls, gates, feeding system and milking equipment.
Feed Systems & Stalls	Per double ten barn, $62,000

Milk, Wash, and Equipment Room

	Average Quality	**Good Quality**
Foundation (20% of total cost)	Reinforced concrete.	Reinforced concrete.
Floors (15% of total cost)	Concrete slab.	Reinforced concrete slab.
Walls (25% of total cost)	8" concrete block with 2" x 6", 16" o.c. framing above or all concrete block.	8" concrete block with 2" x 6" @16" o.c. framing above or all concrete block.
Exterior (10% of total cost)	Stucco or concrete block.	Stucco and masonry veneer.
Roof (10% of total cost)	Wood frame, corrugated iron or aluminum.	Wood frame, or steel beams, steel roofing or tile, skylights, and gutters.
Windows (5% of total cost)	Metal sash on 10% of wall area.	Tile floors and walls in many areas.
Interior (5% of total cost)	Smooth finish plaster, cove base.	Smooth finish plaster, cove base.
Electrical (4% of total cost)	Conduit, average fixtures.	Conduit, better lighting and ample outlets.
Plumbing (6% of total cost)	One wash basin, one water closet, one lavatory, floor drains.	One wash basin, one water closet, ¾ bath, vinyl floors, floor drains.
Square Foot Costs	$78.90 to $88.70	$88.70 to $99.40

Note: Use the percent of total cost to help identify the correct quality classification.

Miscellaneous Dairy Costs

Holding Corral and Wash Area Costs

Components	Cost
Floor or Ramp	Sloping concrete with abrasive finish, $4.70 to $5.30 per square foot.
Wall	5' to 6' high plastered interior, $47.70 to $67.20 per linear foot.
Metal rail fence	Welded pipe, posts 10' o.c. in concrete, top rail and 3 cables, $14.70 to $17.30 per linear foot.
Cable fence	Pipe posts 10' o.c. set in concrete, $11.90 + $.84 per linear foot per cable.
Gates	54" high, pipe with necessary bracing, $202 to $242 ea.
Sprinklers	Hooded rainbird, $204 to $255 each, including plumbing and pump.
Roof	Pipe column supports, average wood frame, corrugated iron or aluminum cover, open sides, $6.80 to $9.40 per square foot.
Typical wash area (without sprinkler)	Sloped concrete floors, 5' concrete block exterior walls, welded pipe interior fences and gates, usual floor drains $32.30/S.F. with roof, $22.50/S.F. without roof.

Milk Line, 3" stainless steel

40-cow conventional barn	$30,590 total $767 per cow
Double ten herringbone barn	$18,400 total $923 per cow
Dual milk pump, add	$3,162
In-line filter, add	$3,445
In-line cold plate, add	$7,724 to $11,230

Refrigeration Compressors

7-1/2 HP	$8,110
10 HP	10,670
15 HP	13,370

Corral Costs

Components	Cost
4" concrete flatwork	$4.60 to 6.90 per square foot.
6" concrete flatwork	$4.80 to 7.20 per square foot.
8" curb	$11.50 per linear foot.
Cable fence	$10.40 per linear foot.
Water tank, concrete typical	$690.00 each.
Steel stanchions	$70.00 each. $30.90 to $33.40 per linear foot.
Steel lockable stanchions	$66.70 to $71.50 each. $35.30 to $39.00 per linear foot.
Light standards	$3,260 each.
Sump pumps, each	
3 HP	$3,620
4 HP	$4,660
10 HP	$7,840
Hay shelters	$7.65 to $8.60 per square foot.
Loafing sheds	$6.70 to $8.40 per square foot.
Typical 80-cow corral	$40,700 to $46,000 total. $505 to $570 per cow.

Refrigerated Holding Tanks

1,530 gallon	$40,400
2,040 gallon	45,500
2,550 gallon	48,800
3,060 gallon	52,900
3,570 gallon	59,480
4,080 gallon	62,200
5,100 gallon	66,100

Vacuum Pumps (air pumps)

Units	Cost
8	$5,650
12	9,600
20	15,600

Poultry Houses

Conventional Lay Cage Type

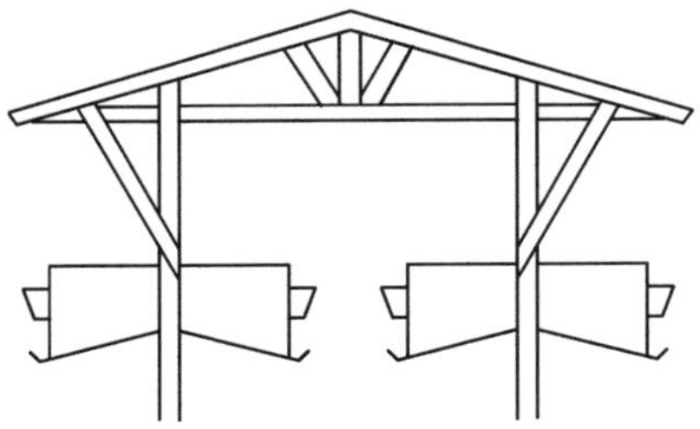

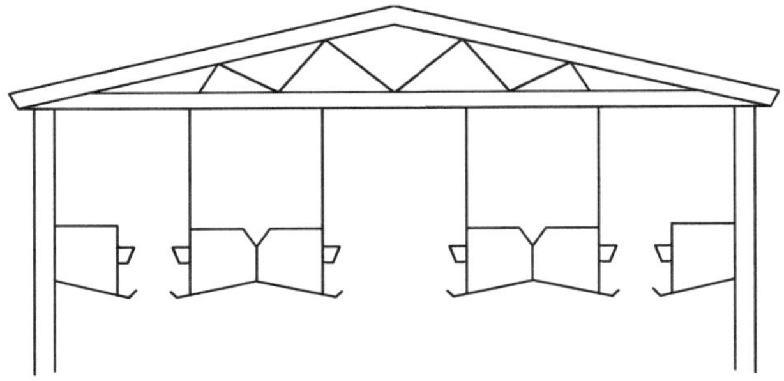

Basic Building

Component	Good Quality	Average Quality	Low Quality
Floors (20% of total cost)	2" concrete.	Dirt with 4' concrete walkways.	Dirt, leveled and compacted.
Foundations (15% of total cost)	Thickened slab.	Concrete piers.	Wood piers.
Frame (20% of total cost)	Light steel or average wood frame.	Average wood frame.	Light wood frame.
Roof Cover (5% of total cost)	Aluminum or corrugated iron.	Light aluminum or composition.	Light aluminum or composition.
Exterior (8% of total cost)	Plywood.	Vinyl curtains.	Wood lath.
Lighting (20% of total cost)	Good system, automatic controls.	Average system, automatic controls.	Minimum system, manual controls.
Plumbing (10% of total cost)	Good system.	Average system.	Fair system.
Insulation (2% of total cost)	Roof only.	None.	None.
Basic Building Cost Per S.F.	$21.00 to $25.20	$13.80 to $17.70	$11.90 to $14.00

Note: Use the percent of total cost to help identify the correct quality classification.

Equipment

Component	Best Quality	Good Quality	Average Quality	Low Quality
Cages (35% of total cost)	12" x 20" double deck.	12" x 20" single deck.	12" x 20" single deck.	12" x 12" single deck.
Water System (20% of total cost)	Automatic cup system.	Automatic cup system.	Simple "V" trough.	Simple "V" trough.
Feed System (30% of total cost)	Automatic system.	"V" trough.	"V" trough.	"V" trough.
Egg Gathering	Manual.	Manual.	Manual.	Manual.
Cooling (15% of total cost)	Pad and fan system.	Pad and fan system.	Simple fogging system.	Simple fogging system.
Cost Per S.F.	$30.70 to $31.40	$19.30 to $22.50	$13.40 to $17.20	$11.40 to $16.00

Note: Use the percent of total cost to help identify the correct quality classification.

Poultry Houses

Modern Controlled Environment Type

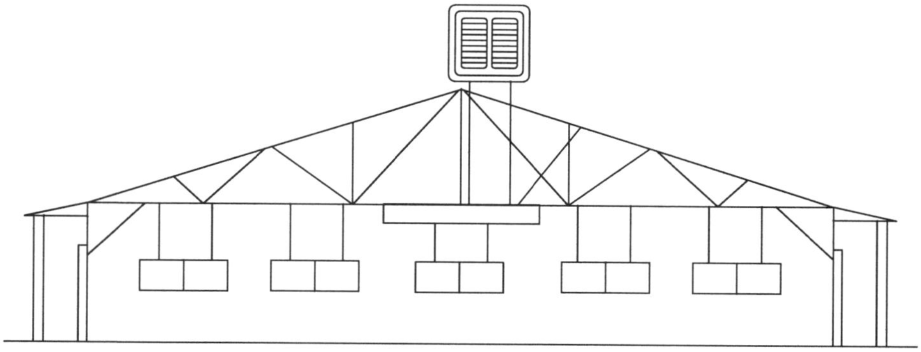

Basic Building	
Foundation (20% of total cost)	Concrete.
Floor (15% of total cost)	3-1/2 concrete slab.
Wall Frame (20% of total cost)	2" x 4" @ 24" o.c.
Roof & Cover (10% of total cost)	Wood trusses with 2" x 4" purlins @24" o.c., corrugated iron or aluminum cover
Exterior (18% of total cost)	Two rib aluminum or corrugated iron.
Interior (7% of total cost)	4" fiberglass roll with aluminum foil facing or 3/4" insulation.
Lighting (5% of total cost)	Fluorescent or LED system.
Plumbing (5% of total cost)	Good basic system.
Basic Building Cost Per S.F.	$42.30 to $46.40

Component	**Equipment Single Deck**	**Stair Step**
Cages (45% of total cost)	12" x 20" single deck.	12" x 20" double deck.
Watering System (30% of total cost)	Automatic cup system.	Automatic cup system.
Feeding System	Manual.	Manual.
Egg Gathering System	Manual.	Manual.
Cooling (25% of total cost)	Evaporative coolers.	Evaporative coolers.
Heating	None.	None.
Building & Equipment Square Foot Cost	$48.00 to $54.90	$52.00 to $58.00

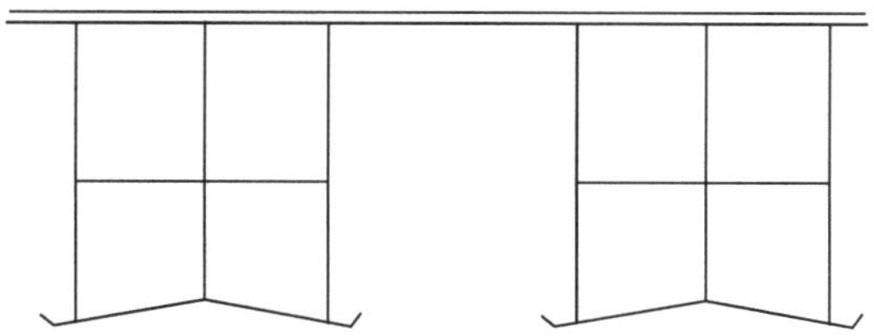

Single-deck cage system

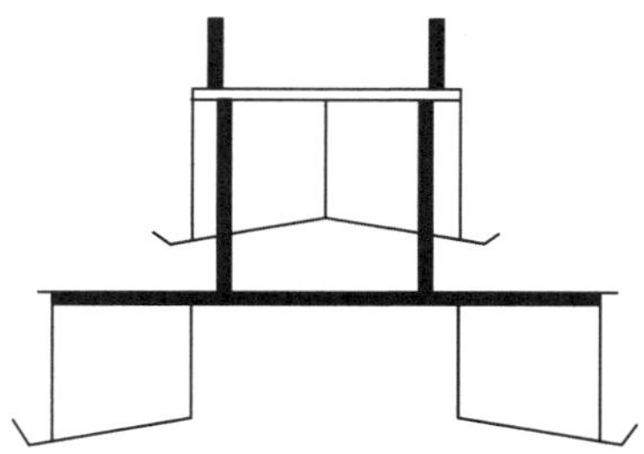

Double-deck cage system

Poultry Houses

High Rise Type

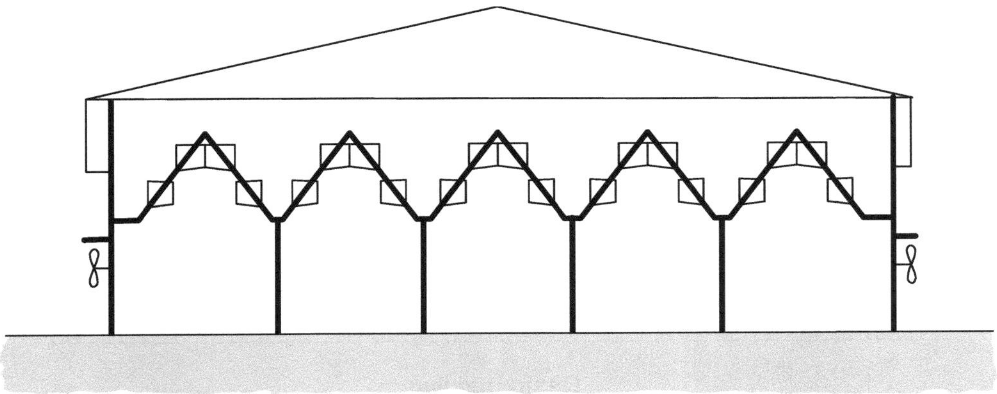

Basic Building

Foundation (25% of total cost)	Concrete piers.
Floors (5% of total cost)	Dirt, leveled and compacted.
Wall Frame (20% of total cost)	2" x 4" @24" o.c.
Roof & Cover (15% of total cost)	Wood trusses with 2" x 4" purlins @24" o.c., corrugated iron or aluminum cover.
Exterior (18% of total cost)	Two rib aluminum or corrugated iron.
Interior (7% of total cost)	3/4" insulation.
Lighting (5% of total cost)	Fluorescent or LED system.
Plumbing (5% of total cost)	Good basic system.
Basic Building Cost Per S.F.	$77.70 to $85.70

Equipment

Component	Flat Deck	Stair Step
Cages (33% of total cost)	12" x 20".	12" x 20".
Watering System (20% of total cost)	Automatic cup system.	Automatic cup system.
Feeding (25% of total cost)	Automatic system.	Automatic system.
Egg Gathering (15% of total cost)	Automatic system.	Automatic system.
Cooling (7% of total cost)	Negative pressure system.	Negative pressure system.
Heating	None.	None.
Building & Equipment Square Foot Cost	$91.50 to $102.80	$98.00 to $111.00

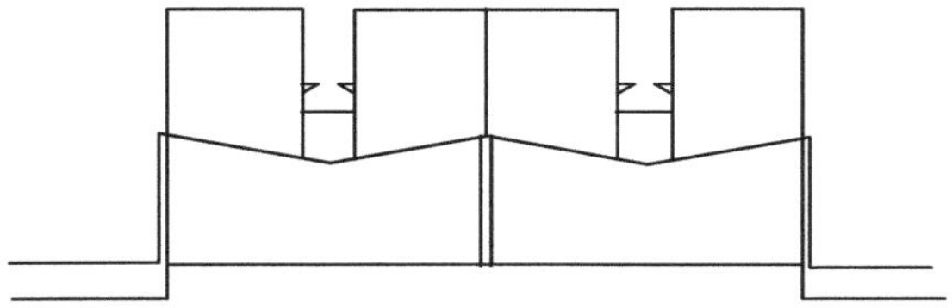

Flat-deck cage system

Poultry Houses

Deep Pit Type

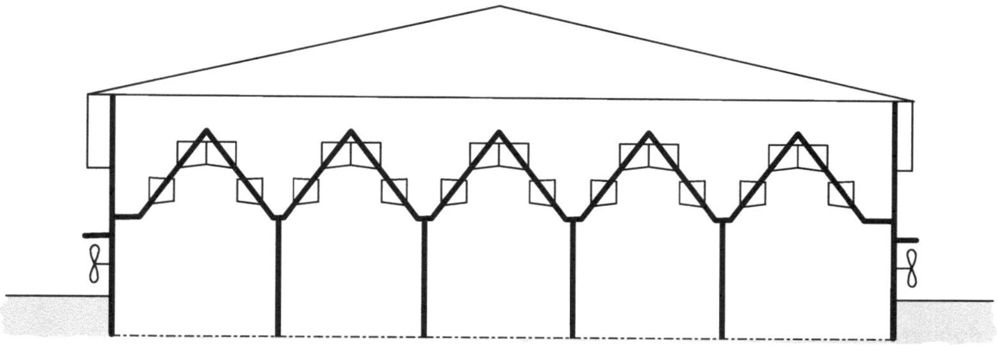

Basic Building

Foundation (25% of total cost)	Concrete piers.
Floors (15% of total cost)	Concrete with waterproof membrane.
Wall Frame (15% of total cost)	2" x 4" @24" o.c.
Roof & Cover (15% of total cost)	Wood trusses with 2" x 4" purlins @24" o.c., corrugated iron aluminum cover.
Exterior (15% of total cost)	Two rib aluminum or corrugated iron.
Interior (5% of total cost)	3/4" insulation.
Lighting (5% of total cost)	Fluorescent or LED system.
Plumbing (5% of total cost)	Good basic system.
Basic Building Cost Per S.F.	$50.30 to $53.00

Equipment

Component	Flat Deck	Stair Step
Cages (33% of total cost)	12" x 20".	12" x 20".
Watering (20% of total cost)	Automatic cup system.	Automatic cup system.
Feeding (25% of total cost)	Automatic system.	Automatic system.
Egg Gathering (15% of total cost)	Automatic system.	Automatic system.
Cooling (7% of total cost)	Negative pressure system.	Negative pressure system.
Heating	None.	None.
Building & Equipment Square Foot Cost	$53.90 to $57.70	$55.70 to $60.20

Equipment Costs

Add these costs to the basic building cost

Component	Serving One Row of Cages	Serving Two Rows of Cages
Automatic feeders	$2.74 per bird	$1.59 per bird
Automatic egg gathering	2.13 per bird	1.18 per bird
Automatic water cup system	12.18 per bird	.99 per bird
	6.60 per cup	4.30 per cup
"V" water trough	.52 per bird	.38 per bird
16" feed trough	.62 per bird	.40 per bird

Foggers

1/2" galvanized pipe	$3.00/linear foot
3/4" galvanized pipe	3.18/linear foot
1" galvanized pipe	3.31/linear foot

Roof sprinklers

$3.39 per linear foot

Evaporative coolers

$970 each. $2.45 per S.F. of building

Fans

36"	$500 each
48"	650 each
50"	720 each

Negative pressure air conditioning system

$2.00 to $2.57 per S.F. of building.

Cooling pads in walls

$1.16 per S.F. of surface.

Heating systems

$2,790 per unit.

Cages, 12" x 20" or 18"

$10.30 each. $2.66 per bird.

Greenhouses

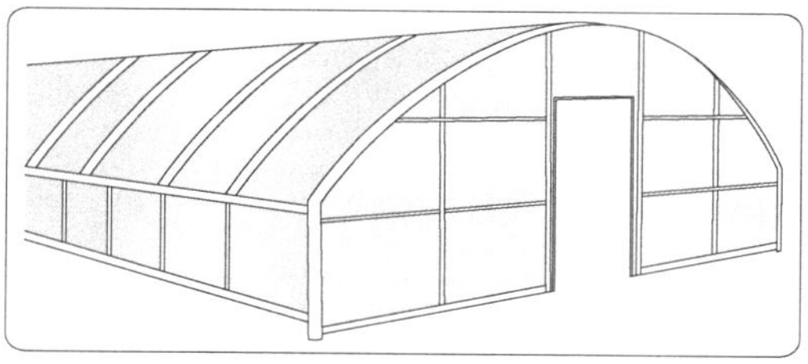

Greenhouse, Class 3

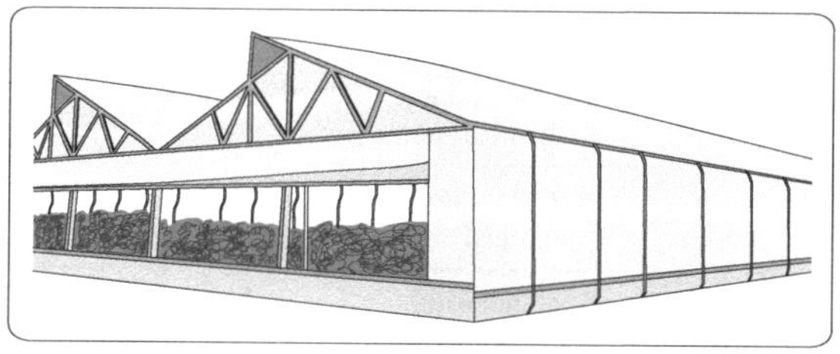

Greenhouse, Class 1

Quality Classification

Component	Class 1 High Quality	Class 2 Average Quality	Class 3 Low Quality
Wall and roof	Heavy steel frame, 8' wall, glass or multi-wall polycarbonate cover.	Galvanized steel frame, 8' wall, double polycarbonate or fiberglass cover.	Light pipe, 4' wall, single light polyethylene cover, fiberglass ends.
Floor	Finished concrete walks, concrete foundation.	Gravel, some plain concrete walks.	Dirt, some gravel.
Interior	Good lighting, running water, roof vents, and exhaust fans.	Average lighting, water, and roof vents.	Light wood frame, 10' eave height.

Note: Use the percent of total cost to help identify the correct quality classification.

Square Foot Area

Quality Class	5,000	7,500	10,000	15,000	20,000	25,000	30,000	35,000	40,000	45,000	50,000
High	35.61	34.35	33.21	31.13	29.16	28.13	27.34	26.56	25.80	25.31	25.19
Average	26.76	25.95	25.06	22.85	21.30	20.60	20.14	19.53	19.34	18.95	18.68
Low	6.30	5.95	5.65	5.51	5.46	5.38	5.27	4.89	4.57	4.37	4.20

Migrant Worker Housing

Quality Classification

	Class 1 Best Quality	Class 2 Prefabricated	Class 3 Good Quality	Class 4 Average Quality	Class 5 Low Quality
Slab Foundation (15% of total cost)	Spread footing around perimeter and thickened slab at partitions.	Spread footing around perimeter.	Thickened around perimeter.	Thickened around perimeter.	Thickened around perimeter.
Floor (10% of total cost)	4" concrete slab reinforced.	4" concrete slab reinforced.	4" concrete slab.	4" concrete slab.	4" concrete slab.
Walls (20% of total cost)	Masonry exterior walls, wood frame interior partitions and ceiling.	Metal building, prefabricated.	2" x 4" studs @16" o.c., 2" x 4" stud partitions.	Box construction 4" x 4" at 48" o.c.	Box construction 2" x 4" at 48" o.c.
Exterior Cover (15% of total cost)	Natural blocks.	Prefabricated metal building.	Average grade red-wood board and batt or horizontal siding or stucco finish.	Fair grade redwood or fir board and batt or horizontal board.	Poor grade of red-wood or fir, vertical or horizontal.
Interior Finish (15% of total cost)	Gypsum wallboard.	None.	Gypsum wallboard.	Plywood partitions.	None.
Roof Framing (10% of total cost)	Rafters, collar beams and ceiling joists.	Prefabricated metal building.	Rafters, collar beams and ceiling joists.	Very simple truss.	Rafters and tie at plate line.
Roofing (5% of total cost)	Composition shingles.	Metal.	Aluminum or wood.	Composition or sheet metal.	Composition or used sheet metal.
Doors (3% of total cost)	1 metal door and metal frame each room.	Metal doors.	1 average door each room.	3 or 4 average doors.	2 or 3 cheap doors.
Windows (2% of total cost)	1 steel sash or aluminum window in each room.	Approximately 20% of floor area.	1 steel or aluminum window in each room.	1 window each room.	Few and small.
Electrical (5% of total cost)	1 good light and plug in each room.	1 light, 1 plug for each 300 S.F.	1 light, 1 plug in each room.	1 pull chain light and plug for each 400 S.F.	1 pull chain light for each 500 S.F.

Note: Use the percent of total cost to help identify the correct quality classification.

Migrant Worker Housing, Class 1

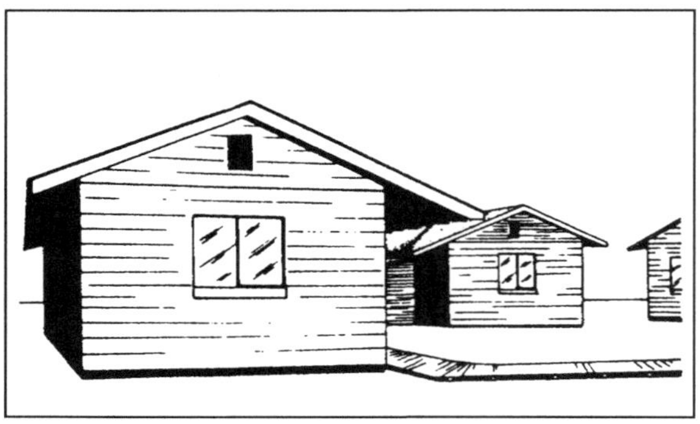

Migrant Worker Housing, Class 4

Square Foot Area

Quality Class	400	600	800	1,000	1,200	1,500	2,000	2,500	3,000
1, Best	77.47	71.60	68.39	66.44	64.97	63.49	61.93	60.92	60.18
2, Prefabricated	71.01	65.52	62.65	60.77	59.49	58.09	56.69	55.74	55.14
3, Good	66.35	61.31	58.59	56.77	55.58	54.31	52.97	52.06	51.51
4, Average	55.84	51.63	49.31	47.80	46.78	45.72	44.55	43.73	43.25
5, Low	45.93	42.42	40.61	38.99	38.47	37.66	36.77	36.12	35.70

Costs do not include any plumbing. Add $970 to $1,140 per fixture.

Miscellaneous Agricultural Structures

Livestock Scales

Type	Size	Capacity	In-Place
Full-capacity beam	16' x 8'	5 ton	$12,470
Printing beam	16' x 8'	5 ton	15,600
Full-capacity beam	22' x 8'	10 ton	16,500
Printing beam	22' x 8'	10 ton	19,950

Additional Costs for Livestock Scales

Types and Size	Cost
Each foot arm is removed from scale	$150/L.F.
Angle iron stock rack for 16' x 8' scale	4,080 ea.
Angle iron stock rack (wood) for 16' x 8'	606 ea.
Angle iron stock rack for 22' x 8' scale	7,110 ea.

Scale pit has 4" concrete walls and slab poured in place. May be poured in or on top of the ground. If on top, compacted ramps and steps to the scale beam are included.

Motor Truck Scales

Five inch reinforced concrete platform. All-steel structure and scale mechanism. Reinforced concrete pit. Motor truck scales are of two general types, the beam type (either manual or type registering) and the full-automated dial type. The construction of both, insofar as the weight-carrying mechanism is concerned is very similar. The method of recording and weight capacity make the cost vary.

Capacity	Platform Size	Total Cost
20 tons	24' x 10'	$25,850
30 tons	34' x 10'	32,670
40 tons	40' x 10'	35,950
50 tons	45' x 10'	40,950
50 tons	50' x 10'	42,700
50 tons	60' x 10'	48,800
50 tons	70' x 10'	53,100

Above costs are for full-capacity beam scales.
Add $860 for the registering type beam.

Septic Tanks

2 bedroom home with 1,200 gallon tank	$2,905
3 bedroom home with 1,500 gallon tank	3,080
4 bedroom home with 2,000 gallon tank	3,570

Bulk Feed Tanks

Size and Type	Cost
5 Ton	$2,674
9 Ton	3,773
10.5 Ton	4,023
13 Ton	4,430
15 Ton	5,212
20 Ton	6,640
25 Ton	7,251
31 Ton	8,361
34 Ton	8,758
40 Ton	10,118
45 Ton	11,602
60 Ton	12,995

Tanks are equipped with a scissor-type opening chute

Domestic Water Systems. Submersible pump, installed at 105' depth. Add the costs below for well (with casing) and pressure tank (if installed.)

	Typical Installation		
	1/2 HP	**3/4 HP**	**1 HP**
Total cost	$2,405	$2,675	$3,140
Pressure tank size	82 gal.	82 gal.	120 gal.
Cost per ft. above or below 105' depth	$2.97	$3.33	$3.72
	1-1/2 HP	**2 HP**	**3 HP**
Total cost	$3,585	$4,555	$5,495
Pressure tank size	220 gal.	220 gal.	315 gal.
Cost per ft. above or below 105' depth	$3.88	$5.59	$7.54

6" wells average $37.47 per foot of depth
8" wells average $45.25 per foot or depth.

Pressure Tank Sizes and Installed Costs

42 gal. 16" dia. x 48" depth	$338 to $ 466
82 gal. 20" dia. x 60" depth	437 to 554
120 gal. 24" dia. x 60" depth	560 to 830
220 gal. 30" dia. x 72" depth	1,360 to 1,550
315 gal. 36" dia. x 72" depth	1,920 to 2,170
525 gal. 36" dia. x 120" depth	2,480 to 2,790

Typical Physical Lives in years for agricultural structures

Building Type	Good	Average	Low
Barns	40	30	20
Dairy barns	25	25	—
Dairy barns, low cost	—	20	20
Storage sheds	40	30	20
Poultry houses, modern	30	25	—
Poultry houses, conventional	—	25	20

To determine the useful life remaining, use the percent good table for residential structures on page 43.

Military Construction Costs

The Office of the Under Secretary of Defense prepared the following square foot guidelines to reflect the cost of military construction for fiscal 2025. The costs are based on construction of permanent facilities built on military bases worldwide. Use the "Construction Cost Indices" at the end of this section to adapt the square foot costs to any other area. The "Size Cost Adjustment Chart" should be used to determine the approximate cost of a structure larger or smaller than the typical size shown.

Included in these costs are all items of equipment which are permanently built-in or attached to the structure, including items with fixed utility connections.

The costs include items such as the following:

- Furniture, cabinets and shelving, built-in.
- Venetian blinds and shades.
- Window screens and screen doors.
- Elevators and escalators.
- Drinking water coolers.
- Telephone, fire alarm and intercom systems.
- Theater seats.
- Pneumatic tube systems.
- Heating, ventilating and air conditioning installations.
- Electrical generators and auxiliary gear.
- Waste disposers such as incinerators.
- Food preparation & serving equipment, built-in.
- Raised flooring.
- Hoods and vents.
- Chapel pews and pulpit.
- Refrigerators, built-in.
- Laboratory furniture, built-in.
- Cranes and hoists, built-in.
- Dishwashers.

The costs listed are the estimated contract award costs, excluding contingencies, supervision, and administration. They include construction to the five-foot line only, but do not include the cost of outside utilities or other site improvements.

Figures listed do not include the cost of piles or other special foundations which are considered as an additional supporting item. The cost of air conditioning is included to the extent authorized by the Construction Criteria Manual.

The costs of equipment such as furniture and furnishings which are loose, portable, or can be detached from the structure without tools are excluded from the unit costs. The cost of permanently attached equipment related directly to the operating function for which the structure is being provided, such as technical, scientific, production and processing equipment, is normally excluded from these costs. The following items are excluded from the costs on the following page.

- Furniture, loose.
- Furnishings, including rugs, loose.
- Filing cabinets and portable safes.
- Office machines, portable.
- Wall clocks, plug-in
- Food preparation and serving equipment, including appliances, portable.
- Training aids and equipment, including simulators.
- Shop equipment.
- Bowling lanes, including automatic pin spotting equipment, score table and players' seating.
- Automatic data processing equipment.
- Communications equipment.
- Photographic equipment, portable.
- Any operational equipment for which installation, mounting and connections are provided in building design and which are detachable without damage to the building or equipment.

Estimating Procedure

Determine the area relationship of the proposed building by dividing the gross area by the typical size as shown in the Square Foot Cost Table. Locate the quotient on the Area Relationship scale and trace vertically to the Factor Line, then trace horizontally to the Cost Relationship scale. This value is then multiplied by the unit cost in the Square Foot Cost Table and factored by the Construction Cost Index to determine the adjusted unit cost for the building.

Military Construction Costs

Facilities	Typical size Gross S.F.	Unit cost per S.F. FY 2025
Administrative office		
Multi-purpose	25,000	277.00
Data processing	21,000	403.00
Aircraft operations		
without tower	25,000	476.00
tower only	4,500	1324.00
Applied instruction building		
General instruction	25,000	319.00
High Tech (Auto-Aid)	25,000	363.00
Barracks, dormitory		
(No Kitchenette equip)	99,500	289.00
Bowling alley, 8 lanes,	7,800	348.00
Chapel center	15,000	448.00
Commissary		
(sales store/equipment)	85,000	292.00
Family housing, U.S.	None	165.00
Family housing, outside U.S.	None	214.00
Family support		
Child development center	15,000	359.00
Education center	10,000	354.00
Youth center	15,000	342.00
Family service center	5,000	326.00
Fire and rescue station (airfield)	8,000	436.00
Fire Station, community	8,000	362.00
Hangars		
Maintenance/Gen. purpose	23,000	332.00
High bay Maintenance	35,000	353.00
Library	12,000	338.00
Main Exchange		
with mall service shops	80,000	237.00
Medical facility		
Station hospital	N/A	519.00
Regional medical center	N/A	562.00
Medical clinic	50,000	388.00
Dental clinic	20,000	495.00
Ambulatory clinic	30,000	457.00

Facilities	Typical size Gross S.F.	Unit cost per S.F. FY 2025
Dining facility		
(with kitchen equipment)	21,500	461.00
Operations building		
General purpose	25,000	290.00
Squadron	36,000	340.00
Physical fitness		
training center	30,000	328.00
Recreation center	20,000	346.00
Reserve Center	20,000	284.00
Satellite Communications center	6,000	1,036.00
Service club	22,500	472.00
School for dependents		
Elementary	None	298.00
Jr. high/middle	None	301.00
High school	None	317.00
Shops		
Vehicle maint. (wheeled)	30,000	294.00
Vehicle maint. (tracked)	25,000	320.00
Aircraft avionics	23,000	290.00
Installation maintenance	31,000	206.00
Parachute and dinghy	8,000	343.00
Aircraft machine shop	20,000	357.00
Storage facility		
Cold storage warehouse		
w/processing	11,000	256.00
Cold storage warehouse	6,000	343.00
General purpose warehouse		
low bay	40,000	149.00
General purpose warehouse		
high bay	100,000	160.00
General purpose magazine		
w/o crane	10,000	420.00
High Explosive magazine	5,000	437.00
Temporary lodging facility	30,000	301.00
Unaccompanied officers quarters	44,000	317.00

Size Cost Adjustment

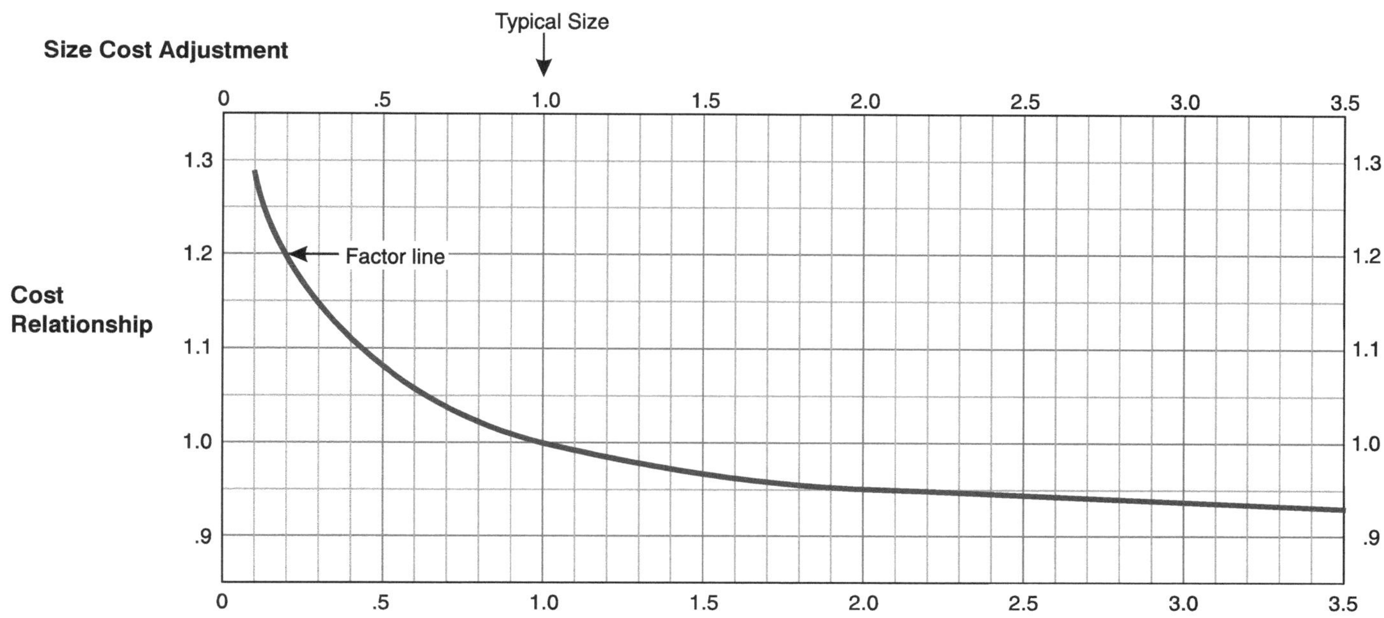

Military Construction Cost Indices

State	Index
Alabama	**0.83**
Mobile	0.84
Montgomery	0.83
Anniston AD	0.83
Fort Rucker	0.81
Maxwell AFB	0.83
Mobile	0.84
Redstone Arsenal	0.85
Alaska	**1.84**
Anchorage	1.67
Clear AFB	2.13
EareckssonAFB	3.44
Eielson AFB	2.02
Elmendorf AFB	1.67
Fairbanks	2.00
Fort Greeley	2.14
Fort Richardson	1.67
Ft Wainwright	2.00
Arizona	**0.99**
Flagstaff	0.97
Tucson	1.01
Davis Monthan AFB	1.01
Fort Huachuca	1.14
Luke AFB	0.98
Yuma MCAS	1.15
Yuma PG	1.15
Arkansas	**0.87**
Fort Smith	0.85
Pine Bluff	0.89
Fort Chaffee	0.85
Little Rock AFB	0.85
Pine Bluff Arsenal	0.89
California	**1.19**
29 Palms MCB	1.35
San Diego	1.19
San Francisco	1.20
Beale AFB	1.22
Camp Pendleton MC	1.19
Centerville Beach	1.10
China Lake NWC	1.29
Concord	1.20
Edwards AFB	1.29
El Centro NAF	1.24
Ft Hunter Liggett	1.25
Fort Irwin	1.29
Los Angeles	1.12
March AFB	1.05
Monterey	1.16
Oakland AB	1.37
Point Magu	1.12
Port Sur	1.16
Port Hueneme	1.12
Riverbank AAP	1.14
San Bruno	1.20
San Clemente Island	1.90
Seal Beach	1.12
Sharpe AD	1.17
Sierra AD	1.27
Stockton	1.14
Travis AFB	1.24
Vandenberg AFB	1.22
Colorado	**1.05**
Buckley AFB	1.02
Colorado Springs	1.09
Denver	1.02
Air Force Academy	1.09
Cheyenne Mountain	1.16
Fort Carson	1.09
Peterson AFB	1.09
Pueblo AD	0.94
Rock Mountain ARS	1.02
Schriever AFS	1.16
Connecticut	**1.08**
Bridgeport	1.05
New London	1.10
New London area	1.10
Delaware	**0.98**
Dover	1.00
Wilmington	0.97
Dover AFB	1.00
Florida	**0.85**
Miami	0.90
Panama City	0.80
Cape Canaveral	0.97
Eglin AFB	0.82
Homestead AFB	0.90
Jacksonville area	0.93
Key West NAS	1.09
Mayport	0.93
McDill AFB	0.88
Orlando	0.84
Panama City area	0.80
Patrick AFB	0.98
Pensacola area	0.83
Tyndall AFB	0.80
Georgia	**0.85**
Albany	0.78
Albany area	0.82
Athens	0.84
Atlanta	0.91
Fort Benning	0.80
Fort Gillem	0.91
Fort Gordon	0.84
Fort McPherson	0.91
Fort Stewart	0.82
Robins AFB	0.82
Kings Bay	0.99
Moody AFB	0.85
Hawaii	**1.71**
Honolulu	1.67
Kaneohe Bay NAS	1.75
Barking Sand	1.74
Fort DeRussy	1.55
Fort Shafter	1.57
Hickam AFB	1.55
Kaneohe MCAS	1.75
Pearl Harbor	1.67
Pohakuloa	1.94
Schofield Barracks	1.66
Tripler AMC	1.68
Wheeler AFB	1.66
Ford Island	1.67
Idaho	**1.07**
Boise	1.00
Mtn Home	1.14
Mtn Home AFB	1.14
Illinois	**1.17**
Belleville	1.11
Chicago	1.22
Great Lakes NTC	1.26
Rock Island Arsenal	1.05
Scott AFB	1.19
Forest Park	1.22
Indiana	**0.97**
Indianapolis	0.99
Logansport	0.96
Crane NWSC	1.06
Fort Ben Harrison	1.04
Grissom AFB	1.02
Iowa	**1.05**
Burlington	1.11
Des Moines	1.00
Iowa AAP	1.16
Kansas	**0.93**
Manhattan	0.96
Wichita	0.91
Fort Leavenworth	1.05
Fort Riley	1.01
Kansas AAP	0.97
McConnell AFB	0.91
Kentucky	**0.94**
Lexington	0.91
Louisville	0.97
Fort Campbell	1.05
Fort Knox	1.07
Louisville NAS	0.97
Louisiana	**0.92**
New Orleans	0.94
Shreveport	0.89
Barksdal AFB	0.89
Fort Polk	0.89
Louisiana AAP	0.89
New Orleans AB	0.94
Maine	**1.06**
Bangor	1.07
Portland	1.05
Brunswick	1.05
Cutler Winter Harbor	1.01
Kittery/Portmouth	1.06
Maryland	**0.98**
Baltimore	0.89
Lexington Park	1.08
Aberdeen PG	0.89
Andrews AFB	0.98
Fort Detrick	0.98
Fort Meade	0.98
Fort Ritchie	0.92
Harry Diamond Lab	0.98
Indian Head	0.96
Patuxent River	1.08
Annapolis	0.98
Bethesda	0.98
Cheltonham	0.98
Chesapeake Beach	0.90
Thurmont	1.08
Massachusetts	**1.11**
Boston	1.15
Fitchburg	1.08
Army M & M Lab	1.13
Fort Devens	1.15
Hanscomb AFB	1.21
Michigan	**1.16**
Detroit	1.18
Marquette	1.15
Detroit Arsenal	1.18
Minnesota	**1.07**
Duluth	1.10
Minneapolis	1.04
Mississippi	**0.86**
Biloxi	0.92
Columbus	0.81
Columbus AFB	0.81
Gulfport area	0.92
Keesler AFB	0.92
Meridian	0.95
Missouri	**0.98**
Kansas City	1.00
Sedalia	0.95
Fort Leonard Wood	1.09
Lake City AAP	1.04
St. Louis AAP	1.04
Whiteman AFB	1.01
Montana	**1.16**
Billings	1.16
Great Falls	1.15
Malmstrom AFB	1.15
Nebraska	**0.94**
Grand Island	0.88
Omaha	1.00
Cornhusker AAP	0.88
Offutt AFB	1.00
Nevada	**1.18**
Hawthorne	1.15
Las Vegas	1.21
Fallon	1.19
Hawthorne AAP	1.15
Nellis AFB	1.25
New Hampshire	**1.04**
Concord	1.05
Portsmouth	1.04
Portsmouth area	1.04
New Jersey	**1.18**
Newark	1.19
Trenton	1.17
Bayonne	1.21
Earle	1.21
Fort Dix	1.17
Fort Monmouth	1.16
Lakehurst	1.21
McGuire AFB	1.17
Picatinny Arsenal	1.23
New Mexico	**0.98**
Alamagordo	0.98
Albuquerque	0.99
Cannon AFB	1.06
Holloman AFB	0.98
Kirtland AFB	0.99
White Sands MR	1.03
New York	**1.08**
Albany	1.02
Buffalo	1.14
New York City	1.51
Fort Drum	1.15
Long Island	1.11
Scotia	1.02
USMA	1.40
Watervliet AFS	1.02
Niagara	1.14
North Carolina	**0.86**
Fayetteville	0.88
Greensboro	0.84
Camp Lejeune	0.94
Cherry Point	0.94
Fort Bragg	0.88
New River	0.94
Pope AFB	0.88
Seymour AFB	0.82
Sunny Point	0.91
North Dakota	**1.05**
Grand Forks	1.01
Minot	1.10
Ohio	**0.99**
Dayton	0.97
Youngstown	1.00
Ravenna AAP	1.03
Wright-Patterson	0.97
Oklahoma	**0.90**
Lawton	0.91
Oklahoma City	0.89
Altus AFB	0.98
Fort Sill	0.91
McAlester AAP	0.84
Tinker AFB	0.89
Vance AFB	0.91
Oregon	**1.10**
Pendleton	1.11
Portland	1.08
Umatilla AD	1.24
Pennsylvania	**1.03**
Philadelphia	1.08
Pittsburgh	0.98
Carlisle Barracks	0.93
Indiantown Gap MR	1.00
Letterkenny AD	0.99
Mechanicsburg	0.93
New Cumberland	0.93
Philadelphia area	1.08
Tobyhanna AD	1.05
Warminster	1.06
Willow Grove	1.06
Rhode Island	**1.06**
Newport	1.09
Providence	1.03
South Carolina	**0.86**
Charleston	0.88
Columbia	0.84
Beaufort	1.04
Charleston AFB	0.88
Fort Jackson	0.84
Shaw AFB	0.82
South Dakota	**0.93**
Rapid City	0.91
Sioux Falls	0.95
Ellsworth AFB	0.91
Tennessee	**0.87**
Chattanooga	0.83
Memphis	0.91
Volunteer AAP	0.90
Arnold AFS	0.87
Memphis NAS	0.99
Texas	**0.82**
San Angelo	0.81
San Antonio	0.83
Brooks AFB	0.83
Camp Bullis	0.83
Corpus Christi	0.90
Dallas	0.88
Dyess AFB	0.97
Fort Bliss	0.91
Fort Hood	0.85
Fort Sam Houston	0.83
Fort Worth	0.88
Goodfellow AFB	0.81
Ingleside NS	0.94
Kingville area	0.91
Lackland AFB	0.83
Laughlin AFB	0.88
Lone Star AAP	0.89
Longhorn AAP	0.80
Randolph AFB	0.83
Red River AD	0.89
Sheppard AFB	0.94
Utah	**1.03**
Ogden	1.04
Salt Lake City	1.02
Dugway PG	1.05
Fort Douglas	1.02
Hill AFB	1.04
Tooele AD	1.05
Vermont	**0.91**
Burlington	0.90
Montpelier	0.92
Virginia	**0.92**
Norfolk	0.94
Richmond	0.92
Camp Peary	0.94
Chesapeake	0.94
Dahlgren	0.98
Fort A.P. Hill	0.90
Fort Belvoir	0.98
Fort Eustis	0.94
Fort Lee	0.94
Fort Monroe	0.94
Fort Myer	0.98
Fort Pickett	0.94
Fort Story	0.94
Quantico	0.98
Radford AAP	0.94
Vint Hill Farms	0.90
Wallops Island	1.12
Langley AFB	0.94
Washington	**1.06**
Bangor	1.16
Everett	1.11
Spokane	1.06
Tacoma	1.06
Bremerton	1.16
Fairchild AFB	1.06
Fort Lewis	1.06
Indian Head	1.16
McChord AFB	1.06
Silverdale	1.16
Whidbey Island	1.27
Yakima Firing Range	1.07
West Virginia	**0.95**
Bluefield	0.93
Charleston	0.97
Sugar Grove	1.43
Wisconsin	**1.13**
Madison	1.14
Milwaukee	1.12
Badger AAP	1.22
Fort McCoy	1.18
Wyoming	**1.01**
Casper	0.98
Cheyenne	1.04
F.E.Warren AFB	1.04
Washington DC	**0.98**
Fort McNair	0.98
Walter Reed AMC	0.98
Bolling AFB	0.98

Index

Index

Practical References for Builders

National Estimator Cloud

Generate professional construction estimates for all residential and commercial construction from your internet browser. Includes 10 Craftsman construction cost databases, over 40,000 labor and material costs for construction, in an easy-to-use format. Cost estimates are well-organized and thoroughly indexed to speed and simplify writing estimates for nearly any residential or light commercial construction project - new construction, improvement or repair. Convert the bid to an invoice - in either QuickBooks Desktop or QuickBooks Online. Access your estimates anywhere with a Web browser. Monthly and one-time billing options available. Visit
https://craftsman-book.com/national-estimator-cloud for more details.

National Construction Estimator

Current building costs for residential, comercial, and industrial construction. Estimated prices for every common building material. Provides manhours, recommended crew, and gives the labor cost for installation.
672 pages, 8½ x 11, $117.50
Revised annually
***eBook (PDF) also available;* $58.75 at https://craftsman-book.com**

Getting Financing & Developing Land

Developing land is a major leap for most builders — yet that's where the big money is made. This book gives you the practical knowledge you need to make that leap. Learn how to prepare a market study, select a building site, obtain financing, guide your plans through approval, and then control your building costs so you can ensure yourself a good profit. Includes a CD-ROM with forms, checklists, and a sample business plan you can customize and use to help you sell your idea to lenders and investors. **232 pages, 8½ x 11, $39.00**
***eBook (PDF) also available;* $19.50 at https://craftsman-book.com**

Construction Forms for Contractors

This practical guide contains 78 useful forms, letters and checklists, guaranteed to help you streamline your office, organize your jobsites, gather and manage records and documents, keep a handle on your subs, reduce estimating errors, administer change orders and lien issues, monitor crew productivity, track your equipment use, and more. Includes accounting forms, change order forms, forms for customers, estimating forms, field work forms, HR forms, lien forms, office forms, bids and proposals, subcontracts, and more. All are also on the CD-ROM included, in *Excel* spreadsheets, as formatted Rich Text that you can fill out on your computer, and as PDFs.
360 pages, 8½ x 11, $48.50
***eBook (PDF) also available;* $24.25 at https://craftsman-book.com**

National Appraisal Estimator

An Online Appraisal Estimating Service. Produce credible single-family residence appraisals - in as little as five minutes. A smart resource for appraisers using the cost approach. Reports consider all significant cost variables and both physical and functional depreciation.

For more information, visit
https://craftsman-book.com/national-appraisal-estimator-online-software.

Paper Contracting: The How-To of Construction Management Contracting

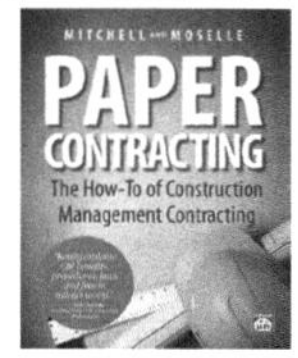

Risk, and the headaches that go with it, have always been a major part of any construction project — risk of loss, negative cash flow, construction claims, regulations, excessive changes, disputes, slow pay — sometimes you'll make money, and often you won't. But many contractors today are avoiding almost all of that risk by working under a construction management contract, where they are simply a paid consultant to the owner, running the job, but leaving him the risk. This manual is the how-to of construction management contracting. You'll learn how the process works, how to get started as a CM contractor, what the job entails, how to deal with the issues that come up, when to step back, and how to get the job completed on time and on budget. Includes a link to free downloads of CM contracts legal in each state.
272 pages, 8½ x 11, $55.50
***eBook (PDF) also available;* $27.75 at https://craftsman-book.com**

Construction Contract Writer

Relying on a "one-size-fits-all" boilerplate construction contract to fit your jobs can be dangerous — almost as dangerous as a handshake agreement. *Construction Contract Writer* lets you draft a contract in minutes that precisely fits your needs and the particular job, and meets both state and federal requirements. You just answer a series of questions — like an interview — to construct a legal contract for each project you take on. Anticipate where disputes could arise and settle them in the contract before they happen. Include the warranty protection you intend, the payment schedule, and create subcontracts from the prime contract by just clicking a box. Includes a feedback button to an attorney on the Craftsman staff to help should you get stumped — *No extra charge.* **$199.95.** Download *Construction Contract Writer* at
https://www.constructioncontractwriter.com

Insurance Restoration Contracting: Startup to Success

Insurance restoration — the repair of buildings damaged by water, fire, smoke, storms, vandalism and other disasters — is an exciting field of construction that provides lucrative work immune to economic downturns. And, with insurance companies funding the repairs, your payment is virtually guaranteed. But this type of work requires special knowledge and equipment, and that's what you'll learn about in this book. It covers fire repairs and smoke damage, water losses and specialized drying methods, mold remediation, content restoration, even damage to mobile and manufactured homes. You'll also find information on equipment needs, training classes, estimating books and software, and how restoration leads to lucrative remodeling jobs. It covers all you need to know to start and succeed as the restoration contractor that both homeowners and insurance companies call on first for the best jobs. **640 pages, 8½ x 11, $69.00**
***eBook (PDF) also available;* $34.50 at https://craftsman-book.com**

Construction Estimating Reference Data eBook

Provides the 300 most useful manhour tables for practically every item of construction. Labor requirements are listed for sitework, concrete work, masonry, steel, carpentry, thermal and moisture protection, doors and windows, finishes, mechanical and electrical. Each section details the work being estimated and gives appropriate crew size and equipment needed.
***Available only as an eBook (PDF);* $29.50 at https://craftsman-book.com**

Markup & Profit: A Contractor's Guide, Revisited

In order to succeed in a construction business, you have to be able to price your jobs to cover all labor, material and overhead expenses, and make a decent profit. But calculating markup is only part of the picture. If you're going to beat the odds and stay in business — profitably, you also need to know how to write good contracts, manage your crews, work with subcontractors and collect on your work. This book covers the business basics of running a construction company, whether you're a general or specialty contractor working in remodeling, new construction or commercial work. The principles outlined here apply to all construction-related businesses. You'll find tried and tested formulas to guarantee profits, with instructions and easy-to-follow examples to help you learn how to operate your business successfully. Includes a link to free downloads of blank forms and checklists used in this book.
336 pages, 8½ x 11, $59.50
***Also available as an eBook (ePub, mobi for Kindle),* $39.95 at https://craftsman-book.com**

Estimating Home Building Costs, Revised

Estimate every phase of residential construction from site costs to the profit margin you include in your bid. Shows how to keep track of manhours and make accurate labor cost estimates for site clearing and excavation, footings, foundations, framing and sheathing finishes, electrical, plumbing, and more. Provides and explains sample cost estimate worksheets with complete instructions for each job phase. This practical guide to estimating home construction costs has been updated with digital *Excel* estimating forms and worksheets that ensure accurate and complete estimates for your residential projects. Enter your project information on the worksheets and *Excel* automatically totals each material and labor cost from every stage of construction to a final cost estimate worksheet. Load the enclosed CD-ROM into your computer and create your own estimate as you follow along with the step-by-step techniques in this book.
336 pages, 8½ x 11, $38.00
***eBook (PDF) also available;* $19.00 at https://craftsman-book.com**

Contractor's Plain-English Legal Guide

For today's contractors, legal problems are like snakes in the swamp — you might not see them, but you know they're there. This book tells you where the snakes are hiding and directs you to the safe path. With the directions in this easy-to-read handbook you're less likely to need a $250-an-hour lawyer. Includes simple directions for starting your business, writing contracts that cover just about any eventuality, collecting what's owed you, filing liens, protecting yourself from unethical subcontractors, and more. For about the price of 15 minutes in a lawyer's office, you'll have a guide that will make many of those visits unnecessary. Includes a CD-ROM with blank copies of all the forms and contracts in the book. **272 pages, 8½ x 11, $49.50**

Craftsman's Construction Installation Encyclopedia

Step-by-step installation instructions for just about any residential construction, remodeling or repair task, arranged alphabetically, from *Acoustic tile* to *Wood flooring*. Includes hundreds of illustrations that show how to build, install, or remodel each part of the job, as well as manhour tables for each work item so you can estimate and bid with confidence. Also includes a CD-ROM with all the material in the book, handy look-up features, and the ability to capture and print out for your crew the instructions and diagrams for any job. **792 pages, 8½ x 11, $65.00**
***eBook (PDF) also available;* $32.50 at https://craftsman-book.com**

Craftsman eLibrary

Craftsman's eLibrary license gives you immediate access to 60+ PDF eBooks in our bookstore for 12 full months! **You pay only one low price. $129.99**
Visit **https://craftsman-book.com** for more details.

Contractor's Survival Manual Revised

The "real skinny" on the down-and-dirty survival skills that no one likes to talk about – unique, unconventional ways to get through a debt crisis: what to do when the bills can't be paid, finding money and buying time, conserving income, transferring debt, setting payment priorities, cash float techniques, dealing with judgments and liens, and laying the foundation for recovery. Here you'll find out how to survive a downturn and the key things you can do to pave the road to success. Have this book as your insurance policy; when hard times come to your business it will be your guide.
336 pages, 8½ x 11, $38.00
***Also available as an eBook (PDF),* $19.00 at https://craftsman-book.com**

Fences & Retaining Walls Revised

Everything you need to know to run a profitable business in fence and retaining wall contracting. Takes you through layout and design, construction techniques for wood, masonry, and chain link fences, gates and entries, including finishing and electrical details. How to build retaining and rock walls. How to get your business off to the right start, keep the books, and estimate accurately. The book even includes a chapter on contractor's math.
416 pages, 8 x 11, $98.75
***Also available as an eBook (PDF, EPUB & MOBI/Kindle);* $49.38 at https://craftsman-book.com**

National Repair & Remodeling Estimator

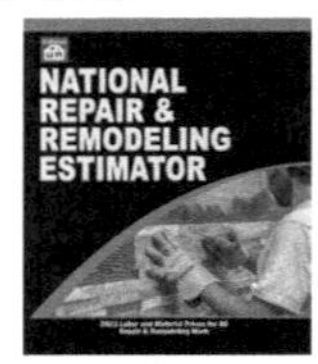

The complete pricing guide for dwelling reconstruction costs. Reliable, specific data you can apply on every repair and remodeling job. Up-to-date material costs and labor figures based on thousands of jobs across the country. Provides recommended crew sizes; average production rates; exact material, equipment, and labor costs; a total unit cost and a total price including overhead and profit. Separate listings for high- and low-volume builders, so prices shown are specific for any size business. Estimating tips specific to repair and remodeling work to make your bids complete, realistic, and profitable.
468 pages, 8½ x 11, $118.50. Revised annually
***Also available as an eBook (PDF),* $59.25 at https://craftsman-book.com**

National Home Improvement Estimator

Current labor and material prices for home improvement projects. Provides manhours for each job, recommended crew size, and the labor cost for removal and installation work. Material prices are current, with location adjustment factors and free monthly updates on the Web. Gives step-by-step instructions for the work, with helpful diagrams, and home improvement shortcuts and tips from experts. **548 pages, 8½ x 11, $118.75. Revised annually**
***Also available as an eBook (PDF),* $59.38 at https://craftsman-book.com**

Building Layout

Shows how to use a transit to locate a building correctly on the lot, plan proper grades with minimum excavation, find utility lines and easements, establish correct elevations, lay out accurate foundations, and set correct floor heights. Explains how to plan sewer connections, level an out-of-level foundation, use a story pole and batterboards, work on steep sites, and minimize excavation costs. **240 pages, 5½ x 8½, $22.00**

Builder's Guide to Accounting Revised

Step-by-step, easy-to-follow guidelines for setting up and maintaining records for your building business. This practical guide to all accounting methods shows how to meet state and federal accounting requirements, explains the new depreciation rules, and describes how the Tax Reform Act can affect the way you keep records. Full of charts, diagrams, simple directions and examples to help you keep track of where your money is going. Recommended reading for many state contractor's exams. Each chapter ends with a set of test questions, and a CD-ROM included FREE has all the questions in interactive self-test software. Use the Study Mode to make studying for the exam much easier, and Exam Mode to practice your skills. **360 pages, 8½ x 11, $61.50**

***eBook (PDF) also available;* $30.75 at https://craftsman-book.com**

Pipe & Excavation Contracting Revised

This popular manual has been updated and improved to bring it more current with modern earthmoving and trenching equipment, refined excavation techniques, stricter safety rules, and improved materials. Shows how to read plans and compute quantities for both trench and surface excavation, figure crew and equipment productivity rates, estimate unit costs, bid the work, and get the bonds you need. Learn how to choose the right equipment for each job, use GPS, how to lay all types of water and sewer pipe, work on steep slopes or in high groundwater, efficiently remove asphalt and rock, and the various pipe, joints and fittings now available. Explains how to switch your business to excavation work when you don't have pipe contracts, and how to avoid the pitfalls that can wipe out your profits on any job.

328 pages, 8½ x 11, $55.00

***eBook (PDF) also available;* $27.50 at https://craftsman-book.com**

Moving to Commercial Construction

In commercial work, a single job can keep you and your crews busy for a year or more. The profit percentages are higher, but so is the risk involved. This book takes you step-by-step through the process of setting up a successful commercial business: finding work, estimating and bidding, value engineering, getting through the submittal and shop drawing process, keeping a stable work force, controlling costs, and promoting your business. Explains the design/build and partnering business concepts and their advantage over the competitive bid process. Includes sample letters, contracts, checklists and forms that you can use in your business, plus a CD-ROM with blank copies in several word-processing formats for both Mac™ and PC computers. **256 pages, 8½ x 11, $42.00**

***eBook (PDF) also available;* $21.00, at https://craftsman-book.com**

Basic Engineering for Builders

This book is for you if you've ever been stumped by an engineering problem on the job, yet wanted to avoid the expense of hiring a qualified engineer. Here you'll find engineering principles explained in non-technical language and practical methods for applying them on the job. With the help of this book you'll be able to understand engineering functions in the plans and how to meet the requirements, how to get permits issued without the help of an engineer, and anticipate requirements for concrete, steel, wood and masonry. See why you sometimes have to hire an engineer and what you can undertake yourself: surveying, concrete, lumber loads and stresses, steel, masonry, plumbing, and HVAC systems. This book is designed to help you, the builder, save money by understanding engineering principles that you can incorporate into the jobs you bid. **400 pages, 8½ x 11, $39.50**

***eBook (PDF) also available;* $19.75 at https://craftsman-book.com**

Steel-Frame House Construction eBook

Framing with steel has obvious advantages over wood, yet building with steel requires new skills that can present challenges to the wood builder. This book explains the secrets of steel framing techniques for building homes, whether pre-engineered or built stick by stick. It shows you the techniques, the tools, the materials, and how you can make it happen. Includes hundreds of photos and illustrations.

***Available only as an eBook (PDF) and software download;* $19.88 at https://craftsman-book.com**

National Renovation & Insurance Repair Estimator

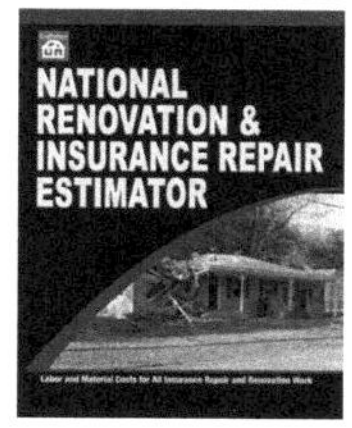

Current prices in dollars and cents for hard-to-find items needed on most insurance, repair, remodeling, and renovation jobs. All price items include labor, material, and equipment breakouts, plus special charts that tell you exactly how these costs are calculated.

488 pages, 8½ x 11, $119.50. Revised annually

***eBook (PDF) also available;* $59.75 at https://craftsman-book.com**

Blueprint Reading for the Building Trades eBook

How to read and understand construction documents, blueprints, and schedules. Includes layouts of structural, mechanical, HVAC and electrical drawings. Shows how to interpret sectional views, follow diagrams and schematics, and covers common problems with construction specifications.

***Available only as an eBook (PDF);* $8.38, at https://craftsman-book.com**

Home Building Mistakes & Fixes

This is an encyclopedia of practical fixes for real-world home building and repair problems. There's never an end to "surprises" when you're in the business of building and fixing homes, yet there's little published on how to deal with construction that went wrong - where out-of-square or non-standard or jerry-rigged turns what should be a simple job into a nightmare. This manual describes jaw-dropping building mistakes that actually occurred, from disastrous misunderstandings over property lines, through basement floors leveled with an out-of-level instrument, to a house collapse when a siding crew removed the old siding. You'll learn the pitfalls the painless way, and real-world working solutions for the problems every contractor finds in a home building or repair jobsite. Includes dozens of those "surprises" and the author's step-by-step, clearly illustrated tips, tricks and workarounds for dealing with them.

384 pages, 8½ x 11, $52.50

***eBook (PDF) also available,* $26.25 at https://craftsman-book.com**

Finish Carpenter's Manual

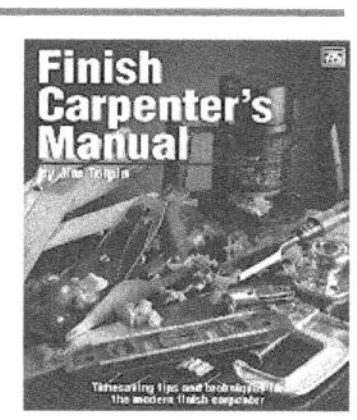

Everything you need to know to be a finish carpenter: assessing a job before you begin, and tricks of the trade from a master finish carpenter. Easy-to-follow instructions for installing doors and windows, ceiling treatments (including fancy beams, corbels, cornices and moldings), wall treatments (including wainscoting and sheet paneling), and the finishing touches of chair, picture, and plate rails. Specialized interior work includes cabinetry and built-ins, stair finish work, and closets. Also covers exterior trims and porches. Includes manhour tables for finish work, and hundreds of illustrations and photos.

208 pages, 8½ x 11, $32.50

Concrete Construction

Just when you think you know all there is about concrete, many new innovations create faster, more efficient ways to do the work. This comprehensive concrete manual has both the tried-and-tested methods and materials, and more recent innovations. It covers everything you need to know about concrete, along with Styrofoam forming systems, fiber reinforcing adjuncts, and some architectural innovations, like architectural foam elements, that can help you offer more in the jobs you bid on. Every chapter provides detailed, step-by-step instructions for each task, with hundreds of photographs and drawings that show exactly how the work is done. To keep your jobs organized, there are checklists for each stage of the concrete work, from planning, to finishing and protecting your pours. Whether you're doing residential or commercial work, this manual has the instructions, illustrations, charts, estimating data, rules of thumb and examples every contractor can apply on their concrete jobs. **288 pages, 8½ x 11, $28.75**

***eBook (PDF) also available;* $14.38 at https://craftsman-book.com**

Electrician's Exam Preparation Guide to the 2017 *NEC*

Need help in passing the apprentice, journeyman, or master electrician's exam? This is a book of questions and answers based on actual electrician's exams over the last few years. Almost a thousand multiple-choice questions exactly the type you'll find on the exam – cover every area of electrical installation: electrical drawings, services and systems, transformers, capacitors, distribution equipment, branch circuits, feeders, calculations, measuring and testing, and more. It gives you the correct answer, an explanation, and where to find it in the latest *NEC*. Also tells how to apply for the test, how best to study, and what to expect on examination day. Includes a certificate for a FREE download of an Interactive Study Center, with all the questions in the book in test-yourself software that makes studying for the exam almost fun! Based on the 2017 *NEC*. **352 pages, 8½ x 11, $67.99**

***Also available as an eBook (PDF),* $33.99 at https://craftsman-book.com**

Also available: Electrician's Exam Preparation Guide 2014, $59.50

***eBook (PDF)* also available; $29.75 at https://craftsman-book.com**

Excavation & Grading Handbook Revised

The foreman's, superintendent's and operator's guide to highway, subdivision and pipeline jobs: how to read plans and survey stake markings, set grade, excavate, compact, pave and lay pipe on nearly any job. Includes hundreds of informative, on-the-job photos and diagrams that even experienced pros will find invaluable. This new edition has been completely revised to be current with state-of-the-art equipment usage and the most efficient excavating and grading techniques. You'll learn how to read topo maps, use a laser level, set crows feet, cut drainage channels, lay or remove asphaltic concrete, and use GPS and sonar for absolute precision. For those in training, each chapter has a set of self-test questions, and a Study Center CD-ROM included has all 250 questions in a simple interactive format to make learning easy and fun. **512 pages, 8½ x 11, $65.00**

eBook *(PDF) also available;* $32.50 at https://craftsman-book.com

Contractor's Guide to QuickBooks by Online Accounting eBook

This book is designed to help a contractor, bookkeeper and their accountant set up and use QuickBooks Desktop specifically for the construction industry. No use re-inventing the wheel, we have used this system with contractors for over 30 years. It works and is now the national standard. By following the steps we outlined in the book you, too, can set up a good system for job costing as well as financial reporting.

***Available only as an eBook (PDF);* $29.99, at https://craftsman-book.com**

Estimating Excavation Revised eBook

How to calculate the amount of dirt you'll have to move and the cost of owning and operating the machines you'll do it with. Detailed, step-by-step instructions on how to assign bid prices to each part of the job, including labor and equipment costs. Also, the best ways to set up an organized and logical estimating system, take off from contour maps, estimate quantities in irregular areas, and figure your overhead. This revised edition includes a chapter on earthwork estimating software. As with any tool, you have to pick the right one. Written by an experienced dirt contractor and instructor of computer estimating software, this chapter covers the program types, explains how they work, gives the basics of how to use them, and discusses what will work best for the type of work you handle. This e-Book is the download version of the book in text searchable, PDF format. Craftsman eBooks are for use in the freely distributed Adobe Reader and are compatible with Reader 6.0 or above.

***Available only as an eBook (PDF);* $21.75, at https://craftsman-book.com**

Building Code Compliance for Contractors & Inspectors

An answer book for both contractors and building inspectors, this manual explains what it takes to pass inspections under the 2009 International Residential Code. It includes a checklist for every trade, covering some of the most common reasons why inspectors reject residential work: footings, foundations, slabs, framing, sheathing, plumbing, electrical, HVAC, energy conservation and final inspection. The requirement for each item is explained, and the code section cited. Knowing in advance what the inspector wants to see gives you an (almost unfair) advantage. To pass inspection, do your own pre-inspection before the inspector arrives. If you're considering a career in code enforcement, this can be your guidebook. **232 pages, 8½ x 11, $32.50**

***eBook (PDF) also available;* $16.25 at https://craftsman-book.com**

Plumber's Handbook Revised, 6th Edition

This new edition explains simply and clearly, in non-technical, everyday language, how to install all components of a plumbing system to comply not only with recent changes in the International Plumbing Code and the Uniform Plumbing Code, but with the requirements of the Americans with Disabilities Act. Originally written for working plumbers to assure safe, reliable, code-compliant plumbing installations that pass inspection the first time, Plumber's Handbook, because of its readability, accuracy and clear, simple diagrams, has become the textbook of choice for numerous schools preparing plumbing students for the plumber's exams. Now, with a set of questions for each chapter, full explanations for the answers, and with a 200-question sample exam in the back, this handbook is one of the best tools available for preparing for almost any plumbing journeyman, master or state-required plumbing contracting exam.

384 pages, 8½ x 11, $67.00

***eBook (PDF) also available;* $33.50 at https://craftsman-book.com**

Commercial Metal Stud Framing

Master the transition from wood to metal stud framing with this comprehensive guide. Written by industry expert Ray Clark, this book offers step-by-step instructions, essential tools, and proven techniques to excel in commercial metal stud framing. Ideal for experienced wood framers, it includes hundreds of job site photos and valuable tips to help you work quickly, accurately, and safely on commercial projects. **208 pages, 8½ x 11, $65.50**

***eBook (PDF) also available;* $32.75, at https://craftsman-book.com**

Paint Contractor's Manual

How to start and run a profitable paint contracting company: getting set up and organized to handle volume work, avoiding mistakes, getting maximum production from your crews and the most value from your advertising dollar. Shows how to estimate all prep and painting. Loaded with manhour estimates, sample forms, contracts, charts, tables and examples you can use. **224 pages, 8½ x 11, $46.50**

***eBook (PDF) also available;* $23.25 at https://craftsman-book.com**

Roofing Construction & Estimating, Revised

Detailed, step-by-step instructions, with photographs and diagrams, for installing, repairing and estimating nearly every type of roof covering available today for residential and commercial structures: asphalt shingles, roll roofing, wood shingles and shakes, clay tile, slate, metal, built-up, elastomeric, TPO and more. Provides guidance on sheathing, synthetic and felt underlayment, as well as tips and tricks from an experienced pro for dealing with those difficult points on a roof that are prone to leaks, such as valleys and roof penetrations. For each roofing type, instructions are provided for estimating material quantities and labor costs, with formulas, easy-to-follow examples and sample estimates for you to test your skill. Use these methods to create reliable estimates that will help insure a profit on every job you take. **448 pages, 8½ x 11, $62.50**

***eBook (PDF) also available,* $31.25 at https://craftsman-book.com**

Build Smarter with Alternative Materials

New building products are coming out almost every week. Some of them may become new standards, as sheetrock replaced lath and plaster some years ago. Others are little more than a gimmick. To write this manual, the author researched hundreds of products that have come on the market in recent years. The ones he describes in this book will do the job better, creating a superior, longer-lasting finished product, and in many cases also save you time and money. Some are made with recycled products — a good selling point with many customers. But most of all, they give you choices, so you can give your customers choices. In this book, you'll find materials for almost all areas of constructing a house, from the ground up. For each product described, you'll learn where you can get it, where to use it, what benefits it provides, any disadvantages, and how to install it — including tips from the author. And to help you price your jobs, each description ends with manhours – for both the first time you install it, and after you've done it a few times.

336 pages, 8½ x 11, $34.75

Renovating & Restyling Older Homes

Any builder can turn a run-down old house into a showcase of perfection — if the customer has unlimited funds to spend. Unfortunately, most customers are on a tight budget. They usually want more improvements than they can afford — and they expect you to deliver. This book shows how to add economical improvements that can increase the property value by two, five or even ten times the cost of the remodel. Sound impossible? Here you'll find the secrets of a builder who has been putting these techniques to work on Victorian and Craftsman-style houses for twenty years. You'll see what to repair, what to replace and what to leave, so you can remodel or restyle older homes for the least amount of money and the greatest increase in value.

416 pages, 8½ x 11, $33.50

How to Succeed With Your Own Construction Business

Everything you need to start your own construction business: setting up the paperwork, finding the jobs, advertising, using contracts, dealing with lenders, estimating, scheduling, finding and keeping good employees, keeping the books, and coping with success. If you're considering starting your own construction business, all the knowledge, tips, and blank forms you need are here. **336 pages, 8½ x 11, $28.50**

***eBook (PDF) also available,* $14.25 at https://craftsman-book.com**

Home Inspection Handbook

Every area you need to check in a home inspection – especially in older homes. Twenty complete inspection checklists: building site, foundation and basement, structural, bathrooms, chimneys and flues, ceilings, interior & exterior finishes, electrical, plumbing, HVAC, insects, vermin and decay, and more. Also includes information on starting and running your own home inspection business.

324 pages, 5½ x 8½, $39.95

***eBook (PDF) also available,* $19.98 at https://craftsman-book.com**

Profits in Buying & Renovating Homes

Step-by-step instructions for selecting, repairing, improving, and selling highly profitable "fixer-uppers." Shows which price ranges offer the highest profit-to-investment ratios, which neighborhoods offer the best return, practical directions for repairs, and tips on dealing with buyers, sellers, and real estate agents. Shows you how to determine your profit before you buy, what "bargains" to avoid, and how to make simple, profitable, inexpensive upgrades.

304 pages, 8½ x 11, $24.75

Wood-Frame House Construction

Step-by-step construction details, from the layout of the outer walls, excavation and formwork, to finish carpentry and painting. Packed with clear illustrations and explanations updated for modern construction methods. Everything you need to know about framing, roofing, siding, interior finishings, floor covering and stairs — your complete book of wood-frame homebuilding.

320 pages, 8½ x 11, $25.50. Revised edition

Handbook of Construction Contracting, Vol. 1

Volume 1: Everything you need to know to start and run your construction business; the pros and cons of each type of contracting, the records you'll need to keep, and how to read and understand house plans and specs so you find any problems before the actual work begins. All aspects of construction are covered in detail, including all-weather wood foundations, practical math for the job site, and elementary surveying. **416 pages, 8½ x 11, $32.75**

Handbook of Construction Contracting, Vol. 2

Volume 2: Everything you need to know to keep your construction business profitable; different methods of estimating, keeping and controlling costs, estimating excavation, concrete, masonry, rough carpentry, roof covering, insulation, doors and windows, exterior finishes, specialty finishes, scheduling work flow, managing workers, advertising and sales, spec building and land development, and selecting the best legal structure for your business.

320 pages, 8½ x 11, $33.75